*UNITED STATES ARMY IN THE KOREAN WAR*

# SOUTH TO THE NAKTONG, NORTH TO THE YALU

(June–November 1950)



*by*

*Roy E. Appleman*

Book 2

*CENTER OF MILITARY HISTORY*

*UNITED STATES ARMY*

*WASHINGTON, D.C., 1992*

# CHAPTER XXI

# August Build-up and September Portents

> The impact of an army, like the total of mechanical co-efficients, is equal to the mass multiplied by the velocity . . . when you are about to give battle concentrate all your strength, neglect nothing; a battalion often decides the day.
>
> NAPOLEON

### *The Far East Air Forces in August*

The Far East Air Forces probably exercised a greater relative influence in August 1950 in determining the outcome of the Korean battles than in any other month of the war. As the number of tactical air control parties increased in late July and during August, the standard practice of the Fifth Air Force was to place one with each U.S. Army regiment and division headquarters and one with each ROK division and corps headquarters. Fighter aircraft in August normally left their Japanese bases at Itazuke and Ashiya on a daily schedule of two planes every fifteen minutes. They reported to the tactical air control center at Taegu where they received specific missions. After receiving them the planes reported to the proper division TACP and then to a regimental TACP for their target assignment. By 23 August, the Fifth Air Force operated twenty-nine T–6 Mosquitoes, all using the Taegu airstrip. The Mosquitoes operated over six stations from dawn to dusk, each plane on station for a 2-hour period before being relieved by another.[1] The pilots of these planes were tactical co-ordinators. They located and controlled close support missions when TACP's did not have visual control.

At the end of the month, eight fighter squadrons were engaged in combat operations in Korea. They were about all that could be supported at the Kyushu bases. In July, FEAF flew 4,635 sorties in close support of ground troops in Korea; in August, 7,397 sorties. An average of 40 sorties daily supported each U.S. division in the August Pusan Perimeter battles. A colorful pilot from Ohio, Maj. Dean E. Hess, who had a record of 63 combat missions in Europe in World War II, had been assigned to train South Korean pilots. He was known to many as "the one man Air Force of the South Korean Army" and by his call sign "MacIntosh One." Hess was grounded by official order near the end of August

---

[1] Hq, X Corps, Analysis of the Air-Ground Operations System, 28 Jun–8 Sep 50. A tactical air control party is described above, p. 95.

after he had flown 95 combat missions in less than two months.[2]

Aviation engineer units available to the Far East Air Forces in July had been badly understrength and deficient in technical training. This slowed the construction of six planned airfields in Korea and, together with the ground reverses, prevented a deployment of fighter planes to bases there.[3]

On 4 August FEAF began B–29 interdiction attacks against all key bridges north of the 37th Parallel in Korea, and on 15 August some light bombers and fighter-bombers joined in the interdiction campaign. This campaign sought the destruction of thirty-two rail and highway bridges on the three main transportation routes across Korea: (1) the line from Sinanju south to P'yongyang and thence northeast to Wonsan on the east coast; (2) the line just below the 38th Parallel from Munsan-ni through Seoul to Ch'unch'on to Chumunjin-up on the east coast; (3) the line from Seoul south to Choch'iwon and hence east to Wonju to Samch'ok on the east coast. The interdiction campaign marked nine rail yards, including those at Seoul, P'yongyang, and Wonsan, for attack, and the ports of Inch'on and Wonsan to be mined. This interdiction program, if effectively executed, would slow and perhaps critically disrupt the movement of enemy supplies along the main routes south to the battlefront.[4]

The Air Force B–29's on 7 August bombed and largely destroyed the P'yongyang Army Arsenal and the P'yongyang railroad yards. On 7, 9, and 10 August they bombed and completely destroyed the large Chosen petroleum refinery at Wonsan. This plant, with its estimated capacity of 250,000 tons, annually produced approximately 93 percent of the North Korean petroleum products. Throughout the month the Air Force bombed the chemical complex in the Hungnam area, the largest in Asia, dropping 1,761 tons of bombs there in the period between 30 July and 19 September. It bombed the Najin docks only 17 miles south of the Siberian border and 105 air miles from Vladivostok. (Najin was an important port of entry for vessels carrying supplies from Vladivostok and it was also a rail center.) The bombers struck the metal-working industry at Songjin with 326 tons of bombs on 28 August, and three days later they heavily damaged the aluminum and magnesium plants at Chinnamp'o with 284 tons of bombs.[5]

The supremacy of the Fifth Air Force in the skies over Korea forced the North Koreans in the first month of the war to resort to night movement of supplies to the battle area. To counter this,

[2] USAF Hist Study 71, pp. 14, 46; 25th Div WD, 21 Aug 50; New York *Times*, August 29, 1950. Hess was an ordained minister in the Campbellite Church of the Disciples of Christ. See *Newsweek*, August 1, 1955, p. 28, for Hess' subsequent service in Korea.

[3] USAF Hist Study 71, pp. 18, 20.

[4] "Air War in Korea," *op. cit.*, IV, No. 2, 19–39; Col Raymond S. Sleeper, "Korean Targets for Medium Bombardment," *Air University Quarterly Review*, IV, No. 3 (Spring, 1951), 21–22.

[5] Sleeper, "Korean Targets for Medium Bombardment," *op. cit.*, pp. 24–26; *Air Intelligence Digest*, September 1950; Hq X Corps, A General Review of U.S. Tactical Air Support in Korea, 28 Jun–8 Sep 50, p. 61; New York *Times*, August 16, 1950.

FLAMING ROCKET *penetrating a loaded boxcar (bottom foreground) and napalm tank (right) exploding at the moment of impact.*

General Stratemeyer ordered nightly visual reconnaissance of the enemy supply routes, beginning on 6 August. On the 8th, Stratemeyer ordered Partridge to increase the night sorties to fifty; by 24 August, Fifth Air Force B–26's alone averaged thirty-five sorties nightly. Late in August the Air Force began flare missions over North Korea. B–29's would release parachute flares at 10,000 feet that ignited at 6,000 feet, whereupon co-operating B–26 bombers attacked any enemy movement discovered in the illuminated area. These M-26 parachute flares from World War II stock functioned poorly, many of them proving to be duds.[6]

Since capturing Seoul, the North Koreans had built two ponton bridges over the Han at that point, one north and one south of the rail and highway bridges. They had also started a new railroad bridge north of the old triple bridge group. The steel cantilever railroad bridge on the west still stood, defying all the efforts of the Far East Air Forces to bring it down. For almost four weeks the Air Force bombed this bridge daily with 1-, 2-, and 4-thousand-pound general purpose bombs with fuze settings, intended to damage both the superstructure and the abutments. On 19 August, nine B–29's of the 19th Group dropped 54 tons of 1,000-pound bombs on the bridge, but it still stood. The same day, Navy carrier-based planes attacked the bridge, scoring eight direct hits, and brought it down. The next day when Air Force planes returned to the bridge they found that three spans had dropped into the river.[7]

Attacks against the Han River ponton bridges at Seoul do not seem to have been successful until FEAF on 27 August ordered the Bomber Command to lay delayed action bombs alongside the bridges, set to detonate at night. This method of attack seems to have caused such heavy casualties among the North Korean labor force trying to keep the pontons in repair that the enemy finally abandoned the effort. These bridges re-

[6] Sleeper, "Korean Targets for Medium Bombardment," *op. cit.*, p. 66; USAF Hist Study 71, p. 45.

[7] GHQ FEC G–3, Opn Rpts 57, 20 Aug, and 62, 25 Aug 50; *Ibid.*, Sitrep, 25 Aug 50; USAF Hist Study 71, pp. 41–44; Sleeper, "Korean Targets for Medium Bombardment," *op. cit.*, p. 65. The Bomber Command made eighty-six sorties with 643 tons of demolition bombs against this bridge.

mained unfinished when the American forces recaptured Seoul.[8]

While it is clear that air power wrought great destruction of enemy equipment and troops during this period of the war, it is not possible to state accurately just how great it really was. Pilot claims are the basis of most estimates of air damage and destruction. Experience has shown that these are subject to many kinds of error. As an example, pilots often mistakenly claimed the destruction of enemy equipment if it remained immobile after attack. It is often impossible for a pilot of a high-speed aircraft to determine if his target is live or not, and three or four different pilots may claim as a "kill" a vehicle already knocked out by ground action. One study revealed a surprisingly great discrepancy between pilot claims and a ground study of destroyed enemy equipment. Pilots claimed to have destroyed ten times as many tanks with rockets as with napalm in the first three months of the Korean War, but a ground survey of destroyed enemy tanks after the Eighth Army breakout from the Pusan Perimeter showed three times as many tanks destroyed with napalm as with rockets. This gives a discrepancy factor of thirty to one in relation to pilot claims. Napalm seldom destroyed a tank by the burst of flame itself. But it did set off chain events that often led to the complete destruction of the tank. The splashing napalm on the bogie wheels set the rubber tires on fire, it heated ammunition to the point where it detonated inside the tank, or it set fuel on fire, and sometimes it splashed into the air intake vents and started fires inside the tank.[9]

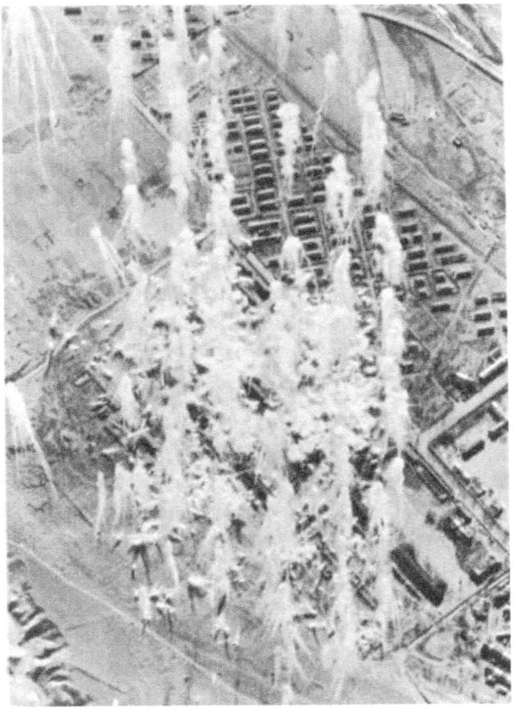

EXPLODING PHOSPHORUS BOMBS *dropped by B–29's over a North Korean barracks create an unusual pattern.*

## *Ground Build-up*

The Far East Command's "Operation Rebuild" by August had assumed the proportions of a gigantic production of ordnance materiel. Before the end of

[8] USAF Hist Study 71, pp. 44–45; Operations Research Office, The Employment of Armor in Korea, ORO–R–1 (FEC), vol. 1, pp. 212–13.

[9] ORO, Close Air Support Operation in Korea, ORO–R–3 (FEC), pp. 37–38, 59, 344–45. Experiments by Eighth Army disclosed that in a napalm burst of two 110-gallon tanks, the intense flame lasted for only approximately twelve seconds and was entirely burned out in twenty seconds. The burst covered an area of about 15,000 square feet; it was considered effective in an area fifty yards square.

1950 it had expanded to employ 19,908 people in eight Japanese shops. In August, 950 2½-ton trucks alone were repaired. During the first three months of the Korean War practically all ammunition the U.N. and South Korean forces used came from rebuild stocks in Japan.

The daily rail and water Red Ball Express from Yokohama to Sasebo to Pusan, beginning on 23 July, operated with increased efficiency in August and demonstrated that it could deliver promptly to Korea any supplies available in Japan. On 5 August, for instance, it delivered 308 measurement tons; on 9 August, 403 tons; on 22 August, 574 tons; and on 25 August, 949 tons. The success of the Red Ball Express cut down the amount of airlift tonnage. This fell from 85 tons on 31 July to 49 tons on 6 August. The express eliminated the need for nearly all airlift of supplies to Korea from Japan. It delivered supplies to Korea in an average time of 60–70 hours, while the airlift delivery varied from 12 hours to 5 days. The Red Ball delivery was not only far cheaper, it was more consistent and reliable.[10]

The drop in air delivery to Korea caused General Partridge, commanding the Far East Air Forces, to complain on 10 August that the Army was not fully using the airlift's 200-ton daily capacity. That day, Eighth Army ordered curtailment of delivery by the Red Ball Express and increased use of the airlift to its maximum capacity. The reason given for this action was a sudden apprehension that the port of Pusan could not process promptly the flow of water-borne supplies. The absurdity of the logistical situation was illustrated the next day, 11 August, when, upon General Partridge's suggestion, two 2½-ton trucks were airlifted in a C–119 from Tachikawa Air Base in Japan to Taegu. The Air Force planned to airlift two trucks daily in this manner. As a result of this development, Eighth Army on 12 August ordered that, effective 15 August, the Red Ball Express be discontinued except on Tuesday and Friday of each week when it would carry cargo difficult for the planes to handle. Under this arrangement airlift tonnage greatly increased. On 16 August, transport planes carried 324 tons of cargo and 595 passengers; on 19 August, 160 tons of cargo and 381 passengers; on 28 August, 398 tons of cargo and 343 passengers; and, on 29 August, 326 tons of cargo and 347 passengers.[11]

After the Russian-built T34 tank appeared on the Korean battlefield, the Department of the Army acted as quickly as possible to correct the imbalance in armor. It alerted three medium tank battalions for immediate movement to Korea. These battalions were the 6th (M46), the 70th (M26 and M4A3), and the 73d (M26). Two of them were the school troop battalions of the Armored School at Fort Knox and of the Infantry School at Fort Benning; the third was the organic battalion of the 1st Armored Division. The Department of the Army notified General MacArthur on 10 July that it planned to ship these battalions

[10] GHQ FEC Sitreps, 8, 11, 23, 27 Aug 50; EUSAK WD, 8 Aug 50.

[11] EUSAK WD, G–4 Jnl, 10 Aug 50 (Rear, Yokohama); *Ibid.*, G–4 Sec, 11–12 Aug 50; GHQ FEC Sitrep, 18 Aug 50; GHQ FEC G–3, Opn Rpts 57, 20 Aug, 66, 29 Aug, and 67, 30 Aug 50.

to the Far East as the quickest way it could devise of getting medium tanks and trained crews to the battlefield. Ships carrying these three tank battalions sailed from San Francisco on 23 July and arrived at Pusan on 7 August. The tank battalions unloaded the next day. The 6th Medium Tank Battalion served as Eighth Army reserve near Taegu in August; the 70th joined the 1st Cavalry Division on 12 August; and the 73d on army orders sent its companies to support various ground operations around the Pusan Perimeter—A Company to Ulsan guarding the eastern main supply route, B Company to Task Force Bradley at Kyongju and Kigye, and C Company to the 27th Infantry in the Bowling Alley north of Taegu. For further reinforcement of Eighth Army, the SS *Luxembourg Victory* departed San Francisco on 28 July with eighty medium tanks in its cargo. Still more armor reinforcements arrived on 16 August, when the 72d Medium Tank Battalion, organic to the 2d Infantry Division, landed at Pusan. The 2d Division also had two regimental tank companies.[12]

During August, therefore, 6 U.S. medium tank battalions landed in Korea, 5 of them in the first eight days of the month. There were, in addition, 4 regimental tank companies and about 30 light tanks for reconnaissance purposes. The tanks in the battalions were about equally divided between M26 Pershings and M4A3 Shermans, except for 1 battalion which had M46 Pattons. The tank battalions averaged 69 tanks. Through 22 August, Eighth Army had lost 20 medium tanks in action.[13] By the third week of August there were more than 500 U.S. medium tanks within the Pusan Perimeter. At the beginning of September American tanks outnumbered the enemy's on the Pusan Perimeter battlefield by at least five to one.

The Korean battle situation in August 1950 caused the Department of the Army to decide to increase its strength there by moving the 3d Infantry Division from the United States. Anticipating future offensive operations in Korea, General MacArthur on 19 August requested troops for two corps headquarters and asked that these two corps be designated I and IX Corps.[14]

Losses in the American divisions fighting in Korea had been so great in the first two months that special steps had to be taken to obtain replacements. On 19 August to help meet this demand, Eighth Army Rear in Japan ordered what it called "Operation Flushout." This required that all units in Japan reassign part of their troops as replacements for use in Korea. By 6 September, 229 officers and 2,201 enlisted men had been reassigned to Korea under this plan. Altogether, during August, 11,115 officer and enlisted replacements arrived in Ko-

[12] Schnabel, FEC, GHQ Support and Participation in the Korean War, ch. III, p. 29; 6th Med Tk Bn Opn Jnl, 23 Jul, 7-8 Aug 50; GHQ FEC Sitrep, 31 Jul 50.

[13] Arty School, Ft. Sill, Employment of Armor in Korea, The First Year, vol. 1, pp. 51-52; GHQ FEC G-3, Opn Rpt 64, 27 Aug 50; ORO, The Employment of Armor in Korea, ORO-R-1 (FEC), vol. 1, p. 167. The tank units in Korea in August were: 1st Marine Tank Battalion; U.S. Army 6th, 70th, 72d, 73d, and 89th Tank Battalions; Regimental tank company with the 5th RCT, three regimental tank companies with the 2d Infantry Division; and the light tanks of the reconnaissance companies of the 2d, 24th, and 25th Infantry, and the 1st Cavalry Divisions.

[14] Schnabel FEC, GHQ Support and Participation in Korean War, ch. III, pp. 24, 33.

rea from Japan and the United States.[15]

The United Nations Command had a supported strength in Korea on 1 September 1950 of nearly 180,000 men, according to figures available at the time. The major organizations reported their personnel strengths as follows:

| | | |
|---|---|---|
| Total | | 179,929 |
| U.S. Eighth Army | | 78,762 |
| 2d Infantry Division | 17,498 | |
| 24th Infantry Division | 14,739 | |
| 25th Infantry Division | 15,007 | |
| 1st Cavalry Division | 14,703 | |
| ROK Army | | 91,696 |
| U.S. 1st Provisional Marine Brigade | | 4,290 |
| British 27th Infantry Brigade | | 1,578 |
| U.S. Fifth Air Force | | 3,603 |

Available for aerial action over Korea and naval action in the waters around it there must be counted an additional 33,651 men in the Far East Air Forces, 330 men of the Royal Australian Air Force, and 36,389 men in the U.S. Naval Forces, Far East.[16]

In early September a distinguished soldier joined the Eighth Army staff. The Department of the Army sent Maj. Gen. Leven C. Allen to Korea to serve as General Walker's chief of staff. General Allen in World War II had been General Omar N. Bradley's army group chief of staff in the European Theater of Operations. He entered on duty at Eighth Army Headquarters in Taegu on 4 September. Colonel Landrum, highly regarded by General Walker, remained as deputy chief of staff.[17]

The Korean War was more than two months old before the first United Nations troops, other than those of the United States, arrived in Korea. Since the Republic of Korea was not a member of the United Nations, the ROK Army was considered an allied force.

The British War Office on 20 August announced that it was dispatching to Korea at once from Hong Kong an infantry force of two battalions. These were regular troops and comprised the 27th Infantry Brigade headquarters, the 1st Battalion of the Middlesex Regiment, and the 1st Battalion of the Argyll and Sutherland Highlanders Regiment. Both regimental organizations dated from the American Revolution—the Middlesex Battalion from 1775, the Argylls from 1776. Since that time they had seen service in many parts of the world, including Wellington's Peninsular Campaign, India, and South Africa. Brigadier Basil A. Coad commanded the force. The British troops sailed from Hong Kong for Pusan, 1,300 miles to the north, on 25 August with bagpipes playing "Auld Lang Syne" and "The Campbells Are Coming." Five ships, including the cruiser *Ceylon* and the carrier *Unicorn*, carried the British to Pusan where they docked on 29 August. En route at sea, the rumor had spread among the troops that the North Koreans were only five miles from Pusan. Instead of the anticipated sound of gunfire the British soldiers found relative quiet in the port city. Debarking at once, the 27th In-

[15] Japan Logistical Comd, Hist Sec, Logistical Problems and Their Solutions, 25 Aug–31 Aug 50, p. 6 (15 Feb 52); GHQ FEC, Ann Narr Hist Rpt, 1 Jan–31 Oct 50, p. 32.

[16] EUSAK WD and PIR 50, 31 Aug 50; *Ibid.*, Sep 50 Summ; GHQ FEC Sitrep, 1 Sep 50. A postwar tabulation, ROK and UN Ground Forces Strength in Korea, 31 July 1950–31 July 1953, prepared by COA, 7 Oct 54 (copy in OCMH), shows a strength of 90,092 for Eighth Army, 126,580 for the ROK Army, and 4,468 for the 1st Provisional Marine Brigade.

[17] EUSAK WD, 4 Sep 50, sec 23, GO 52.

fantry Brigade, on Eighth Army orders, moved by train that night to an assembly area near Kyongsan, ten miles southeast of Taegu.[18]

On 24 August, General MacArthur established the Japan Logistical Command (JLC) as a major organization of the Far East Command. It relieved Eighth Army Rear of all responsibilities concerning posts, camps, and stations in Japan and assumed responsibility as well for the logistical support of all U.N. forces in Korea, except those specifically delegated to other commands. Other organizational changes came on 27 August when General MacArthur designated the Far East Air Forces and the U.S. Naval Forces, Far East, officially as part of the United Nations Command, thus clarifying his relationship to them as Commander in Chief of the United Nations Command. This action served as a precedent for subsequent attachments of other U.N. air and naval forces to the United Nations Command.[19]

The movement of refugees through the front lines and their removal from the battle area was a constant source of worry to the military authorities in August. Between 12 and 19 August, the 25th Division helped the ROK police screen and remove more than 50,000 refugees from its battlefront area between Chindong-ni and the Nam River. Altogether, the 25th Division evacuated 120,335 refugees from its sector during August. In mid-August, the 24th Division estimated there were 100,000 refugees in its southern sector seeking an opportunity to cross the Naktong River. On 24 August, about 300,000 refugees, assembled in collecting points near Yongsan and Changnyong, began moving under ROK police control to areas away from the front lines. They were warned not to stray from their assigned routes of travel lest they be mistaken for guerrillas and shot. The 1st Cavalry Division and the ROK divisions eastward had similar experiences with refugees. In all cases, the ROK police, working in collaboration with the local army commanders, screened the refugees and moved them away from the combat area as quickly as possible.[20]

Occasionally, guerrillas would attack trains in rear areas of the Pusan Perimeter, usually in the Yongch'on-Kyongju area in the east or along the lower Naktong in the Samnangjin area. These attacks generally resulted in only a few persons wounded and minor damage to rail equipment. The most successful guerrilla attack behind the lines of the Pusan Perimeter occurred on 11 August against a VHF radio relay station on Hill 915, eight miles south of Taegu. A guerrilla force, estimated at 100 men, at 0515 attacked the 70 ROK police guarding the station and its American operators. They drove off the ROK police and set fire to the buildings. American casualties were 2 killed, 2 wounded, and 3 missing. When a ROK police force re-

[18] EUSAK WD, 29 Aug 50; *Ibid.*, G–3 Sec, Troop Control, 29 Aug 50; *Ibid.*, POR 139, 28 Aug 50; Maj Gen B. A. Coad, "The Land Campaign in Korea," *Journal of the Royal United Service Institution* (February, 1952), vol. XCVII, No. 585 (lecture given 29 Oct 51); New York *Times*, August 26, 1950; New York *Herald Tribune*, August 30, 1950, Bigart dispatch from Pusan.

[19] GHQ FEC, Ann Narr Hist Rpt, 1 Jan–21 Oct 50, ch. II.

[20] 25th Div WD, 11–19, 31 Aug 50; 2d Div Public Info Off file, Aug 50.

occupied the area later in the day the guerrillas had disappeared.[21]

How to use South Korean manpower to the greatest advantage became one of the most important problems early in the conduct of the war. An immediate need was for more troops to oppose and stop the advancing North Koreans. A longer range need was to build up the manpower of the allied forces to the point where they could drive the enemy back across the 38th Parallel. The program adopted was threefold: (1) fill the five ROK divisions to full strength with replacements; (2) activate new ROK divisions; and (3) attach large numbers of South Korean recruits to American units (a novel expedient).

As part of its projected expansion program the ROK Army opened training schools and centers for officers and replacements. On 14 July it opened the 1st Replacement Training Center at Taegu. This center at first operated on a 10-day schedule to receive and send out 1,000 men daily. The 2d Replacement Training Center at Pusan opened on 20 August. Its capacity was 500 daily, half that of the Taegu center. On 15 August, the ROK Army activated the Ground General School at Tongnae, near Pusan, which received its first class on 23 August. This school was principally a center for training infantry second lieutenant replacements. Its normal capacity was 250 candidates a week. After the pressing needs of the Pusan Perimeter battles had passed, all these schools lengthened their courses of training.[22]

On 10 August, General MacArthur, having received the necessary authority from the Department of the Army, authorized General Walker to increase the strength of the ROK Army to any practicable number.[23] Walker on 18 August requested authority to activate and equip five new ROK divisions at the rate of one a month beginning in September. The divisions were to have a strength of 10,500 men. General MacArthur denied General Walker this authority because of other needs for the available equipment, but he did concur in the recommendation to activate new divisions and service units and so reported to the Department of the Army. On 19 August the strength of the ROK tactical troops was 61,152; service troops, 23,672; total strength of the ROK Army, 84,824 men. The reported strength of ROK tactical organizations was as follows: [24]

| | |
|---|---|
| Total | 76,842 |
| I ROK Corps Headquarters | 1,275 |
| Capital Division | 16,376 |
| 8th Division | 9,106 |
| II ROK Corps Headquarters | 499 |
| 1st Division | 10,482 |
| 6th Division | 9,300 |
| ROK's Headquarters | 2,159 |
| 3d Division | 7,154 |
| P'ohang Task Force | 575 |
| Task Force Kim | 4,025 |
| Special type troops | 14,641 |
| Training Center and Headquarters Company | 1,250 |

The pay scale of the ROK Army in won per month was as follows, with the

[21] EUSAK WD, Aug 50, 25 Aug; *Ibid.*, Sig Sec and G–3 Sec, 11 Aug 50; *Ibid.*, Aug 50, 11 Aug, p. 59.

[22] ORO, Close Air Support Operations in Korea, ORO–R–3 (FEC), pp. 27–28, 42.

[23] Schnabel, FEC, GHQ Support and Participation in the Korean War, ch. IV, pp. 29, 33; GHQ FEC Sitrep, 10 Aug 50.

[24] EUSAK WD, G–4, 19 Aug 50; Schnabel, FEC, GHQ Support and Participation in the Korean War, ch. IV, pp. 31–33; ORO, Utilization of Indigenous Manpower in Korea, ORO–R–4 (FEC), Table XXII, p. 65.

exchange rate of 4,000 won to one U.S. dollar:

| | |
|---|---|
| Pvt | 3,000 |
| Pfc | 3,600 |
| Cpl | 4,500 |
| Sgt | 5,400 |
| SSgt | 6,000 |
| MSgt | 24,900 |
| WO | 29,700 |
| 2d Lt | 30,900 |
| 1st Lt | 33,300 |
| Capt | 35,700 |
| Maj | 41,700 |
| Lt Col | 46,500 |
| Col | 51,300 |
| Gen | 60,000 |

Of the five planned new divisions, the ROK Army intended to reactivate first the ROK 7th Division and second the ROK 11th Division. General MacArthur warned that the new divisions could be equipped only from stocks delivered from the United States. The ROK Army did not wait upon the planned schedule, but was in the process of reactivating the ROK 7th Division at the end of August, forming at least one battalion in each of the 3d, 5th, and 8th Regiments. Task Force Min as an organization disappeared from the ROK Army Order of Battle and became instead the 1st and 2d Battalions of the 5th Regiment, ROK 7th Division.[25]

The attachment by 10 September of a U.S. battalion of 105-mm. howitzers to each of the six ROK divisions then in action considerably increased their combat effectiveness. But even with this new artillery, it must be noted that the ROK divisions had only approximately one-fourth the artillery support of that of the American divisions in Eighth Army.[26] It should not have been surprising that sometimes the ROK divisions did not perform as satisfactorily as the U.S. divisions.

A proper ration for the ROK soldier finally evolved after experimentation and testing, and was adopted in September. It provided 3,165 calories a day for an active 130-pound man. The ration included canned fish, field biscuit, barley, rice, kelp, and tea. Supplemental items were furnished from American stocks. This diet gave promise of improving the physical stamina of the ROK soldier.[27]

### *Korean Augmentation to the United States Army*

Concurrent with the steps taken in August to rebuild the ROK Army, the Far East Command planned to incorporate 30,000 to 40,000 ROK recruits in the four American divisions in Korea and the one still in Japan but scheduled to go to Korea. This was admittedly a drastic expedient to meet the replacement requirement in the depleted American ground forces. As early as 10 August, Eighth Army began planning for the Korean augmentation, but it was not until 15 August that General MacArthur ordered it—General Walker was to increase the strength of each company and battery of United States troops by 100 Koreans. The Koreans legally would be part of the ROK Army and would be

[25] GHQ FEC G–3, Opn Rpts 61, 24 Aug, and 65, 28 Aug 50; Schnabel, FEC, GHQ Support and Participation in the Korean War, ch. IV, pp. 32–33.

[26] EUSAK WD, 11 Sep 50, Arty Rpt, sec 5. Artillery was assigned to the ROK divisions as follows: 10th FA Bn to Capital Div; 17th FA Bn to 1st Div; 11th FA Bn to 3d Div; 16th FA Bn to 6th Div; 18th FA Bn to 7th Div; 50th FA Bn to 8th Div.

[27] EUSAK WD, sec 17, Quartermaster Rpt, 13 Sep 50.

paid and administered by the South Korean Government. They would receive U.S. rations and special service items. The Far East Command initially expected that each ROK recruit would pair with a United States soldier.[28]

Before the augmentation program began there had been a few cases in which American unit commanders had used volunteer South Koreans unofficially to strengthen their forces. One of the first of these officers, if not the first, was Colonel Clainos, commanding officer of the 1st Battalion, 7th Cavalry Regiment. About the first of August, just after Eighth Army had retired behind the Naktong River, four Korean officers and 133 men from the South Korean police at Taegu voluntarily joined Clainos' battalion on the unofficial basis that they would receive arms and food to the best of Colonel Clainos' ability. A Lieutenant Chung, a Tokyo-trained Korean wearing a Japanese samurai sword, marched his unit to the 1st Battalion. Colonel Clainos attached Lieutenant Chung to his staff and the other three officers to A, B and C Companies, respectively. He then attached two Korean policemen to each rifle squad in the companies. Nine days after these Koreans joined the 1st Battalion they took part in the battle at Triangulation Hill, after the North Korean crossing of the Naktong in the 1st Cavalry Division sector. Two of them were killed in this action, and seven wounded. Of the wounded, all refused evacuation except one who could not walk.[29]

The U.S. 7th Infantry Division in Japan was far understrength, having contributed key personnel to the 24th, 25th, and 1st Cavalry Divisions in succession when they mounted out for Korea. In an effort to rebuild this division, the first Korean augmentation recruits were assigned to it rather than to the divisions in Korea. The first three platoons of 313 recruits left Pusan by ship the morning of 16 August and arrived in Japan the afternoon of the 18th. Once started, the shipments of recruits left Pusan at the rate of nearly 2,000 daily. The final shipment arrived at Yokohama on the 24th and debarked the next day, making a total of 8,625 Korean officers and men for the division. The South Korean Government at first obtained many of these recruits directly from the streets of Pusan and Taegu. In the contingents shipped to Japan, schoolboys still had their schoolbooks; one recruit who had left home to obtain medicine for his sick wife still had the medicine with him.[30]

On 20 August, the American divisions in Korea received their first augmentation recruits—the 24th and 25th Divisions, 250 each; the 2d and 1st Cavalry Divisions, 249 each. For the next week each of the divisions received a daily average of 250 Korean recruits. On the 29th and 30th, the 1st Cavalry Division got an average of 740, and the 24th Division, 950 recruits daily. Near the end of August the plan changed so that every fourth day each division would receive 500 men until it had a total of 8,300 Korean recruits. Except for the first

[28] EUSAK WD, 10 and 20 Aug 50; *Ibid.*, Off Asst CofS, G–1, 15 Aug 50; GHQ FEC Sitreps, 10 and 19 Aug 50.

[29] Ltr, Clainos to author, n.d., but recd May 54.

[30] EUSAK WD, G–3 Sec, 19 Aug 50; *Ibid.*, 20 Aug 50; *Ibid.*, G–1 Sec Rpt, 24 Aug 50; ORO, Close Air Support Operations in Korea, ORO–R–4 (FEC), p. 54.

groups, the recruits received five days' training at the Kup'o-ri Training Center near Pusan, which was opened 20 August.[31]

Even though it initially had been the intention of the Far East Command to pair Korean augmentation recruits with American soldiers in a "buddy system," this did not work out uniformly in practice in the Eighth Army. The 1st Cavalry and the 2d Infantry Divisions used the buddy system, with the American responsible for the training of the recruit in use of weapons, drill, personal hygiene, and personal conduct. Two regiments of the 25th Division used the system, while the third placed the recruits in separate platoons commanded by American officers and noncommissioned officers. General Church directed the 24th Division to place all its augmentation recruits in separate squads and platoons commanded by selected Korean officers and noncommissioned officers. These Korean squads and platoons were attached to American units.[32]

Capt. Robert K. Sawyer who, as a 2d lieutenant, commanded a platoon of these new augmentation recruits in the Reconnaissance Company, 25th Division, has given a fair appraisal of the typical Korean recruit in the United States Army in August and September 1950:

When a fresh batch arrived our First Sergeant ran them through a brief schooling on methods of attack, and they were ready for us. Recon Company's ROK contingent ate with us (our menu plus a huge, steaming plate of rice), but otherwise was a force apart.

SOUTH KOREAN RECRUIT *with an American soldier.*

About sixty ROK's were assigned to each Recon platoon, under the command of an American Lieutenant, as support for the Recon platoon leader. In other words, each Recon platoon had two U.S. officers; one for the Americans, the other for the ROK's. I had the latter job for a few weeks. On some occasions I controlled forces consisting of nearly one hundred ROK's, plus ten or twelve GI's scattered throughout for control. At other times I had a fifty-fifty combination. Sometimes the Americans predominated.

It is difficult for me to evaluate the Koreans who augmented our ranks. All in all, however, I was not impressed by my charges and was happy to see the last of them. Mere recruits, they simply had not had time to become soldiers, and I used them for little more than carrying ammunition and rations. On the occasions I had to use them for fighting I spread my GI's

[31] 25th Div WD, 20 Aug 50; 1st Cav Div WD, 20 Aug 50; GHQ FEC Sitrep, 21 Aug 50; ORO, Close Air Support Operations in Korea, ORO-R-4 (FEC), Table XVII, pt. 1, pp. 52, 54-56, 60-62; Lt Ed. E. Balforth, "Getting Our ROK's," *U.S. Combat Forces Journal* (February, 1951), p. 23.

[32] EUSAK WD, Memo from Asst CofS G-1, 5 Sep 50, sub: Korean Augmentation; 19th Inf WD, 29 Aug 50; 35th Inf WD, 20 Aug 50; 24th Div WD, G-1 Hist Rpt, 3-4 Sep 50; 25th Div WD, 4 Sep 50.

around and prayed that nothing of consequence would happen.

My ROK's were always hungry, and never did understand that the cardboard box of C rations was meant for one day's subsistence. Often, an hour after doling out the one-box-per-man I have heard my interpreter ask me for more 'chop-chop.' The Koreans had already eaten their entire day's supply! Invariably they fell asleep when on guard, requiring constant checking by the Americans. And to make matters worse, most Koreans I have observed love to greet the morning sun with a song. This habit did not always fit into our security plan.

In one action I had spread my ROK's in a half circle position, with GI's posted here and there along the line for control. Late in the morning one lone sniper fired at us, and immediately my ROK's went to pieces. Hysterical, they lay on the ground with faces pressed into the earth, weapons pointed in the general direction of the enemy, firing madly, wasting ammunition, completely out of hand. There was only one way to straighten out the situation, so my GI's and I went from ROK to ROK, kicking them and dragging them bodily to where they could see. We eventually succeeded in quieting them down, and when the enemy attacked us later in the day my ROK's held pretty well.[33]

The buddy system of using the Korean augmentation recruits gradually broke down and was abandoned. Most American soldiers did not like the system. Most units found they could employ the recruits, organized in ROK squads and platoons with American officers and noncommissioned officers in charge, to best advantage as security guards, in scouting and patrolling, and in performing various labor details. They were particularly useful in heavy weapons companies where the hand-carrying of machine guns, mortars, and recoilless rifles and their ammunition over the rugged terrain was a grueling job. They also performed valuable work in digging and camouflaging defensive positions.[34]

There also began in August the extensive use of Korean civilians with A-frames as cargo carriers up the mountains to the front lines. This method of transport proved both cheaper and more efficient than using pack animals. American units obtained the civilian carriers through arrangements with the ROK Army. Soon the American divisions were using Korean labor for nearly all unskilled work, at an average of about 500 laborers and carriers to a division.

The U.S. divisions in Korea never received the number of Korean augmentation recruits planned for them. In September the divisions began to take steps to halt further assignments. In the middle of the month, the 24th Division requested Eighth Army not to assign to it any more such troops until the division asked for them. As one observer wrote, "The Koreans haven't had time to learn our Army technique. An American doughboy hated to have his life dependent on whether his Oriental buddy knew enough to give him covering fire at the right moment." [35] The language barrier, the difference in loyalties, the

[33] Sawyer, Notes on Korea, Aug–Nov 50, prepared for author in 1952.

[34] 25th Div WD, 31 Aug 50; 1st Cav Div WD, 29 Aug 50; Maj Norman F. J. Allen, Korean Army Troops, U.S.A. (KATUSA), student MS, Advanced Inf Course, Class 2, Inf School, Ft. Benning, Ga., 1952–53. This is an interesting case study of augmentation in one rifle company. The Korean carrier received a wage of 500 won daily (12½ cents) and a rice ration.

[35] Hal Boyle, "The AP Reports on the Buddy System," *U.S. Combat Forces Journal* (February, 1951), p. 23; 24th Div WD, G–1 Stf Sec Summ, 26 Aug–28 Sep 50.

lack of training in the recruits, and their relative combat ineffectiveness all put great strain on the attempt to integrate the Koreans. It was not strange that as fast as American units obtained American replacements they dispensed with their Korean replacements. By winter, the buddy system had been quietly dropped.

### *Eighth Army Realignment and Extension Eastward*

The last of the 2d Division's regiments, the 38th, known as "The Rock of the Marne" and commanded by Col. George B. Peploe, landed at Pusan on 19 August. The next day, 20 August, Eighth Army issued an operational directive ordering the 2d Infantry Division to relieve the 24th Division as soon as the 38th Regiment closed on Miryang. The 2d Division completed relief of the 24th Division in its Naktong River positions on 24 August and Keiser, 2d Division commander, assumed responsibility for the sector at 1800 that date.

The strength of the 24th Division on 25 August was approximately 10,600 men. It needed about 8,000 replacements as well as quantities of arms, equipment, and vehicles to bring it up to war strength. The 19th Infantry and the 11th Field Artillery Battalion were attached to the 2d Division as a reserve force; the 21st Infantry became Eighth Army reserve; the rest of the division assembled in the vicinity of Kyongsan, twelve miles southeast of Taegu.[36]

General Walker, after discussing the matter with General Church on 26 August, ordered the 34th Infantry reduced to paper status and its personnel and remaining equipment transferred to the 19th and 21st Regiments. At the same time, Eighth Army also reduced to paper status the 63d Field Artillery Battalion, which had been in support of the 34th Infantry, and transferred its troops and equipment to the newly activated C Batteries of the 11th, 13th, and 52d Field Artillery Battalions. The effective dates for the transfer were 26 August for the artillery and 31 August for the infantry. The troops of the 1st Battalion, 34th Infantry, were transferred to the 19th Infantry as its newly activated 3d Battalion and the men of the 3d Battalion, 34th Infantry, were transferred to the 21st Infantry as its newly activated 2d Battalion. Out of the nearly 2,000 men who originally entered Korea with the 34th Infantry on 3 July, there were 184 left in the regiment at the end of August—the rest either had been killed, wounded, or were missing in action. Colonel Beauchamp was reassigned to the command of his old regiment, the 32d Infantry of the 7th Division.[37]

Simultaneously with this action, General Walker transferred the 5th Regimental Combat Team to the 24th Division as its third regiment. The 5th Regimental Combat Team at this time numbered about 3,500 men. The 6th Medium Tank Battalion with about 650 men also was to be attached to the 24th

[36] 24th Div WD, 20–26 Aug 50; 2d Div POR 64, 24 Aug 50; EUSAK WD, G–3 Sec, 24 Aug 50.

[37] EUSAK WD, GO 37, 26 Aug 50; *Ibid.*, GO 37, 28 Aug 50; 34th Inf WD, 26 Aug 50; 24th Div WD, G–4 Opn Highlights, 26 Aug–29 Sep 50; 24th Div WD, 31 Aug 50; 19th Inf WD, Pers Summ 26 Aug–28 Sep 50; 34th Inf WD, Summ, 22 Jul–26 Aug 50; Interv, author with Beauchamp, 1 Aug 52. The 34th Infantry from 5 July to 23 August had suffered 1,714 casualties—98 KIA, 569 WIA, 773 MIA, and 274 nonbattle casualties.

Division. The division still needed approximately 4,000 replacements.[38]

The 27th and 35th Regiments of the 25th Division had received their third battalions early in August with the transfer to them of the two battalions of the 29th Infantry. The 1st Cavalry Division on 26 August received the third battalions for its regiments in organizations sent from the United States. It also received 3 provisional artillery batteries to provide the third firing battery for 3 battalions of artillery. At the end of August, therefore, the 4 U.S. divisions in Korea had finally built up their regiments to the normal 3 battalions.[39]

In Eighth Army a confused order of battle had prevailed generally throughout August. Battle conditions frequently had compelled the army to separate battalions and regiments from their parent organizations and send them posthaste to distant points of the Pusan Perimeter to bolster a threatened sector. All divisions except the 1st Cavalry at various times were broken up by this process. At the end of August, Eighth Army made an effort to unscramble the disorder. It ordered the 23d Infantry on 28 August to leave the Taegu front and return to 2d Division control at Miryang; it ordered the 27th Infantry on 30 August to rejoin the 25th Division at Masan; and it ordered the 5th Regimental Combat Team north from the Masan area to join the 24th Division.[40]

The course of battle in the ROK eastern sector of the Perimeter and the enemy advance down the Sangju-Taegu road during August caused General Walker near the end of the month to decide on a shift of the boundary eastward between the American and ROK troops. He considered the existing boundary near the Sangju-Taegu road a source of military weakness. On 26 August he ordered a new boundary line slanting southeast from a point two miles north of the Walled City of Ka-san to a point east of and below Taegu. This placed the Sangju-Taegu road and the former zone of the ROK 1st Division in the American zone. The 1st Cavalry Division was to move eastward into the ROK 1st Division zone, and the U.S. 2d Division at the same time was to extend its zone northward into the 1st Cavalry zone. The shift of units was to take place as soon as practicable, but no later than 30 August.[41]

Pursuant to the army directive, General Gay on 28 August ordered the 7th Cavalry Regiment to occupy the left (west) part of the ROK 1st Division sector and the 8th Cavalry Regiment to occupy the right (east) part. This shift placed the 7th and 8th Cavalry Regiments in mountainous terrain north of Taegu. The supply of these units now became much more difficult than it had been along the Naktong. On 29 August, the 3d Battalion, 23d Infantry, relieved

---

[38] 24th Div WD, 29 Aug 50; Interv, author with Church, 25 Sep 52. Given his choice, General Church chose to have the 5th RCT as his third regiment rather than rebuild the 34th Infantry with replacements.

[39] EUSAK WD, 26 Aug 50; 5th and 8th Cav Regts WD, 26 Aug 50; 7th Cav WD, 27 Aug 50; EUSAK GO 182, 29 Aug 50. The 3d Bn, 7th Regt, 3d Inf Div, from Fort Devens, Mass., became the 3d Bn, 8th Cav; the 2d Bn, 30th Inf Regt, 3d Div, from Fort Benning became the 3d Bn, 7th Cav, and a battalion from Camp Carson became the 3d Bn, 5th Cav Regt.

[40] EUSAK WD, 30 Aug 50; GHQ FEC G–3 Opn Rpt 65, 28 Aug 50.

[41] EUSAK WD, Opn Dir, 26 Aug 50. For a description of the Walled City of Ka-san, see pages 422–23, below.

the 7th Cavalry Regiment in the southern part of the division sector, and the 7th Cavalry in turn relieved the ROK 13th Regiment and part of the 12th. When the newly arrived 3d Battalion of the 5th Cavalry assumed responsibility on 30 August for the generally quiet 14,000-yard sector of the 8th Cavalry Regiment, the result was a 32,000-yard front for the 5th Cavalry. The 8th Cavalry Regiment then moved to the sectors of the ROK 11th and part of the 12th Regiments. The 1st Cavalry Division completed the relief of the ROK 1st Division at 1300 on 30 August, whereupon the ROK division moved to its new sector just eastward of the new boundary. The contemplated shift of the 2d Division zone of responsibility northward proved impracticable because the area could be supplied only over the road net from Taegu, and Eighth Army reestablished the old boundary between the two divisions, effective 30 August. To defend this old 7th Cavalry sector, Eighth Army attached the 3d Battalion, 23d Infantry, to the 1st Cavalry Division.[42]

On 30 August the 714th Transportation Railway Operating Battalion arrived in Korea and became responsible for operating the approximately 500 miles of rail lines within the Pusan Perimeter.[43] The rail lines usually carried supplies from Pusan to a division railhead. From there they were trucked forward to regiment and battalion.

August was a month of heavy casualties for Eighth Army. Battle casualties in its four divisions were for the 24th Division, 1,941; 25th Division, 1,800; 1st Cavalry Division, 1,503; and the 9th Regiment of the 2d Division, 827. Nonbattle casualties were high in all units, many of them caused by heat exhaustion; the 9th Regiment alone had 419 nonbattle casualties. Loss among officers was very heavy.[44]

During the same period, battle losses had been far greater in the ROK Army than in United States forces, but nonbattle casualties were fewer. On some days ROK battle losses were wholly disproportionate to American. As extreme examples, on 6 August American battle losses were 74, the ROK 1,328; on 21 August the American battle losses were 49, the ROK, 2,229.[45]

As is customary in most army and theater zones of military action, Eighth Army had prepared plans to meet all eventualities anticipated as probable. In early August, General MacArthur outlined to General Walker a defense line closer to Pusan than the Naktong River line. He wanted this line prepared for occupancy in the event Eighth Army could not stop the North Koreans at the Naktong. On 11 August, General Walker verbally instructed Brig. Gen. Garrison H. Davidson, an Engineer officer, to lay out this secondary defense line. Davidson, after looking over the ground, recommended to General Walker that because of better defensive terrain the line should be somewhat farther back toward Pusan than General MacArthur had indicated. General Walker replied that the

[42] 1st Cav Div WD, 28–29 Aug 50; EUSAK POR 144, 29 Aug and POR 147, 30 Aug 50; 5th Cav WD, 27 Aug 50.

[43] Mossman and Middleton, Logistical Problems and Their Solutions, p. 53.

[44] 25th Div WD, 26 and 31 Aug 50; 1st Cav Div WD, Aug 50 Summ; 2d Div WD, G– Sec, 9 Jul–Aug 50; Ltr, Asst CofS to CG 2d Div; 24th Div WD, AG Sec, 1–26 Aug 50, 23 Aug.

[45] GHQ FEC G–3 Opn Rpts 39–67, 2–30 Aug 50.

line would be constructed where General MacArthur had indicated it should go. General Davidson began laying out the line with very few resources. He received some help from Brig. Gen. Crump Garvin and the 2d Logistical Command at Pusan and from the 2d and 25th Divisions. This line, known as the Davidson Line, began on the east coast at Sodong-ni, approximately eight miles north of Ulsan, and extended generally west along high ground to a point northeast of Miryang, then curved down the ridge east of Muan-ni, turned south across the Naktong River and anchored on the high ground northeast of Masan. General Walker would not approve Davidson's recommendation to remove all houses from in front of the line to clear a field of fire. Davidson succeeded in laying a trace of the line on the ground, cleared fields of fire except for houses, ordered material for fortifications, and was able to have a few positions dug before he reported to the 24th Division as assistant division commander on the first of September.[46]

While General Walker had many capable staff officers at his Eighth Army headquarters at this time, perhaps none was more valuable to him than Col. John A. Dabney, Assistant Chief of Staff, G–3, who had joined the Army in Korea during July. Dabney was quiet and unassuming, possessed of a good mind, sound professional knowledge, persistent in his search of facts, and blessed with a fine judgment in evaluating combat information. He showed common sense throughout the critical Naktong battles of the Perimeter, and was a trusted and valued adviser to General Walker and his chief of staff.

At the beginning of September the United Nations had a large numerical superiority of men in the line divisions and in army reserve. In the skies over the battlefield and in the coastal waters guarding the Perimeter flanks, United Nations aerial and naval might was virtually uncontested. Approximately 600 American medium tanks mounting 90-mm. and 76-mm. guns were in the battle area on 1 September, as contrasted with probably not more than 100 North Korean Russian-built medium T34 tanks mounting 85-mm. guns. Eighth Army also had overwhelming superiority in artillery and mortar fire.

The Eighth Army intelligence officer on 30 August estimated that the twelve known enemy rifle divisions had an effective strength of 82,590 men, with combat effectiveness varying between 27 percent for the *13th* and *15th Divisions* to 96 percent for the *7th* and 100 percent for the *2d Division*. His estimate gave the North Korean divisions a loss of 26,820 men in August against a gain of only 14,770 replacements.[47] This estimate, as noted below, was not entirely correct.

### *The North Korean Plan*

In their action against the Perimeter in August the plan and tactics of the North Koreans showed no departure from those that had characterized their

[46] Interv, author with Maj Gen Garrison H. Davidson, 28 Jan 54; Ltr, Dabney to author, 19 Jan 54.

[47] EUSAK WD and PIR 49, 30 Aug 50; *Ibid.*, Sep 50 Summ. On 1 September the United Nations held 1,753 North Korean prisoners—1,372 captured by ROK forces, 381 by Eighth Army.

advance south of the Han River in July. Their divisions simply followed the American and ROK forces on all avenues leading south and closed with them as soon as possible. Enemy action followed the familiar pattern of frontal holding attack, envelopment of the flank, and infiltration to the rear. These tactics had paid high dividends during July when the North Koreans were numerically superior to the forces opposing them and when there was no continuous and connected defense line across the width of Korea. When Eighth Army and the ROK Army withdrew into the Pusan Perimeter in early August and there stabilized a line in relatively connected although thinly held defense positions, these tactics failed for the first time in the war to accomplish their desired result.

The battle line on both flanks rested on the sea. U.N. naval forces secured these flanks. Flanking operations and a tactical decision by grand maneuver were now impossible. Success could come to the North Korean command now in only one way—by frontal attack and penetration of the Perimeter defense followed by immediate exploitation.

Generals MacArthur and Walker applied classical principles of defense in the Pusan Perimeter battles—interior lines of communications for movement of supplies and reinforcements, superior artillery fire power to break the offensive spirit of enemy soldiers and reduce their numbers, and a strong air force which is ideally suited for operational defense because it can intervene quickly in adding its fire power to turn the tide of battle.

The North Korean Army strength during August fell below the combined strength of the U.S. Eighth and the ROK Armies. It is certain also that its combat effectiveness at the first of September was considerably below what it had been a month earlier. While its numbers may have been as large, its trained troops, tanks, and heavy weapons were fewer. Many of the recruits that filled the North Korean divisions in September had no small arms.

The North Korea People's Army had shown a remarkable ability to maintain transport to its front lines over long lines of communications despite heavy and constant air attacks. This accomplishment is one of the outstanding feats of the North Korean war effort in the Pusan Perimeter period. The United Nations air effort failed to halt military rail transport. Ammunition and motor fuel, which took precedence over all other types of supply, continued to arrive at the front, though in diminished quantity. There was still a considerable resupply of heavy weapons, such as tanks, artillery, and mortars, at the front in early September, although a steady decline in artillery can be traced from the middle of August. There was a sufficient supply of small arms ammunition, but a shortage of small arms themselves became apparent by mid-August and continued to worsen with each passing week. Rear areas were able to fill only about one third of the requisitions from the front for small arms in mid-August and resupply ceased entirely about the middle of September. New trucks were almost impossible to obtain. There was no resupply of clothing. At best there were rations for only one or two meals a day. Most units had to live at least partially off the country. By 1 September the food situation was so bad in the North Korean Army at the front that most of the soldiers showed a loss of stamina with resulting impaired combat effectiveness.[48]

The North Koreans directed the Pusan Perimeter battles from their *Front Headquarters* in Kumch'on. Marshal Choe Yong Gun, the North Korean Minister of Defense, was Deputy Commander of the North Korean Armed Forces. He had formerly been associated with the Chinese Communist *8th Route Army*. In command of the *Front Headquarters* during August and September was General Kim Chaek. His chief of staff was Lt. Gen. Kang Kon until the latter was killed near Andong by a land mine explosion on 8 September.

The *II Corps* from its headquarters at Mun'gyong directed the action from north of Taegu eastward to the coast. Lt. Gen. Kim Mu Chong, a graduate of the Whampoa Military Academy under Chiang Kai-shek and a Communist veteran of the Chinese wars, commanded the *II Corps*. He had accompanied Mao Tse Tung on the "Long March" and reportedly was the only one of thirty Koreans to survive that march.

The *I Corps*, which had captured Seoul in the early days of the war, had direct charge under the *Front Headquarters* for the western half of the enemy arc around the Perimeter, from Waegwan south to the Korea Strait. Lt. Gen. Kim Ung, a spectacular soldier, commanded the *I Corps*. Kim had gone from Korea to the Whampoa Military Academy and eventually served with the Communist *8th Route Army* in North China where reportedly he became a brigade or division commander. He was generally considered the ablest of the North Korean field commanders. He was energetic and harsh, feared rather than loved by his subordinates. His *I Corps* headquarters was at Chonju.[49]

With time running against it, the North Korean High Command prepared a massive co-ordinated offensive all around the Pusan Perimeter for the first of September. As the North Korea People's Army prepared for its great effort, it brought 13 infantry divisions, 1 armored division, 2 armored brigades, and miscellaneous security forces into the line. On the *I Corps* front, reaching from opposite Taegu southward along the Naktong River, in line from north to south, were the *10th, 2d, 4th, 9th, 7th,* and *6th Infantry Divisions*. Elements of the *105th Armored Division* and the newly arrived *16th Armored Brigade* supported these troops. The *16th Armored Brigade,* really a regiment, had forty-three new T34 tanks when it left P'yongyang in August to take part in the September offensive. Back of the *6th Division* was the *104th Security Brigade*. Deployed along the *II Corps* front from northwest of Taegu eastward to the coast and in line from west to east were the *3d, 13th, 1st, 8th, 15th, 12th,* and *5th Infantry Divisions*. Elements of the *105th Armored Division* and the newly arrived *17th Armored Brigade* supported this corps. The *17th Armored Brigade,* also actually

[48] DA Intel Rev, Dec 50, 175, pp. 36–38, ATIS Interrog Rpts, Issue 3, Rpt 895, p. 214, Maj Kim Song Won, CO *19th Regt, 13th Div*.

[49] GHQ FEC, History of the North Korean Army, pp. 41, 84, 91–94, 98; ATIS Enemy Docs, Issue 9, p. 66, ltr, Kim Man Hwa to Col Lee Hak Ku, CofS N.K. *13th Div,* 8 Sep 50; New York *Times,* August 25 and September 11, 1950.

a regiment, had forty new tanks when it left P'yongyang. Most, if not all, of the tanks in the two brigades apparently arrived in P'yongyang on or about 23 August, coming from the Russians by way of Manchuria. Trained crews were immediately assigned to the tanks. The two armored brigades each had two battalions; each battalion was composed of four tank companies. The two new armored brigades moved to the front by rail at night.[50]

Other than the *17th Armored Brigade,* the *II Corps* had no new units along the northern and eastern front for the September offensive. In the *I Corps* sector were two new and previously uncommitted infantry formations to strengthen the assault forces there. The *9th Infantry Division,* formed around the old *3d Border Constabulary,* arrived in the Hyopch'on area from Seoul (less its *3d Regiment* which remained at Inch'on) on or about 25 August. The *7th Infantry Division,* in the Chinju-Masan area, had not been committed except for two battalions that fought briefly against ROK marines at T'ongyong.[51]

The North Korean force assembled at the front on 1 September for the assault against the Pusan Perimeter numbered about 98,000 men. Perhaps a third were raw recruits, most of them forcibly conscripted in South Korea and hastened to the front with little or no training and with few weapons. It is believed that the major organizations had personnel strength approximately as follows:[52]

| | |
|---|---|
| Total | 97,850 |
| 1st Infantry Division | 5,000 |
| 2d Infantry Division | 6,000 |
| 3d Infantry Division | 7,000 |
| 4th Infantry Division | 5,500 |
| 5th Infantry Division | 7,000 |
| 6th Infantry Division | 10,000 |
| 7th Infantry Division | 9,000 |
| 8th Infantry Division | 6,500 |
| 9th Infantry Division | 9,350 |
| 10th Infantry Division | 7,500 |
| 12th Infantry Division | 5,000 |
| 13th Infantry Division | 9,000 |
| 15th Infantry Division | 7,000 |
| 104th Security Brigade | 2,000 |
| 105th Armored Division | 1,000 |
| 16th Armored Brigade | 500 |
| 17th Armored Brigade | 500 |

Planning for the massive attack was under way for at least the last ten days of August since the N.K. Army operational order for the *I Corps* attack was issued on or about 20 August. The enemy plan indicated five major groupings of assault units and objectives:

1. *6th* and *7th Divisions* to break through the U.S. 25th Division to Masan in the south.

2. *9th, 4th, 2d,* and *10th Divisions* to break through the U.S. 2d Division to Miryang and the Pusan-Taegu railroad and highway by way of Changnyong and Yongsan.

3. *3d, 13th,* and *1st Divisions* to break through the U.S. 1st Cavalry Division and the ROK 1st Division to Taegu.

[50] GHQ FEC, History of the North Korean Army; ATIS Res Supp Interrog Rpts, issues for the N.K. divisions previously cited; ORO, Employment of Armor in Korea, ORO-R-1 (FEC), vol. 1, p. 165, app. F (Apr 51); EUSAK WD, 9 Sep 50, PW Rpts, 2d Lt Won Hong Ki and Sgt Choi Soong Moon; *Ibid.,* PIR 55, 5 Sep 50.

[51] ATIS Res Supp Interrog Rpts, Issue 100 (N.K. *9th Div*), p. 49; *Ibid.,* Issue 99 (N. K. *7th Div*), p. 35.

[52] These figures are based on information derived from enemy material: ATIS Res Supp Interrog Rpts, Issues 3, 4, 94, 96, 99, 100, 104 (N.K. *1st, 2d, 3d, 4th, 5th, 6th, 7th, 8th, 9th, 10th, 12th, 13th Divs);* GHQ FEC, History of the North Korean Army.

4. *8th* and *15th Divisions* to break through the ROK 8th and 6th Divisions to Hayang and Yongch'on in the lateral corridor east of Taegu.

5. *12th* and *5th Divisions* to break through the ROK Capital and 3d Divisions to P'ohang-dong, Yonil Airfield, and the Kyongju corridor to Pusan.

Assault groupings 1 and 2 of *I Corps* were to begin their co-ordinated attacks at 2330, 31 August; assault groupings 3, 4, and 5 of *II Corps* were to attack at 1800, 2 September.[53]

---

[53] GHQ FEC, History of the North Korean Army, pp. 57, 65, 73; ATIS Res Supp Interrog Rpts, Issue 100 (N.K. *6th Div*) pp. 40–42; *Ibid.*, Issue 3 (N.K. *15th Div*), p. 44; ATIS Enemy Docs, Issue 9, p. 65, notebooks belonging to Col Lee Hak Ku, CofS *13th Div*, gives summary of order from Mu Chong, CG N.K. *II Corps; Ibid.*, p. 69, ltr from Choe Fam, CO *21st Regt, 13th Div,* to Col Lee Hak Ku, 2 Sep 50.

# CHAPTER XXII

# Perimeter Battle

Every battle has a turning point when the slack water of uncertainty becomes the ebb tide of defeat or the flood water of victory.

ADMIRAL CHARLES TURNER JOY

For most of the men who fought the battles of the Pusan Perimeter in early September 1950, it was a period of confusion. So many actions went on simultaneously that only a wide-screen view could reveal the situation as the commander had to cope with it in its totality. Since this panoramic approach is not feasible, the story in this chapter will follow the battles from the east coast near P'ohang-dong westward to Taegu and the Naktong River for the first two weeks of September. The next chapter will follow the battles for the same period of time in the southern part of the Pusan Perimeter.

It is necessary to keep in mind that not one of the battles in this phase of the war was an isolated event, but that everywhere over the extent of the Perimeter other battles of equal, greater, or lesser intensity were being waged. As an example of their impact, on 3 September 1950 General Walker faced at least five distinct and dangerous situations on the Perimeter—an enemy penetration in the east at P'ohang-dong, severance of the lateral corridor at Yongch'on between Taegu and P'ohang-dong, alarming enemy gains in the mountains north of Taegu, the threat posed by North Korean units slicing through the defenses of the Naktong Bulge area of the lower Naktong, and enemy penetration behind the greater part of the 25th Division in the Masan area in the extreme south. In addition, at this time in the east the ROK II Corps was on the point of collapse; above Taegu the 1st Cavalry Division withdrew closer to that city; and in the south disaster threatened the U.S. 2d and 25th Divisions.

## *Action in the East—Task Force Jackson*

Although the N.K. *II Corps'* general attack in the north and east was planned for 2 September, the enemy *12th Division,* now numbering about 5,000 men, started earlier to move forward from the mountain fastnesses where it had reorganized after its defeat in the Kigye and P'ohang-dong area. (*Map 14*) The division was low in food supply, weapons, and ammunition, and its men were in a state of lowered morale. On 26 August, American and ROK officers in the P'ohang-dong–Kigye area with great optimism congratulated each other on having repulsed what they thought was the last serious threat to the Pusan Perimeter. In their view the North Koreans were now on the defensive and the war might end by Thanksgiving.

Nearest to the N.K. *12th Division* was the ROK Capital Division. At 0400, 27 August, a North Korean attack overran one company of the ROK 17th Regiment, Capital Division, north of Kigye. This caused the whole regiment to give way. Then the 18th Regiment on the right fell back because of its exposed flank. The 17th Regiment lost the town of Kigye, and the entire Capital Division fell back three miles to the south side of the Kigye valley. This enemy blow fell with startling impact on Eighth Army in the predawn hours of 27 August.[1]

At the briefing at Eighth Army headquarters in Taegu on Sunday, 27 August, General Walker showed his concern over this development. One of those present was Maj. Gen. John B. Coulter who had arrived in Korea about a month earlier. Half an hour after the briefing ended, General Walker called General Coulter to him and said, "I can't get reliable reports. I want you to go to the eastern front and represent me. I am sending a regiment from the 24th Division to help." [2]

Coulter flew to Kyongju at once, arriving there at noon. Walker in the meantime formally appointed Coulter Deputy Commander, Eighth Army, placing him in command of the ROK I Corps, the U.S. 21st Infantry, the 3d Battalion, 9th Infantry, and the 73d Medium Tank Battalion, less C Company. General Coulter designated these units Task Force Jackson and established his headquarters in the same building in Kyongju in which the ROK I Corps commander and the KMAG officers had their command post. He assumed command of Task Force Jackson at 1200, 27 August.[3]

When he arrived at Kyongju that Sunday, General Coulter found the ROK I Corps disintegrating rapidly and in low morale. Coulter talked to the ROK commanders and their staffs about the terrible effect of their failure to stop the North Koreans and the danger it posed for the entire Pusan Perimeter. General Walker had instructed him to issue his orders to the ROK I Corps commander or his chief of staff in the form of advice, which Coulter did. Coulter had the mission of eliminating the enemy penetration in the Kigye area and of seizing and organizing the high ground extending from north of Yongch'on northeasterly to the coast at Wolp'o-ri, about twelve miles north of P'ohang-dong. This line passed ten miles north of Kigye. Coulter was to attack at once with Task Force Jackson, his immediate objective being to gain the first high ground north of Kigye. The U.S. 21st Infantry Regiment on the morning of 27 August was moving to a position north of Taegu, when General Walker revoked its orders and instructed Colonel Stephens to turn the regiment around and proceed as rapidly

---

[1] EUSAK WD, G-3 Sec, Msg 0655 from KMAG, 27 Aug 50; *Ibid.*, Aug 50 Summ, p. 77; GHQ FEC G-3 Opn Rpt 64, 27 Aug 50; New York *Herald Tribune*, August 28, 1950, Bigart dispatch of 27 August.

[2] Interv, author with Lt Gen John B. Coulter, 3 Apr 53.

[3] *Ibid.*, EUSAK WD, G-3 Sec, 27 Aug 50; GHQ FEC G-3 Opn Rpt 64, 27 Aug 50; *Ibid.*, Sitrep, 28 Aug 50; Capt. George B. Shutt, Operational Narrative History of Task Force Jackson, MS in National Archives Record Service.

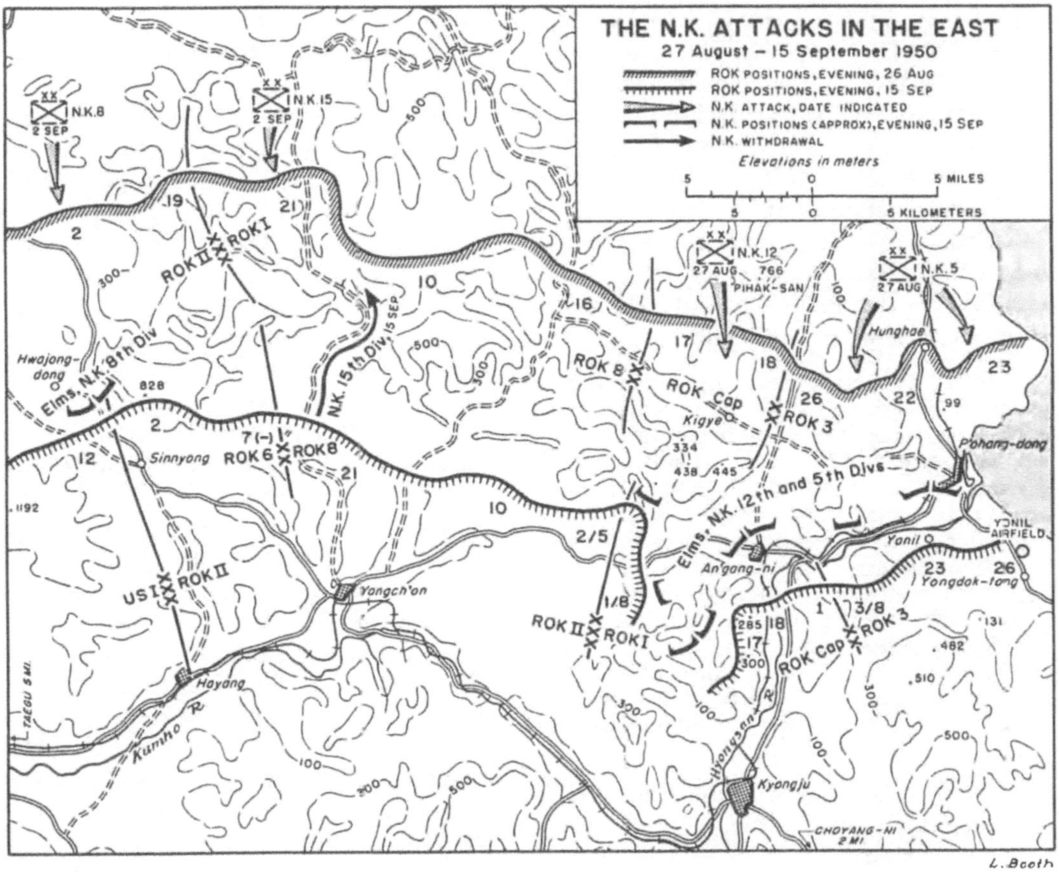


MAP 14

as possible to Kyongju and report to General Coulter. The regiment departed Taegu at 1000 and arrived at Kyongju that afternoon. Coulter immediately sent the 3d Battalion north to An'gang-ni where it went into a position behind the ROK Capital Division.[4]

General Coulter's plan to attack on 28 August had to be postponed. The ROK I Corps commander told him he could not attack, that there were "too many enemy, too many casualties, troops tired." Also, the N.K. *5th Division* above P'ohang-dong had begun to press south again and the ROK 3d Division in front of it began to show signs of giving way. On the 28th, Colonel Emmerich, the KMAG adviser to the ROK 3d Division, at a time he deemed favorable, advised Brig. Gen. Kim Suk Won, the ROK division commander, to counterattack, but General Kim refused to do so. The next day Kim said he was going to move his command post out of

[4] Ltr, Coulter to author, 7 Jul 53; Ltr, Stephens to author, 14 May 53; EUSAK WD, G–3 Sec, 27 Aug 50; 21st Inf WD, 27 Aug 50; 24th Div WD, 27 Aug 50.

P'ohang-dong. Emmerich replied that the KMAG group was going to stay in P'ohang-dong. Upon hearing that, Kim became hysterical but decided to stay for the time being to avoid loss of face. That day, 28 August, General Walker issued a special statement addressed to the ROK Army, and meant also for the South Korean Minister of Defense. He called on the ROK's to hold their lines in the Perimeter, and said:

> It is my belief, that the over-extended enemy is making his last gasp, while United Nations forces are daily becoming stronger and stronger. The time has now come for everyone to stand in place and fight, or advance to a position which will give us greater tactical advantage from which the counter-offensive can be launched. If our present positions are pierced, we must counterattack at once, destroy the enemy and restore the positions.
>
> To you officers and soldiers of the Army of the Republic of Korea, I ask that you rise as one and stop the enemy on your front in his tracks.[5]

The ROK disorganization was so great in the face of continued enemy pressure that Task Force Jackson could not launch its planned co-ordinated attack. Colonel Stephens' 21st Infantry was in an assembly area two miles north of An'gang-ni and ready for an attack the morning of the 28th, but during the night the ROK 17th Regiment lost its position on the high ridge northward at the bend of the Kigye valley, and the attack was canceled. The ROK's regained their position in the afternoon but that night lost it again. At the same time, elements of the enemy *5th Division* penetrated the ROK 3d Division southwest of P'ohang-dong. General Coulter directed Colonel Stephens to repel this penetration. During the 29th, B Company, 21st Infantry, supported by a platoon of tanks of B Company, 73d Medium Tank Battalion, successfully counterattacked northwest from the southern edge of P'ohang-dong for a distance of a mile and a half, with ROK troops following. The American units then withdrew to P'ohang-dong. That night the ROK's withdrew, and the next day an American infantry-tank force repeated the action of the day before. Colonel Stephens now received orders to take over from the ROK 3d Division a sector extending 1,000 yards north and 3,000 yards northwest of P'ohang-dong.[6]

Also on the 29th, the ROK Capital Division, with American tank and artillery support, recaptured Kigye and held it during the night against enemy counterattacks, only to lose it finally at dawn. American air attacks continued at an increased tempo in the Kigye area. On 31 August, the aircraft carrier USS *Sicily* alone launched 38 sorties. ROK troops reported finding the bodies of many North Koreans, apparently killed by air attack. They also found many suits of white clothing scattered on the ground, abandoned when enemy soldiers changed into uniforms.

Coincidentally with this air action in the Kigye area, U.S. naval vessels continued their efforts to help stop the N.K. *5th Division* on the east coast. A

[5] Ltr, Coulter to author, 7 Jul 53; EUSAK WD, G-3 Sec, 28 Aug 50; 21st Inf WD, 28 Aug 50; GHQ FEC Sitrep, 28 Aug 50; Ltr, Emmerich to Farrell, 29 Aug 50, recommending relief of Gen Kim Suk Won; New York *Times*, August 28, 1950.

[6] Ltr, Stephens to author, 14 May 53, and accompanying map of 21st Inf positions; EUSAK WD, G-3 Sec, Msg 1710 from TF Jackson; 21st Inf WD, 29 Aug 50.

cruiser and two destroyers concentrated their fire power on the Hunghae area five miles north of P'ohang-dong where the enemy division's troop assembly and forward supply center were located. On 29 and 30 August the three vessels fired almost 1,500 5-inch shells at enemy targets there in support of the ROK 3d Division. Despite this aerial and naval support, on the last day of August the battle continued to go against the ROK forces both at Kigye and P'ohang-dong.[7]

Aerial observation on 1 September disclosed that North Koreans were moving southward in the mountains above Kigye and P'ohang-dong. The next day another major enemy attack was forming north and northwest of Kigye. In the afternoon, KMAG advisers with the Capital Division estimated that 2,500 enemy soldiers had penetrated a gap between the ROK 17th and 18th Regiments.

At the same time, enemy pressure built up steadily north of P'ohang-dong, where the N.K. *5th Division* fed replacements on to Hill 99 in front of the ROK 23d Regiment. This hill became almost as notorious as had Hill 181 near Yongdok earlier because of the almost continuous and bloody fighting there for its control. Although aided by U.S. air attacks and artillery and naval gunfire, the ROK 3d Division was not able to capture this hill, and suffered many casualties in the effort. On 2 September Colonel Stephens' 21st Infantry attacked northwest from P'ohang-dong in an effort to help the ROK's recapture Hill 99. A platoon of tanks followed the valley road between P'ohang-dong and Hunghae. Stephens assigned K Company Hill 99 as its objective. The 21st Infantry made very slow progress in this attack, and in some quarters none at all. Casualties were heavy. By 1525 that afternoon K Company could account for only thirty-five men. The company was unable to take Hill 99 from the well dug-in North Koreans who threw showers of hand grenades to repel all efforts to reach the top. Two tanks of the 6th Tank Battalion were lost in this attack, one in an enemy mine field and another because of a thrown track. At dusk an enemy penetration occurred along the boundary between the ROK Capital and 3d Divisions three miles east of Kigye.[8]

The next morning, an hour and a half after midnight, the N.K. *12th Division*, executing its part of the co-ordinated N.K. *II Corps* general attack, struck the Capital Division on the high hill masses south of the Kigye valley. This attack threw back the ROK 18th Regiment on the left in the area of Hills 334 and 438, and the ROK 17th Regiment on the right in the area of Hill 445. By dawn of 3 September the enemy penetration there had reached the vital east-west corridor road three miles east of An'gang-ni. As a result of this 5-mile enemy gain during the night the Capital Division all but collapsed.[9]

---

[7] Interv, author with Coulter, 3 Apr 53; GHQ FEC G-3 Opn Rpts 64-68, 27-31 Aug 50; *Ibid.*, Sitreps, 28 Aug-2 Sep 50; EUSAK WD, G-3 Sec, 31 Aug 50; *Ibid.*, POR 149, 31 Aug 50; 21st Inf WD, 31 Aug 50.

[8] Ltr and marked map, Stephens to author, 14 May 53; EUSAK WD, G-3 Jnl, 2 Sep 50; 21st Inf WD, 2 Sep 50; 6th Tk Bn WD, 1-3 Sep 50; Emmerich, MS review comments, 30 Nov 57.

[9] EUSAK WD, G-3 Jnl, 0845 3 Sep 50; *Ibid.*, PIR 53, 3 Sep 50; GHQ FEC G-3 Opn Rpt 71, 3 Sep 50; *Ibid.*, Sitrep, 3 Sep 50; ROK Army Hq, MS review comments, 11 Jul 58.

This dire turn of events forced General Coulter to withdraw the 21st Infantry at once from the line northwest of P'ohang-dong and concentrate it forthwith in the vicinity of Kyongju. The 2d Battalion, commanded by Lt. Col. Gines Perez, had joined the regiment as its third battalion on 31 August, but General Coulter had held it in Task Force reserve at An'gang-ni. That battalion now took up a horseshoe-shaped defense position around the town, with some elements on high ground two miles eastward where they commanded the Kyongju–P'ohang-dong highway. The rest of the regiment closed into an assembly area north of Kyongju. At the same time, General Walker started the newly activated ROK 7th Division toward the enemy penetration. Its 5th Regiment closed at Yongch'on that afternoon, and the 3d Regiment, less its 1st Battalion, closed at Kyongju in the evening. Walker also authorized Coulter to use the 3d Battalion, 9th Infantry; the 9th Infantry Regimental Tank Company; and the 15th Field Artillery Battalion as he deemed advisable. These units, held at Yonil Airfield for its defense, had not previously been available for commitment elsewhere. The two antiaircraft batteries of automatic weapons (D Battery, 865th AAA Battalion, and A Battery, 933d AAA Battalion) were not to be moved from the airfield except in an emergency.[10]

During the day (3 September), Colonel Emmerich at P'ohang-dong sent General Coulter a message that the ROK 3d Division commander was preparing to withdraw from P'ohang-dong. Coulter went immediately to the ROK I Corps commander and had him issue an order that the ROK 3d Division would not withdraw. Coulter checked every half hour to see that the division stayed in its P'ohang-dong positions.

That night, 3–4 September, the ROK I Corps front collapsed. Three enemy tanks overran a battery of ROK artillery and then scattered two battalions of the newly arrived ROK 5th Regiment. Following a mortar preparation, the North Koreans entered An'gang-ni at 0220. An hour later the command post of the Capital Division withdrew from the town and fighting became increasingly confused. By 0400 American tanks ceased firing because remnants of the Capital Division had become hopelessly intermingled with enemy forces. Colonel Perez said, "We couldn't tell friend from foe." At daylight, G Company, 21st Infantry, discovered that it was alone in An'gang-ni, nearly surrounded by the enemy. ROK troops had disappeared. At 1810, G Company withdrew from the town and dug in along the road eastward near the rest of the 2d Battalion at the bridge over the Hyongsan-gang. North Koreans held the town and extended southward along the railroad.[11]

Receiving orders from Colonel Stephens to withdraw the 2d Battalion and join the regiment above Kyongju, Colonel Perez had to fight his way through an enemy roadblock on the east side of

[10] Interv, author with Coulter, 3 Apr 53; Ltr, Stephens to author, 14 May 53; EUSAK WD, G–3 Jnl 0840, 1420 3 Sep 50; *Ibid.*, G–3 Sec and Br for CG, 3 Sep 50; New York *Herald Tribune*, September 6, 1950, Bigart dispatch.

[11] EUSAK WD, G–3 Jnl, 4 Sep 50; 21st Inf WD, 28 Aug–28 Sep 50 Summ, for 3–4 Sep; GHQ FEC Sitrep, 4 Sep 50; New York *Herald Tribune*, September 6, 1950, Bigart dispatch.

ASSAULT TROOPS OF COMPANY K, *21st Infantry, under mortar fire on Hill 99, 2 September.*

the Hyongsan-gang three miles southeast of An'gang-ni. When he got through he discovered that G Company was missing. Colonel Stephens ordered Perez to turn around and get G Company. The 2d Battalion fought its way back north and found G Company at the bridge. Reunited, the battalion fought its way out again, with tanks firing down the road ahead of the column and into the hills along the sides. Enemy fire knocked the tracks off three Patton tanks. Friendly artillery then destroyed them to prevent enemy use. The 2d Battalion arrived in the Kyongju area shortly before noon.[12]

[12] 21st Inf WD, 5 Sep 50; 24th Div WD, 4 Sep 50; Ltr, Stephens to author, 14 May 53.

By noon, 4 September, enemy units had established roadblocks along the Kyongju–An'gang-ni road within three miles of Kyongju. A 2-mile-wide gap existed between the ROK 3d and Capital Divisions in the P'ohang-dong area. But the big break in the United Nations line was in the high mountain mass west of the Hyongsan valley and southwest of An'gang-ni. In this area northwest of Kyongju there was an 8-mile gap between the Capital Division and the ROK 8th Division to the west. From that direction the enemy posed a threat, General Coulter thought, to the railroad and the road net running south through the Kyongju corridor to Pusan. He was not equally concerned about enemy advances in the P'ohang-dong coastal area. Faced with this big gap on his left flank, Coulter put Stephens' 21st Infantry in the broad valley and on its bordering hills northwest of Kyongju to block any enemy approach from that direction.[13]

The situation at Kyongju during the evening of 4 September was tense. The ROK corps commander proposed to evacuate the town. He said that the North Koreans were only three miles away on the hills to the north, and that they would attack and overrun the town that night. General Coulter told him that he would not move his command post—that they were all staying in Kyongju. And stay they did. Coulter put four tanks around the building where the command posts were located. Out on the roads he stationed KMAG officers to round up ROK stragglers and get them into positions at the edge of the town. One KMAG major at pistol point stopped ROK troops fleeing southward. Most of his staff at Kyongju found Coulter irritable and hard to please, but they also say that he went sleepless and was determined to hold Kyongju.[14]

That night radio conversations between tankers on the road just north of Kyongju, overheard at Coulter's headquarters, told of knocking North Koreans off the tanks. The expected North Korean attack on Kyongju, however, never came. The enemy turned east, crossed the highway a few miles north of the town, and headed toward Yonil Airfield. The next day the Air Force, attacking enemy gun positions four miles north of Kyongju along the road, found enemy targets at many points within the triangle Kigye–Kyongju–P'ohang-dong.

North of P'ohang-dong the situation worsened. At 0200 5 September Colonel Emmerich hastened to Yonil Airfield where he conferred with Lt. Col. D. M. McMains, commanding the 3d Battalion, 9th Infantry, stationed there, and informed him of the situation in P'ohang-dong. Emmerich obtained a platoon of tanks and returned with them to the town. He placed the tanks in position and awaited the expected enemy armored attack. At 0530 he received information that elements of the ROK 22d Regiment had given way. Enemy troops entered this gap and just before 1100 the American tanks in P'ohang-dong were under heavy enemy machine gun fire. Five N.K. self-propelled guns approached and began firing. At a range of one city block the tanks knocked out the lead gun, killing three crew members. In the ensuing exchange of fire the other four withdrew. Emmerich then directed air strikes and artillery fire which destroyed the other four guns. But, nevertheless, that afternoon at 1435 the order came to evacuate all materiel and supplies from the Yonil airstrip.[15]

That night, 5–6 September, events reached a climax inside P'ohang-dong.

---

[13] Interv, author with Coulter, 3 Apr 53; Ltr, Stephens to author, 14 May 53; EUSAK WD, Br for CG, 4 Sep 50; 6th Med Tk Bn WD, 4 Sep 50.

[14] Interv, author with Coulter, 3 Apr 53; Interv, author with Col John F. Greco, 12 Aug 51 (Greco was Coulter's G–2 at Kyongju); Interv, author with Maj Wm. C. Hungate, Jr., 28 Jul 51; Interv, author with Maj George W. Flagler, 28 Jul 51 (both Hungate and Flagler were at Kyongju with Coulter); Shutt, History of Task Force Jackson.

[15] Interv, author with Coulter, 3 Apr 53; EUSAK WD, G–3 Sec, 5 Sep 50; *Ibid.*, Br for CG, 5 Sep 50; *Ibid.*, PIR 55, 5 Sep 50; 21st Inf WD, 5 Sep 50; GHQ FEC Opn Rpt 74, 6 Sep 50; *Ibid.*, Sitrep, 6 Sep 50; Shutt, History of Task Force Jackson; Emmerich, MS review comments, 30 Jan 57.

At midnight, after ten rounds of enemy mortar or artillery fire struck near it, the ROK 3d Division command post moved to another location. Enemy fire that followed it to the new location indicated observed and directed fire. The ROK division commander and his G–2 and G–3 "got sick." The division withdrew from P'ohang-dong, and on 6 September this coastal town was again in enemy hands. The ROK Army relieved both the ROK I Corps and the 3d Division commanders.[16]

Because the big gap between the ROK Capital and 8th Divisions made it impossible for I Corps at Kyongju to direct the action of the 8th Division, the ROK Army at 1030, 5 September, transferred that division to the control of the ROK II Corps, and attached to it the 5th Regiment of the ROK 7th Division. This shift of command came just as the N.K. *15th Division* penetrated the ROK 8th Division lines to enter Yongch'on in the Taegu–P'ohang-dong corridor. From west of An'gang-ni the ROK 3d Regiment drove toward Yongch'on, still trying to close the gap.[17]

The startling gains of the North Koreans in the east on 4 September caused General Walker to shift still more troops to that area. The day before, he had ordered the 24th Division to move from its reserve position near Taegu to the lower Naktong River to relieve the marines in the Naktong Bulge area of the 2d Division front. It bivouacked that night in a downpour of rain on the banks of the Naktong near Susan-ni. On the morning of the 4th, before it could begin relief of the marines, the 24th received a new order to proceed to Kyongju. General Davidson, the assistant division commander, proceeding at once by jeep, arrived at Kyongju that evening. Division troops and the 19th Infantry started at 1300 the next day, 5 September, and, traveling over muddy roads, most of them arrived at Kyongju just before midnight. General Church had arrived there during the day. All division units had arrived by 0700, 6 September.[18]

General Coulter knew that the N.K. *15th Division* had crossed the Taegu lateral corridor at Yongch'on and was heading in the direction of Kyongju. On the 6th, he ordered the 21st Infantry to attack the next day up the valley and bordering hills that lead northwest from Kyongju into the high mountain mass in the direction of Yongch'on. When it attacked there on 7 September the 21st Infantry encountered virtually no opposition.

At 1230 Eighth Army redesignated Task Force Jackson as Task Force Church, and half an hour later General Coulter departed Kyongju for Taegu to resume his planning duties. General Church was now in command on the eastern front. That afternoon, 7 September, General Church canceled General Coulter's order for the 21st Infantry to attack into the mountains. He felt it was a useless dispersion of troops and he wanted the regiment concentrated near Kyongju. Church made still another change in the disposition of the

[16] Interv, author with Emmerich, 5 Dec 51; Interv, author with Coulter, 3 Apr 53.

[17] GHQ FEC Sitrep, 5 Sep 50; *Ibid.*, G–3 Opn Rpts 73–74, 5–6 Sep 50.

[18] Ltr, Davidson to author, 18 Feb 54; 24th Div WD, 5–6 Sep 50; EUSAK WD, G–3 Jnl, 1800 5 Sep 50; EUSAK Opn Ord for CG 24th Div and CG TF Jackson, 5 Sep 50.

task force. On the 8th he moved its command post from Kyongju to the vicinity of Choyang-ni, four miles southward. He believed the command post could be more easily defended there in the open if attacked than in a town, and that traffic congestion near it would be less.[19]

Fighting continued between the North Koreans and the ROK Capital Division on the hills bordering the valley from An'gang-ni to Kyongju. The 3d Battalion, 19th Infantry, became involved there just after midnight, 8–9 September. An enemy force attacked K Company and drove it from Hill 300, a defensive position midway between An'gang-ni and Kyongju. North Koreans held the hill during the 9th against counterattack. Farther north, on the left side of the valley, the ROK 17th Regiment attacked and, with the support of the U.S. 13th Field Artillery Battalion, captured Hill 285 and held it against several enemy counterattacks. On the opposite side of the valley (east) the ROK 18th Regiment made limited gains. These battles took place in drenching typhoon rains. Low-hanging clouds allowed very little air support. The rains finally ceased on 10 September.[20]

In this second week of September elements of the N.K. *5th Division* had spread out over the hills west, southwest, and south of P'ohang-dong. One North Korean force, estimated to number 1,600 men, reached Hills 482 and 510, four to five miles southwest of Yonil Airfield. Facing this enemy force were two regiments of the ROK 3d Division, which held a defensive position on the hills bordering the west side of the valley south of the airfield. Enemy pressure threatened to penetrate between the two ROK regiments.

On the evening of 9 September, General Church formed Task Force Davidson to eliminate this threat to Yonil. The airfield itself had not been used since the middle of August except for emergency landing and refueling of planes, but evacuation of Air Force equipment, bombs, and petroleum products was still in progress. General Davidson commanded the task force, which was composed of the 19th Infantry, less the 3d Battalion; the 3d Battalion, 9th Infantry; the 13th Field Artillery Battalion; C Battery, 15th Field Artillery Battalion; A Company, 3d Engineer Combat Battalion; the 9th Infantry Regimental Tank Company; two batteries of antiaircraft automatic weapons; and other miscellaneous units.[21]

The enemy having cut off all other approaches from the Kyongju area, the task force spent all of 10 September making a circuitous southern approach to its objective. It arrived in its assembly area at Yongdok-tong, one mile south of Yonil Airfield, at 1900 that evening. General Davidson early that morning had flown on ahead from Kyongju to Yongdok-tong. Colonel Emmerich was there to meet him when his light plane landed on the road. On the flight over, Davidson looked for but did not see any

[19] Ltrs, Church to author, 3 May and 26 Jul 53; Ltr, Davidson to author, 18 Feb 54; 24th Div WD, 7–8 Sep 50; EUSAK WD, G–3 Jnl, 7 Sep 50; *Ibid.*, POR 171, 7 Sep 50.

[20] 24th Div WD, 9 Sep 50; 3d Engr C Bn WD, Sep 50, Summ.

[21] Interv, author with Davidson, 28 Jan 54; Ltr, Davidson to author, 18 Feb 54; 19th Inf WD, 10 Sep 50; 24th Div WD, 9 Sep 50; EUSAK WD, G–3 Sec, 10 Sep 50; 3d Engr C Bn Unit Hist, 6 Aug–28 Sep 50.

enemy soldiers. Emmerich told Davidson the North Koreans had driven the ROK's from Hill 131. This hill was on the southern side of the boundary between the two ROK regiments holding the Yonil defensive position. Davidson and Emmerich agreed that the ROK's would have to recapture Hill 131 during the night and that then the task force would attack through the ROK 3d Division to capture the main enemy positions on Hill 482. They thought that if the task force could establish the ROK's on Hill 482 the latter should be able to hold it and control the situation themselves thereafter. Emmerich took Davidson to meet the ROK 3d Division commander. Davidson told him that he was in command in that area and informed him of his plan for the attack. That night the ROK's did succeed in recapturing Hill 131 and restoring their lines there. In this attack the ROK 3d Engineer Battalion fought as infantry, and under the leadership and guidance of Capt. Walter J. Hutchins, the KMAG adviser to the battalion, contributed heavily to the success.[22]

The next morning, 11 September, the 19th Infantry passed through the left-hand ROK regiment just south of Hill 131 and, with the 1st Battalion leading, attacked west. At 0930 it captured without opposition the first hill mass two miles west of the line of departure. The 2d Battalion then passed through the 1st Battalion and continued the attack toward Hill 482 (Unje-san), a mile westward across a steep-sided gorge. There, North Koreans held entrenched positions, and their machine gun fire checked the 2d Battalion for the rest of the day. The morning of 12 September four Australian pilots struck the enemy positions with napalm, and an artillery preparation followed the strike. The 2d Battalion then launched its attack and secured the rough and towering Hill 482 about noon. In midafternoon, ROK forces relieved Task Force Davidson on the hill mass, and the latter descended to the valley southwest of Yongdok-tong for the night. During the day, General Walker had visited the task force's command post two or three times. On 13 September, Task Force Davidson returned to Kyongju.[23]

While this action was in progress near Yonil Airfield, the week-long battle for Hill 300 north of Kyongju came to an end. A regiment of the ROK 3d Division captured the hill on 11 September. In midafternoon the 3d Battalion, 19th Infantry, relieved the ROK's there. Scattered over Hill 300 lay 257 counted enemy dead and great quantities of abandoned equipment and weapons, some of it American. In this fighting for Hill 300, the U.S. 3d Battalion, 19th Infantry, lost eight lieutenants and twenty-nine enlisted men killed.[24]

Tuesday, 12 September, may be considered as the day when the North Korean offensive in the east ended. By that date, the N.K. *12th Division* had been virtually destroyed and the *5th Division* was trying to consolidate its

[22] Ltr, Davidson to author, 18 Feb 54; 24th Div WD, 10 Sep 50; EUSAK WD, G–3 Sec, 10 Sep 50; Emmerich, MS review comments, 30 Nov 57.

[23] Ltr, Davidson to author, 18 Feb 54; 24th Div WD, 11–12 Sep 50; *Ibid.*, Opn Summ, 26 Aug–28 Sept 50; EUSAK POR 186, 12 Sep 50; EUSAK WD, G–3 Jnl, 1046 12 Sep 50.

[24] 21st Inf WD, 11 Sep 50; 24th Div WD, 26 Aug–28 Sep 50, p. 40; 24th Div Arty WD, 12 Sep 50. Ltr, Stephens to author, 14 May 53. The 3d Battalion, 19th Infantry, was attached to the 21st Infantry during this action.

survivors near P'ohang-dong. Aerial observers reported sighting many enemy groups moving north and east.[25]

The ROK 3d Division followed the withdrawing *5th Division,* and the Capital Division advanced against the retreating survivors of the enemy *12th Division.* On 15 September some elements of the Capital Division reached the southern edge of An'gang-ni. Reports indicated that enemy troops were retreating toward Kigye. With the enemy threat in the east subsiding, Eighth Army dissolved Task Force Church, effective at noon 15 September, and the ROK Army resumed control of the ROK I Corps. Eighth Army also ordered the 24th Division to move to Kyongsan, southeast of Taegu, in a regrouping of forces. The 21st Infantry Regiment had already moved there on the 14th. The 19th Infantry was to remain temporarily at Kyongju in Eighth Army reserve.[26]

In the eastern battles during the first two weeks of September, the ROK troops, demoralized though they were, did most of the ground fighting. American tanks, artillery, and ground units supported them. Uncontested aerial supremacy and naval gunfire from offshore also supported the ROK's, and probably were the factors that tipped the scales in their favor. After the initial phase of their September offensive, the North Koreans labored under what proved to be insurmountable difficulties in supplying their forward units. The North Korean system of supply could not resolve the problems of logistics and communication necessary to support and exploit an offensive operation in this sector of the front.

### *Enemy Breakthrough at Yongch'on*

In the high mountains between the Taegu sector on the west and the Kyongju–east coast sector, two North Korean divisions, the *8th* and *15th,* stood ready on 1 September to attack south and sever the Taegu–P'ohang-dong corridor road in the vicinity of Hayang and Yongch'on, in co-ordination with the North Korean offensive in the Kigye-P'ohang area. Hayang is 12, and Yongch'on 20, air miles east of Taegu. The N.K. *8th Division* was astride the main Andong-Sinnyong-Yongch'on road 20 air miles northwest of Yongch'on; the *15th* was eastward in the mountains just below Andong, 35 air miles north of Yongch'on on a poor and mountainous secondary road. The objective of the *8th Division* was Hayang; that of the *15th* was Yongch'on, which the enemy division commander had orders to take at all costs. Opposing the N.K. *8th Division* was the ROK 6th Division; in front of the N.K. *15th Division* stood the ROK 8th Division.[27]

In ten days of fighting the N.K. *8th Division* gained only a few miles, and not until 12 September did it have possession of Hwajong-dong, 14 air miles northwest of Yongch'on. In this time it

[25] ATIS Res Supp Interrog Rpts, Issue 106 (N.K. *12th Div*), p. 70; *Ibid.,* Issue 96 (N.K. *5th Div*), p. 43. There are many individual enemy interrogation reports in ATIS Interrogation Reports, Issues 6 and 7 (N.K. Forces), describing the condition of these two divisions at this time.

[26] 21st Inf WD, Summ, Sep 50; EUSAK WD, G–3 Jnl, 1125 14 Sep 50; *Ibid.,* Br for CG, 14 Sep 50; EUSAK PIR 64 and 65, 14–15 Sep 50; *Ibid.,* POR 192 and 195, 14–15 Sep 50; GHQ FEC G–3 Opn Rpt 83, 15 Sep 50.

[27] ATIS Res Supp Interrog Rpts, Issue 4 (N.K. *8th Div*), p. 25; *Ibid.,* Issue 3 (N. K. *15th Div*), p. 44.

lost nearly all the twenty-one new tanks of the *17th Armored Brigade* that were supporting it. Just below Hwajong-dong, towering mountains close in on the road, with 3,000-foot-high Hill 928 (Hwa-san) on the east and lesser peaks 2,000 feet high on the west. At this passage of the mountains into the Taegu corridor, the ROK 6th Division decisively defeated the enemy *8th Division* and, in effect, practically destroyed it. Of these battles around Hwajong-dong an enemy diarist wrote on 2 September, "Today we opened a general attack"; after 6 September, "We underwent extremely desperate battles. With no place to hide or escape from the fierce enemy artillery bombardment our main force was wiped out.' On 8 September he wrote, "We suffered miserably heavy casualties from fierce enemy air, artillery, and heavy machine gun attacks. Only 20 remain alive out of our entire battalion." [28]

On the next road eastward above Yongch'on, the N.K. *15th Division* launched its attack against the ROK 8th Division on 2 September. Although far understrength, with its three regiments reportedly having a total of only 3,600 men, it penetrated in four days to the lateral corridor at Yongch'on. North of the town one regiment of the ROK 8th Division panicked when an enemy tank got behind its lines. Elements of the enemy division were in and south of Yongch'on by midafternoon 6 September. The North Koreans did not remain in the town, but moved to the hills south and southeast of it overlooking the Taegu-Kyongju-Pusan road. On 7 September some of them established a roadblock three and a half miles southeast of Yongch'on, and other elements attacked a ROK regiment a mile south of the town. During the day, however, the ROK 5th Regiment of the newly activated 7th Division, attacking from the east along the lateral corridor, cleared Yongch'on itself of enemy and then went into a defensive position north of the town. But the next day, 8 September, additional elements of the *15th Division* arrived before Yongch'on and recaptured it. That afternoon the 11th Regiment of the ROK 1st Division arrived from the Taegu front and counterattacked the enemy in and near the town. This action succeeded in clearing the enemy from most of Yongch'on, but some North Koreans still held the railroad station southeast of it. Still others were an unknown distance southeast on the road toward Kyongju.[29]

There, in the hills southeast and east of Yongch'on, the enemy *15th Division* came to grief. Its artillery regiment foolishly advanced ahead of the infantry, expended its ammunition, and, without support, was then largely destroyed by ROK counterbattery fire. The artillery commander lost his life in the action. After the ROK 5th and 11th Regiments arrived in the vicinity of Yongch'on to

[28] *Ibid.*, Issue 4 (N.K. *8th Div*), p. 25; ATIS Interrog Rpts, Issue 4, Rpts 922 and 923, pp. 47 and 51; ATIS Enemy Documents, Nr 28, p. 7, diary of Pak Han Pin, *83d Regt, 8th Div;* EUSAK WD, 9 Sep 50, PW Interrog Rpt, 2d Lt Wong Hong Ki; *Ibid.*, PIR 53, 3 Sep 50; New York *Herald Tribune*, September 7, 1950.

[29] EUSAK WD, G–3 Jnl, 1230 3 Sep 50, and an. to PIR 53, 3 Sep 50, and PIR 56, 6 Sep 50; *Ibid.*, Br for CG, 7 Sep 50; GHQ FEC Sitrep, 8–9 Sep 50; *Time* Magazine, September 18, 1950, J. Bell article; New York *Herald Tribune*, September 7, 1950, Bigart dispatch.

reinforce the demoralized 8th Division, ROK battle action was so severe against the enemy units that they had no chance to regroup for co-ordinating action. On 9 and 10 September ROK units surrounded and virtually destroyed the N.K. *15th Division* southeast of Yongch'on on the hills bordering the Kyongju road. The North Korean division chief of staff, Col. Kim Yon, was killed there together with many other high-ranking officers. The part played by KMAG officers in rounding up stragglers of the ROK 8th Division and in reorganizing its units was an important factor in the successful outcome of these battles. On 10 September, the ROK 8th Division cleared the Yongch'on-Kyongju road of the enemy, capturing 2 tanks, 6 howitzers, 1 76-mm. self-propelled gun, several antitank guns, and many small arms. The capture of the self-propelled gun is a revealing story in itself. The driver drove the gun, followed by a truckload of enemy infantry, from the southeast through the ROK lines to Yongch'on, where he stopped and was quietly eating dinner with ROK troops when he came under suspicion and had to make a dash for it, hotly pursued by groups of ROK's. He surrendered four miles northward to a lone ROK soldier with the explanation that he could not drive the vehicle and shoot at the same time.[30]

Advancing north of Yongch'on after the retreating survivors of the N.K. *15th Division,* the ROK 8th Division and the 5th Regiment of the ROK 7th Division encountered almost no resistance. On 12 September, elements of the two ROK organizations were eight miles north of the town. On that day they captured 4 120-mm. mortars, 4 antitank guns, 4 artillery pieces, 9 trucks, 2 machine guns, and numerous small arms. ROK forces now also advanced east from Yongch'on and north from Kyongju to close the big breach in their lines.[31]

Perhaps the most critical period of the fighting in the east occurred when the N.K. *15th Division* broke through the ROK 8th Division to Yongch'on. The enemy division at that point was in a position to turn west toward Taegu and take Eighth Army and the 1st Cavalry Division there in the rear, or to turn east and southeast and take Task Force Jackson in the rear or on its left flank. It tried to do the latter. But General Walker's quick dispatch of the ROK 5th and 11th Regiments from two widely separated sectors of the front to the area of penetration resulted in destroying the enemy force before it could exploit its breakthrough. General Walker's prompt judgment of the reinforcements needed to stem the North Korean attacks in the Kyongju-P'ohang and the Yongch'on areas, and his rapid shifting of these reinforcements to the threatened sectors from other fronts, constitute a notable command achievement in the battles of the Pusan Perimeter.[32]

---

[30] EUSAK WD, POR 180, 10 Sep 50; *Ibid.*, 14 Sep 50, interrog of Lee Yong Sil; *Ibid.*, 9, 11, 12 Sep 50, interrogs of 1st Lt Kim Yong Chul, Cpl So Yong Sik, Capt Pak Chang Yong; ATIS Res Supp Interrog Rpts, Issue 106 (N.K. Arty), p. 62; *Ibid.*, Issue 3 (N.K. *15th Div*), p. 44; GHQ FEC Sitrep, 11 Sep 50; New York *Times*, September 11 and 13, 1950. There are scores of interrogations of prisoners from the N.K. *15th Division* in ATIS Interrogation Reports, Issues 4, 5, and 9.

[31] EUSAK WD, 14 Sep 50; GHQ FEC Sitrep, 13 Sep 50; New York *Times*, September 13, 1950.

[32] Ltr, Stephens to author, 14 May 53; Interv, author with Maj Gen Edwin K. Wright (FEC G–3 at the time), 7 Jan 54; Ltr, Landrum to author, recd 28 Jun 54; Interv, author with Stebbins, 4 Dec 53.

*Back on Taegu*

While four divisions of the N.K. *II Corps* attacked south in the P'ohang-dong, Kyongju, and Yongch'on sectors, the remaining three divisions of the corps—the *3d, 13th,* and *1st,* in that order from west to east—were to execute their converging attack on Taegu from the north and northwest. The *3d Division* was to attack in the Waegwan area northwest of Taegu, the *13th Division* down the mountain ridges north of Taegu along and west of the Sangju-Taegu road, and the *1st Division* along the high mountain ridges just east of the road. (*Map 15*)

Defending Taegu, the U.S. 1st Cavalry Division had a front of about thirty-five miles. General Gay outposted the main avenues of entry into his zone and kept his three regiments concentrated behind the outposts. At the southwestern end of his line General Gay initially controlled the 3d Battalion, 23d Infantry, 2d Division, which had been attached to the 1st Cavalry Division. On 5 September the British 27th Brigade, in its first commitment in the Korean War, replaced that battalion. Next in line northward, the 5th Cavalry Regiment defended the sector along the Naktong around Waegwan and the main Seoul highway southeast from there to Taegu. Eastward, the 7th Cavalry Regiment was responsible for the mountainous area between that highway and the hills bordering the Sangju road. The 8th Cavalry Regiment, responsible for the latter road, was astride it and on the bordering hills.[33]

Greatly concerned at the beginning of September over the North Korean attack and penetration of the southern sector of the Pusan Perimeter in the 2d and 25th Divisions' zone, General Walker on 1 September ordered the 1st Cavalry Division to attack north or northwest in an effort to divert to that quarter some of the enemy strength in the south. General Gay's initial decision upon receipt of this order was to attack north up the Sangju road, but his staff and regimental commanders all joined in urging that the attack instead be against Hill 518 in the 7th Cavalry zone, and they talked him out of his original intent. Only two days before, Hill 518 had been in the ROK 1st Division zone and had been considered an enemy assembly point. The 1st Cavalry Division, accordingly, prepared for an attack in the 7th Cavalry sector and for diversionary attacks by two companies of the 3d Battalion, 8th Cavalry, on the 7th Cavalry's right flank. This left the 8th Cavalry only one rifle company in reserve. The regiment's 1st Battalion was on the hill mass to the west of the Bowling Alley and north of Tabu-dong; its 2d Battalion was astride the road.

This planned attack against Hill 518 chanced to coincide with the defection and surrender on 2 September of Maj. Kim Song Jun, the S–3 of the N.K. *19th Regiment, 13th Division.* He reported that a full-scale North Korean attack was to begin at dusk that day. The N.K. *13th Division,* he said, had just taken in 4,000 replacements, 2,000 of them without weapons, and was now back to a strength of approximately 9,000 men. Upon receiving this intelligence, General Gay alerted all front-line units to

---

[33] 1st Cav Div WD, 1 Sep 50; *Ibid.,* G–3 Jnl, 2 Sep 50.

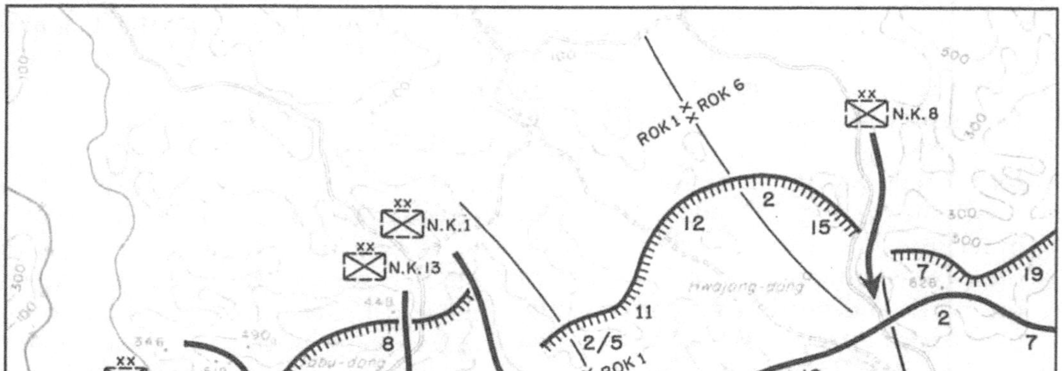


MAP 15

be prepared for the enemy attack.[34]

Complying with Eighth Army's order for what was in effect a spoiling attack against the North Koreans northwest of Taegu, General Gay on 1 September ordered the 7th Cavalry to attack the next day and seize enemy-held Hill 518. Hill 518 (Suam-san) is a large mountain mass five miles northeast of Waegwan and two miles east of the Naktong River. It curves westward from its peak to its westernmost height, Hill 346, from which the ground drops abruptly to the Naktong River. Situated north of the lateral Waegwan–Tabu-dong road, and

[34] 1st Cav Div WD, 1–2 Sep 50; GHQ FEC, History of the N.K. Army, p. 73; EUSAK WD, 5 Sep 50, ATIS Interrog Rpt 895, Maj Kim Song Jun; ATIS Interrog Rpts, Issue 3, pp. 214ff.; ATIS Res Supp Interrog Rpts, Issue 104 (N.K. *13th Div*), p. 67; Ltr and attached notes, Gay to author, 17 Jul 53; Ltr, Col Harold K. Johnson to author, n.d., but recd in Aug 54.

1st Cavalry Observation Post *overlooking the enemy-held Hill 518 complex northeast of Waegwan, 1 September.*

about midway between the two towns, it was a critical terrain feature dominating the road between the two places. After securing Hill 518, the 7th Cavalry attack was to continue on to Hill 314. Air strikes and artillery preparations were to precede the infantry attack on 2 September. Forty pieces of artillery, four-fifths of that available to the 1st Cavalry Division, were to support the attack.[35]

[35] Ltr, Gay to author, 17 Jul 53; 77th FA Bn Hist, Sep 50; 7th Cav WD, 1-2 Sep 50. The 77th FA Bn was in direct support of 7th Cav. To assist in firing support for the regiment were A Btry, 61st FA Bn (105-mm.); B and C Btrys, 9th FA Bn (155-mm.); and one plat, B Btry, 82d FA Bn (155-mm.)

On the morning of 2 September the Air Force delivered a 37-minute strike against Hills 518 and 346. The artillery then laid down its concentrations on the hills, and after that the planes came over again napalming and leaving the heights ablaze. Just after 1000, and immediately after the final napalm strike, the 1st Battalion, 7th Cavalry, attacked up Hill 518.

The plan of regimental attack unfortunately brought a minimum of force against the objective. While the 1st Battalion made the attack, the 2d Battalion was in a blocking position on its left (west) and the newly arrived 3d Battalion, in its first Korean operation, was to be behind the 2d Battalion and in an open gap between that battalion and Hill 518. The 1st Battalion moved up through ROK forces and, from high ground, was committed along a narrow ridge line, attacking from the southeast in a column of companies. This in turn resolved itself in a column of platoons, and finally in a column of squads. The final effect, therefore, was that of a regimental attack amounting to a one-squad attack against a strongly held position.

The attack was doomed to failure from the start. The heavy air strikes and the artillery preparations had failed to dislodge the North Koreans. From their positions they delivered mortar and machine gun fire on the climbing infantry, stopping the weak, advanced force short of the crest. In the afternoon the battalion withdrew from Hill 518 and attacked northeast against Hill 490, from which other enemy troops had fired in support of the North Koreans on Hill 518.

The next day at noon, the newly arrived 3d Battalion resumed the attack against Hill 518 from the south, over unreconnoitered ground, and, as did the 1st Battalion the day before, in a column of companies that resolved itself in the end into a column of squads. Again the attack failed. Other attacks failed on 4 September. An enemy forward observer captured on Hill 518 said that 1,200 North Koreans were dug in on the hill and that they had 120-mm. and 82-mm. mortars with ammunition.[38]

While these actions were in progress on its right, the 2d Battalion, 5th Cavalry Regiment, on 4 September attacked and captured Hill 303. The next day it had the utmost difficulty in holding the hill against enemy counterattacks. By 4 September it had become quite clear that the N.K. *3d Division* in front of the 5th and 7th Cavalry Regiments was itself attacking, and that, despite continued air strikes, artillery preparations, and infantry efforts on Hill 518, it was infiltrating large numbers of its troops to the rear of the attacking United States forces. That day the I&R Platoon reported that enemy soldiers held Hill 464, a high hill mass opposite Hill 518 on the south side of the Waegwan–Tabu-dong road, and that it had to destroy its radio and machine gun to keep them from falling into enemy hands. That night large enemy forces came through the gap between the 3d Battalion on the southern slope of Hill 518 and the 2d Battalion westward. For a time those in the 3d Battalion command post thought the attack was going to turn east and overrun them but, instead, the North Koreans turned west and occupied Hill 464 in force. By 5 September, although it was not yet known by the 7th Cavalry, Hill 464 to its rear probably had more North Koreans on it than Hill 518 to its front. North Koreans cut the Waegwan–Tabu-dong road east of the regiment so that its communications with friendly units now were only to the west. During the day the 7th Cavalry made a limited withdrawal on Hill 518. Any hope that the regiment could capture the hill vanished. One American officer described the situation north of Taegu at this time with the comment, "I'll be damned

---

[38] 7th Cav Regt WD, 4 Sep 50; 1st Cav Div WD, 5 Sep 50; 77th FA Bn Hist, 2 Sep 50; EUSAK WD, G–3 Jnl, 0400 5 Sep 50; Lt Col James B. Webel (Capt and S–3, 7th Cav, Sep 50), MS review comments, 15 Nov 57.

if I know who's got who surrounded." [37]

On the division right, Tabu-dong was in enemy hands, on the left Waegwan was a no-man's land, and in the center strong enemy forces were infiltrating southward from Hill 518. The 7th Cavalry Regiment in the center could no longer use the Waegwan–Tabu-dong lateral supply road behind it, and was in danger of being surrounded. After discussing a withdrawal plan with General Walker and Colonel Collier, General Gay on 5 September issued an order for a general withdrawal of the 1st Cavalry Division during the night to shorten the lines and to occupy a better defensive position. The movement was to progress from right to left beginning with the 8th Cavalry Regiment, then the 7th Cavalry in the Hill 518 area, and finally the 5th Cavalry in the Waegwan area. This withdrawal caused the 3d Battalion, 8th Cavalry, to give up a hill it had just attacked and captured near the Tabu-dong road on the approaches of the Walled City of Ka-san. In the 7th Cavalry sector the 1st, 3d, and 2d Battalions were to withdraw in that order, after the withdrawal of the 1st Battalion, 8th Cavalry, on their right. The 2d Battalion, 5th Cavalry, on Hill 303 north of Waegwan was to cover the withdrawal of the 7th Cavalry and hold open the escape road.[38]

[37] 5th Cav Regt WD, 4 Sep 50; 7th Cav Regt WD, 4–5 Sep 50; 1st Cav Div WD, 4–5 Sep 50; 77th FA Bn Hist, 5–6 Sep 50; Webel, MS review comments, 15 Nov 57; New York *Times*, September 5, 1950.

[38] 1st Cav Div WD, 5 Sep 50; 7th Cav Regt Opn Ord 14, 051840 Sep 50; Ltr, Gay to author, 17 Jul 53; Brig Gen Marcel G. Crombez, Notes for author, 28 Jun 55.

### *Crisis in Eighth Army Command*

At this time, about 5 September, as the 7th Cavalry Regiment was forced into a withdrawal, and enemy penetrations in the south had opened the way to Pusan, a crisis developed in appraisals and decisions called for in the Eighth Army command. Everywhere around the Perimeter the North Koreans were penetrating the defense positions and in some places making spectacular gains. It was a question whether the Eighth Army and the ROK's could hold anything like the Pusan Perimeter based on the line of the Naktong. The ROK Army and most of the American divisions appeared to be near the breaking point. Should the United Nations line be withdrawn to the Davidson Line? That question was under debate in Eighth Army headquarters. The decision to withdraw to that line seemed near as the North Koreans captured P'ohang-dong and drove to the edge of Kyongju in the east, reached Yongch'on in the Taegu lateral corridor, captured Waegwan, Tabu-dong, and Ka-san north of Taegu, drove through the old Naktong Bulge area to Yongsan, and in the south split the U.S. 25th Division and poured into its rear areas almost to the edge of Masan. (The Naktong Bulge and Masan penetrations have not yet been described, but they had already taken place as part of the North Korean co-ordinated attack.)

General Walker discussed the issue of withdrawing to the Davidson Line one night with his principal staff officers, most of the division commanders, and General Coulter, his deputy commander in the east. Colonel Dabney, Eighth Army G–3, told General Walker that for once he did not know what to recommend, that the decision was a hard one to make, but that he hoped the Army could stay. He pointed out that North Korean penetrations in the past had waned after a few days and that they might do so again. Upon orders from Colonel Landrum, Dabney started the G–3 Section that evening working on preparing withdrawal orders for Eighth Army. The staff section worked all night long on them. They were published and ready for issuing at 0500 in the morning, but they were held in the G–3 Section pending General Walker's personal order to put them into effect. The order was not given. At some time during the night Walker reached the decision that Eighth Army would not withdraw.[39]

But at this time Eighth Army headquarters did leave Taegu. The tactical situation had deteriorated so much on the afternoon of 4 September that the 1st Cavalry Division ammunition supply point in Taegu loaded nearly all its ammunition on rail cars on Eighth Army orders and prepared for a hasty evacuation southward. The Army transportation officer placed an embargo on all rail shipments north of Samnangjin on the main line, and north and east of Kyongju on the east line. The next morning, 5 September, General Walker reached the decision to move the main army headquarters back to the old Fisheries College between Pusan and Tongnae, north of Pusan, and it made the move during the day. The ROK Army headquarters moved to Pusan. The ROK Army headquarters opened at Pusan at 0800 and Eighth Army headquarters at 1600, 6 September. Walker himself and a few staff officers remained in Taegu as an advanced echelon of the army command post, constituting a tactical headquarters. The principal reason General Walker moved Eighth Army headquarters to Pusan was for the greater protection of the army signal communication equipment. Had the Eighth Army teletype equipment been destroyed or captured by the enemy there was no other similar heavy equipment in the Far East to replace it. The Army's operations would have been seriously handicapped had this signal equipment been lost or damaged.[40]

At this time, General Garvin issued verbal orders to service troops in the 2d Logistical Command at Pusan to take

[39] Ltr, Dabney to author, 19 Jan 54; Notes, Landrum to author, recd 28 Jun 54; Interv, author with Wright, 7 Jan 54; Interv, author with Stebbins, 4 Dec 53; Interv, author with Tarkenton, 3 Oct 52; Interv, author with Col Robert G. Fergusson, 2 Oct 52; Interv, author with Bullock, 28 Jan 54; Collier, MS review comments, 10 Mar 58.

[40] Notes, Landrum to author, recd 28 Jun 54; Interv, author with Allen, 15 Dec 53; Collier, MS review comments, 10 Mar 58; FEC CofS files, Summ of conversation, Hickey with Landrum, 060900 Sep 50; GHQ FEC Sitrep, 7 Sep 50; EUSAK WD, G–3 Jnl, 2030 6 Sep 50; 1st Cav Div, Ordnance Act Rpt, Sep 50.

Communication between Eighth Army and the Far East Command would have suffered most if this equipment had been lost or damaged, not the tactical control of units in Korea under Eighth Army. The Marc 2, 4-van unit for teletype, and a 1,200-line switchboard could not have been replaced—there was only one each in Korea and Japan. The big teletype unit with 180 lines to Pusan and the trunk cables constituted the critical items in U.N. signal communication at this time. Interv, author with Col William M. Thames (Deputy Sig Off, FEC, 1950), 17 Dec 53; Interv, author with Col Thomas A. Pitcher (Acting EUSAK Sig Off, Sep 50), 16 Dec 53; Interv, author with Lt Col William E. Kaley, 16 Dec 50.

defensive positions on the hills bordering the port city and within the city itself if and when the tactical situation required it.[41]

What the South Korean civilian estimate of the situation was at this time can be surmised from the fact that about 5 September prominent Koreans started to leave Pusan for the island of Tsushima, midway in the Korean Strait between Korea and Japan. Operators of small 10- to 20-ton vessels smuggled them across to the island. Wealthy and influential Chinese residing in the Pusan area were planning to leave for Formosa, the first group expecting to depart about 8 September. They, too, were to be smuggled away in small vessels.[42]

This period in early September 1950 tested General Walker as perhaps no other did. Walker was generally an undemonstrative man in public, he was not popular with the press, and he was not always popular with his troops. He could be hard and demanding. He was so at this time. When many of his commanders were losing confidence in the ability of Eighth Army to stop the North Koreans he remained determined that it would. On one occasion in early September he told one of his division commanders in effect, "If the enemy gets into Taegu you will find me resisting him in the streets and I'll have some of my trusted people with me and you had better be prepared to do the same. Now get back to your division and fight it." He told one general he did not want to see him back from the front again unless it was in a coffin.[43]

By day, General Walker moved around the Perimeter defense positions either by liaison plane or in his armored jeep. The jeep was equipped with a special iron handrail permitting him to stand up so that he could observe better while the vehicle was in motion, and generally it was in rapid motion. In addition to his .45 automatic pistol, he customarily carried a repeating shotgun with him, because, as he told a fellow officer, "I don't mind being shot at, but these —— are not going to ambush me."[44] Walker was at his best in Korea in the Pusan Perimeter battles. Famous previously as being an exponent of armored offensive warfare, he demonstrated in August and September 1950 that he was also skilled in defensive warfare. His pugnacious temperament fitted him for directing the fighting of a bitter holding action. He was a stouthearted soldier.

### *The 7th Cavalry's Withdrawal Battle*

It was in this crisis that the 7th Cavalry began its withdrawal northwest of Taegu. In his withdrawal instructions for the 7th Cavalry, Col. Cecil Nist, the regimental commander, ordered, "The 2d Battalion must clear Hill 464 of enemy tonight." This meant that the 2d Battalion must disengage from the enemy to its front and attack to its rear

[41] 2d Logistical Comd Activities Rpt, Sep 50, G-3 and Trans Secs. Operation Plan 4, dated 10 September, confirmed these orders.

[42] EUSAK WD, PIR 57, an. 2, 441st Counter Intelligence Corps Agent Rpt, 7 Sep 50.

[43] Notes, Landrum to author, recd 28 Jun 54; Interv, author with Maj Gen Leven C. Allen, 15 Dec 53; Interv, author with Bullock, 28 Jan 54; Interv, author with Lt Col Paul F. Smith, 1 Oct 52; Collier, MS review comments, Mar 58.

[44] Ltr, Wright to author, 12 Feb 54.

GENERAL WALKER CROSSING THE NAKTONG *in his armored jeep with handrail.*

to gain possession of Hills 464 and 380 on the new main line of resistance to be occupied by the regiment. Since efforts to gain possession of Hill 464 by other elements had failed in the past two or three days this did not promise to be an easy mission.

Heavy rains fell during the night of 5–6 September and mud slowed all wheeled and tracked vehicles in the withdrawal. The 1st Battalion completed its withdrawal without opposition. During its night march west, the 3d Battalion column was joined several times by groups of North Korean soldiers who apparently thought it was one of their own columns moving south. They were made prisoners and taken along in the withdrawal. Nearing Waegwan at dawn, the battalion column was taken under enemy tank and mortar fire after daybreak and sustained about eighteen casualties.

The 2d Battalion disengaged from the enemy and began its withdrawal at 0300, 6 September. The battalion abandoned two tanks, one because of mechanical failure and the other because it was stuck in the mud. The battalion moved to the rear in two main groups: G Company to attack Hill 464 and the rest of the battalion to seize Hill 380, half a mile farther south. The North Koreans quickly discovered that the 2d Battalion was withdrawing and attacked it. The battalion commander, Maj. Omar T. Hitchner, and his S-3, Capt. James T. Milam, were killed. In the vicinity of Hills 464 and 380 the battalion discovered at daybreak that it was virtually surrounded by enemy soldiers. Colonel Nist thought that the entire battalion was lost.[45]

Moving by itself and completely cut off from all other units, G Company, numbering only about eighty men, was hardest hit. At 0800, nearing the top of Hill 464, it surprised and killed three enemy soldiers. Suddenly, enemy automatic weapons and small arms fire struck the company. All day G Company maneuvered around the hill but never gained its crest. At midafternoon it received radio orders to withdraw that night. The company left six dead on the hill and, carrying its wounded on improvised litters of ponchos and tree branches, it started down the shale slopes of the mountain in rain and darkness. Halfway down, a friendly artillery barrage killed one of the noncommissioned officers, and a rock thrown by one of the exploding shells hit Capt. Herman L. West, G Company commander, inflicting a painful back injury. The company scattered but Captain West reassembled it. Cautioning his men to move quietly and not to fire in any circumstances, so that surrounding enemy troops might think them one of their own columns, West led his men to the eastern base of Hill 464 where he went into a defensive position for the rest of the night.[46]

On the division left, meanwhile, the 2d Battalion, 5th Cavalry, on Hill 303 came under heavy attack and the battalion commander wanted to withdraw. Colonel Crombez, the regimental commander, told him he could not do so until the 7th Cavalry had cleared on its withdrawal road. This battalion suffered heavy casualties before it abandoned Hill 303 on the 6th to the enemy.[47]

While G Company was trying to escape from Hill 464, the rest of the 2d Battalion was cut off at the eastern base of Hill 380, half a mile southward. Colonel Nist organized all the South Korean carriers he could find before dark and loaded them with water, food, and ammunition for the 2d Battalion, but the carrier party was unable to find the battalion. At dawn on 7 September the men in G Company's perimeter at the eastern base of Hill 464 saw in the dim light four figures coming down a trail toward them. Soon recognizing them as North Koreans, the men killed them. This rifle fire brought answering fire from enemy troops in nearby positions. At this time, Captain West heard what he recognized as fire from American weapons on a knob to his west. Thinking that it might be from the Weapons Platoon which had become separated from him during the night, he led his company in that direction. He was right; soon the company was reunited.

[45] 7th Cav Regt WD, 6–7 Sep 50; Capt Robert M. Ballard, Action of G Company on Hill 464, an. to 7th Cav Regt WD, 6 Sep 50; Webel, MS review comments, 15 Nov 57; New York *Herald Tribune*, September 8, 1950, Bigart dispatch.

[46] Ballard, Action of G Company on Hill 464; Capt. Russell A. Gugeler, *Combat Actions in Korea*, ch. 4, "Attack to the Rear," pp. 41–42.

[47] 5th Cav Regt WD, 5–6 Sep 50; Notes, Crombez for author, 28 Jul 55.

The Weapons Platoon, led by Lt. Harold R. Anderegg, had undergone a strange experience. After becoming separated from the rest of the company, three times during the night it encountered North Koreans on the trail it was following but in each instance neither side fired, each going on its way. At dawn, the platoon came upon a group of foxholes on a knoll. Enemy soldiers were occupying some of them. In a swift action which apparently surprised and paralyzed the North Koreans, the platoon killed approximately thirteen and captured three enemy soldiers. From the body of an officer the men took a brief case containing important documents and maps. These showed that Hill 464 was an assembly point for part of the N.K. *3d Division* in its advance from Hill 518 toward Taegu.[48]

Later in the day (7 September), Capt. Melbourne C. Chandler, acting commander of the 2d Battalion, received word of G Company's location on Hill 464 from an aerial observer and sent a patrol which guided the company safely to the battalion at the eastern base of Hill 380. The battalion, meanwhile, had received radio orders to withdraw by any route as soon as possible. It moved southwest into the 5th Cavalry sector. At one point it escaped ambush by turning aside when North Koreans dressed in American uniforms waved helmets and shouted, "Hey, this way, G.I.! "[49]

East of the 2d Battalion, the enemy attacked the 1st Battalion in its new position on 7 September and overran the battalion aid station, killing four and wounding seven men. That night the 1st Battalion on division order was attached to the 5th Cavalry Regiment. The rest of the 7th Cavalry Regiment moved to a point near Taegu in division reserve. During the night of 7–8 September the 5th Cavalry Regiment on division orders withdrew still farther below Waegwan to new defensive positions astride the main Seoul-Taegu highway. The enemy *3d Division* was still moving reinforcements across the Naktong. Observers sighted fifteen barges loaded with troops and artillery pieces crossing the river two miles north of Waegwan on the evening of the 7th. On the 8th the North Korean communiqué claimed the capture of Waegwan.[50]

The next day the situation grew worse for the 1st Cavalry Division. On its left flank, the N.K. *3d Division* forced the 1st Battalion, 5th Cavalry, to withdraw from Hill 345, three miles east of Waegwan. The enemy pressed forward and the 5th Cavalry was immediately locked in hard, seesaw fighting on Hills 203 and 174. The 1st Battalion, 7th Cavalry, before it left that sector to rejoin its regiment, finally captured the latter hill after four attacks.

Only with the greatest difficulty did the 5th Cavalry Regiment hold Hill 203 on 12 September. Between midnight and 0400, 13 September, the North Koreans attacked again and took Hill 203 from E Company, Hill 174 from L Company, and Hill 188 from B and F Companies. In an afternoon counterattack

[48] Ballard, Action of G Company on Hill 464; 7th Cav Regt WD, 7 Sep 50; Gugeler, *Combat Actions in Korea*, pp. 39–45.

[49] Ballard, Action of G Company on Hill 464; 7th Cav Regt WD, 6–7 Sep 50.

[50] 1st Bn, 7th Cav WD, 7 Sep 50; 1st Cav Div WD, 7 Sep 50; 7th Cav Regt WD, 7th Sep 50, Opn Ord 15; GHQ FEC Sitrep, 8 Sep 50.

the regiment regained Hill 188 on the south side of the highway, but failed against Hills 203 and 174 on the north side. On the 14th, I Company again attacked Hill 174, which had by now changed hands seven times. In this action the company suffered 82 casualties. Its 2d Platoon with 27 Americans and 15 ROK's at the start had only 11 Americans and 5 ROK's when it reached its objective. Even so, the company held only one side of the hill, the enemy held the other, and grenade battles between the two continued for another week. The battalions of the 5th Cavalry Regiment were so low in strength at this time as to be scarcely combat effective. This seesaw battle continued in full swing only eight air miles northwest of Taegu.[51]

### *Troopers in the Mountains—Walled Ka-san*

Hard on the heels of Major Kim's warning that the North Korean attack would strike the night of 2 September, the blow hit with full force in the Bowling Alley area north of Taegu. It caught the 8th Cavalry Regiment defending the Sangju road badly deployed in that it lacked an adequate reserve. The North Koreans struck the 2d Battalion, 8th Cavalry, the night of 2–3 September on Hill 448 west of the Bowling Alley and two miles north of Tabu-dong, and overran it. On the right, E Company, although not under attack, was cut off and had to withdraw by a roundabout way. Lt. Col. Harold K. Johnson, commanding officer of the 3d Battalion, placed I Company in a blocking position just north of Tabu-dong astride the road. There, two enemy tanks and some enemy infantry struck it at 0200 in the morning of 3 September. In this action, I Company suffered many casualties but repelled the enemy attack. The overrun 2d Battalion withdrew through the 3d Battalion which had assembled hastily in a defensive position south of Tabu-dong. During the day, elements of the N.K. *1st Division* forced the 8th Cavalry I&R Platoon and a detachment of South Korean police from the Walled City of Ka-san on the crest of Hill 902, four miles east of Tabu-dong. On 3 September, therefore, Eighth Army lost to the enemy both Tabu-dong and Hill 902, locally called Ka-san, the dominant mountaintop ten miles north of Taegu.[52]

The North Koreans now concentrated artillery north of Hill 902 and, although its fire was light and sporadic, it did cause minor damage in the 99th Field Artillery positions. This sudden surge of the enemy southward toward Taegu caused concern in Eighth Army headquarters. The Army ordered a ROK battalion from the Taegu Replacement Training Center to a position in the rear of the 8th Cavalry, and the 1st Cavalry Division organized Task Force Allen, to be commanded by Assistant Division Commander Brig. Gen. Frank

[51] EUSAK WD, Br for CG, 130800 Sep 50; *Ibid.*, G–3 Sec, 15 Sep 50; 1st Cav Div WD, 9–10, 13–15 Sep 50; 5th Cav Regt WD, 13–15 Sep 50; 7th Cav Regt WD, 10 Sep 50; New York *Times*, September 9, 1950 (London rebroadcast of Moscow broadcast); I Corps WD, 15 Sep 50, Hist Narr, p. 5; ATIS Interrog Rpts, Issue 6 (N.K. Forces), Rpt 1157, Che Nak Hwan, p. 127; Allen, Korean Army Troops, USA, pp. 6–7.

[52] 8th Cav Regt WD, 3 Sep 50; 1st Cav Div WD, 3 Sep 50; EUSAK WD, G–3 Sec, 3 Sep 50; *Ibid.*, PIR 53, 3 Sep 50; Ltr, Johnson to author, recd Aug 54.

RUINS OF ANCIENT FORTRESS *and stone wall on the crest of Ka-san (Hill 902).*

A. Allen, Jr. This task force comprised two provisional battalions formed of division headquarters and technical service troops, the division band, the replacement company, and other miscellaneous troops. It was to be used in combat should the North Koreans break through to the edge of the city.[53]

Eighth Army countered the North Korean advance down the Tabu-dong road by ordering the 1st Cavalry Division to recapture and defend Hill 902. This hill, ten miles north of Taegu, gave observation all the way south through Eighth Army positions into the city, and, in enemy hands, could be used for general intelligence purposes and to direct artillery and mortar fire. Hill 902 was too far distant from the Tabu-dong road to dominate it; otherwise it would have controlled this main communication route. The shortage of North Korean artillery and mortar ammunition nullified in large part the advantages the peak held as an observation point.

Actually, there was no walled city on the crest of Ka-san. Ka-san, or Hill 902, the 3,000-foot-high mountain which differs from most high peaks in this part of Korea in having an oval-shaped semilevel area on its summit. This oval is a part of a mile-long ridgelike crest, varying from 200 to 800 yards in width, which slopes down from the peak at 902

[53] 1st Cav Div WD, 3 Sep 50, Ordnance Stf Sec; Ltr, Gay to author, 17 Jul 53.

D Company, 8th Engineer Combat Battalion, *position on the summit of Hill 755.*

meters to approximately 755 meters at its southeastern end. On all sides of this ridge crest the mountain slopes drop precipitously. In bygone ages Koreans had built a thirty-foot-high stone wall around the crest and had converted the summit into a fortress. One man who fought in the shadow of the wall commented later, "It looked to me like they built that wall just to keep the land from sliding down." Most of the summit in 1950 was covered with a dense growth of scrub brush and small pine trees. There were a few small terraced fields. Koreans knew Ka-san as the Sacred Mountain. Near the northern end of the crest still stood the Buddhist Poguk Temple.

When the 1st Cavalry Division on 29 August assumed responsibility for the old ROK 1st Division sector north of Taegu it sent a patrol from the I&R Platoon to the top of Ka-san. There the patrol found 156 South Korean police. There was some discussion between General Gay and Eighth Army about whether the 1st Cavalry Division or the ROK 1st Division should have the responsibility for the mountain. General Gay maintained that his understrength division with a 35-mile front was already overextended and could not extend eastward beyond the hills immediately adjacent to the Tabu-dong road. Uncertainty as to final responsibility for Ka-san ended on the afternoon of 3 September after North Koreans had seized the mountain. The Eighth Army G-3 Section telephoned Col. Ernest V. Holmes, Chief of Staff, 1st Cavalry Division, and told him that the 1st Cavalry Division had responsibility for the Walled City. Holmes replied he believed that General Gay, who was then absent from the headquarters, would not like the decision, but that pending his return he would send a company of engineers to Ka-san. When General Gay returned to his command post he said that if the army had ordered the responsibility it had to be complied with, and he approved Holmes' decision to send a company to the mountain.[54]

After his telephone conversation with Eighth Army, Colonel Holmes ordered Lt. Col. William C. Holley, commanding officer of the 8th Engineer Combat Battalion, to report to Col. Raymond D. Palmer, commanding the 8th Cavalry Regiment. That afternoon Colonel Palmer in his command post on the Tabu-dong road outlined to Holley and the commanding officers of D Company, 8th Engineer Combat Battalion, and E Company, 8th Cavalry, his attack plan to regain control of Ka-san. The Engineer company, commanded by 1st Lt. John T. Kennedy, was to lead the attack, E Company following. Once the force had gained the crest and E Company had established itself in defensive positions, the Engineer company was to come off the mountain. Luckily, many of the men in D Company had been infantrymen in World War II.[55]

That evening, D Company loaded into trucks and in a driving rain traveled north, eventually turning off the main road to the designated assembly area. On the way they met two truckloads of South Korean police going south, some of them wounded. These were the police who, together with the detachment of the I&R Platoon, had been driven off Ka-san that afternoon. After waiting in the rain awhile for orders, the Engineer company turned around and went back to camp.

The next morning (4 September) at breakfast, D Company received orders to move immediately as infantry to Ka-san. One platoon had to forego its breakfast. The company carried no rations since E Company, 8th Cavalry, was to bring food and water later. The Engineer troops arrived at their assembly area near the village of Kisong-dong two miles east of the Tabu-dong road, where Colonel Holley set up a communications command post. Sniper fire came in on the men as they moved up the trail half a mile to the base of Ka-san's steep slope. Word was given to the company that there were about seventy-five disorganized enemy troops on Ka-san. But actually, during the afternoon and evening of 3 September, the N.K. *2d Battalion, 2d Regiment, 1st Division*, had occupied the summit of Ka-san.[56]

The Engineer company started its attack up the mountain about noon, 4

[54] Ltr and notes, Gay to author, 17 Jul 53; Interv, author with Holmes, 27 Oct 53; 1st Cav Div WD, 29 Aug 50.

[55] Ltr, Capt John T. Kennedy to author, 2 Apr 52; Interv, author with Holley, 20 Feb 52; Interv, author with 1st Sgt Cornelius C. Kopper (D Co, 8th Engr C Bn, in 1950), 20 Feb 52.

[56] Ltr, MSgt James N. Vandygriff to author, 19 May 53 (Vandygriff was Plat Sgt, 2d Plat, D Co, 8th Engr C Bn, Sep 50); Ltr, Capt Thomas T. Jones to author, 21 Jun 53 (Plat Ldr, 3d Plat, D Co, 8th Engr C Bn, Sep 50); EUSAK WD, 11 Sep 50, interrog of Kim Choe Ski; *Ibid.*, G–3 Jnl, 0620 4 Sep 50; GHQ FEC Sitrep, 4 Sep 50.

September, following a trail up a southern spur. The 1st Platoon was in the lead, single file, followed by the 2d and 3d Platoons. Colonel Palmer considered the mission so important that he and his S–2, Capt. Rene J. Guiraud, accompanied the engineers. Platoon Sgt. James N. Vandygriff, 2d Platoon, D Company, in a brief conversation with Colonel Holley as he went ahead of the latter on his way up the trail, said he thought it was a suicide mission.

Less than a mile up the trail, D Company came under machine gun fire from its right front, which inflicted several casualties. Lieutenant Kennedy rejected Vandygriff's request to take a squad and knock out the gun, so the file got past the line of fire as best it could until BAR fire from the 3d Platoon silenced the weapon. Farther up the trail another enemy machine gun fired from the right along the trail and held up the advance until radio-adjusted artillery fire silenced it.

The file of men, with Lt. Robert Peterson of the I&R Platoon as guide, left the traillike road, which dead-ended, dropped over into a ravine on the left, and continued the climb. Enemy mortar fire killed two men and wounded eight or ten others in this phase of the ascent. At this time the 2d Platoon leader collapsed from a kidney ailment and command passed to Sergeant Vandygriff. Vandygriff led his platoon, now at the head of the company, on up the gully and finally, about 1700, came through a tunnel under a small ridge and the stone wall into the bowl-shaped summit of Hill 755, the southern arm of the Hill 902 crest. The 2d and 3d Platoons soon arrived, in that order. When he was within fifty feet of the top, Colonel Palmer received radio orders from General Gay to come off the mountain; Gay had not known that Palmer had accompanied the attack until he telephoned Holley trying to locate him.[57]

Lieutenant Kennedy quickly placed the approximately ninety men of his company in position facing in an arc from west to northeast; the 2d Platoon took the left flank near the stone wall, the 1st Platoon took the center position on a wooded knoll, and the 3d Platoon the right flank at the edge of a woods. Just as he reached the top, 2d Lt. Thomas T. Jones, commanding the 3d Platoon, saw and heard three North Korean mortars fire, approximately 1,000 yards away on a grassy ridge to the right (east). He suggested to Lieutenant Kennedy several times that he request artillery fire on these mortars, but Kennedy did not act on the suggestion. Kennedy established his command post inside the tunnel behind the 2d Platoon position. The D Company position was entirely within the area enclosed by the stone wall, which was nearly intact except on the northeast near the 3d Platoon position where it had crumbled and was covered with brush and trees. Lieutenant Jones pointed out to his platoon sergeant and squad leaders where he wanted them to take position at the edge of the woods facing the enemy mortars he had seen on the grassy ridge beyond. He then remained a few minutes

[57] Ltr, Jones to author, 21 Jun 53; Ltr, Vandygriff to author, 19 May 53; Ltr and notes, Gay to author, 17 Jul 53; Interv, author with Holley, 20 Feb 52; Ltr, Kennedy to author, 2 Apr 52; Interv, author with Guiraud, 21 Apr 54.

in conversation with Lieutenant Kennedy.[58]

A few minutes later Jones joined his 3d Squad men at the edge of the woods. They told him that the platoon sergeant and the rest had continued on toward the narrow grassy ridge. Just then one of the squad called Jones to the edge of the woods and pointed out ten or twelve well-camouflaged North Korean soldiers, one of them carrying a machine gun, coming down the narrow ridge toward them from the mortar position. Apparently this group was a security force for the mortars because they dropped to the ground about one-third of the way down the ridge.

Jones decided he had better bring back his other two squads to form a solid line and, expecting to be gone only a few minutes, he left his SCR–300 radio behind. That, as he said later, was his big mistake. Jones found one squad but the other had gone on farther and was not visible. While he studied the terrain and waited for a messenger he had sent to bring back that last squad, North Koreans attacked the main company position behind him. Judging by the firing and yelling, Jones thought North Koreans were all over the wooded bowl between him and the rest of the company. When the firing ended, all he could hear was North Korean voices. Jones never got back to his 3d Squad. He and the rest of the platoon dropped down off the ridge into a gully on the left, the two squads separated but for a time within sight of each other.

That night Jones and the eight men with him stayed in the ravine just under the crest. Without his radio he could not communicate with the rest of the company which he thought had been destroyed or driven off the hill. The next day when American fighter planes strafed the hilltop it confirmed his belief that no D Company men were there. Some of the men in the advanced squad made their way to safety, but North Koreans captured Jones and the eight men with him near the bottom of Ka-san on 10 September as they were trying to make their way through the enemy lines. This account of the 3d Platoon explains why—except for the 3d Squad which rejoined D Company that evening—it was out of the action and off the crest almost as soon as it arrived on top, all unknown to Lieutenant Kennedy and the rest of the company at the time.[59]

Half an hour after D Company had reached Hill 755, an estimated enemy battalion launched an attack down the slope running south to Hill 755 from the crest of Hill 902. The main attack hit Vandygriff's 2d Platoon just after Vandygriff had set up and loaded his two machine guns. These machine guns and the protection of the 15-foot wall on its left enabled D Company to turn back this attack, which left one dead and three wounded in the 2d Platoon. That night, enemy mortar and small arms fire harassed the company and there were several small probing attacks. Having

[58] Ltr, Jones to author, 21 Jun 53; Ltr, Kennedy to author, 2 Apr 52; Ltr, Vandygriff to author, 19 May 53.

[59] Ltrs, Jones to author, 26 May, 21 and 30 Jun 53; 1st Lt Thomas T. Jones, "Two Hundred Miles to Freedom," *The Military Engineer* (September–October 1951), pp. 351–54. The North Koreans, strangely enough, released Jones and three other soldiers later near Ch'unch'on in central Korea, where they entered the lines of the ROK 6th Division.

no communication with the 3d Platoon, Kennedy sent a patrol to its supposed position. The patrol reported back that it could find no one there but had found the rocket launchers and two light machine guns.[60]

It rained most of the night, and 5 September dawned wet and foggy on top of Hill 755. Just after daylight in a cold drizzle the North Koreans attacked. The engineers repulsed this attack but suffered some casualties. Enemy fire destroyed Vandygriff's radio, forcing him to use runners to communicate with Kennedy's command post. Ammunition was running low and three C–47 planes came over to make an airdrop. Kennedy put out orange identification panels, then watched the enemy put out similarly colored one. The planes circled, and finally dropped their bundles of ammunition and food—to the enemy. Immediately after the airdrops, two F–51 fighter planes came over and attacked D Company. It was obvious that the enemy panels had misled both the cargo and fighter planes. The fighters dropped two napalm tanks within D Company's perimeter, one of which fortunately failed to ignite; the other injured no one. The planes then strafed right through the 2d Platoon position, but miraculously caused no casualties. Soon after this aerial attack, enemy burp gun fire wounded Kennedy in the leg and ankle.[61]

Sometime between 1000 and 1100 the advanced platoon of E Company, 8th Cavalry Regiment, arrived on top of Hill 755 and came into D Company's perimeter. Some of the engineers fired on the E Company men before the latter identified themselves. The E Company platoon went into position on the right of Vandygriff, and Kennedy turned over command of the combined force to the E Company commander. Kennedy then assembled twelve wounded men and started down the mountain with them. The party was under small arms fire most of the way. A carrying party of Korean A-frame porters led by an American officer had started up the mountain during the morning with supplies. Enemy fire, killing several of the porters, turned it back.[62]

The day before, E Company had been delayed in following D Company to Hill 755. Soon after the Engineer company had started up the trail on the 4th, E Company arrived at Colonel Holley's command post at the base of the mountain. Enemy mortar fire was falling on the trail at the time and the company commander said he could not advance because of it. Holley radioed this information to Colonel Palmer who designated another company commander and said, "Tell him to come on through." This second officer broke his glasses on a rock and informed Holley that he could not go on. Holley put him on the radio to Palmer who ordered him to continue up the hill. Soon thereafter this officer was wounded in the leg. Holley then designated a third officer, who started up the mountain with E Company that evening about 2000. Enemy fire stopped the company 500 yards

[60] Ltr, Vandygriff to author, 19 May 53; Ltrs, Kennedy to author, 2 Apr 52 and 4 Jun 52.

[61] Ltrs, Kennedy to author, 2 Apr and 4 Jun 52; Ltr, Vandygriff to author, 19 May 53; GHQ FEC Sitrep, 5 Sep 50.

[62] Ltr, Vandygriff to author, 19 May 53; Ltrs, Kennedy to author, 2 Apr and 4 Jun 52.

short of the crest before dawn. It was this same company that the N.K. *13th Division* had cut off when it launched its attack the evening of 2 September and overran the 2d Battalion north of Tabu-dong. Tired and dispirited from this experience and their roundabout journey to rejoin the regiment, E Company men were not enjoying the best of morale.[63]

Shortly after the E Company platoon joined Vandygriff, the North Koreans attacked again. The E Company infantrymen had brought no mortars with them—only small arms. In this situation, Vandygriff took a 3.5-inch rocket launcher and fired into the North Koreans. They must have thought that it was mortar or 75-mm. recoilless rifle fire for they broke off the attack. Vandygriff checked his platoon and found it was nearly out of ammunition. He then instructed his men to gather up all the weapons and ammunition from enemy dead they could reach, and in this manner they obtained for emergency use about 30 to 40 rifles, 5 burp guns, and some hand grenades.

In the course of gathering up these enemy weapons, Vandygriff passed the dug-in position of Pfc. Melvin L. Brown, a BAR man in the 3d Squad. Brown was next to the wall on the extreme left of the platoon position at a point where the wall was only about six or seven feet high. At the bottom of the wall around Brown's position lay about fifteen or twenty enemy dead. Vandygriff asked Brown what had happened. The latter replied, "Every time they came up I knocked them off the wall." Earlier in the day, about 0800, Kennedy had visited Brown and had seen five enemy dead that Brown had killed with BAR fire. Subsequently Brown exhausted his automatic rifle ammunition, then his few grenades, and finally he used his entrenching tool to knock the North Koreans in the head when they tried to climb over the wall. Brown had received a flesh wound in the shoulder early in the morning, but had bandaged it himself and refused to leave his position.[64]

At 1330 General Gay ordered the 8th Cavalry Regiment to withdraw its men off Ka-san. Gay decided to give up the mountain because he believed he had insufficient forces to secure and hold it and that the enemy had insufficient ammunition to exploit its possession as an observation point for directing artillery and mortar fire. It is not certain that this order actually reached anyone on the hill. Colonel Holley could not reach anyone in D Company, 8th Engineer Combat Battalion.[65]

Rain started falling again and heavy fog closed in on the mountain top so that it was impossible to see more than a few yards. Again the enemy attacked the 2d Platoon and the adjacent E Company infantrymen. One of the engineers was shot through the neck and Vandygriff sent him to the company command post. In about thirty minutes he returned. "What's wrong?" asked Vandygriff. Barely able to talk from his wound

[63] Interv, author with Holley, 20 Feb 52; Ltrs, Kennedy to author, 2 Apr and 4 Jun 52; Ltr, Vandygriff to author, 19 May 53; Ltr, Jones to author, 21 Jun 53.

[64] Ltr, Vandygriff to author, 19 May 53; Ltrs, Kennedy to author, 2 Apr and 4 Jun 52.

[65] Interv, author with Holley, 20 Feb 52; Ltr, Gay to author, 17 Jul 53.

and shock, the man replied that there was no longer a command post, that he could not find anyone and had seen only enemy dead. Vandygriff now went to the infantry sergeant who was in command of the E Company platoon and asked him what he intended to do. The latter replied, in effect, that he was going to take his platoon and go over the wall.

Vandygriff went back to his own platoon, got his squad leaders together and told them the platoon was going out the way it came in and that he would give the wounded a 30-minute start. Enemy fire was falling in the platoon area now from nearly all directions and the situation looked hopeless. Sgt. John J. Philip, leader of the 3d Squad, started to break up the weapons that the platoon could not take out with them. Vandygriff, noticing that Brown was not among the assembled men, asked Philip where he was. The latter replied that he didn't know but that he would try to find out. Philip returned to the squad's position and came back fifteen minutes later, reporting to Vandygriff that Brown was dead. Asked by Philip if he should take the identification tags off the dead, Vandygriff said, "No," that he should leave them on because they would be the only means of identification later. Vandygriff put his platoon in a V formation and led them off the hill the same way they had come up, picking up four wounded men on the way down.[66]

At the base of the mountain, Colonel Holley and others in the afternoon saw E Company men come down from the top and, later, men from the engineer company. Each group thought it was the last of the survivors and told confused, conflicting stories. When all remaining members of D Company had been assembled, Colonel Holley found that the company had suffered 50 percent casualties; eighteen men were wounded and thirty were missing in action.[67]

Among the wounded carried off the mountain was an officer of D Company, 8th Engineer Combat Battalion. Enemy machine gun fire struck him in the leg just before he jumped off a high ledge. Two men carried him to the bottom and at his request left him in a Korean house, expecting to come back in a jeep for him. A little later, other members escaping off the mountain heard his screams. Two weeks passed before the 1st Cavalry Division recaptured the area. They found the officer's body in the house. The hands and feet were tied, the eyes gouged, a thumb pulled off, and the body had been partly burned. Apparently he had been tied, tortured, and a fire built under him.

Soldiers of the ROK 1st Division captured a North Korean near Ka-san on 4 September who said that about 800 of his fellow soldiers were in the Walled City area with three more battalions following them from the north. The Engineer company had succeeded only in establishing a perimeter briefly within the enemy-held area. By evening of 5

[66] Ltr, Vandygriff to author, 19 May 53. Department of the Army General Order 11, 16 February 1951, awarded the Medal of Honor posthumously to Pfc. Melvin L. Brown.

[67] Ltr, Jones to author, 21 Jun 53; Intervs, author with Holley and Kopper, 20 Feb 52; Maj Hal D. Steward, "Engineers Fight as Infantry," *Ad Cosantoir* (August, 1951), pp. 366–67.

September, Ka-san was securely in enemy hands with an estimated five battalions, totaling about 1,500 enemy soldiers, on the mountain and its forward slope. A North Korean oxtrain carrying 82-mm. mortar shells and rice reportedly reached the top of Ka-san during the day. The ROK 1st Division captured this oxtrain a few days later south of Ka-san.[68]

When Lieutenant Jones went back up the mountain as a prisoner on 10 September he saw at least 400–500 enemy soldiers on the ridge. A Mosquito spotter plane flew over and he felt sure it would sight the large number of enemy troops and call in fighter planes for strafing attacks. But, he said, "The pilot of the plane took one look and went away which amazed me, except that the minute they heard the plane the North Koreans all either hit the ground or squatted and ducked their heads, which attested to the effectiveness of the leaves, branches, etc., that almost every man had stuck in the string netting on the back of his shirt and the top of his cloth hat." [69]

Now, with Ka-san firmly in their possession, the N.K. *13th* and *1st Divisions* made ready to press on downhill into Taegu.

On the 6th, the day after the American troops were driven off Ka-san, an enemy force established a roadblock three miles below Tabu-dong and other units occupied Hill 570, two miles southwest of the Walled City and overlooking the Taegu road from the east side. The next morning five tanks of the 16th Reconnaissance Company prepared to lead an attack against the roadblock. The enemy troops were in a rice field west and on the hills east of the road. General Gay was at the scene to watch the action. He ordered the reconnaissance company commander to launch the attack into the rice fields at maximum speed, saying, "I don't want a damn tank moving under 25 miles per hour until you are on top of those men." [70] The tank attack speedily disposed of the enemy in the rice field, but the infantry spent several hours clearing the hills on the east side of the road.

Enemy artillery during 7 September shelled batteries of the 9th and 99th Field Artillery Battalions, forcing displacement of two batteries during the day. U.S. air strikes and artillery kept both Hills 902 and 570 under heavy attack. Even though the 1st Cavalry Division fell back nearly everywhere that day, General Walker ordered it and the ROK II Corps to attack and seize Hill 902 and the Walled City, the time of the attack to be agreed upon by the commanders concerned. He directed the ROK 1st Division and the 1st Cavalry Division to select a boundary between them and to maintain physical contact during the attack.[71]

On the morning of the 8th, Lt. Col. Harold K. Johnson's 3d Battalion, 8th Cavalry, after executing a withdrawal during the night from its former position, tried to drive the enemy from Hill 570. The three peaks of this mountain

[68] EUSAK WD, PIR 55, 5 Sep 50; *Ibid.*, G–3 Jnl, 0750 and 1900 5 Sep 50; *Ibid.*, interrogation of Kim Choe Ski, 11 Sep 50.

[69] Ltr, Jones to author, 30 Jun 53.

[70] Ltr, Gay to author, 17 Jul 53; 1st Cav Div WD, 7 Sep 50.

[71] EUSAK WD, G–3 Sec and G–3 Jnl, 1330 7 Sep 50; *Ibid.*, Br for CG, 7 Sep 50.

mass were under clouds, making it impossible to support the infantry attack with air strikes or artillery and mortar fire. Johnson placed all three of his rifle companies in the assault against the three peaks; two of them reached their objectives, one with little opposition, the other catching enemy soldiers asleep on the ground. But enemy counterattacks regained this second peak. The main enemy force on Hill 570 was on the third and highest of the three peaks and held it firmly against the L Company attack. The I Company commander and the L Company executive officer were killed, as were several noncommissioned officers. The Eighth Army Intelligence Section estimated that 1,000 enemy soldiers were on Hill 570, only eight air miles north of Taegu, and on 8 September it stated that the continued pressure against the eastern flank of the 1st Cavalry Division sector "represents what is probably the most immediate threat to the U.N. Forces." [72]

That same day, 8 September, the 1st Cavalry Division canceled a planned continuation of the attack against Hill 570 by the 3d Battalion, 7th Cavalry Regiment, when enemy forces threatened Hills 314 and 660, south and east of 570.

In the midst of this enemy drive on Taegu, an ammunition shortage became critical for the U.N. forces. The situation was such that General MacArthur on 9 September sent messages urging that two ammunition ships then en route to Yokohama and Pusan carrying 172,790 rounds of 105-mm. shells, with estimated arrival time 11 September, proceed at maximum speed consistent with the safety of the vessels. Eighth Army on 10 September reduced the ration of 105-mm. howitzer ammunition from fifty to twenty-five rounds per howitzer per day, except in cases of emergency. Carbine ammunition was also in critical short supply. The 17th Field Artillery Battalion, with the first 8-inch howitzers to arrive in Korea, could not engage in the battle for lack of ammunition.[73]

The N.K. *1st Division* now began moving in the zone of the ROK 1st Division around the right flank of the 1st Cavalry Division. Its *2d Regiment*, about 1,200 strong, advanced six air miles eastward from the vicinity of the Walled City on Hill 902 to the towering 4,000-foot-high mountain of P'algong-san. It reached the top of P'algong-san about daylight on 10 September, and a little later new replacements, prodded by burp guns from behind, made a wild charge toward the ROK positions. The ROK's turned back the charge, killing or wounding about two-thirds of the attacking force.[74]

The 1st Cavalry Division now had most of its combat units concentrated on its right flank north of Taegu. The 3d Battalion, 7th Cavalry, attached to the 8th Cavalry Regiment, was behind that regiment on Hills 181 and 182

---

[72] Johnson, Review notes for author on draft chapter, Aug 54; 8th Cav Regt WD, 8 Sep 50; EUSAK PIR 58, 8 Sep 50; 7th Cav Opn Ord 16 and WD overlay, 8 Sep 50.

[73] Schnabel, FEC, GHQ Support and Participation in the Korean War, ch. 4, pp. 22–23; 24th Div WD, G–4 Summ, 25–26 Aug and 10–11 Sep 50; 159th FA Bn Unit Rpt, 1–30 Sep 50; GHQ FEC Sitrep, 9 Sep 50.

[74] EUSAK WD, 14 Sep 50, interrog rpt of Kim Yong Gi; ATIS Interrog Rpts, Issue 6, Rpt 1103, p. 8; ATIS Res Supp Interrog Rpts, Issue 3 (N.K. *1st Div*), p. 36.

astride the Tabu-dong road only 6 air miles north of Taegu. The rest of the 7th Cavalry Regiment (the 1st Battalion rejoined the regiment during the day) was in the valley of the Kumho River to the right rear between the enemy and the Taegu Airfield, which was situated 3 miles northeast of the city. The 5th Cavalry was disposed on the hills astride the Waegwan road 8 air miles northwest of Taegu. On its left the entire 8th Engineer Combat Battalion was in line as infantry, with the mission of holding a bridge across the Kumho River near its juncture with the Naktong east of Taegu.[75]

The fighting north of Taegu on 11 September in the vicinity of Hills 660 and 314 was heavy and confused. For a time, the 1st Cavalry Division feared a breakthrough to the blocking position of the 3d Battalion, 7th Cavalry. The rifle companies of the division were now very low in strength. On 11 September, for instance, E Company, 5th Cavalry, in attacking Hill 203 on the division left toward Waegwan had only 3 officers and 63 men. The day before, C Company, 7th Cavalry, had only 50 men. Colonel Johnson stated later that any company of the 3d Battalion, 8th Cavalry, that had 100 men during this period was his assault company for the day.[76]

### *Hill 314*

While the 3d Battalion, 8th Cavalry, again vainly attacked Hill 570 on 11 September, enemy soldiers seized the crest of Hill 314 two miles southeast of it and that much closer to Taegu. Actually, the two hill masses are adjacent and their lower slopes within small arms range of each other. The North Koreans drove the 16th Reconnaissance Company from the hill and only the ROK 5th Training Battalion, previously hurried into the line from Taegu in a supporting position, prevented the enemy from gaining complete control of this terrain feature. This ROK battalion still held part of the reverse slope of Hill 314 when the 3d Battalion, 8th Cavalry, hurried to the scene from its fruitless attacks on Hill 570 and tried to retake the position. The ROK battalion twice had attacked and reached the crest but could not hold it, and had dug in on the lower southern slopes. The 3d Battalion, 7th Cavalry, command post had to fight off infiltrating enemy on 12 September as it issued its attack order and prepared to attack through the 8th Cavalry lines against Hill 314.

This attack on the 12th was to be part of a larger American and ROK counterattack against the N.K. *13th* and *1st Divisions* in an effort to halt them north of Taegu. The 2d Battalion, 7th Cavalry, relieved the ROK units on Hill 660, east of Hill 314, and had the mission of securing that hill. Farther east the ROK 1st Division had the mission of attacking from P'algong-san toward the Walled City on Hill 902. The point nearest Taegu occupied by enemy forces at this time was Hill 314. Some called it the "key to Taegu." Although this may be an exaggeration, since other hills, like links in a chain, were possibly equally important, the enemy *13th Division* valued its possession and had concentrated about 700 soldiers on it. The

---

[75] Ltr, Gay to author, 17 Jul 53; 1st Cav Div WD, 11 Sep 50.

[76] 1st Cav Div WD, 11 Sep 50; 1st Bn, 7th Cav Regt WD, 10 Sep 50; 5th Cav Regt WD, Sep 50 Narr; Review notes, Johnson to author, Aug 54.

North Koreans meant to use it, no doubt, in making the next advance on Taegu. From it, observation reached to Taegu and it commanded the lesser hills southward rimming the Taegu bowl.

Hill 314 is actually the southern knob of a 500-meter hill mass which lies close to the east side of Hill 570 and is separated from that hill mass only by a deep gulch. The hill mass is shaped like an elongated teardrop, its broad end at the north. The southern point rises to 314 meters and the ridge line climbs northward from it in a series of knobs to 380 and, finally, to 500 meters. The ridge line from the 314-meter to the 500-meter point is a mile in length. All sides of the hill mass are very steep.[77]

Lt. Col. James H. Lynch's 3d Battalion, 7th Cavalry, on the eve of its attack against Hill 314 numbered 535 men, less its rear echelons. The battalion, which had been organized at Fort Benning, Ga., from the 30th Infantry Regiment of the 3d Division, had arrived in Korea at the end of August. The ill-fated action of the 7th Cavalry at Hill 518, begun nine days earlier, had been its first action. This was to be its second. The battalion attack plan this time differed radically from that employed against Hill 518 and was a direct development of that failure. The key aspect of the Hill 314 attack plan was to mass as many riflemen as possible on top of the narrow ridge line, by attacking with two companies abreast along the ridge, and not to repeat the mistakes of Hill 518 where the fire power of only a platoon, and at times of only a squad, could be brought to bear against the enemy. Because of the ammunition shortage there was no artillery preparation on Hill 314, but there was an air strike before Colonel Lynch's battalion, with L Company on the left and I Company on the right, at 1100, 12 September, started its attack. The point of departure was the front lines of the 3d Battalion, 8th Cavalry, on the lower slope of the hill.[78]

Enemy 120-mm. mortar fire was falling on and behind the line of departure as the battalion moved out. For 500 yards it encountered only sporadic small arms and machine gun fire; then enemy rifle fire became intense and preregistered mortar fire came down on the troops, pinning them to the ground. On the left, men in L Company could see approximately 400 North Koreans preparing to counterattack. They radioed for an air strike but the planes were on the ground refueling. Fortunately, they were able to repulse the counterattack with combined artillery, mortar, and small arms fire. The air strike came in at 1400, blanketing the top and the north slope of the ridge.

By this time enemy mortar fire had caused many casualties, and elements of L and I Companies became intermingled. But, in contrast to the action on Hill 518, the men continued the attack largely of their own volition after many of the officers had become casualties. The example of certain officers,

[77] Ltr, Johnson to author, recd Aug 54; 1st Cav Div WD, 12 Sep 50; 7th Cav Regt WD, 12 Sep 50; Attack Ord, 3d Bn, 7th Cav Regt, attached to 1st Lt Morris M. Teague, Jr.'s, Narrative and Supporting Documents Concerning Hill 314, in the 7th Cav Regt WD.

[78] Teague, Hill 314 Narr; Ltr, Gay to author, 17 Jul 53; Webel, MS review comments, 15 Nov 57; 7th Cav Regt WD, 12 Sep 50.

however, pointed the way. The commanding officer of I Company, 1st Lt. Joseph A. Fields, reorganized his company under mortar fire without regard to his own safety after the company had suffered 25 percent casualties; 1st Lt. Marvin H. Haynes led a small group which killed or drove off enemy troops that had overrun part of L Company; and Capt. Robert W. Walker, commanding officer of L Company, continued his superb personal leadership. Fields was wounded, Haynes killed. MSgt. Roy E. McCullom, the weapons platoon leader of I Company, organized his men as riflemen, and though wounded three times, in shoulders and right arm, he led them on until he received a fourth wound in the head. Wounded by mortar fragments, 2d Lt. Marshall G. Engle, I Company, refused evacuation twice, telling litter teams to go farther forward and get the more critically injured. Engle lay on the hill for twelve hours, far into the night, receiving another mortar wound during that time before a litter team finally evacuated him.[79]

Fifteen minutes after the air strike, the 3d Battalion resumed its attack toward the crest. As it neared it the North Koreans came out of their positions in a violent counterattack and engaged at close quarters. Some men gained the crest but enemy mortar and machine gun fire drove them off. They reached it a second time but could not hold it. Another air strike hit the enemy. Then, a third time, Captain Walker led a group of men of L and I Companies to the top. When Walker reached the crest he shouted back, "Come on up here where you can see them! There are lots of them and you can kill them." The men scrambled up a 60-degree slope for the last 150 yards to the top, where they closed with the North Koreans and overran their positions. Walker and the remaining men of the two companies secured the hill at 1530 and then Walker reorganized the two companies jointly under his command. There were fewer than forty effectives left in L Company and about forty in I Company; the latter had lost all its officers.[80]

General Gay caused a special study to be made of this action, so outstanding did he consider it to be. He found that the 3d Battalion suffered 229 battle casualties in the first two hours, most of them incurred during the second hour of the attack. Of these, 38 Americans were killed and 167 wounded, the remainder were attached South Koreans. The battalion aid station reported treating 130 casualties. Other wounded were treated at the 8th Cavalry aid station. Many men with minor wounds did not ask for medical attention until the battle had ended, and there were only five cases of combat shock in contrast to the eighteen on Hill 518. Enemy mortar fire caused 80 percent of the casualties.[81]

Colonel Lynch's battalion held Hill

[79] 1st Cav Div WD, 12 Sep 50; 7th Cav Regt WD, 12 Sep 50; Medical Log, Hill 314, attached to 3d Bn, 7th Cav Unit Rpt, 12 Sep 50; Teague, Hill 314 Narr.

[80] Medical Log with 3d Bn, 7th Cav, Unit Rpt; Teague, Hill 314 Narr.

[81] Medical Log with 3d Bn, 7th Cav, Unit Rpt; Ltr, Gay to author, 17 Jul 53; Webel, MS review comments, 15 Nov 57; Interv, author with Robert Best, ORO analyst, 3 Apr 53. Best made a study of the Hill 314 action and stated that the 3d Battalion casualties were 30 Americans killed, 119 wounded, and 10 ROK's killed or wounded. General Gay and Colonel Webel say these figures are inaccurate.

314 for the next six days and gathered up a large amount of enemy equipment and ammunition. The enemy soldiers on Hill 314 wore American uniforms, helmets, and combat boots. Many of them had M1 rifles and carbines. Two hundred of their number lay dead on the hill. Of the other 500 estimated to have been there, prisoners said most of them had been wounded or were missing.

Several atrocity cases came to light during the action on Hill 314. Capt. James B. Webel found the first one on the afternoon of the 12th while the final action on the hill was taking place. He came upon an American officer who had been bound hand and foot, gasoline poured over him, and burned. A 5-gallon can lay close to the body. Two days later members of the battalion found on the hill the bodies of four other American soldiers with their hands tied. The bodies bore evidence that the men had been bayoneted and shot while bound.[82]

After the capture of Hill 314 on 12 September, the situation north of Taegu improved. On 14 September the 2d Battalion, 8th Cavalry, attacked and, supported by fire from Hill 314, gained part of Hill 570 from the N.K. *19th Regiment, 13th Division.*

Across the army boundary on the right, the ROK 1st Division continued its attack northwest and advanced to the edge of the Walled City. The ROK 11th Regiment seized Hill 755 about dark on 14 September, and small elements of the ROK 15th Regiment reached the stone ramparts of the Walled City area at the same time. The ROK's and North Koreans fought during the night and on into the 15th at many points along the high mountain backbone that extends southeast from the Walled City to Hills 755 and 783 and on to P'algong-san. Prisoners taken by the ROK's estimated that there were about 800 North Koreans on this high ridge. The ROK 1st Division later estimated that approximately 3,000 enemy were inside the Walled City perimeter and about 1,500 or 2,000 outside it near the crest. It appears that at this time the bulk of the N.K. *1st Division* was gradually withdrawing into the Walled City and its vicinity. Indications were that the N.K. *13th Division* also was withdrawing northward. Aerial observers on the afternoon of 14 September reported that an estimated 500 enemy troops were moving north from Tabu-dong. But, while these signs were hopeful, General Walker continued to make every possible preparation for a final close-in defense of Taegu. As part of this, fourteen battalions of South Korean police dug in around the city.[83]

The fighting continued unabated north of Taegu on the 15th. The 2d Battalion, 8th Cavalry, still fought to gain control of Hill 570 on the east side of the Tabu-dong highway. On the other side, the 3d Battalion, 8th Cavalry, attacked Hill 401 where an enemy force had penetrated in a gap between the 8th and 5th Cavalry Regiments. The fighting on Hill 401 was particularly severe. Both sides had troops on the

[82] 7th Cav Regt WD, 12 and 15 Sep 50; Teague, Hill 314 Narr; Webel, MS review comments, 15 Nov 57; 1st Cav Div WD, 14 Sep 50.

[83] I Corps WD, Sep 50, Hist Narr, pp. 5–6; EUSAK WD, G–3 Jnl, 14 Sep 50; GHQ FEC G–3 Opn Rpts 82 and 83, 14–15 Sep 50; *Ibid.*, Sitrep, 15 Sep 50; EUSAK WD, an. to PIR 64, 14 Sep 50.

mountain when night fell. In this action, SFC Earl R. Baxter, at the sacrifice of his life, covered the forced withdrawal of his platoon (2d Platoon, L Company), killing at least ten enemy soldiers in close combat before he himself was killed by an enemy grenade.[84]

While the N.K. *II Corps* was striving to capture Taegu and penetrate behind Eighth Army toward Pusan by way of the P'ohang-Kyongju corridor, the N.K. *I Corps* along the lower Naktong and in the south had unleashed simultaneously a violent offensive to bring the entire Pusan Perimeter under assault. Of the entire Perimeter, the parts tactically most vulnerable to enemy action lay along the lower Naktong, and accordingly they promised the greatest dividends strategically to successful North Korean attack. There the battle in early September rose to great intensity and for a period the outcome hung in the balance.

[84] Ltr, Johnson to author, recd Aug 54; 1st Cav Div WD, 15 Sep 50. General Order 328, 20 May 1951, awarded the Distinguished Service Cross posthumously to Sergeant Baxter. EUSAK WD.

## CHAPTER XXIII

# North Korean Breakthrough in the South

> Exact knowledge of the terrain regulates the dispositions of the troops and the order of battle. . . . Knowledge of the country is to a general what a rifle is to an infantryman and what the rules of arithmetic are to a geometrician.
>
> FREDERICK THE GREAT, *Instructions for His Generals*

The dog days of August had given way to September. Casualties during the next two weeks were to be the greatest of the Korean War. To the men of Eighth Army, these were to be the worst of "the days along the Naktong." And, as if to envelop this deadly clash of arms with a misery of nature's own making, the elements brought to the battlefield blackened skies and torrential rains. It was the end of the summer monsoon season.

Aerial reconnaissance in the last week of August had disclosed to Eighth Army exceptional enemy activity behind the lines opposite the U.S. 2d and 25th Divisions in the southern part of the Pusan Perimeter. Ominously, the enemy had built three new underwater bridges across the Nam River in front of the 35th Infantry in the 25th Division sector. Aerial bombing only temporarily and partially destroyed these bridges, for they could be repaired overnight.

Eighth Army intelligence credited the North Koreans with having moved one or two new divisions and about twenty tanks to the Hyopch'on area on the west side of the Naktong River opposite the U.S. 2d Division. On 28 August the Eighth Army intelligence officer warned that a general attack "may be expected at any time along the 2d Division and 25th Division front," aimed at severing the Taegu-Pusan railroad and highway and capturing Masan.[1]

With this tense situation as the setting, the N.K. *I Corps* before midnight 31 August started its great offensive. As the final hours of August gave way to the first hours of September, North Korean soldiery crossed the lower Naktong at a number of points in a well-planned attack. From Hyongp'ung southward to the coast, in the zones of the U.S. 2d and 25th Divisions, the enemy's greatest effort struck in a single massive co-ordinated attack.

In the southern part of its sector, where the U.S. 25th Division held the U.N. line, the N.K. *I Corps* planned a crushing blow, co-ordinating it with an

---

[1] GHQ FEC Sitrep, 1 Sep 50; EUSAK PIR's 46–50, 27 Aug–1 Sep 50; EUSAK WD, entry for 30 Aug 50, and Aug 50 Summ; 25th Div WD, 28 Aug 50.

attack against the 2d Division just to the north. The North Korean *6th* and *7th Divisions* prepared for the breakthrough effort against the 25th Division after receiving their attack orders about 20 August. The operation order called for the N.K. *I Corps* to assault all along the line at 2200, 31 August. The *6th Division,* farthest south on the enemy right flank, was to attack through Haman, Masan, and Chinhae and capture Kumhae, on the west side of the Naktong River delta fifteen miles from Pusan, by 3 September. The division zone of attack was to be south of the Chinju–Komam-ni (Saga)–Masan highway. The *7th Division,* next in line north of the *6th Division,* was to attack north of the Masan highway, wheel left to the Naktong, and wait for the *6th Division* on its right and the *9th* on its left to join it. Part of the *7th Division* was concentrated in the Uiryong area west of the Nam River. This plan pitted the *6th Division* against the 24th Infantry and the *7th Division* against the 35th Infantry.[2]

On 24 August, Maj. Gen. Pang Ho San, commanding general of the N.K. *6th Division,* much decorated for the exploits of his division thus far, issued an order calculated to improve troop morale. He said the mission of the division was "to liberate Masan and Pusan within a few days." He demanded stricter discipline and more perseverance than ever before, and stated that tactics must adjust to the changes "this epoch-making conflict has introduced into the art of warfare." He summed up the battle lessons:

> Our experience in night combat up to now shows that we can operate only four or five hours in the dark since we start night attacks between 2300 and 2400 hours, and, therefore, if the battle continues until dawn, we are likely to suffer losses. From now on, use daylight hours for full combat preparation, and commence attacks soon after sunset. Concentrate your battle actions mostly at night and capture enemy base positions. From midnight on, engage enemy in close combat by approaching to within 100 to 150 meters of him. Then, even with the break of dawn, the enemy planes will not be able to distinguish friend from foe, which will enable you to prevent great losses. This is the most valuable battle experience we have gained from the Chinju operation.[3]

### *Midnight Near Masan*

On 31 August 1950 the 25th Division held a front of almost thirty miles, beginning in the north at the Namji-ri bridge over the Naktong River and extending westward on the hills south of the river to the Nam's confluence with it. *(Map V)* It then bent southwest up the south side of the Nam to where the Sobuk-san mountain mass tapered down in its northern extremity to the river. There the line turned south along rising ground to 850-foot-high Sibidang-san (Hill 276), crossed the saddle on its south face through which passed the Chinju-Masan railroad and highway, and continued southward, climbing to 2,200-foot-high Battle Mountain (Hill 665) and on to 2,400-foot-high P'il-bong (Hill 743). From P'il-bong the line dropped down

---

[2] GHQ FEC, History of the N.K. Army, p. 63; ATIS Res Supp Interrog Rpts, Issue 100 (N.K. *6th Div*), pp. 39–40; *Ibid.*, Issue 99 (N. K. *7th Div*), p. 35.

[3] ATIS Res Supp Interrog Rpts, Issue 100 (N.K. *6th Div*), pp. 41–42; GHQ FEC, History of the N.K. Army, p. 97.

MOUNTAIN MASS WEST OF HAMAN. *The town is in the center foreground; the 2d Battalion, 24th Regiment, position is on the crest of the second ridge west of Haman.*

spur ridge lines to the southern coastal road near Chindong-ni.

Colonel Fisher's 35th Infantry held the northern part of the division line, approximately 26,000 yards of it from the Namji-ri bridge to the Chinju-Masan highway. The regiment was responsible for the highway. Colonel Fisher considered his weakest and most vulnerable point to be a 3-mile gap along the Naktong River between most of F Company on the west and its 1st Platoon to the east. This platoon guarded the Namji-ri cantilever steel bridge on the division extreme right at the boundary with the 2d Division across the Naktong.

South of the highway, Colonel Champney's 24th Infantry held the high country west of Haman up to and including Battle Mountain and P'il-bong. Colonel Throckmorton's 5th Infantry held the southern spur of Sobuk-san to the coastal road at Chindong-ni. From Chindong-ni some ROK Marine units continued the line to the southern coast. General Kean's 25th Division command post was at Masan; Colonel Fisher's 35th Infantry command post was on the east side of the Chirwon–Chung-ni road about midway between the two towns; Colonel Champney's 24th Infantry command post was at Haman; and Colonel Throckmorton's command post was at Chindong-ni.[4]

In the left center of the 25th Division line, Lt. Col. Paul F. Roberts' 2d Bat-

[4] 25th Div, 35th, 24th, and 5th Inf WD's, 31 Aug 50; Interv, author with Fisher, 5 Jan 52; Interv, author with Champney, 22 Jul 51; Interv, author with Throckmorton, 20 Aug 52.

talion, 24th Infantry, held the crest of the second ridge west of Haman, a little more than a mile from the town. From Chungam-ni, in enemy territory, a secondary road zigzagged to Haman along the shoulders of low hills and across paddy ground, running generally east a mile south of the main Chinju-Masan road. It came through Colonel Roberts' 2d Battalion position in a pass a little more than a mile directly west of Haman.

Late in the afternoon of 31 August, observers with G Company, 24th Infantry, noticed a lot of activity a mile to their front. They called in two air strikes that hit this enemy area at twilight. Artillery also took it under fire. All line units were alerted for a possible enemy attack.[5]

Shortly before midnight the North Koreans struck, first hitting F Company on the north side of the pass on the Chungam-ni–Haman road. The ROK troops in the pass left their positions and fell back on G Company south of the pass. The North Koreans captured a 75-mm. recoilless rifle in the mouth of the pass and turned it on American tanks, knocking out two of them. They then overran a section of 81-mm. mortars at the east end of the pass. South of the pass, at dawn, 1st Lt. Houston M. McMurray found that only 15 out of 69 men remained with him, 8 from his own 1st Platoon, G Company, and 7 ROK's of a group he had taken into his position during the night. The enemy attacked his position at first light. They came through an opening in the barbed wire, supposedly covered by a BAR, but the BAR men had fled. Throwing grenades and spraying the area with burp gun fire, the North Koreans quickly overran the position.[6]

Farther up the slope, enemy tank fire hit E Company at midnight. The company commander, 1st Lt. Charles Ellis, an able and courageous officer, ran over to his left flank when he heard a noise there. He found that his 3d Platoon was leaving its position. Ellis threatened the platoon leader, saying he would shoot him if he did not get back in position, and fired a shot between his feet to impress him. Ellis then went to his right flank and found that platoon also leaving its position. During the night everyone in E Company ran off the hill except Ellis and eleven men. Several E Company men in fleeing their position had run through their own mine field and were killed.

It is worthwhile to anticipate a bit and tell the fate of Ellis and his small group of men who stood their ground. Enemy fire pinned them down after daylight. When three or four of the group tried to run for it, enemy machine gun fire killed them. Ellis and the rest stayed in their holes on the hill for two days, repelling several attacks in that time. Ellis was then able to withdraw southward up the mountain to the 3d Battalion's position. In his withdrawal, Ellis, discovering a man who had been injured earlier in a mine explosion, entered the mine field to rescue him.[7]

[5] 24th Inf WD, 31 Aug 50; Interv, author with Corley, 6 Nov 51; Fisher, MS review comments, Jan 58.

[6] 24th Inf WD, 1 Sep 50; EUSAK IG Rpt, 24th Inf Regt, Sep 50, testimony of Lt McMurray; Col John T. Corley, MS review comments, 22 Jul 53.

[7] Interv, author with Champney, 22 Jul 51; Interv, author with Corley, 6 Nov 51; EUSAK IG Rpt, testimony of Capt Charles Ellis, E Co, 24th Inf.

The fact is that shortly after the enemy attack started most of the 2d Battalion, 24th Infantry, fled its positions. The enemy passed through the line quickly and overran the 2d Battalion command post, killing many men there and destroying much equipment. Haman was then open to direct attack. As the enemy encircled Haman, Colonel Roberts, the 2d Battalion commander, ordered an officer to take remnants of the battalion and establish a roadblock at the south edge of the town. Although the officer directed a large group of men to accompany him, only eight did so. The 2d Battalion was no longer an effective fighting force.[8]

Colonel Champney at 0400, 1 September, moved the 24th Regiment command post from Haman two miles northeast to a narrow defile on the New Engineer Road. At this time, an enemy group attacked C Battery, 159th Field Artillery Battalion, a mile north of Haman. Two tanks of the 89th Tank Battalion helped defend the battery until the artillerymen could pull out the howitzers and escape back through Haman and then eastward over this recently improved trail.[9]

The enemy assault did not strike the southern part of the line held by Corley's 3d Battalion, 24th Infantry, and Colonel Throckmorton's 5th Infantry. That part of the line, however, did receive artillery and mortar fire and some diversionary light attacks. About 0200, 1 September, men in an outpost on the right flank of Colonel Corley's battalion watched an estimated 600 enemy soldiers file past at a distance of 100 yards, going in the direction of Haman. Viewed during the night from the high ground of the 3d Battalion, Haman seemed to be in flames. At dawn, men in the battalion saw an estimated 800 enemy troops enter the town.[10]

When the enemy attack broke through the 2d Battalion, Colonel Champney ordered the 1st Battalion, about three miles south of Haman on the Chindong-ni road, to counterattack and restore the line. Colonel Roberts, a superior battalion commander, assembled all the men of the disorganized 2d Battalion he could find—about forty—to join in this counterattack, which got under way at 0730. But it was of short duration. Upon contact with the enemy, the 1st Battalion broke and fled to the rear. Thus, shortly after daylight the scattered and disorganized men of the 1st and 2d Battalions of the 24th Infantry had fled to the high ground two miles east of Haman. The better part of two regiments of the N.K. *6th Division* poured into and through the 3-mile-wide Haman gap.[11]

Meanwhile, action-packed events were taking place simultaneously to the north, on the right side of the 25th Division line. Half an hour before midnight, 31 August, an enemy self-propelled high-velocity gun from across the Nam fired shells into the position of G Company, 35th Infantry, overlooking the river.

[8] Interv, author with Corley, 6 Nov 51; Corley, MS review comments, 22 Jul 53; EUSAK IG Rpt, testimony of 1st Lt John L. Herren.

[9] 24th Inf WD, 1 Sep 50; 159th FA Bn WD, Sep 50; Barth MS, p. 16; Interv, author with Champney, 22 Jul 51.

[10] 3d Bn, 24th Inf WD, 1 Sep 50; Corley, MS review comments, 22 Jul 53.

[11] 25th Div WD, 1 Sep 50; Interv, author with Corley, 6 Nov 51; EUSAK IG Rpt, testimony of Col Roberts.

Within a few minutes, enemy artillery had taken under fire all front-line rifle companies of the regiment from the Namji-ri bridge west. Under cover of this fire a reinforced regiment of the N.K. *7th Division* crossed the Nam River and attacked F and G Companies, 35th Infantry. Other enemy soldiers crossed the Nam on an underwater bridge in front of the paddy ground north of Komam-ni and near the boundary between the 2d Battalion, led by Lt. Col. John L. Wilkins, Jr., holding the river front and Lt. Col. Bernard G. Teeter's 1st Battalion holding the hill line that stretched from the Nam River to Sibidang-san and the Chinju-Masan highway.

In the low ground between these two battalions at the river ferry crossing site, Colonel Fisher had placed about 300 ROK police. He expected them to hold there long enough in case of a major attack to serve as a warning device. Guns from the flanking hills there could cover the low ground with fire. Back of Komam-ni he held the 3d Battalion ready for use in counterattack to stop an enemy penetration should it occur.

Unexpectedly, the ROK police companies near the ferry scattered at the first enemy fire. Half an hour after midnight enemy troops streamed through this hole in the line, some turning left to take G Company in flank and rear, and others turning right to attack C Company, which was on a spur of ground west of the Komam-ni road. The I&R Platoon and elements of C and D Companies formed a defense line along the dike at the north edge of Komam-ni where tanks joined them at daybreak. But the enemy did not drive for the Komam-ni road fork four miles south of the river as Colonel Fisher expected him to do; instead, he turned east into the hills behind Fisher's 2d Battalion.[12]

The position of B Company, 35th Infantry, on 1,100-foot-high Sibidang-san, flanking the Masan road two miles west of Komam-ni and giving observation over all the surrounding country, was certain to figure prominently in the enemy's attack. It was a key position in the 25th Division line. The enemy's preparatory barrage there lasted from 1130 to midnight. Under cover of it two battalions of the N.K. *13th Regiment, 6th Division,* moved up within 150 yards of the American foxholes. At the same time, enemy tanks, self-propelled guns, and antitank guns moved toward Komam-ni on the road at the foot of Sibidang-san. An American Sherman tank there destroyed a T34 just after midnight, and a 3.5-inch bazooka team destroyed a self-propelled gun and several 45-mm. antitank guns.

On the crest of Sibidang-san, an antipersonnel mine field stopped the first enemy infantry assault. Others followed in quick succession. They were met and turned back with the fire of all weapons. By 0230 the B Company riflemen were stripping machine gun ammunition belts for their rifles. The 1st Platoon of C Company, at the base of the mountain behind B Company, met the emergency by climbing Sibidang-san in forty-five minutes with an ammunition resupply for the company. Just before dawn the enemy attack subsided. Daylight dis-

[12] The Distinguished Unit Citation was awarded the 35th Infantry Regiment for action on 1–6 September 1950. Supporting Docs, AG files. 25th Div WD, 1 Sep 50; EUSAK WD, G–3 Jnl, 0120 1 Sep 50; 2d Bn, 35th Inf, Narr of Act, 31 Aug–1 Sep 50; 35th Inf WD, 31 Aug 50; I&R Plat Unit Hist, 1 Sep 50; Interv, author with Fisher, 5 Jan 52.

closed a great amount of abandoned enemy equipment scattered on the slope just below the crest, including thirty light and three heavy machine guns. Among the enemy dead lay the body of the commanding officer of the N.K. *13th Regiment*.[13]

At daybreak, 1 September, a tank-led relief force of C Company headquarters troops cleared the road to Sibidang-san and resupplied the 2d Platoon, B Company, with ammunition just in time for it to repel a final North Korean assault, killing seventy-seven and capturing twenty-one of the enemy.

Although Colonel Fisher's 35th Infantry held all its original positions, except that of the forward platoon of G Company, it nevertheless was in a dangerous situation. Approximately 3,000 North Korean soldiers were behind its lines. The farthest eastern penetration reached the high ground just south of Chirwon overlooking the north-south road there.

### *Agok*

On the 35th Regiment's right flank, in the 9th Infantry, 2d Division, sector across the Naktong, the enemy also made deep penetrations. (*Map VI*) There, in the southern part of the U.S. 2d Division zone, the 9th Infantry Regiment held a sector more than 20,000 yards long, including the bulge area of the Naktong where heavy fighting had taken place earlier in August. The rifle companies on the river line here had frontages of 3,000 to 4,000 feet, and, like the units to the north and south of them, they held only key hills and observation points.

As August neared its end, men on these hills could see minor enemy activity across the river, which they interpreted as North Koreans organizing the high ground on the west side of the Naktong against a possible American attack. There was moderate enemy infiltration into the 9th Infantry forward positions, but to the men in the front line this appeared to be only normal patrol action.

Opposite, on the west side of the Naktong, General Pak Kyo Sam, commanding the N.K. *9th Division*, issued his operational order to the division on 28 August. Its mission in the forthcoming attack was stated in part as follows:

> To outflank and destroy the enemy by capturing the Miryang and Samnangjin areas, thereby cutting off his [Eighth Army] route of withdrawal between Taegu and Pusan, is the mission of this division.[14]

The North Koreans apparently did not know on the eve of their attack that the U.S. 2d Division had replaced the 24th Division in this sector of the front, since they named the latter division in the attack order as being opposite it in the attack zone.

On the left and southern flank of the 9th Infantry river line, just above the junction of the Nam River with the Naktong, A Company was dug in on a long finger ridge paralleling the Naktong that terminates in Hill 94 at the Kihang ferry site. The river road from Namji-ri running west along the Naktong passes the southern tip of this ridge

[13] 35th Inf WD, 31 Aug 50; 35th Inf DUC award supporting docs, AG files.

[14] ATIS Enemy Documents, Issue 4, pp. 16–20, *9th Inf Div* Opn Ord concerning Naktong River crossing, signed by CG Pak Kyo Sam.

and crosses to the west side of the river at the ferry. A small village of a few huts, called Agok, lay at the base of Hill 94 and 300 yards from the river. Two medium tanks of A Company, 72d Tank Battalion and two antiaircraft vehicles of D Battery, 82d AAA Battalion, one mounting twin 40-mm. guns and the other four .50-caliber machine guns, together with two rifle squads of A Company, 9th Infantry, held a roadblock near the ferry and close to Agok. On the evening of 31 August, A Company, in accordance with orders just received, moved from its ridge positions overlooking Agok and the river to new positions along the river below the ridge line.[15]

That evening Sgt. Ernest R. Kouma took a Pershing tank to Agok to replace one that had developed gun trouble. Kouma placed his tank on the west side of Agok about forty yards from the Kihang ferry. At 2000 a heavy fog covered the river. An hour later dogs started barking on the far side of the Naktong, and continued to bark in the otherwise unbroken silence until enemy mortar shells began falling on the American-held side of the river at 2200. Fifteen minutes later a heavy enemy mortar preparation struck A Company's positions. American mortars and artillery began firing counterbattery. Some of the A Company men reported they heard noises on the opposite side of the river and splashes in the water.[16]

Suddenly at 2230 the fog lifted and Kouma saw that an enemy bridge, already two-thirds completed, was being laid across the river directly in front of him. He ordered his tank gunner to lay the 90-mm. cannon on the bridge and he himself went to the .50-caliber machine gun mounted behind the tank cupola. Kouma's gunner opened fire on the bridge and the bridging party; the other tank and the two antiaircraft vehicles joined in the action. After about a minute of this heavy fire the bridge collapsed, and after another two minutes the ponton boats used to hold the bridge in place broke loose. Machine gun fire then sank many of them. Except for the barking of the dogs across the river and an occasional mortar round, silence once again reigned as Kouma's guns fell silent after the destruction of the bridge.

At 2300 this quiet suddenly gave way to a small arms fight which flared around the left side of A Company north of the tanks. This gunfire had lasted only two or three minutes when the A Company roadblock squads near the tanks received word by field telephone that the company was withdrawing to the original ridge positions and that they should do likewise. Someone in the outpost shouted, "We are moving out, tankers!"[17] Then, as Kouma tells it:

> The infantry outpost had hardly left when I spotted seven men running towards me from the direction of where Able Company's CP formerly was located. I halted them and noticed that they were wearing the division patch. [The Indianhead of 2d Division, which the newly augmented Koreans wore on their herringbone twill as did regular members of the division. Com-

[15] Ltr, Capt Albert J. Fern, Jr. (Plat Ldr, 2d Plat, A Co, 9th Inf, 31 Aug 50) to author, 1 Apr 56; Ltr, Maj Robert L. Cody (S-3, 1st Bn, 9th Inf, Aug-Sep 50) to author, 18 Nov 55.

[16] Ltr, MSgt Ernest R. Kouma to author, 1 May 53; Ltr, SFC Oscar V. Berry to author with location sketch map, 24 Apr 53 (Berry commanded the second tank at the Kihang ferry); Ltr, Fern to author, 1 Apr 56.

[17] Ltr, Kouma to author, 1 May 53; Ltr, Fern to author, 1 Apr 56.

pany A had some of these South Koreans.] One of them spoke excellent English. All seven came next to my tank . . . three of them crawled on the deck of the tank and informed me that a large force had crossed the river farther down approaching my position and that most of Able Company were killed or captured. At the time I had the idea that they were part of the 9th Infantry. During this time I was on top of the turret checking my 50 cal. machine gun. At a given signal they leaped from the tank and began throwing grenades on the tank and about the same time a steady spray of machine gun and rifle fire began hitting the tanks and AA guns from the crest of the high bluff about 150 yards to my right. My gunner at once took them under fire as well as SFC Berry's and the AA guns. I got back in the turret and threw about 7 or 8 grenades over the house as well as inside the house through the door which faced us.[18]

In this exchange, enemy grenades and fire wounded Kouma twice. Enemy soldiers now attacked the tanks and the antiaircraft vehicles from the rear. The group approaching the quad-50 knew the password and overran the vehicle, killing all crew members except one who escaped. Several men in the second, the dual 40-mm. gun vehicle (M19), were wounded but this tracked vehicle escaped to the rear. The two tanks were alone. They quickly changed their positions, driving out from under the cliffs and near the village to open ground with clear fields of fire for 200 yards in every direction. There they repelled repeated attacks, some enemy soldiers reaching within twenty yards of the tanks before they turned back leaving their dead and wounded. About 0130, SFC Oscar V. Berry informed Kouma his tank engine was overheating and that he was going to withdraw. A mile to the rear Berry's tank engine caught fire and he abandoned the tank. Kouma maintained his position throughout the night. With the coming of daylight the enemy attempts to destroy the tank by infantry attack ceased. At 0730 Kouma started back toward friendly lines and got through safely, firing into enemy positions on the way.[19]

In the attack against A Company, the North Koreans happened to strike the 1st Platoon, which was near Agok, but they did not find the 2d Platoon northward, commanded by 2d Lt. Albert J. Fern, Jr. Fern could tell by the sound of combat that C Company on his right and that part of A Company on his left were under heavy attack. Two stragglers from C Company soon told him the North Koreans had overrun that company. The A Company commander, 1st Lt. Adam B. Rodriguez, quickly found it necessary to abandon his command post in Agok and withdraw up the ridge to his original positions, ordering his subordinate units to do likewise. Fern's 2d Platoon had a skirmish with a small group of North Koreans in the dark in going up the slope. On top, the company reassembled and went into perimeter defense positions. For them the rest of the night passed quietly.

The N.K. *9th Division*'s infantry crossing of the Naktong and attack on its east side near midnight quickly overran the positions of C Company, north of A Company. There the North Koreans as-

[18] Ltr, Kouma to author, 1 May 53.

[19] Ltrs, Kouma and Berry, 1 May 53 and 24 Apr 53. Department of the Army General Order 38, 4 June 1951, awarded the Medal of Honor to Sergeant Kouma.

saulted with unusual force, to the accompaniment of green flares and blowing of whistles. The company held its positions only a short time and then attempted to escape. Many of the men moved southward, a few of them coming into A Company's ridge line positions near Agok during the night. Most of C Company moved all the way to the 25th Division positions south of the Naktong. On 2 September that division reported that 110 men of C Company had come into its lines.[20]

### *Task Force Manchu Misfires*

Five miles north of Agok and A Company's position, B Company, 9th Infantry, held a similar position on Hill 209 overlooking the Paekchin ferry crossing of the river. This ferry was located at the middle of the Naktong Bulge where the Yongsan road came down to the Naktong and crossed it. The U.S. 2d Division, as it chanced, had planned an important reconnaissance action to start from there the night of 31 August, the very night that the N.K. *I Corps* offensive rolled across the river.

Near the end of the month two reconnaissance patrols from the 9th Infantry had crossed to the west side of the Naktong and from a hill position watched enemy tank and troop activity at a place approximately two miles west of the river, which they suspected was a division command post. Information obtained later indicated it was in fact the command post of the N.K. *9th Division*. On 25 August, Col. John G. Hill outlined projected "Operation Manchu," which was to be a company-sized combat patrol to cross the river, advance to the suspected enemy command post and communications center, destroy it, capture prisoners, and gain information of enemy plans.[21]

The 9th Infantry Regiment had planned Task Force Manchu on orders from the 2d Division, which in turn had received instructions from Eighth Army for aggressive patrolling. Colonel Hill selected three possible crossing sites for the operation. General Keiser decided on the one at the Paekchin ferry. The 9th Infantry reserve, E Company, reinforced with one section of light machine guns from H Company, was to be the attack force. The 1st Platoon, 2d Engineer Combat Battalion, was to transport it across the river in assault boats the night of 31 August. Two heavy weapons companies, D and H, were each to furnish one section of heavy machine guns, one section of 81-mm. mortars, and one section of 75-mm. recoilless rifles for supporting fires. A platoon of 4.2-inch mortars was also to give support.[22]

After dark on the evening of 31 August, 1st Lt. Charles I. Caldwell of D Company and 1st Lt. Edward Schmitt of H Company, 9th Infantry, moved their men and weapons to the base of Hill 209, which was within B Company's defense sector and overlooked the Paekchin ferry crossing of the Naktong River. The raid-

[20] 9th Inf WD, 2 Sep 50; Ltr, Fern to author, 1 Apr 56.

[21] Interv, author with Hill, 30 Jun 53; Cody, Operation Manchu, student MS, Advanced Inf Off Course, Class 2, Inf School, Ft. Benning, 1952–53.

[22] Interv, author with Hill, 30 Jun 50; Cody, Operation Manchu; Ltr, Capt Lee E. Beahler (CO D Co, 2d Engr C Bn, Aug–Sep 50) to author, 10 Jun 53.

ing force, E Company, was still in its regimental reserve position about two miles west of Yongsan, getting ready with the engineer platoon to move to the crossing site. Colonel Hill, the regimental commander, went forward in the evening with the 4.2-inch mortar platoon to its position at the base of Hill 209 where the mortarmen prepared to set up their weapons.

Schmitt and Caldwell took their section leaders up the hill and showed them where they wanted the weapons set up. The first of the carrying parties soon followed them. It was now a little after 2100, and dark.

The closest front line unit was B Company on top of Hill 209, approximately a mile north of the river road which curved around the hill's southern base. The regimental chaplain, Capt. Lewis B. Sheen, had gone forward in the afternoon to B Company to hold services.[23] On top of Hill 209, Chaplain Sheen and men in B Company after dark thought they could hear a swishing sound in the water below them. By straining their eyes and staring through field glasses for a long time into the near darkness, they made out a long line of North Korean soldiers wading the river.

The first enemy crossing at the Paekchin ferry caught the Heavy Mortar Platoon wholly unaware in the act of setting up its weapons. It also caught most of the D and H Company men at the base of Hill 209, only a little more than half a mile from the crossing site. The North Koreans killed or captured many of them. Colonel Hill was there, but escaped to the rear just before midnight, together with several others, when the division canceled Operation Manchu. His S–3, who was with him, delayed a bit and never got out. The first heavy weapons carrying party was on its way up the hill when the enemy engulfed the men below. It hurried on to the top where the advance group waited and there all hastily dug in on a small perimeter. This group was unmolested during the night.

Word of the enemy crossing that had caught the support elements of Task Force Manchu flat-footed had been received at the 2d Division headquarters. This news, together with the heavy enemy barrages that had developed all along the river, caused the division to cancel Operation Manchu five minutes before midnight.

From approximately 2130 until shortly after midnight the N.K. *9th Division* crossed the Naktong at a number of places and climbed the hills quietly toward the 9th Infantry river line positions. Then, when the artillery barrage preparation lifted, the North Korean infantry were in position to launch their assaults. These began in the northern part of the regimental sector and quickly spread southward. Chaplain Sheen in the B Company perimeter heard cries of "Manzai!" northward and saw many flares light the sky in that direction. At the river crossing below him he could hear enemy troops working on a bridge. By this time the sounds of enemy tanks and trucks and shouting men came up from the river. And from the hills upstream the men in B Company heard, ever so often, after a flurry of small arms

[23] Interv, author with Hill, 15 Apr 53; Ltr, Capt Charles I. Caldwell to author, 29 May 53, together with sketch map of positions of D and H Co units of TF Manchu, 31 Aug–4 Sep 50. Hill 209 is Hill 210 on the revised map of Korea, AMS 4, 1950.

fire, a massed shout which they interpreted as the North Korean capture of another position.[24]

At 0200, B Company's turn came. A truck stopped at the bottom of the hill, a whistle sounded, then came a shouted order, and enemy soldiers started climbing the slope. The hills on both sides of B Company were already under attack as was also Hill 311, a rugged terrain feature a mile and a half back from the river and apparently the enemy's principal immediate objective. The North Koreans apparently were not aware of the Task Force Manchu group lower down on the hill for it remained unmolested during the night. But higher up on Hill 209 the enemy drove B Company from its position, inflicting very heavy casualties on it. Chaplain Sheen led one group of soldiers back to friendly lines on 4 September.[25]

Approximately at 0300, 1 September, the 9th Infantry Regiment ordered its only reserve, E Company, which was to have been the striking force of Task Force Manchu, to move west along the Yongsan–Naktong River road and take a blocking position at the pass between Cloverleaf Hill and Obong-ni Ridge, about three miles from the river and six miles from Yongsan. This was the critical terrain where so much heavy fighting had taken place in the first battle of the Naktong Bulge. Fighting began at the pass at 0230 when an American medium tank of A Company, 72d Tank Battalion, knocked out a T34 at Tugok (Morisil). E Company never reached its blocking position. A strong enemy force surprised and delivered heavy automatic fire on it at 0330 from positions astride the road east of the pass. The company suffered heavy casualties, the killed including the company commander and General Keiser's aide who had accompanied the force. With the critical parts of Cloverleaf Hill and Obong-ni Ridge in enemy hands before dawn of 1 September, the best defensive terrain between Yongsan and the river was lost. The 2d Division now had to base its defense of Yongsan on relatively poor defensive terrain, the low jumbled hills at the western edge of the town.[26]

### *The North Koreans Split the U.S. 2d Division*

North of the 9th Infantry sector of the 2d Division front along the Naktong, the 23d Regiment on 29 August had just relieved the 3d Battalion of the 38th Infantry Regiment, which in turn had only a few days before relieved the 21st Infantry Regiment of the 24th Division. On 31 August, therefore, the 23d Regiment was in a new sector of which it had only a limited knowledge. It took over a 16,000-yard Naktong River front without its 3d Battalion which had been attached to the U.S. 1st Cavalry Division. Colonel Freeman, the regimental commander, deployed the 1st Battalion on the high ground along the river with the three companies abreast. Actually,

[24] Cody, Operation Manchu; 9th Inf WD, Sep 50, Sheen MS, From Encirclement to Safety; 9th Inf WD, 1 Sep 50, Rpt at 0040 to 9th Regt.

[25] Sheen, From Encirclement to Safety; Ltr, Caldwell to author, 29 May 53; Interv, author with Hill, 30 Jun 53.

[26] Interv, author with Hill, 30 Jun 53; 72d Tk Bn WD, 1 Sep 50; 9th Inf WD, 1 Sep 50; EUSAK WD, G–3 Jnl, 0343 1 Sep 50; Cody, Operation Manchu.

the 1st Battalion, under Lt. Col. Claire E. Hutchin, Jr., little more than outposted the hills with platoons and squads. He placed the 2d Battalion in a reserve position approximately eight miles in the rear of the 1st Battalion and in a position where it commanded the road net in the regimental sector. On the last day of the month the 2d Division moved E Company south to a reserve position in the 9th Infantry sector.[27]

Two roads ran through the regimental sector from the Naktong River to Changnyong. The main road bent south along the east bank of the river to Pugong-ni and then turned northeast to Changnyong. A northern secondary road curved around marshland and lakes, the largest of which was Lake U-p'o, to Changnyong. In effect, the 1st Battalion of the 23d Regiment guarded these two approach routes to Changnyong.

The forty-two men of the 2d Platoon, B Company, 23d Infantry, led by 1st Lt. William M. Glasgow held outpost positions on seven hills covering a 2,600-yard front along the east bank of the Naktong north of Pugong-ni. Across the river in the rice paddies they could see, in the afternoon of 31 August, two large groups of enemy soldiers. Occasionally artillery fire dispersed them.

Just before dusk turned to darkness, Glasgow and the men in his 1st Squad saw "a large and bizarre torchlight parade" come out of the hills and proceed toward the river. Glasgow immediately reported the spectacle to the battalion command post. The artillery forward observer, who estimated the crowd to number 2,000 people, thought they were refugees. When the matter was referred to Colonel Freeman, he immediately ordered the artillery to fire on the torchbearers. With each bursting shell some of the torches disappeared but others took their places and the procession continued unchecked toward the river bank.[28]

At 2100 the first shells of what proved to be a two-hour enemy artillery and mortar preparation against the American river positions jarred the fascinated Glasgow and his companions from their absorbed contemplation of the torchlight scene. As the enemy barrage rolled on, North Korean infantry crossed the river and climbed the hills in the darkness under cover of its fire. At 2300 the barrage lifted. A green flare signaled the North Korean assault. A few minutes later enemy grenades showered into Glasgow's position. After a short fight at close quarters, Glasgow and his men ran off the hill toward the rear. Similar assaults took place elsewhere along the battalion outpost line.

On the regimental left along the main Pugong-ni–Changnyong road enemy soldiers completely overran C Company by 0300, 1 September. Capt. Cyril S. Bartholdi, the company commander, and most of his men were lost. Only seven men of C Company could be accounted for, and

[27] 23d Inf WD, Aug 50 Summ; Freeman, Highlights of the Combat Activities of the 23d Infantry Regiment from 5 August to 30 September 1950, MS, copy in OCMH; Maj Gen Paul L. Freeman, Jr., MS review comments, 30 Oct 57.

[28] Glasgow, Platoon Leader in Korea, pp. 100ff; Glasgow, "Through Hell and Out," *Bluebook* Magazine (August, 1951), pp. 71–77 (reproduces that part of above MS covering experiences of 1–7 September 1950); 23d Inf WD, Aug 50 Summ; Interv, author with Lt Col Frank Meszar (S–3, 23d Inf, Sep 50), 15 May 53; Freeman, MS review comments, 30 Oct 57.

three days later, after all the stragglers and those cut off behind enemy lines had come in, there were fewer than twenty men in the company.[29]

As the enemy attack developed during the night, Colonel Hutchin succeeded in withdrawing a large part of the battalion, less C Company, to his command post just north of Lake U-p'o and the hills there covering the northern road into Changnyong, three miles east of the river and five air miles west of the town. B Company lost heavily in this action.

When word of Colonel Hutchin's plight and of the disaster that had overtaken C Company reached regimental headquarters, Colonel Freeman obtained the release of G and F Companies from 2d Division reserve and sent the former to help Hutchin and the latter on the southern road toward Pugong-ni and C Company. Maj. Lloyd K. Jensen, executive officer of the 2d Battalion, accompanied F Company down the Pugong-ni road. This force was unable to reach C Company, but Major Jensen collected stragglers from it and seized high ground astride this main approach to Changnyong near Ponch'o-ri above Lake Sanorho, and went into a defensive position there. The 2d Division released E Company to the regiment and the next day it joined F Company to build up what became the main defensive position of the 23d Regiment in front of Changnyong. Lt. Col. James W. Edwards took command of this 2d Battalion position.[30]

Enemy troops during the night passed around the right flank of Colonel Hutchin's northern blocking position and reached the road three miles behind him near the division artillery positions. The 23d Infantry Headquarters and Service Companies and other miscellaneous regimental units finally stopped this enemy penetration near the regimental command post five miles northwest of Changnyong.

Before the morning of 1 September had passed, reports coming in to 2d Division headquarters made it clear that North Koreans had penetrated to the north-south Changnyong-Yongsan road and cut the division in two; the 38th and 23d Infantry Regiments with the bulk of the division artillery in the north were separated from the division headquarters and the 9th Infantry Regiment in the south. General Keiser decided that this situation made it advisable to control and direct the divided division as two special forces. Accordingly, he placed the division artillery commander, Brig. Gen. Loyal M. Haynes, in command of the northern group. Haynes' command post was seven miles north of Changnyong. Task Force Haynes became operational at 1020, 1 September.[31] Southward, in the Yongsan area, General Keiser placed Brig. Gen. Joseph S. Bradley, Assistant Division Commander,

[29] 2d Div WD, JA Stf Sec Hist Rpt, 1 Sep–31 Oct 50, p. 5; 23d Inf WD, Narr Summ, Sep 50; Interv, author with Meszar, 15 May 53; EUSAK WD, 21 Sep 50, ADVATIS Interrog Rpts, Sr Lt Lee Kwan Hyon, Med Off, *17th Regt, 2d Div;* Glasgow, Platoon Leader in Korea.

[30] Interv, author with Meszar, 15 May 53; Freeman MS; Freeman, MS review comments, 30 Oct 57.

[31] 2d Div Arty WD, Narr Summ, Sep 50. Task Force Haynes remained operational until 1300, 15 September 1950. Units forming the task force were the following: 23d Inf; 38th Inf; 37th FA Bn; C Btry, 503d FA Bn; Btrys A, B, C, 82d AAA AW Bn (SP); and C Co, 72d Tk Bn.

AREA OF MAIN DEFENSIVE POSITION *in front of Changnyong.*

in charge of the 9th Infantry Regiment, the 2d Engineer Combat Battalion, most of the 72d Tank Battalion, and other miscellaneous units of the division. This southern grouping was known as Task Force Bradley.

All three regiments of the enemy *2d Division*—the *4th, 17th,* and *6th,* in line from north to south—crossed during the night to the east side of the Naktong River into the 23d Regiment sector. The enemy *2d Division,* concentrated in the Sinban-ni area west of the river, had, in effect, attacked straight east across the river and was trying to seize the two avenues of advance into Changnyong above and below Lake U-p'o. The water area of this lake and the surrounding marshland varied according to the season and the amount of rainfall. On 31 August 1950, Lake U-p'o was a large body of water although in most places only a few feet deep.[32]

### *General Walker's Decisions on 1 September*

At daybreak of 1 September, General Keiser at 2d Division headquarters in Muan-ni, seven air miles east of Yongsan on the Miryang road, knew that his division was in the midst of a crisis. A massive enemy attack was in progress and had made deep penetrations everywhere in his sector except in the north in the zone of the 38th Infantry. The N.K. *9th Division* had effected major crossings of the Naktong at two principal points

[32] Interv, author with Meszar, 15 May 53; 2d Div WD, Aug 50 (G-3 Stf Sec Rpt is incorporated with Div Narr Summ—all G-3 supporting documents were destroyed or lost through enemy action at Kunu-ri in November 1950); 23d Inf WD, Narr Summ, Sep 50.

against the 9th Infantry; the *2d Division,* three major crossings against the 23d Infantry; and the *10th Division* had crossed more troops in the Hill 409 area near Hyongp'ung in the 38th Infantry sector. At 0810 General Keiser telephoned Eighth Army headquarters and reported the situation as he then understood it, indicating that the heaviest and deepest enemy penetrations were in the 9th Infantry sector.

The picture of the situation darkened as the morning hours passed. Liaison planes rose from the division strip every hour to observe the enemy's progress and to locate 2d Division front-line units. Communication from division and regimental headquarters to nearly all the forward units was broken. Beginning at 0930 and continuing throughout the rest of the day, the light aviation section of the division artillery located front-line units cut off by the enemy, and made fourteen drops of ammunition, food, water, and medical supplies. As information slowly built up at division headquarters it became apparent that the North Koreans had punched a hole six miles wide and eight miles deep in the middle of the division line and made lesser penetrations elsewhere. The front-line battalions of the 9th and 23d Regiments were in various states of disorganization and some companies had virtually disappeared. General Keiser hoped he could organize a defense along the Changnyong-Yongsan road, five to eight miles east of the Naktong River, and prevent enemy access to the passes eastward leading to Miryang and Ch'ongdo.[33]

On its part, the Eighth Army staff had sufficient information soon after daybreak of 1 September to realize that a big enemy attack was under way in the south. At 0900 General Walker requested the Air Force to make a maximum effort along the Naktong River from Toksong-dong, just above the 2d Division boundary, southward and to a depth of ten to fifteen miles west of the river. He wanted the Air Force to isolate the battlefield and prevent enemy reinforcements and supplies from moving across the river in support of the North Korean spearhead units. The Far East Command requested the Navy to join in the air effort, and the Seventh Fleet, pursuant to NAVFE orders, turned back from its strikes in the Inch'on-Seoul area and sped southward at full steam toward the southern battle front. General Walker came to the 2d Division front at noon and ordered a "stand or die" defense. He had already ordered ground reinforcements to the Yongsan area.[34]

For a few hours during the morning of 1 September, General Walker weighed the news coming in from his southern front, wavering in a decision as to which part of the front most needed his Pusan Perimeter reserves. Since midnight the N.K. *I Corps* had broken his Pusan Perimeter in two places—the N.K. *2d* and *9th Divisions* in the U.S. 2d Division sector, and the *7th* and *6th Divisions* in the U.S. 25th Division sector, below the junction of the Nam and Naktong Rivers. In the 2d Division sector enemy troops were at the edge of Yongsan, the

[33] EUSAK WD, G–3 Jnl, 1 Sep 50; *Ibid.,* PIR 51, 1 Sep 50; 2d Div WD, Hq Co, Aviation Sec, 1 Sep 50; Ibid., vol. II, Summ, 1 Sep–31 Oct 50, p. 6.

[34] EUSAK WD, G–3 Jnl, 1 Sep 50; 2d Div WD, vol. II, Summ, 1 Sep–3 Oct 50, pp. 6–7; Memo for Dept of Navy, Hist Sec, 1950, in OCMH.

gateway to the corridor leading twelve air miles eastward to Miryang and the main Pusan-Mukden railroad and highway.

Walker had a critical decision to make. He had in reserve three understrength infantry regiments and the 2-battalion British 27th Infantry Brigade which was not yet completely equipped and ready to be placed in line. Even so, this was an unusually large reserve for Eighth Army in the summer of 1950. The three U.S. regiments available to Walker were the 5th Marines at Changwon, six miles northeast of Masan, preparing for movement to the port of Pusan; the 27th Regiment of the 25th Division which had arrived at Masan only the night before at 2030 to relieve the 5th Regimental Combat Team, which was then to join the 24th Division in the Taegu area; and the 19th Infantry Regiment of the 24th Division, then with that division's headquarters at Kyongsan southeast of Taegu. Walker alerted both the 24th Division headquarters, together with its 19th Regiment, and the 1st Provisional Marine Brigade to move at a moment's notice; the 24th Division either to the 2d or 25th Division fronts, and the marines to an unannounced destination.

As the morning passed, General Walker decided that the situation was most critical in the Naktong Bulge area of the 2d Division sector. There the North Koreans threatened Miryang and with it the lifeline of the entire Eighth Army position. There, for the moment at least, was the most critical spot of the far-flung battlefield. An hour before noon General Walker ordered General Craig to prepare the marines to move at once. Just after noon the action order came and the marines made ready to depart at 1330. They were going back to the bulge area.[35]

---

[35] EUSAK WD, G-3 Sec, 1 Sep 50; 2d Div WD, 1 Sep 50; Interv, author with Hill, 30 Jun 53; Transcription and Summ of fonecon, Walker with Hickey, Deputy CofS, FEC, 020935 Sep 50, CofS files FEC.

CHAPTER XXIV

# The North Korean Great Naktong Offensive

> If men make war in slavish observance of rules, they will fail. . . . War is progressive, because all the instruments and elements of war are progressive.
>
> ULYSSES S. GRANT

The situation on Eighth Army's southern front was chaotic by midday of 1 September. The North Koreans at one place had crossed at the Kihang ferry, captured Agok, engaged Kouma's tanks, and scattered A Company, 9th Infantry of the 2d Division, at its positions from Agok northward. (*See Map VI.*) Lieutenant Rodriguez succeeded in withdrawing most of A Company to its old positions on the ridge line back of the river. From there at daylight the men could see enemy soldiers on many of the ridges surrounding them, most of them moving east. After several hours, Lieutenant Fern, 2d Platoon leader, sent a patrol down the hill to Agok to obtain supplies abandoned there during the night. The patrol encountered a small enemy group in the village, killed three men and sustained two casualties, but returned with much needed water, rations, and ammunition.

Later in the morning enemy barges crossed the Naktong below A Company but they were out of range. Rodriguez sent a squad with a light machine gun to the southern tip of the ridge overlooking Agok to take these enemy troops under fire. About halfway down, the squad came upon a critically wounded Negro soldier. Around him lay ten dead North Koreans. The wounded man was evacuated to the company command post but died that afternoon. When the squad reached the tip of the ridge they saw that an enemy force occupied houses at its base. They reported this to Lieutenant Fern, who called for artillery fire through the forward observer. This artillery fire was delivered within a few minutes and was on target. The North Koreans broke from the houses, running for the river. At this the light machine gun at the tip of the ridge took them under fire, as did another across the Naktong to the south in the 25th Division sector. Proximity fuze artillery fire decimated this group. Combined fire from all weapons inflicted an estimated 300 casualties.

In the afternoon, light aircraft dropped food and ammunition to the company; only part of it was recovered.

The 1st Battalion ordered Rodriguez to withdraw the company that night.

Lieutenant Fern's 2d Platoon led the A Company withdrawal immediately after dark, moving eastward along the ridge crest. At the eastern tip the platoon started down. Near the bottom the leading men saw a column of about 400 North Koreans marching on the road some 200 yards below them with a number of machine guns mounted on wheels. Rodriguez ordered the company to circle back up the ridge and away from the road. Fern was to bring up the rear and carry with him the wounded, two of whom were litter cases. Transporting the wounded over the rough terrain in the darkness was a slow and difficult task and gradually Fern's platoon fell behind the others. By the time he reached the base of the ridge he had lost contact with the rest of the company.

At this juncture a furious fire fight erupted ahead of Fern. Enemy machine gun fire from this fight struck among the 2d Platoon and pinned it down. For their safety, Fern decided to send the wounded back into the ravine they had just descended, and put them in charge of Platoon Sgt. Herbert H. Freeman and ten men. Several stragglers from the advanced elements of the company joined Fern and reported that Rodriguez and the rest of the company had run into a sizable enemy force and had scattered in the ensuing fight. Lieutenant Rodriguez and most of the company were killed at close range. In this desperate action, Pfc. Luther H. Story, a weapons squad leader, so distinguished himself by a series of brave deeds that he was awarded the Medal of Honor. Badly wounded, Story refused to be a burden to those who might escape, and when last seen was still engaging enemy at close range. Of those with Rodriguez, approximately ten men escaped to friendly lines.

Fern decided shortly before dawn that he must try to escape before daylight. He sent word by a runner back to Freeman, who should have been about 500 yards in the rear, to rejoin the platoon. The runner returned and said he could not find Freeman. There had been no firing to the rear, so Fern knew that Freeman had not encountered enemy troops. Two men searched a second time for Freeman without success. Fern then decided that he would have to try to lead those with him to safety.

A heavy ground fog, so thick that one could hardly see twenty-five yards, developed in the early morning of 2 September and this held until midmorning. Under this cloak of concealment Fern's group made its way by compass toward Yongsan. From a hill at noon, after the fog had lifted, the men looked down on the battle of Yongsan which was then in progress. That afternoon Fern brought the nineteen men with him into the lines of the 72d Tank Battalion near Yongsan.

Upon reporting to Lt. Col. John E. Londahl, Fern asked for permission to lead a patrol in search of Sergeant Freeman's group. Londahl denied this request because every available man was needed in the defense of Yongsan. As it turned out, Freeman brought his men to safety. Upon moving back from Fern's platoon during the night battle, he had taken his group all the way back up to the top of the ridge. They had stayed there in seclusion all day, watching many enemy groups moving about in all directions below them. Freeman assumed that most of A Company had been killed or captured. For five days and nights he maintained his squad and the four wounded behind enemy lines, finally guiding them all safely to friendly lines.[1]

### *The End of Task Force Manchu*

It will be recalled that the North Koreans who crossed near the middle of the Naktong Bulge in front of B Company, 9th Infantry, surprised the advanced support elements of Task Force Manchu at the base of Hill 209 where the Yongsan road came down to the Naktong. Some elements of the two Heavy Weapons Companies, D and H, had already started to climb the hill to emplace their weapons there when the North Korean surprise river crossing caught most of the support elements and the Heavy Mortar Company at the base of the hill. This crossing was about five miles north of the enemy crossing that had all but destroyed A Company near the division's southern boundary.

The perimeter position taken by the men of D and H Companies, 9th Infantry, who had started up the hill before the North Koreans struck, was on a southern knob (about 150 meters high) of Hill 209, half a mile south across a saddle from B Company's higher position. As the night wore on, a few more men reached the perimeter. In addition to the D and H Company men, there were a few from the Heavy Mortar Platoon and one or two from B Company. Altogether, there were approximately 60 to 70 men, including 5 officers, in the group—an actual count was never made. An inventory of the weapons and equipment disclosed that the group had 1 SCR–300 radio; 2 heavy machine guns, 1 operable; 2 light machine guns; 1 BAR; about 20 M1 rifles; and about 40 carbines or pistols. Lieutenant Schmitt assumed command of the group.[2]

During the night Lieutenant Schmitt established radio communication with the 1st Battalion, 9th Infantry, and received promises of help on the morrow. When daylight came Schmitt and his group saw that they were surrounded by enemy. One force occupied the higher knob half a mile above them, formerly held by B Company. Below them, North Koreans continued crossing the river and moving supplies forward to their combat units, some of them already several miles eastward.

Enemy troops were not long in discovering the Task Force Manchu group. They first attacked it at 1400 that afternoon, and were repulsed. That night an estimated company attacked three times, pressing the fight to close quarters, but failed each time to penetrate the tight perimeter. Daylight of the second day disclosed many enemy dead on the steep slopes outside the perimeter.

By that morning (2 September) the

[1] Ltr, Fern to author, 1 Apr 56; Ltr, Cody to author, 18 Nov 55. Department of the Army General Order 70, 2 August 1951, awarded the Medal of Honor to Private Story. General Order 187, 5 December 1950, awarded the Distinguished Service Cross to Sergeant Freeman. EUSAK.

[2] Ltr, Caldwell to author, 29 May 53; Interv, author with Hill, 30 Jun 53; Hill, MS review comments, 2 Jan 58; 9th Inf WD, Sep 50, Incl B, Col Charles C. Sloane, Jr., Hill 209 (1138–1386), with sketch map; *Ibid.*, app., 1st Lt Raymond J. McDoniel, Notes (this document misspells "McDoniel" as "McDaniel"); Sheen, From Encirclement to Safety. The officers on the hill were Lt Schmitt, CO H Co; Lt McDoniel, Plat Ldr D Co; Lt Paul E. Kremser, Plat Ldr H Co; Lt Caldwell, Plat Ldr D Co; and Lt Edmund J. Lilly III, Plat Ldr B Co.

need for hand grenades was desperate. About 0900 MSgt. Travis E. Watkins of H Company shot and killed two enemy soldiers 50 yards outside the northeast edge of the perimeter. He jumped from his hole to get the weapons and grenades of the dead men; 20 yards from them three hidden enemy soldiers jumped to their feet and opened fire on him. Watkins killed them and gathered weapons, ammunition, and insignia from all five before returning to the perimeter. An hour later a group of six enemy soldiers gained a protected spot 25 yards from a machine gun position of the perimeter and began throwing hand grenades into it. Although already wounded in the head, Watkins rose from his hole to engage them with rifle fire. An enemy machine gun immediately took him under fire and hit him in the left side, breaking his back. Watkins in some manner managed to kill all six of the nearby enemy soldiers before he sank into his hole paralyzed from the waist down. Even in this condition, Watkins never lost his nerve, but shouted encouragement to his companions. He refused any of the scarce rations, saying that he did not deserve them because he could no longer fight.[3]

In the afternoon of 2 September Schmitt succeeded in radioing a request to the 1st Battalion for an airdrop of supplies. A division liaison plane attempted the drop, but the perimeter was so small and the slopes so steep that virtually all the supplies went into enemy hands. The men in the perimeter did, however, recover from a drop made later at 1900 a case of carbine ammunition, 2 boxes of machine gun ammunition, 11 hand grenades, 2½ cases of rations, part of a package of medical supplies, and 21 cans of beer. Pfc. Joseph R. Ouellette, H Company, left the perimeter to retrieve an airdrop of water cans but found on reaching them that they were broken and empty. Like Watkins, he distinguished himself by leaving the perimeter to gather weapons, ammunition, and grenades from the enemy dead. On one such occasion an enemy soldier suddenly attacked Ouellette, who killed the North Korean in hand-to-hand combat.[4]

In helping to recover the airdropped supplies on the evening of 2 September, Lieutenant Schmitt was wounded but continued to exercise his command, encouraging the diminishing group by his example. That same afternoon, the North Koreans sent an American prisoner up the hill to Schmitt with the message, "You have one hour to surrender or be blown to pieces." Failing in frontal infantry attack to reduce the little defending force, the enemy now obviously meant to take it under observed and registered mortar fire.[5]

Forty-five minutes later enemy antitank fire came in on the knob and two machine guns from positions northward and higher on the slope of Hill 209 swept the perimeter. Soon, enemy mor-

[3] Sworn affidavit, SSgt Grover L. Bozarth and Sgt Ralph G. Lillard, H Co, 9th Inf, 15 Sep 50, Yongsan, recommending Watkins for Medal of Honor, DA AG files.

[4] McDoniel, Notes cited n. 2, Sep 50; Ltr, Caldwell to author, 29 May 53. Department of the Army General Order 25, 25 April 1951, awarded the Medal of Honor posthumously to Private Ouellette.

[5] EUSAK WD, 9 Sep 50, an. 1 to PIR 59; 9th Inf WD, 8 Sep 50, account of Lt McDoniel; Sloane, Hill 209, Sep 50. General Order 54, 6 February 1951, awarded the Distinguished Service Cross posthumously to Lieutenant Schmitt. EUSAK.

tars emplaced on a neighboring high finger ridge eastward registered on Schmitt's perimeter and continued firing until dark. The machine gun fire forced every man to stay in his hole. The lifting of the mortar fire after dark was the signal for renewed enemy infantry attacks, all of which were repulsed. But the number of killed and wounded within the perimeter was growing, and food, water, and ammunition were needed. There were no medical supplies except those carried by one aid man.

The third day, Sunday, 3 September, was the worst of all. The weather was terrifically hot. There was no water, and only one can of C rations per man. Ammunition was almost gone. Since the previous afternoon, enemy mortar barrages had alternated with infantry assaults against the perimeter. Survivors later estimated there were about twenty separate infantry attacks—all repulsed. Two enemy machine guns still swept the perimeter whenever anyone showed himself. Dead and dying were in almost every foxhole or lay just outside. Mortar fragments destroyed the radio and this ended all communication with friendly units. Artillery fire and air strikes requested by Schmitt never came. Some enemy soldiers worked their way close to the perimeter and threw grenades into it. Six times Ouellette leaped from his foxhole to escape grenades thrown into it. Each time the enemy fired on him from close range. In this close action Ouellette was killed. Most of the foxholes of the perimeter received one or more direct mortar hits in the course of the continuing mortar fire. One of these killed Lieutenant Schmitt on 3 September. He had given his men heroic leadership and had inspired them by his example throughout three days and nights of the ordeal. The command passed now to 1st Lt. Raymond J. McDoniel of D Company, senior surviving officer.[6]

In the evening, relief came in the form of rain. McDoniel spread out two blankets recovered with airdropped supplies the day before, and wrung from them enough water to fill a 5-gallon can. The men removed their clothing and wrung water from them to fill their canteens.

The fourth night passed. At daylight on the morning of 4 September only two officers, McDoniel and Caldwell, and approximately half the men who had assembled on the hill, were alive. Some men had broken under the strain and in a state of shock had run from their holes and were killed. As the day passed, with ammunition down to about one clip per man and only a few grenades left and no help in sight, McDoniel decided to abandon the position that night. He told Caldwell that when it got dark the survivors would split into small groups and try to get back to friendly lines. That evening after dark the North Koreans tried to get their men to assault the perimeter again, but, despite shouted orders of "Manzai!" only a few grenades fell inside the perimeter—apparently the enemy soldiers had had enough and refused to charge forward.

At 2200, McDoniel and Caldwell and twenty-seven enlisted men slipped off the hill in groups of four. One poignant scene etched itself on the minds of Sergeant Watkins' comrades. Watkins, still alive in his paralyzed condition, refused efforts of evacuation, saying that he did

---

[6] Ltr, Caldwell to author, 29 May 53; McDoniel, Notes, Sep 50.

not want to be a burden to those who had a chance to get away. He asked only that his carbine be loaded and placed on his chest with the muzzle under his chin. He smiled a last farewell to his buddies and wished them well when they started off the hill.[7]

McDoniel and Caldwell started off the hill together, their plan being to make their way to the river and follow it downstream. At the road they encountered so much enemy activity that they had to wait about an hour for the supply-carrying parties, tanks, and artillery to clear so that they could cross. Once across the road the two men found themselves in the middle of a North Korean artillery battery. They escaped unobserved and hid in a field near the river at daybreak. That night the two men became separated when they ran into an enemy outpost. The next morning two enemy soldiers captured Caldwell, removed his boots and identification, smashed him on the head with a rock, and threw him over a cliff into the Naktong River. Caldwell, not critically injured, feigned death and escaped that night. Four days later, on 10 September, he entered the lines of the 72d Tank Battalion.

Of the twenty-nine men who came off the hill the night of 4 September, twenty-two escaped to friendly lines—many of them following the Naktong downstream, hiding by day and traveling by night, until they reached the lines of the 25th Division.[8]

Members of Task Force Manchu who escaped from Hill 209 brought back considerable intelligence information of enemy activity in the vicinity of the Paekchin ferry crossing site. At the ferry site the enemy had put in an underwater ford. A short distance downstream, each night half an hour after dark they placed a metal floating bridge across the river and took it up before dawn the next morning. Carrying parties of 50 civilians guarded by four soldiers crossed the river continuously at night at a dogtrot, an estimated total of 800–1,000 carriers being used at this crossing site.[9]

### *The Battle of Yongsan*

On the morning of 1 September the *1st* and *2d Regiments* of the N.K. *9th Division* (the *3d Regiment* had been left at Inch'on), in their first offensive of the war, stood only a few miles short of Yongsan after a successful river crossing and penetration of the American line. At that point the chances of the division accomplishing its assigned mission must have looked favorable to its commanding general, Pak Kyo Sam.

As the N.K. *9th Division* approached Yongsan, its *1st Regiment* was on the

[7] Ltr, Caldwell to author, 29 May 53; McDoniel, Notes, Sep 50; Bozarth and Lillard Affidavit; Ltr, MSgt Robert S. Hall (1st Bn, 9th Inf, Aug–Sep 50—Hall maintained morning rpts) to author, 1 Jun 54. Department of the Army General Order 9, 16 February 1951, awarded the Medal of Honor to Sergeant Watkins posthumously.

[8] McDoniel, Notes, Sep 50; New York *Times*, September 9, 1950. Three weeks later, when the N.K. *9th Division* had been driven back across the Naktong, a party of 9th Infantry men climbed to the tragic perimeter on Hill 209. They found most of the dead had been blown to pieces in the foxholes, and it was often difficult to tell whether two or three men had occupied a particular hole. There were approximately thirty American dead at the site, fifteen of whom could be identified. Sloane, Hill 209, Sep 50.

[9] EUSAK WD, PIR 59, an. 1, 9 Sep 50.

north and its *2d Regiment* on the south. The division's attached support, consisting of one 76-mm. artillery battalion from the *I Corps*, an antiaircraft battalion of artillery, two tank battalions of the *16th Armored Brigade*, and a battalion of artillery from the *4th Division*, gave it unusual weapon support. Crossing the river behind it came the *4th Division*, a greatly weakened organization, far understrength, short of weapons, and made up mostly of untrained replacements. A captured enemy document referred to this grouping of units that attacked from the Sinban-ni area into the Naktong Bulge as "the main force" of *I Corps*. Elements of the *9th Division* reached the hills just west of Yongsan during the afternoon of 1 September.[10]

On the morning of 1 September, with only the shattered remnants of E Company at hand, the 9th Infantry had virtually no troops to defend Yongsan. General Keiser in this emergency attached the 2d Engineer Combat Battalion to the regiment. The 72d Tank Battalion and the 2d Division Reconnaissance Company also were assigned positions close to Yongsan. Colonel Hill planned to place the engineers on the chain of low hills that arched around Yongsan on the northwest.

Capt. Frank M. Reed, commanding officer of A Company, 2d Engineer Combat Battalion, led his company westward on the south side of the Yongsan–Naktong River road; Lt. Lee E. Beahler with D Company of the 2d Engineer Battalion was on the north side of the road. Approximately two miles west of Yongsan an estimated 300 enemy troops engaged A Company in a fire fight. Two quad-50's and one twin-40 gun carrier of the 82d AAA Battalion supported Reed's men in this action, which lasted several hours. Meanwhile, Lieutenant Beahler protested his position because of its long frontage and exposed flanks. With the approval of General Bradley, he moved his Engineer company to the hill immediately south of and overlooking Yongsan. A platoon of infantry went into position behind him. Captain Reed was now ordered to fall back with his company to the southeast edge of Yongsan on the left flank of Beahler's company. There, A Company went into position along the road; on its left was C Company of the Engineer battalion, and beyond C Company was the 2d Division Reconnaissance Company. The hill occupied by Beahler's D Company was in reality the western tip of a large mountain mass that lay southeast of the town. The road to Miryang came south out of Yongsan, bent around the western tip of this mountain, and then ran eastward along its southern base. In its position, D Company not only commanded the town but also its exit, the road to Miryang.[11]

North Koreans had also approached Yongsan from the south. The 2d Division Reconnaissance Company and tanks of the 72d Tank Battalion opposed them in a sharp fight. In this action, SFC

[10] GHQ FEC, History of the North Korean Army, p. 68; ATIS Res Supp Interrog Rpts, Issue 100 (N.K. *9th Div*), p. 49, and Issue 94 (N.K. *4th Div*), pp. 49–50; ATIS Interrog Rpts, Issue 4, p. 118, Rpt 949, 1st Lt So Chung Kun (captured 3 Sep 50, Yongsan); EUSAK WD, 14 Sep 50, an. to PIR 64, and 9 Sep 50, POW Interrog Rpt of 1st Lt So Chung Kun, *9th Div*, and Interrog Rpt of Cha Sook Wha, interpreter, *16th Regt, 4th Div*.

[11] Ltrs, Capt Lee E. Beahler to author, 10 Jun and 1 Jul 53 (sketch map of D Co positions with ltr of 1 Jul); Ltr, Reed to author, 20 Jul 53; Interv, author with Hill, 30 Jun 53.

Charles W. Turner of the Reconnaissance Company particularly distinguished himself. He mounted a tank, operated its exposed turret machine gun, and directed tank fire which reportedly destroyed seven enemy machine guns. Turner and this tank were the objects of very heavy enemy fire which shot away the tank's periscope and antennae and scored more than fifty hits on it. Turner, although wounded, remained on the tank until he was killed. That night North Korean soldiers crossed the low ground around Yongsan and entered the town from the south.[12]

About 0300, 2 September, D Company of the 2d Engineer Battalion alerted A Company that a long line of white-garbed figures was moving through Yongsan toward its roadblock. Challenged when they approached, the white figures opened fire—they were enemy troops. Four enemy tanks and an estimated battalion of North Koreans were in Yongsan.

The North Koreans now attempted a breakthrough of the Engineer position. After daylight, they were unable to get reinforcements into the fight since D Company commanded the town and its approaches. In this fight, which raged until 1100, the engineers had neither artillery nor mortar support. D Company remedied this by using its 9 new 3.5-inch and 9 old 2.36-inch rocket launchers against the enemy infantry. The fire of the 18 bazookas plus that from 4 heavy and 4 light machine guns and the rifles, carbines, and grenades of the company inflicted very heavy casualties on the North Koreans, who desperately tried to clear the way for a push eastward to Miryang. Tanks of A and B Companies, 72d Tank Battalion, at the southern and eastern edge of Yongsan shared equally with the engineers in the honors of this battle. Lieutenant Beahler was the only officer of D Company not killed or wounded in this melee, which cost the company twelve men killed and eighteen wounded. The edge of Yongsan and the slopes of the hill south of the town became a shambles of enemy dead and destroyed equipment.[13]

While this battle raged during the morning at Yongsan, Colonel Hill reorganized about 800 men of the 9th Infantry who had arrived in that vicinity from the overrun river line positions. Among them were F and G Companies, which were not in the path of major enemy crossings and had succeeded in withdrawing eastward. They had no crew-served weapons or heavy equipment. In midafternoon (2 September) tanks and the reorganized 2d Battalion, 9th Infantry, attacked through A Company, 2d Engineer Combat Battalion, into Yongsan, and regained possession of the town at 1500. Later, two bazooka teams from A Company, 2d Engineer Combat Battalion, knocked out three T34 tanks just west of Yongsan. American ground and air action destroyed other enemy tanks during the day southwest of the town. By evening the North Koreans had been driven into the hills

[12] Department of the Army General Order 10, 16 February 1951, awarded the Medal of Honor to Sergeant Turner posthumously.

[13] Ltrs, Beahler to author, 10 Jun and 1 Jul 53; Ltr, Reed to author, 20 Jul 53; Interv, author with Hill, 30 Jun 53; EUSAK WD, G–3 Jnl, Msg 1525, 2 Sep 50; *Ibid.*, PIR 52, 2 Sep 50. General Order 59, 8 February 1951, awarded the Distinguished Service Cross to Lieutenant Beahler for heroic leadership in this action. EUSAK.

westward. In the evening, the 2d Battalion and A Company, 2d Engineer Combat Battalion, occupied the first chain of low hills half a mile beyond Yongsan, the engineers west and the 2d Battalion northwest of the town. For the time being at least, the North Korean drive toward Miryang had been halted.[14]

At 0935 that morning (2 September), while the North Koreans were attempting to destroy the Engineer troops at the southern edge of Yongsan and clear the road to Miryang, General Walker talked by telephone with Maj. Gen. Doyle O. Hickey, Deputy Chief of Staff, Far East Command, in Tokyo. He described the situation around the Perimeter and said the most serious threat was along the boundary between the U.S. 2d and 25th Divisions. He described the location of his reserve forces and his plans for using them. He said he had started the marines toward Yongsan but had not yet released them for commitment there and he wanted to be sure that General MacArthur approved his use of them, since he knew that this would interfere with other plans of the Far East Command. Walker said he did not think he could restore the 2d Division lines without using them. General Hickey replied that General MacArthur had the day before approved the use of the marines if and when Walker considered it necessary. A few hours after this conversation General Walker, at 1315, attached the 1st Provisional Marine Brigade to the 2d Division and ordered a co-ordinated attack by all available elements of the division and the marines, with the mission of destroying the enemy east of the Naktong River in the 2d Division sector and of restoring the river line. The marines were to be released from 2d Division control just as soon as this mission was accomplished.[15]

A conference was held that afternoon at the 2d Division command post attended by Colonel Collier, Deputy Chief of Staff, Eighth Army, General Craig and Maj. Frank R. Stewart, Jr., of the Marine Corps, and General Keiser and 2d Division staff officers. A decision was reached that the marines would attack west the next morning at 0800 (3 September) astride the Yongsan–Naktong River road; the 9th Infantry, B Company of the 72d Tank Battalion, and D Battery of the 82d AAA Battalion would attack northwest above the marines and attempt to re-establish contact with the 23d Infantry; the 2d Engineer Combat Battalion, remnants of the 1st Battalion, 9th Infantry, and elements of the 72d Tank Battalion would attack on the left flank, or south, of the marines to re-establish contact with the 25th Division. Eighth Army now ordered the 24th Division headquarters and the 19th Infantry to move to the Susan-ni area, eight air miles south of Miryang and fifteen miles east of the confluence of the Nam and the Naktong Rivers. There it was to prepare to enter the battle in either the 2d or 25th Division zone. Colonel Fisher, commanding officer of the 35th Infantry, 25th Division, each morning flew along the Naktong River east of the

[14] Ltrs, Beahler to author, 10 Jun and 1 Jul 53; Ltr, Reed to author, 20 Jul 53; Interv, author with Hill, 30 Jun 53; Cody, Operation Manchu; EUSAK WD, G–3 Jnl, 2 Sep 50; *Ibid.*, Br for CG, 2 Sep 50; GHQ FEC Sitrep, 2 Sep 50.

[15] Transcription and summ of fonecon, Walker with Hickey, 0935 2 Sep 50, CofS files, FEC; EUSAK WD, G–3 Sec, Opn Ord 021315 Sep 50.

U.N. Troops Cross Rice Paddies *to attack west of Yongsan.*

Namji-ri bridge to see if North Koreans had crossed from the 2d Division zone.[16]

At 1900 the evening of 2 September, Colonel Hill returned to his command post east of Yongsan where he conferred with Colonel Murray, commanding the 5th Marines, and told him that his line of departure for the attack the next morning was secure. The troops holding this line on the first hills west of Yongsan were: G Company, 9th Infantry, north of the road running west through Kogan-ni to the Naktong; A Company, 2d Engineer Combat Battalion, southward across the road; and, below the engineers, F Company, 9th Infantry. Between 0300 and 0430, 3 September, the 5th Marines moved to forward assembly areas—the 2d Battalion north of Yongsan, the 1st Battalion south of it. The 3d Battalion established security positions southwest of Yongsan along the approaches into the regimental sector from that direction.[17]

During the night, A Company of the engineers had considerable fighting with North Koreans and never reached its objective. At dawn 3 September, Reed led A Company in an attack to gain the high ground which was part of the designated Marine line of departure. The company fought its way up the slope to within 100 yards of the top, which was held by the firmly entrenched enemy. At this

[16] EUSAK WD, G–3 Sec, and Br for CG, 2 Sep 50; 9th Inf WD, 3 Sep 50, Opn Ord 11 and accompanying overlay, 030300 Sep 50; 1st Prov Mar Brig SAR, 2 Aug–6 Sep 50, p. 15; 2d Div Arty WD, entry 12, 2335 2 Sep 50; Fisher, MS review comments, 7 Nov 57.

[17] Interv, author with Hill, 30 Jun 53; Ltr, Beahler to author, 10 Jun 53; 1st Prov Mar Brig SAR, 2 Aug–6 Sep 50, entry for 3 Sep, p. 15; 5th Mar SAR, 3 Sep 50; 2d Bn, 5th Mar, SAR, addendum 1, 3 Sep 50; Montross and Canzona, *The Pusan Perimeter*, pp. 217–20.

point Captain Reed caught an enemy-thrown grenade and was wounded by its fragments as he tried to throw it away from his men. The company with help from Marine tank fire eventually gained its objective, but this early morning battle for the line of departure delayed the planned attack.[18]

The Marine attack started at 0855 across the rice paddy land toward enemy-held high ground half a mile westward. The 1st Battalion, south of the east-west road, gained its objective when enemy soldiers broke under air attack and ran down the northern slope and crossed the road to Hill 116 in the 2d Battalion zone. Air strikes, artillery concentrations, and machine gun and rifle fire of the 1st Battalion now caught enemy reinforcements in open rice paddies moving up from the second ridge and killed most of them. In the afternoon, the 1st Battalion advanced to Hill 91.

North of the road the 2d Battalion had a harder time, encountering heavy enemy fire when it reached the northern tip of Hill 116, two miles west of Yongsan. The North Koreans held the hill during the day, and at night D Company of the 5th Marines was isolated there. In the fighting west of Yongsan Marine armor knocked out four T34 tanks, and North Korean crew members abandoned a fifth. That night the marines dug in on a line generally two miles west of Yongsan. The 2d Battalion had lost 18 killed and 77 wounded during the day, most of them in D Company. Total Marine casualties for 3 September were 34 killed and 157 wounded. Co-ordinating its attack with that of the marines, the 9th Infantry advanced abreast of them on the north.[19]

Just before midnight, the 3d Battalion, 5th Marines, received orders to pass through the 2d Battalion and continue the attack in the morning. That night torrential rains made the troops miserable. The enemy was strangely quiet. September 4 dawned clear.

The counterattack continued at 0800, 4 September, at first against little opposition. North of the road the 2d Battalion quickly completed occupation of Hill 116, from which the North Koreans had withdrawn during the night. South of the road the 1st Battalion occupied what appeared to be a command post of the N.K. *9th Division*. Tents were still up and equipment lay scattered about. Two abandoned T34 tanks in excellent condition stood there. Tanks and ground troops advancing along the road found it littered with enemy dead and destroyed and abandoned equipment. By nightfall the counterattack had gained another three miles.[20]

That night was quiet until just before dawn. The North Koreans then launched an attack against the 9th Infantry on the right of the marines, the heaviest blow striking G Company. It had begun to rain again and the attack came in the midst of a downpour. In bringing his platoon from an outpost position to the relief of the company, SFC Loren R. Kaufman encountered an encircling en-

[18] Ltr, Reed to author, 20 Jul 53; Ltr, Beahler to author, 10 Jun 53; 5th Mar SAR, 3 Sep 50.

[19] Entries for 3 Sep, Marine sources cited n. 20; 1st Bn, 5th Mar, SAR, 3 Sep 50; Montross and Canzona, *The Pusan Perimeter*, pp. 220–22; Geer, *The New Breed*, p. 94.

[20] 1st Prov Mar Brig SAR, 2 Aug–6 Sep 50, pp. 15–16; 1st Bn, 5th Mar SAR, 4 Sep 50; Geer, *The New Breed*, p. 96; Montross and Canzona, *The Pusan Perimeter*, pp. 227–29.

emy force on the ridge line. He bayoneted the lead enemy scout and engaged those following with grenades and rifle fire. His sudden attack confused and dispersed this group. Kaufman led his platoon on and succeeded in joining hard-pressed G Company. In the ensuing action Kaufman led assaults against close-up enemy positions and, in hand-to-hand fighting, he bayoneted four more enemy soldiers, destroyed a machine gun position, and killed the crew members of an enemy mortar. American artillery fire concentrated in front of the 9th Infantry helped greatly in repelling the North Koreans in this night and day battle.[21]

That morning (5 September), after a 10-minute artillery preparation, the American troops moved out in their third day of counterattack. It was a day of rain. As the attack progressed, the marines approached Obong-ni Ridge and the 9th Infantry neared Cloverleaf Hill—their old battleground of August. There, at midmorning, on the high ground ahead, they could see enemy troops digging in. The marines approached the pass between the two hills and took positions in front of the enemy-held high ground.

At 1430 approximately 300 enemy infantry came from the village of Tugok and concealed positions, striking B Company on Hill 125 just north of the road and east of Tugok. Two enemy T34 tanks surprised and knocked out the two leading Marine Pershing M26 tanks. Since the destroyed Pershing tanks blocked fields of fire, four others withdrew to better positions. Assault teams of B Company and the 1st Battalion with 3.5-inch rocket launchers rushed into action, took the tanks under fire, and destroyed both of them, as well as an armored personnel carrier following behind. The enemy infantry attack was quite savage and inflicted twenty-five casualties on B Company before reinforcements from A Company and supporting Army artillery and the Marine 81-mm. mortars helped repel it.[22]

September 5 was a day of heavy casualties everywhere on the Pusan Perimeter. Army units had 102 killed, 430 wounded, and 587 missing in action for a total of 1,119 casualties. Marine units had 35 killed, 91 wounded, and none missing in action, for a total of 126 battle casualties. Total American battle casualties for the day were 1,245 men. Col. Charles C. Sloane, Jr., who had commanded part of Task Force Bradley, resumed command of the 9th Infantry, relieving Colonel Hill.[23]

During the previous night, at 2000, 4 September, General Walker had ordered the 1st Provisional Marine Brigade released from operational control of the 2d Division effective at midnight, 5 September. He had vainly protested against releasing the brigade, believing he needed it and all the troops then in Korea if he were to stop the North Korean offensive against the Pusan Perimeter. At 0015, 6 September, the marines

[21] Department of the Army General Order 61, 2 August 1951, awarded the Medal of Honor to Sergeant Kaufman.

[22] 9th Inf WD, 5 Sep 50; 1st Prov Mar Brig SAR, 5 Sep 50; 1st Bn, 5th Mar SAR, 5 Sep 50; Geer, *The New Breed*, pp. 97–98; Montross and Canzona, *The Pusan Perimeter*, pp. 234–37.

[23] GHQ FEC Sitrep, 5 Sep 50; 9th Inf WD, 5 Sep 50, GO 11.

began leaving their lines at Obong-ni Ridge and headed for Pusan.[24]

The American counteroffensive of 3–5 September west of Yongsan, according to prisoner statements, resulted in one of the bloodiest and most terrifying debacles of the war for a North Korean division. Even though remnants of the division, supported by the low strength *4th Division,* still held Obong-ni Ridge, Cloverleaf Hill, and the intervening ground back to the Naktong on 6 September, the division's offensive strength had been spent at the end of the American counterattack. The *9th* and *4th* enemy divisions were not able to resume the offensive.[25]

Once again the fatal weakness of the North Korean Army had cost it victory after an impressive initial success—its communications and supply were not capable of exploiting a breakthrough and of supporting a continuing attack in the face of massive air, armor, and artillery fire that could be concentrated against its troops at critical points.

### *The 23d Infantry in Front of Changnyong*

North of the 9th Infantry and the battles that ebbed and flowed in the big bulge of the Naktong and around Yongsan, the 23d Infantry Regiment after daylight of 1 September found itself in a very precarious position. Its 1st Battalion had been driven from the river positions and isolated three miles westward. Approximately 400 North Koreans now overran the regimental command post, compelling Colonel Freeman to withdraw it about 600 yards. There, approximately five miles northwest of Changnyong, the 23d Infantry Headquarters and Headquarters Company, miscellaneous regimental units, and regimental staff officers checked the enemy in a 3-hour fight. Capt. Niles J. McIntyre of the Headquarters Company played a leading role.[26]

The infallible sign of approaching enemy troops could be seen in Changnyong itself during the afternoon of 2 September—at 1300 the native population began leaving the town. A little later a security force of 300 local police under the command of Maj. Jack T. Young and Capt. Harry H. White withdrew into the hills eastward when two groups of enemy soldiers approached from the northwest and southwest. North Koreans were in Changnyong that evening.[27]

With his communications broken southward to the 2d Division headquarters and the 9th Infantry, General Haynes during the day decided to send a tank patrol down the Yongsan road in an effort to re-establish communication. Capt. Manes R. Dew, commanding officer of C Company, 72d Tank Battalion, led the tanks southward. They had to fight their way down the road through enemy roadblocks. Of the three tanks that started, only Dew's tank got through to Yongsan. There, Captain Dew

---

[24] 2d Div Narr Summ, 1 Sep–31 Oct 50, p. 14; EUSAK WD, 4 Sep 50, Opn Ord at 042000. The removal of the Marine brigade from the Naktong front will be discussed further in the next chapter.

[25] ATIS Res Supp Interrog Rpts, Issue 100 (N.K. *9th Div*), p. 52.

[26] Freeman, MS review comments, 30 Oct 57; Highlights of the Combat Activities of the 23d Infantry Regiment from 5 August to 30 September 1950, MS, prepared in the regiment, copy in OCMH.

[27] 2d Div Arty WD, 2 Sep 50; EUSAK WD, 7 Sep 50, an. to PIR 57.

delivered an overlay of Task Force Haynes' positions to General Bradley.[28]

Still farther northward in the zone of the 38th Infantry the North Koreans were far from idle. After the enemy breakthrough during the night of 31 August, General Keiser on 1 September had ordered the 2d Battalion, 38th Infantry, to move south and help the 23d Regiment establish a defensive position west of Changnyong. In attempting to do this, the battalion found enemy troops already on the ridges along the road. They had in fact penetrated to Hill 284 overlooking the 38th Infantry command post. This hill and Hill 209 dominated the rear areas of the regiment. At 0600, 3 September, an estimated 300 North Koreans launched an attack from Hill 284 against Colonel Peploe's 38th Regiment command post. Colonel Peploe organized all officers and enlisted men present, including members of the mortar and tank companies and attached antiaircraft artillery units, to fight in the perimeter defense. Peploe requested a bombing strike which was denied him because the enemy target and his defense perimeter were too close to each other. But the Air Force did deliver rocket and strafing strikes.

This fight continued until 5 September. On that day Capt. Ernest J. Schauer captured Hill 284 with two platoons of F Company after four efforts. He found approximately 150 enemy dead on the hill. From the crest he and his men watched as many more North Koreans ran into a village below them. Directed artillery fire destroyed the village. Among the abandoned enemy matériel on the hill, Schauer's men found twenty-five American BAR's and submachine guns, a large American radio, thirty boxes of unopened American fragmentation and concussion grenades, and some American rations.[29]

Meanwhile, during these actions in its rear, the 1st Battalion, 23d Infantry, was cut off three miles westward from the nearest friendly units. On 1 September Colonel Hutchin had received instructions from the regiment to withdraw to the Changnyong area. At 1400 he sent a tank-infantry patrol to see if his withdrawal road was open. It reported that an estimated enemy battalion held the mountain pass just eastward of the battalion's defense perimeter. Upon receiving this report Colonel Hutchin requested permission by radio to remain in his present position and from there try to obstruct the movement of North Korean reinforcements and supplies. That evening Colonel Freeman approved this request, and thus began the 1st Battalion's 3-day stand as an island in a sea of enemy. During this time C–47 planes supplied the battalion by airdrops.[30]

The 2d Division, however, did not leave Colonel Hutchin to his own devices in his isolated perimeter position. Instead, on the morning of 1 September, it started the 3d Battalion, 38th Infantry, in an attack westward from the 23d Regiment command post near Mosan-ni to open the enemy-held road to the 1st Battalion. On the second day of the fighting at the enemy-held pass, the relief force, under Maj. Everett S. Stewart, the

[28] 2d Div Arty WD, 2 Sep 50; Interv, author with Hill, 30 Jun 53.

[29] 38th Inf Comd Rpt, Narr Summ, Sep–Oct 50; EUSAK PIR 58, 8 Sep 50; 2d Div PIR 14, 7 Sep 50.

[30] 23d Inf WD, Narr Summ, Sep 50.

battalion executive officer and temporarily acting battalion commander, broke through the enemy roadblock with the help of air strikes and artillery and tank fire. The advanced elements of the battalion joined Hutchin's battalion at 1700, 2 September. That evening, North Koreans strongly attacked the 3d Battalion, 38th Infantry, on Hill 209 north of the road and opposite Hutchin's battalion, driving one company from its position.[31]

On 4 September, General Haynes changed the boundary between the 38th and 23d Infantry Regiments, giving the northern part of the 23d's sector to the 38th Infantry, thus releasing Colonel Hutchin's 1st Battalion for movement southward to help the 2d Battalion defend the southern approach to Changnyong. The 1st Battalion, 23d Infantry, about 1,100 men strong when the enemy attack began, was now down to a strength of approximately 600 men.

The 23d Infantry now made plans to concentrate all its troops on the position held by its 2d Battalion on the Pugong-ni–Changnyong road. Colonel Hutchin succeeded in moving the 1st Battalion there and took a place on the left flank of the 2d Battalion. At the same time the regimental command post moved to the rear of this position. In this regimental perimeter, the 23d Infantry fought a series of hard battles. Simultaneously it had to send combat patrols to its rear to clear infiltrating enemy from Changnyong and from its supply road.

The N.K. *2d Division* made a desperate effort against the 23d Infantry's perimeter in the predawn hours of 8 September, in an attempt to break through eastward. This attack, launched at 0230 and heavily supported with artillery, penetrated F Company. It was apparent that unless F Company's position could be restored the entire regimental front would collapse. When all its officers became casualties, 1st Lt. Ralph R. Robinson, adjutant of the 2d Battalion, assumed command of the company. With North Koreans rapidly infiltrating his company's position and gaining its rear, Robinson in the darkness made his way through them 500 yards to A Company's position. There he obtained that company's reserve platoon and brought it back to F Company. He accomplished the dangerous and difficult task of maneuvering it into the gap in F Company's lines in darkness and heavy rain.[32]

The enemy attack tapered off with the coming of daylight, but that night it resumed. The North Koreans struck repeatedly at the defense line. This time they continued the fighting into the daylight hours of 9 September. The Air Force then concentrated strong air support over the regimental perimeter and gave invaluable aid to the ground troops. Casualties came to the aid stations from the rifle companies in an almost steady stream during the morning. All available men from Headquarters Company and special units were formed into squads and put into the fight at the most critical points. At one time, the regimental reserve was down to six men. When the enemy attack finally ceased

[31] Interv, author with Lt Col Everett S. Stewart, 19 May 53; 23d Inf WD, G–3 Jnl, entry 131, 1715 1 Sep 50. General Order 196, 14 December 1950, awarded the Distinguished Service Cross to Colonel Hutchin. 2d Div.

[32] 2d Div Arty WD, 2 Sep 50; EUSAK WD, 7 Sep 50, an. to PIR 57.

shortly after noon the 23d Regiment had an estimated combat efficiency of only 38 percent.[33]

This furious night and day battle cost the enemy division most of its remaining offensive strength. The medical officer of the *17th Regiment, 2d Division,* captured a few days later, said that the division evacuated about 300 men nightly to a hospital in Pugong-ni, and that in the first two weeks of September the *2d Division* lost 1,300 killed and 2,500 wounded in the fighting west of Changnyong.[34]

Even though its offensive strength was largely spent by 9 September, the enemy division continued to harass rear areas around Changnyong with infiltrating groups as large as companies. Patrols daily had to open the main supply road and clear the town.

### *A North Korean Puzzle*

While the N.K. *2d Division* was making its great effort near the middle of the U.S. 2d Division line, a sister organization, the N.K. *10th Division,* on its left to the north failed to give the assistance that was expected of it in the co-ordinated corps attack. And therein lies one of the greatest North Korean failures of the war to exploit an opportunity. The singular behavior of this enemy force puzzled American commanders at the time, although they were thankful that it took the pattern it did.

The N.K. *10th Division* was the northernmost major organization of the N.K. *I Corps.* A large part of it occupied Hill 409 in a deep fold of the Naktong River just west of Hyongp'ung. Elements of this division streamed off Hill 409 the night of 31 August–1 September and struck the 1st Battalion, 38th Infantry, which formed the extreme right flank of the U.S. 2d Division. Holding the town of Hyongp'ung was C Company, which withdrew from it under enemy attack during the night of 2–3 September. Beginning with 3 September, Hyongp'ung for two weeks was either in enemy hands or a no man's land.[35]

North and east of the Hill 409 and Hyongp'ung area lay a virtually roadless, high mountain area having no fixed U.N. defensive positions. This, too, was a no man's land in early September. Four miles north of Hyongp'ung was the Yongp'o bridge across the Naktong and the 1st Cavalry Division boundary. The Yongp'o bridge site was defended by the 3d Battalion, 23d Infantry, attached to the 1st Cavalry Division for that purpose, until 0410, 5 September, when the British 27th Infantry Brigade relieved it and went into the line there. This, as previously noted, was the British brigade's first commitment in the Korean War.[36]

During the first two weeks of September large numbers of the enemy *10th*

---

[33] 23d Inf WD, Narr Summ, Sep 50; Interv, author with Meszar, 15 May 53; Highlights of Combat Activities of 23d Inf.

[34] EUSAK WD, 21 Sep 50, ADVATIS Interrog Rpts, Sr Lt Lee Kwan Hyon, Med Off, N.K. *17th Regt, 2d Div;* ATIS Interrog Rpts, Issue 6, p. 81, Kim Il Chin and Issue 7, p. 3, Yu Tong Gi; 23d Inf Comd Rpt, Sep 50, Narr Summ, p. 10.

[35] 38 Inf Comd Rpt, Sep–Oct 50, Narr Summ; EUSAK WD, G–3 Jnl, 4–5 Sep 50. Records of the 38th Infantry for September 1950 were lost in the withdrawal from Kunu-ri, 30 November, and the command report compiled later from recollections of regimental personnel lacks precise information on time, place, and overlay data for this period.

[36] GHQ FEC Sitrep, 5 Sep 50; *Ibid.*, G–3 Opn Rpt 70, 2 Sep 50.

*Division* came off Hill 409 and roamed the mountain mass northeast of Hyong-p'ung in the gap between the U.S. 2d Division and the British 27th Brigade. This caused Eighth Army concern for the safety of Taegu. Gradually, ROK police and British combat patrols forced the North Koreans back to Hill 409. On 6 September, the day after they went into the line, the British had a taste of what the Korean War was like. A combat patrol of the Argylls under Capt. Neil A. Buchanan encountered an enemy unit and had to make its escape, leaving behind, on his own orders, Captain Buchanan badly wounded and, at his side, his wounded batman. Neither was seen again. The British company nearest Hill 409 was so isolated that airdrops of ice to it replaced carrying water cans up the hill.[37]

Had the enemy *10th Division* thrown its full weight into a drive eastward, south of Taegu, it might well have precipitated a major crisis for Eighth Army. It could have moved either northeast toward Taegu or southeast to help the *2d Division,* next in line below it, but it did neither. Its relative inactivity in the vicinity of Hill 409 when its companion divisions were engaged in desperate combat above and below it is something of a mystery. Captured enemy material and statements of prisoners indicate that its mission may have been to stay on Hill 409 until the N.K. *II Corps* had captured Taegu, but they indicated, also, that the division command was inept. The *10th Division* caused General Walker much concern at this time. He and his staff found it puzzling to reconcile the division's favorable position with its inactivity. General Walker charged Colonel Landrum, now Deputy Chief of Staff, Eighth Army, to watch the situation closely and inform him daily on it. At least twice daily Landrum insisted on a summary from the Army G–3 of activities in front of the N.K. *10th Division.*[38]

### *The 35th Infantry—The Rock of the Nam*

On the 25th Division's right flank and north of the Haman breakthrough, the 35th Infantry Regiment at daylight, 1 September, still held all its positions except the low ground between Komam-ni and the Nam River, which the two companies of ROK police had abandoned at midnight. (*See Map V.*) In a counterattack after daylight, K Company and tanks had partially regained control of this area, but not completely. Large numbers of North Koreans, by this time, however, were behind the battle positions of the 35th Infantry as far as the Chirwon-ni and Chung-ni areas, six miles east of Komam-ni and the front positions. The North Koreans continued to cross the Nam River after daylight on 1 September in the general area of the gap between the 1st and 2d Battalions. Aerial observers saw an estimated four companies crossing there and directed proximity (VT) fuze fire of the 64th Field

[37] EUSAK WD, G–3 Jnl, 1330, 10 Sep 50; GHQ FEC Sitrep, 13 Sep 50; Lt Col G. I. Malcolm of Poltallock (London: Thomas Nelson & Sons, Ltd., 1952), *The Argylls in Korea,* pp. 11–12; Coad, The Land Campaign in Korea.

[38] ATIS Res Supp Interrog Rpts, Issue 104 (N.K. *10th Div*), p. 49; Interv, author with Stephens, 8 Oct 51; Interv, author with Brig Gen George B. Peploe, 12 Aug 51; Ltr and notes, Landrum to author, recd 28 Jun 54.

BATTLE TROPHY. *Men of the 35th Infantry display a North Korean flag captured in the Sibidang-san area on 5 September.*

Artillery Battalion on the crossing force, which destroyed an estimated three-fourths of it. Fighter planes then strafed the survivors. Aerial observers saw another large group in the open at the river later in the day and directed artillery proximity fuze fire on it with an estimated 200 enemy casualties.[39]

The enemy *I Corps* plan of attack below the Nam River, as indicated by the North Korean action, seemed to be for its *6th Division* to push east along the main Chinju–Komam-ni–Masan highway through the 1st Battalion, 35th Infantry, and at the same time for major elements of its *7th Division* to swing southeast behind the 2d Battalion, 35th Infantry, and cut the Chirwon road. This road crossed the Naktong River over the cantilever steel bridge at Namji-ri from the 2d Division zone and ran south through Chirwon to join the main Masan highway eight miles east of Komam-ni near the village of Chung-ni, four miles northwest of Masan. These two

[39] 25th Div WD, 1 Sep 50; 64th FA Bn WD, 1 Sep 50.

avenues of approach—the Komam-ni–Masan highway and the Chirwon road converging at Chung-ni—formed the axes of the enemy attack plan.

Engineer troops counterattacking up the secondary road toward Chirwon during 1 September made slow progress, and enemy troops stopped them altogether in the early afternoon. The 35th Infantry was now surrounded by enemy forces of the N.K. *6th* and *7th Divisions,* with an estimated three battalions of them behind its lines. Speaking later of the situation, Colonel Fisher, the regimental commander—a professional soldier, trained at West Point, and a regimental commander in World War II—said, "I never intended to withdraw. There was no place to go. I planned to go into a regimental perimeter and hold." [40] His regiment demonstrated its competency to do this in the September battle along the Nam, winning a Distinguished Unit Citation for its performance there.

On that first day of the enemy thrust, a critical situation existed in the 25th Division sector. Because of it, General Walker flew to General Kean's command post at Masan. In the ensuing discussion there, Kean asked Walker for authority to commit the remainder of the 27th Infantry Regiment (Walker had already released one battalion to Kean's control for use in the 24th Infantry sector) against the large enemy groups behind the 35th Infantry. Walker refused. By midafternoon, however, Kean felt that the situation was so critical that he ordered the 2d Battalion, commanded by Colonel Murch, to attack behind the 35th Infantry. A large part of the division artillery was under direct infantry attack and he felt it mandatory upon himself to commit the 2d Battalion, 27th Infantry. He gave this order on his own authority as the responsible commander on the ground, notwithstanding General Walker's earlier refusal. At a later date when General Walker knew all the facts, he approved General Kean's action.[41]

During the predawn hours of 1 September, when the N.K. *7th Division* troops had swung left after crossing the Nam River to roll up that flank, widen the gap, drive the American troops from their hill positions overlooking the Nam River, and secure a broad bridgehead for the division, the first American unit they encountered was G Company, 35th Infantry, at the north shoulder of the gap. While some enemy units peeled off to attack G Company, others continued on and engaged E Company, two miles downstream from it, and still others attacked scattered units of F Company all the way to its 1st Platoon, which guarded the Namji-ri bridge. There, at the extreme right flank of the 25th Division, this platoon drove off an enemy force after a sharp fight. By 2 September, E Company in a heavy battle had destroyed most of an enemy battalion.

Of all the 2d Battalion units, G Company received the hardest blows. Before dawn of 1 September enemy troops had G Company platoons on separate hills under heavy assault. Shortly after 0300 they overran the 3d Platoon, Heavy Mortar Company, and drove it from its position. These mortarmen climbed Hill

[40] Interv, author with Fisher, 5 Jan 52; 35th Inf WD, 1 Sep 50.

[41] Ltr, Kean to author, 22 Apr 53; Barth MS, p. 27.

179 and on its crest joined the 2d Platoon of G Company.

Meanwhile, the 3d Platoon of G Company, on a low hill along the Nam four miles from its juncture with the Naktong, was also under close-in attack. After daylight, Capt. LeRoy E. Majeske, G Company commanding officer, requested artillery concentrations and air strikes, but the latter were slow in coming. At 1145, the enemy had almost reached the crest of the hill, and only the narrow space covered by the air identification panel separated the two forces. A few minutes later Majeske was killed, and 2d Lt. George Roach, commanding the 3d Platoon, again reported the desperate situation and asked for an air strike. The Air Force delivered the strike on the enemy-held side of the hill, and this checked the assaults. But by this time many enemy troops had captured and occupied foxholes in the platoon position and from them they threw grenades into other parts of the position. One of the grenades killed Lieutenant Roach early in the afternoon. SFC Junius Poovey, a squad leader, now assumed command. In this close fight, one of the heroes was Cpl. Hideo Hashimoto, a Japanese-American, who edged himself forward and threw grenades into the enemy holes, some of them only ten to fifteen feet away. By 1800, Sergeant Poovey had only 12 effectives left in the platoon; 17 of the 29 men still living were wounded. With ammunition almost gone, Poovey requested and received authority to withdraw into the main G Company position. After dark, the 29 men, 3 of them carried on stretchers, escaped by timing their departure from the hill with the arrival of friendly tanks which engaged the enemy and diverted attention from the beleaguered men on top. The group reached the G Company position on Hill 179 half an hour before midnight.[42]

While G Company held its positions on Hill 179 on 2 September against enemy attack, Colonel Murch's 2d Battalion, 27th Infantry, started an attack northwest toward it at 1700 from the Chung-ni area. The battalion made slow progress against formidable enemy forces. The night was extremely dark and the terrain along the Kuhe-ri ferry road was mountainous. After fighting all that night the battalion, the next day at 1500, reached a position 1,000 yards south of the original defensive positions of G Company, 35th Infantry. A co-ordinated attack by armor, artillery, air, and infantry got under way and by 1800 the battalion had re-established the battle line. In this attack the 2d Battalion, 27th Infantry, killed 275 enemy and recovered a large part of the equipment G Company had lost earlier.

Colonel Murch's battalion remained on the regained positions during the night of the 3d. The next morning Murch received orders to attack to the rear and clear the alternate route on the western edge of the battalion zone. At 0800 G Company, 35th Infantry, relieved Murch on the regained positions

[42] 35th Inf WD, 1–2 Sep 50; 2d Bn 35th Inf WD, 1 Sep, and Narr story of action Sep 50; 25th Div WD, 1–2 Sep 50; New York *Herald Tribune*, September 3, 1950, Homer Bigart dispatch; New York *Times*, September 4, 1950, W. H. Lawrence dispatch.

In grenade fighting on slopes the practice of "cooking the grenade" developed. In order to avoid allowing enemy troops time to pick up and throw a grenade back, soldiers pulled the pin, released the handle in the grip for a brief period, and then threw the grenade.

2D BATTALION, 27TH INFANTRY, *on the recaptured supply road.*

and the latter started his attack back up the supply road. While this was in progress, word came that North Koreans had again driven G Company from its newly re-established position. Murch turned around, attacked, and once more restored the G Company positions. By noon of 4 September, Murch again turned over these positions to G Company and resumed his attack to the rear along the road in the gap between the 1st and 2d Battalions, 35th Infantry. Almost immediately he was in contact with enemy forces. Soon North Korean machine guns were firing on Murch's men from three directions. Torrential rains fell and observation became poor. By this time, Murch's battalion was running short of ammunition. Murch ordered the battalion to withdraw about 500 yards to favorable terrain so that he could try to effect a resupply.

But this was not easy to do. He had cleared the supply route two days previously in his attack to the G Company position but now it was closed again. With several thousand North Korean soldiers behind the 35th Infantry front, it was like pulling one's thumb from a pail of water—the space filled again immediately. Murch requested air supply and the next morning, 5 September, eight transport planes accomplished the resupply and the 2d Battalion, 27th Infantry, was ready to resume its attack to the rear. By evening that day it had cleared the supply road and adjacent terrain of enemy soldiers for a distance of 8,000 yards to the rear of G Company's front-line positions. There Murch received orders to halt and prepare to attack northeast to link up with Colonel Check's 1st Battalion, 27th Infantry.[43]

After Murch had left the Chung-ni area on 2 September in his attack toward G Company, enemy infiltrators attacked the 24th Infantry command post and several artillery positions. To meet this new situation, General Kean, again acting on his own authority as the responsible commander on the ground, ordered the remaining battalion of the 27th Infantry (technically still the 3d Battalion, 29th Infantry), commanded by Lt. Col. George H. DeChow, to attack and destroy the enemy operating there. General Kean notified Eighth Army of his action at 1250, 2 September.[44]

After an early morning struggle on 3 September against several hundred North Koreans in the vicinity of the artillery positions, DeChow's battalion launched its attack at 1500 over the high, rugged terrain west of the "Horseshoe," as the deep curve in the Masan road was called, four miles east of Komam-ni. Its mission was to seize and secure the high ground dominating the Horseshoe, and then relieve the pressure on the 24th Infantry rear. Initially only one artillery piece was in position to support the attack. After the battalion advanced some distance, an enemy force, estimated at the time to number more than 1,000 men, counterattacked it and inflicted heavy casualties, which included thirteen officers. The K Company commander, 1st Lt. Elwood F. James, was killed while leading an assault. Additional tanks moved up to help secure the exposed right flank and rear, and air strikes helped to contain the enemy force. The battalion finally succeeded in taking the high ground.[45]

The next morning, 4 September, instead of continuing the attack toward the 24th Infantry command post, DeChow, on changed orders, attacked straight ahead into the Komam-ni area where enemy troops were fighting in the artillery positions. This attack got under way at 0900 in the face of severe enemy small arms fire. In the afternoon, heavy rains slowed the attack, but after an all-day battle, I and K Companies, with the help of numerous air strikes, captured the high ground dominating the Komam-ni crossroads. Numerous casualties in the battalion had led General Kean to attach C Company, 65th Engineer Combat Battalion, to it. The next day, 5 September, the 3d Battalion turned its attack across rugged terrain toward Haman and drove through to the vicinity of the 24th Infantry command post. In its attack, the 3d Battalion counted more than 300 enemy dead in the area it traversed.[46]

The series of events that caused General Kean to change the direction of DeChow's attack toward Komam-ni began at 0100, 3 September. The 1st Battalion, 35th Infantry, protruded farther westward at this time than any other

[43] EUSAK WD, G–3 Jnl, 1125 and 1410, 3 Sep 50; 2d Bn, 27th Inf, WD, Unit Rpt, Sep 50; 27th Inf WD, Unit Rpt, Sep 50; Murch, MS review comments, 2 Jan 58.

[44] EUSAK WD, G–3 Jnl, 1250 2 Sep 50; 25th Div WD, 2 Sep 50; Ltr, Kean to author, 22 Apr 53. The 3d Bn, 29th Inf, became operational as the 3d Bn, 27th Inf, by 25th Div GO 134, 10 Sep 50. The 1st Bn, 29th Inf became operational as the 3d Bn, 35th Inf, the same date. EUSAK GO 49, 2 Sep 50, authorized the transfer.

[45] DeChow, MS review comments, Jul 53; Interv, author with Flynn (3d Bn, 27th Inf, Sep 50), 5 Nov 53; Barth MS, pp. 28–29.

[46] DeChow, MS review comments, Jul 53; Barth MS, pp. 28–29.

unit of the U.N. forces in Korea. Back of its positions on Sibidang-san the main supply route and rear areas were in enemy hands, and only in daylight and under escort could vehicles travel the road. On Sibidang-san the battalion had held its original positions after the heavy fighting of pre-dawn 1 September, completely surrounded by barbed wire, booby traps, and flares, with all supporting weapons inside its tight perimeters. The battalion had the advantage of calling by number for previously zeroed and numbered protective fires covering all approaches, which were quickly delivered. An hour after midnight an unusually heavy enemy assault struck the battalion. The fight there continued until dawn 3 September, when the 1st Battalion, 35th Infantry, counted 143 enemy dead in front of its positions, and on that basis estimated that the total enemy casualties must have been about 500 men.[47]

In this night battle the 64th Field Artillery Battalion gave invaluable support to the 1st Battalion and became directly involved itself in the fighting. About fifty North Koreans infiltrated before dawn to A Battery's position and delivered a banzai-type assault. Enemy soldiers employing submachine guns overran two artillery-machine gun perimeter positions, penetrating to the artillery pieces at 0300. There, Capt. Andrew C. Anderson and his men fought hand-to-hand with the North Koreans. Some of the guns fell temporarily into enemy hands and one North Korean scrawled on a howitzer tube, "Hurrah for our Company!" But the artillerymen threw the North Koreans out, aided greatly by the concentrations of fire from C Battery, 90th Field Artillery Battalion, which were placed within fifty yards of the battery and sealed off enemy reinforcements. In defending its guns in this night battle, A Battery lost seven men killed and twelve wounded—about 25 percent of its strength.[48]

The day before, the 159th Field Artillery Battalion also had distinguished itself in defending its guns in close fighting.

Fighting in support of the Nam River front in the northern part of the 25th Division sector were five batteries of the 159th and 64th Field Artillery Battalions (105-mm. howitzers) and one battery of the 90th Field Artillery Battalion (155-mm. howitzers), for a total of thirty-six guns. One 155-mm. howitzer, called by Colonel Fisher "The Little Professor," fired from Komam-ni on the Notch back of Chungam-ni, through which funneled much of the N.K. *6th Division*'s supplies. Another forward artillery piece kept the Iryong-ni bridge over the Nam under fire. The 25th Division artillery estimated it killed approximately 1,825 North Korean soldiers during the first three days of September.[49]

In this critical time, the Fifth Air Force added its tremendous fire power to that of the division artillery in support of the ground force. On 3 September, General Kean, speaking of the action during the past two days, said, "The close air support rendered by Fifth Air Force again saved this division as they

[47] 35th Inf WD, 3 Sep 50; 1st Bn, 35th Inf, Unit Rpt, Sep 50; EUSAK WD, G–3 Jnl, 0445, 3 Sep 50.

[48] 64th FA Bn WD, 3 Sep 50; Barth MS, p. 29; 159th FA Bn WD, Sep 50; EUSAK WD, G–3 Jnl, 0445 3 Sep 50; File supporting DUC, 35th Inf Regt, DA, AG files.

[49] 25th Div WD, 4 Sep 50; Barth MS, pp. 22, 31.

have many times before." [50] This view was supported by General Walker in an interview in November. Speaking then to a U.S. Air Force Evaluation Group, General Walker said, "I will gladly lay my cards right on the table and state that if it had not been for the air support that we received from the Fifth Air Force we would not have been able to stay in Korea." [51]

It is not possible here to follow in detail the confused ebb and flow of battle behind the 35th Infantry. Battalions, companies, and platoons, cut off and isolated, fought independently of higher control and help except for airdrops which supplied many of them. Airdrops also supplied relief forces trying to reach the front-line units. Tanks and armored cars ran the gantlet to the isolated units with supplies of food and ammunition and carried back critically wounded on the return trips.

In general, the 35th Infantry fought in its original battle line positions, while at first one battalion, and later two battalions, of the 27th Infantry fought toward it through the estimated 3,000 North Koreans operating in its rear areas.

In the confused fighting in the rear areas there were several cases of North Korean atrocities. One of the worst occurred when a group of company mess parties in jeeps pulling trailers with hot breakfast were following tanks toward the front lines. About a mile and a half from G Company, 35th Infantry, the column came under enemy fire in a defile. The tanks went on through, but most of the other vehicles under Capt. Robert E. Hammerquist, 2d Battalion S-3, turned back. At least one of the mess parties, however, pressed on after the tanks. Some of this group were captured. One of its members hid in a haystack and later escaped. He told of hearing the torture and murder of one man. He heard agonized screams, recognized the man's voice, and could hear him saying between sobs, "You might as well kill me now." Later when the area was cleared of enemy this man's body was found castrated and the fingers cut off.[52] Many soldiers of the 25th Division later saw the bodies of Americans lying in a ditch in the 35th Infantry area, their hands tied and their feet cut off. Still others saw dead Americans with their tongues cut out. Members of the N.K. *7th Division* apparently perpetrated these atrocities.[53]

During the September offensive enemy action in rear areas of the 25th Division carried right to Masan. Guerrilla activity increased, with the most tragic single incident taking place during the night of 3–4 September. That night about fifteen guerrillas, including one woman, attacked a radio relay station near Changwon, only four miles from Masan. They surprised a group of seven Americans and two South Koreans inside a tent on a hilltop. The guerrillas

[50] "Air War in Korea," *Air University Quarterly Review*, vol. IV, No. 3 (Spring, 1951), 61.

[51] Interv, USAF Evaluation Board with Lt Gen Walton Walker, 25 Nov 50. See also New York *Times*, September 3, 1950, for General Collins' statement quoting Walker.

[52] Interv, author with Maj Joe B. Lamb, CO 2d Bn, 35th Inf, 4 Sep 51; Intervs, author with 2d Lt Dillon Snell and 1st Lt Charles J. Hoyt, 2d Bn, 35th Inf, 4 Sep 51; Ltr, Hammerquist to author, 17 Apr 53; 35th Inf Unit Hist, 3 Sep 50.

[53] Interv, author with Lamb, 4 Sep 51; Interv, author with Sawyer (Recon Co, 25th Div, Sep 50), 27 Jun 51.

tied up the Americans, took documents from files, gathered up all weapons, and then the woman shot every one of the prisoners with a tommy gun. Two wounded Americans lived to tell the story.[54]

Even in Masan, General Kean faced a dangerous situation. The town was a nest of Communist sympathizers and agents. At the peak of the enemy offensive, Han Gum Jo, manager of the Masan branch of the Korean Press Association, confessed that he was chief of the South Korean Labor Party in Masan and that he funneled information to the enemy through a Pusan headquarters. The chief of guards of the Masan prison was the head of a Communist cell and seven of his guards were members. This and other counterintelligence information came to light at a time when desperate fighting was in progress only a few miles away. General Kean considered the situation so menacing that he ordered Masan evacuated of all people except the police, public officials, railroad and utility workers, and necessary laborers and their families. Evacuation was to be completed in five days. On 10 and 11 September alone the 25th Division evacuated more than 12,000 people by LST from Masan.[55]

Although the 25th Division generally was under much less enemy pressure after 5 September, there were still severe local attacks. On 6 September Colonel Check's 1st Battalion, 27th Infantry, moved north from the Haman area to join Murch's 2d Battalion in the clean-up of enemy troops back of the 35th Infantry and below the Nam River. Caught between the 35th Infantry on its hill positions along the river and the attacking 27th Infantry units, large numbers of North Koreans were killed. Sixteen different groups reportedly were dispersed with heavy casualties during the day. By morning of 7 September there was clear evidence that survivors of the N.K. *7th Division* were trying to escape across the Nam River. The 25th Division buried more than 2,000 North Korean dead, killed between 1 and 7 September behind its lines. This number did not include those killed in front of its positions. About 9 September Colonel Fisher traveled over these rear areas where fighting had been intense. He was astonished at the number of North Korean dead that littered the fields. Speaking of that occasion he has said, "The area of Trun in the Falaise Gap in Europe couldn't match it. Flies were so thick in some areas it limited vision." [56]

Heavy rains caused the Nam and Naktong Rivers to rise more than two feet on 8 and 9 September, thereby reducing the danger of new enemy crossings. At this juncture one of the ironies of the Korean War occurred. On the 9th, American jet planes (F-82's) mistakenly bombed the Namji-ri bridge over the Naktong and with one 500-pound bomb destroyed the 80-foot center span. Only the bridges north of the juncture of the Nam with the Naktong were supposed to be subject to aerial attack at this time. Lieutenant Vickery's 1st Platoon of F

[54] EUSAK WD, G-3, Coordinating Protection Lines of Communications Rear Areas, 4 Sep 50; New York *Herald Tribune*, September 4, 1950; 25th Div WD, 4 Sep 50.

[55] 25th Div WD, 3, 7, 11 and 15 Sep 50.

[56] EUSAK WD, G-3 Jnl 0720, 5 Sep 50; 27th Inf WD, 5 Sep 50; 2d Bn, 27th Inf, Unit Rpt, Sep 50; 25th Div WD, 6 Sep 50; Barth MS, p. 28; Fisher, MS review comments, Jan 58.

Company, 35th Infantry, had effectively defended the bridge—the link between the U.S. 2d and 25th Divisions—throughout the enemy offensive. The platoon had become so closely identified with this bridge that in the 25th Division it was called "Vickery's Bridge." Vickery had placed one squad on the north side of the bridge. From the south side it was supported by the rest of the platoon, a tank, and one 105-mm. howitzer, fondly called "Peg O' My Heart."

Some of the local commanders thought that had the North Koreans bypassed this bridge and crossed the Naktong farther east there would have been nothing between them and Pusan. However, North Korean attacks against Vickery's men were a nightly occurrence. The approaches to the bridge on the north side were mined. At one time there were about 100 North Korean dead lying in that area. One morning a pack of dogs were tearing the bodies when one of the animals set off a mine. That scattered the pack and the dogs in their wild flight set off more mines. Pieces of dog went flying through the air like rocks.[57]

### *Counterattack at Haman*

In the middle of the 25th Division line, south of the 35th Infantry, the enemy breakthrough at Haman became a terrifying fact to the division headquarters after daylight, 1 September. General Kean, commanding the division, telephoned Eighth Army headquarters and requested permission to commit, at once, the entire 27th Infantry Regiment, just arrived at Masan the previous evening and still held in Eighth Army reserve. General Walker denied this request, but did release one battalion of the regiment to General Kean's control.[58]

General Kean immediately dispatched Colonel Check's 1st Battalion, 27th Infantry—which had been alerted as early as 0200—from its assembly area near Masan toward Haman, to be attached to the 24th Infantry upon arrival at Colonel Champney's command post. The 1st Platoon of the 27th Regiment's Heavy Mortar Company; a platoon of B Company, 89th Tank Battalion; and A Battery, 8th Field Artillery Battalion, reinforced Check's battalion. Check with his battalion arrived at Champney's 24th Infantry command post two miles east of Haman at 1000.[59]

The scene there was chaotic. Vehicles of all descriptions, loaded with soldiers, were moving down the road to the rear. Many soldiers on foot were on the road. Colonel Champney tried repeatedly but in vain to get these men to halt. The few enemy mortar shells falling occasionally in the vicinity did no damage except to cause the troops of the 24th Infantry and intermingled South Koreans to scatter and increase their speed to the rear. The road was so clogged with this frightened, demoralized human traffic that Colonel Check had to delay his counterattack. In the six hours he waited at this point, Check observed that none of the retreating troops of the 1st and 2d Battalions, 24th Infantry, could be assem-

[57] 35th Inf WD, S-2 and S-3 Jnls, item 15, 9 Sep 50; 35th Inf Unit Hist, Sep 50; EUSAK WD, G-3 Jnl, 1015, 10 Sep 50; Fisher, MS review comments, 7 Nov 57, and Jan 58.

[58] Ltr, Kean to author, 22 Apr 53; Barth MS, p. 27.

[59] 27th Inf WD, 1 Sep 50; Interv, author with Check, 6 Feb 53.

bled as units. Sgt. Jack W. Riley of the 25th Military Police Company tried to help clear the road. Men ran off the mountain past him, some with shoes off, half of them without weapons, and only a few wearing helmets. He shouted for all officers and noncommissioned officers to stop. None stopped. One man who appeared to have some rank told him, "Get out of the way." Riley pulled back the bolt of his carbine and stopped the man at gun point, and then discovered that he was a first sergeant. Asked why they would not stay in and fight, several in the group that Riley succeeded in halting simply laughed at him and answered, "We didn't see any MP's on the hill." At 1600, the 2d Battalion, 24th Infantry, assembling in the rear of the 27th Infantry, could muster only 150 to 200 men.[60]

At 1445, General Kean's orders for an immediate counterattack to restore the 24th Infantry positions arrived at Champney's command post. Check quickly completed his attack plan. For half an hour the Air Force bombed, napalmed, rocketed, and strafed Haman and adjacent enemy-held ridges. Fifteen minutes of concentrated artillery barrages followed. Haman was a sea of flames. Check's infantry moved out in attack westward at 1630, now further reinforced by a platoon of tanks from A Company, 79th Tank Battalion. Eight tanks, mounting infantry, spearheaded the attack into Haman. North Koreans in force held the ridge on the west side of the town, and their machine gun fire swept every approach—their "green tracers seemed as thick as the rice in the paddies." Enemy fire destroyed one tank and the attacking infantry suffered heavy casualties. But Check's battalion pressed the attack and by 1825 had seized the first long ridge 500 yards west of Haman; by 2000 it had secured half of the old battle position on the higher ridge beyond, its objective, one mile west of Haman. Two hundred yards short of the crest on the remainder of the ridge, the infantry dug in for the night.[61]

All day air strikes had harassed the enemy and prevented him from consolidating his gains and reorganizing for further co-ordinated attack. Some of the planes came from the carriers *Valley Forge* and *Philippine Sea,* 200 miles away and steaming toward the battlefield at twenty-seven knots. The crisis for the 25th Division was not lessened by Eighth Army's telephone message at 1045 that the 27th Infantry was to be alerted for a possible move north into the 2d Division sector.

West of Haman the North Koreans and Check's men faced each other during the night without further battle, but the North Koreans, strangely for them, kept flares over their position. In the rear areas, enemy mortar fire on the 24th Regiment command post caused Colonel Champney to move it still farther to the rear.

In the morning, under cover of a heavy ground fog, the North Koreans struck Check's battalion in a counterattack. This action began a hard fight which lasted all morning. Air strikes using na-

[60] 2d Bn, 24th Inf WD, 1 Sep 50; EUSAK IG Rpt, 24th Inf, Sep 50, testimony of Check, Riley, and Roberts.

[61] 1st Bn, 27th Inf WD, 1 Sep 50; A Co, 79th Hv Tk Bn WD, Sep 50; 24th Inf WD, 1 Sep 50; 25th Div WD, 1 Sep 50; EUSAK IG Rpt, Check testimony; *Newsweek,* September 11, 1950, pp. 18–20.

COMMAND POST *of the 27th Infantry under a bridge east of Haman.*

palm burned to death many North Koreans and helped the infantry in gaining the ridge. At noon, the 1st Battalion, 27th Infantry, at last secured the former positions of the 2d Battalion, 24th Infantry, and took over the same foxholes that unit had abandoned two nights before. Its crew-served weapons were still in place. During 2 September, the Air Force flew 135 sorties in the 25th Division sector, reportedly destroying many enemy soldiers, several tanks and artillery pieces, and three villages containing ammunition dumps.[62]

Early the next morning, 3 September, the North Koreans heavily attacked Check's men in an effort to regain the ridge. Artillery, mortar, and tank fire barrages, and a perfectly timed air strike directed from the battalion command post, met this attack. Part of the battalion had to face about and fight toward its rear. After the attack had been repulsed hundreds of enemy dead lay about the battalion position. A prisoner estimated that during 2–3 September the four North Korean battalions fighting Check's battalion had lost 1,000 men.[63]

Colonel Check's battalion held the ridge until dark on 4 September, then

[62] 1st Bn, 27th Inf WD, 2 Sep 50; 24th Inf WD, 2 Sep 50; 25th Div WD, 2 Sep 50; EUSAK IG Rpt, testimony of Check and Capt Don K. Hickman, Ex Off, 1st Bn, 27th Inf; New York *Herald Tribune,* September 2, 1950, Bigart dispatch.

[63] 1st Bn, 27th Inf WD, Sep 50 Opn Rpt, 3 Sep 50; 27th Inf WD, 3 Sep 50, and Rpt of captured documents; Barth MS, p. 27.

the 1st Battalion and F Company of the 2d Battalion, 24th Infantry, which had reorganized in the rear, relieved it. The 1st Battalion, 27th Infantry, thereupon moved back into a secondary defensive position a mile and a half east of Haman. Colonel Champney moved his command post back into Haman, placing it at the base of a hill 300 yards west of the center of the town.[64]

That night there was a repetition of the earlier disgraceful episode. Before dawn, 5 September, an enemy force of two companies, only half-armed, moved against Haman. A part of this force approached the hill at the western edge of Haman where H Company was posted as security for the 24th Regimental command post situated at its base. The H Company men left their post without firing a shot, abandoning two new machine guns. Men in the regimental command post had their first intimation that enemy troops were in the vicinity when the North Koreans opened fire on them with the captured machine guns. A small group of North Koreans infiltrated into Haman within 100 yards of the command post, where members of the I&R Platoon drove them off in a grenade battle. In the course of this action, an enemy grenade blew up an ammunition truck. The exploding shells and resulting fires gave the impression from a distance that a heavy fight was in progress.

About twenty enemy soldiers approached, undiscovered, close enough to the 1st Battalion, 24th Infantry, command post west of Haman to throw grenades and fire burp guns into it. Perhaps 45 soldiers of the battalion command group and 20 South Korean recruits were in position there at the time. The enemy was driven off at dawn, but Maj. Eugene J. Carson, battalion executive officer, then discovered that he had on position with him only 30 men, 7 of them wounded. Looking back down the hill, Carson saw approximately 40 men get up out of the rice paddies and go over to a tank at a roadblock position. These men reported to the regiment that they had been driven off the hill. Three tanks near the command post helped clear the town of North Koreans.[65]

At the time of this enemy infiltration, a white officer and from 35 to 40 Negro soldiers left their position south of Haman at a roadblock and fled to the rear until they reached Colonel Check's 1st Battalion, 27th Infantry, command post a mile and a half away. There, at 0500 this officer said 2,000 North Koreans had overrun his position and others near Haman, including the 24th Regiment command post. Check reported this story to General Kean, and then sent a platoon of tanks with a platoon of infantry toward Haman to find out what had happened. Some of his officers, meanwhile, had stopped about 220 soldiers streaming to the rear. Colonel Check ordered these men to follow his tank and infantry patrol back into Haman. Some of them did so only when threatened with a gun. The tank-led column entered Haman unopposed, where they found

[64] 27th Inf WD, 4 Sep 50; 24th Inf WD, 4 Sep 50; Interv, author with Champney, 22 Jul 51; Corley, MS review comments for author, 22 Jul 53.

[65] Interv, author with Champney, 22 Jul 51; EUSAK WD, 14 Sep 50, Interrog Rpt, Yun Che Gun; 24th Inf WD, 4–5 Sep 50; EUSAK IG Rpt, testimony of Champney, Roberts, and Carson.

the 24th Regiment command post intact and everything quiet.[66]

The next day, 6 September, a sniper severely wounded Colonel Champney while the latter was inspecting his front-line positions west of Haman. Champney was evacuated at once. Colonel Corley, commanding officer of the 3d Battalion, succeeded to the command of the regiment.[67] Corley, known as "Cash Pays the Rent" because that was a favorite saying of his, became a highly regarded commander of the "Deuce-Four" Regiment. He was destined to fight in four campaigns of the Korean War, winning a Distinguished Service Cross, three Silver Stars, and the Legion of Merit to add to the decorations he had already won as a much-decorated battalion commander of World War II. This 36-year-old energetic West Point combat leader was soon well-known throughout the regiment.

### *Battle Mountain and Sobuk-san*

Although the enemy *6th* and *7th Divisions* had massed their troops for the attempted breakthrough of the U.S. 25th Division positions along the Nam and Naktong Rivers as already related, the *6th Division* did not altogether ignore the mountain backbone stretching southward toward the coast. Enemy artillery and mortar fire fell on Battle Mountain, P'il-bong, and Sobuk-san during the period of the enemy offensive and there were strong local attacks and patrol actions. The 1st Battalion, 5th Infantry, never succeeded in gaining possession of the highest peak of Sobuk-san, which would have given observation into the valley below and into the enemy's rear areas. The instability of the 24th Infantry at this time made it necessary for General Kean to order Colonel Throckmorton to send his only regimental reserve, E Company, north into the 24th Infantry sector along the Haman road to protect the right flank of the 5th Regimental Combat Team. In this position, Capt. William Conger, E Company commander, collected stragglers from the 24th Infantry every night and the next morning sent them back to their units. Even the Navy entered the battle in this part of the line, for its destroyers standing off the south coast gave illumination at night by directing their searchlights against low-hanging clouds on Sobuk-san. One destroyer was on station almost continuously, supporting the ground action with the fire of six 5-inch guns. An artillery aerial observer directed this naval gunfire through the fire direction center.[68]

On 7 September, a North Korean attack succeeded once again in driving ROK and American troops from Battle Mountain. The 25th Division ordered Colonel DeChow to retake the peak. DeChow, who had just counterattacked through the rear areas of the 24th Infantry to the vicinity of Haman, prepared his 3d Battalion, 27th Infantry, for the attempt. Companies K and B of the 24th Infantry were to follow him and secure the crest if he regained it. For three days,

[66] EUSAK IG Rpt, testimony of Check, Hickman, and Capt James D. Hunsaker, S–3, 1st Bn, 27th Inf.

[67] 24th Inf WD, 6 Sep 50; 25th Inf WD, 6 Sep 50; Interv, author with Champney, 22 Jul 51.

[68] 25th Div WD, 9 Sep 50; EUSAK WD, 5 Oct 50, Arty Sec, Arty Info Bul 8, 3 Oct 50; Throckmorton, Notes for author, 17 Apr 53.

7, 8, and 9 September, the 3d Battalion counterattacked up Battle Mountain. On the 9th, Capt. William Mitchell led his I Company to the top and engaged in hand-to-hand combat with the North Koreans. L Company followed to the crest but the dug-in enemy drove both companies off and back down the slope. An estimated two companies of enemy troops held the crest of Battle Mountain and two more companies protected their flanks. DeChow's 3d Battalion suffered heavy casualties in these three days of fighting. On the afternoon of the 9th the American counterattack force dropped back to the high ground which it had recaptured on the 7th, 1,000 yards east of Battle Mountain. Artillery, mortars, and air strikes pounded the enemy position on Battle Mountain. During this impasse, word came from the 25th Division for the battalion to move to the vicinity of Masan.[69]

With the failure of the 3d Battalion, 27th Infantry, to hold the high knob on Battle Mountain after its attacks on 8–9 September, Colonel Corley, the 24th Infantry commander, on the evening of the 9th decided to give up the attempt. He had K Company, 24th Infantry, and C Company, 65th Engineer Combat Battalion, dig in on the hill east of and lower than Battle Mountain, surrounded them with barbed wire and mine fields, and placed registered artillery and mortar fires on all enemy approaches to the position. He planned to contain the enemy on Battle Mountain by artillery and mortar fire. The North Koreans on Battle Mountain attacked the lower American defensive position many times on subsequent nights, but all their attacks were driven off. Thus, finally, after a month of almost constant battle the North Koreans gained and held possession of the crest of Battle Mountain. The defensive fires of the 24th Regiment and attached artillery, however, contained them there and they were unable to exploit the possession of this battle-torn peak.[70]

With Battle Mountain in their possession, the North Koreans set out to gain control of P'il-bong, a towering peak 250 feet higher than Battle Mountain and an air mile to the southeast. In the predawn hours of 14 September an enemy force of 400–500 men attacked I and L Companies, 24th Infantry, on P'il-bong. Several attacks were repulsed, but because of men leaving their positions L Company's strength dwindled from 100 to 40 men. Only the determined leadership of Maj. Melvin R. Blair, a replacement officer who had just assumed command of the battalion, held these men in the fight. With the remnant of L Company, Blair withdrew toward I Company's position on the crest of P'il-bong, only to find that this company under a relatively minor attack had, unknown to him, left the hill. A wounded North Korean sniper, hidden along the trail, shot Blair in the leg. Blair refused to be evacuated, but he could not hold P'il-bong with the handful of men remaining with him and it was lost.[71]

Just as soon as the crisis passed for the 25th Division, General Walker

[69] 25th Div WD, 8–9 Sep 50; 24th Inf WD, 9 Sep 50; EUSAK WD, G–3 Jnl, 1245, 9 Sep 50; DeChow, MS review comments, Jul 53.

[70] Interv, author with Corley, 6 Nov 51; 24th Inf WD, 9 Sep 50.

[71] Interv, author with Corley, 6 Nov 51; 3d Bn, 24th Inf WD, 14 Sep 50; EUSAK IG Rpt, testimony of Corley.

ordered it on 7 September to release the 5th Regimental Combat Team within twenty-four hours. The continuing crisis north of Taegu made it mandatory for Walker to build up his reserve there. That evening the 1st and 2d Battalions, 27th Infantry, moved from the Nam River battlefield to relieve the 5th Regimental Combat Team on the Masan front. Colonel Michaelis assumed command of the regimental zone at 1500, 9 September. The 3d Battalion, 27th Infantry, broke off its counterattacks on Battle Mountain that day, rejoined the regiment, and took its place in the southern end of the line on 11 September. Meanwhile, the 5th Regimental Combat Team began moving to Samnangjin on the 10th, the last train with its units clearing Masan at 1600 the next day. Upon arrival at Samnangjin, it passed to Eighth Army reserve.[72]

About the time the all-out North Korean assault on the Pusan Perimeter had been turned back and the 27th Infantry was relieving the 5th Regimental Combat Team in the line west of Masan, the "beer issue" came to a head and evoked strong reactions from the men who were fighting the Korean battles. Free beer had been provided U.S. soldiers on much the same basis as candy bars and cigarettes. It had been purchased with appropriated money and issued at intervals as supplementary to the food ration. Various temperance, church, and social groups, and some individuals in the United States protested the issue of beer to the soldiers. The controversy even reached the floor of Congress, with one Congressman who favored the free beer ration saying, "Water in Korea is deadlier than bullets." The pressure was sufficient to cause the Army through the Far East Command to order that 12 September would be the last day free beer could be issued to the troops. A typical infantryman's comment was, "Those organizations or whatever they are have nothing to do with us. We are doing the fighting over here and it gets pretty bad. One can of beer never hurt nobody." But henceforth, Eighth Army troops could obtain beer purchased only with nonappropriated funds and issued through the post exchanges.[73]

The defensive battles on the Masan front during August and early September brought to a head a problem that had bothered General Kean ever since the 25th Division entered the Korean War; in a larger sense, it was a problem that had concerned Eighth Army as well. Two of the division's regiments, the 27th and the 35th, had performed well in Korea. Not so the 24th Infantry, the division's third regiment. Ever since its entrance into combat in the Sangju area in July the Negro regiment had given a poor performance, although there were some exceptions and many individual acts of heroism and capable performance of duty. The unstable nature of the regiment was demonstrated in the fighting on Battle Mountain during August. Then, on the night of 31 August–1 September two battalions evaporated in the face of the enemy, and a large part of them repeated this performance four nights later. General Kean placed his

---

[72] EUSAK WD, G–3 Sec, 8–9 Sep 50; 27th Inf WD, 7–9 Sep 50; *Ibid.*, Unit Rpt, Sep 50.

[73] See 25th Div WD, 11 Sep 50; New York *Times*, September 14, and October 16, 1950; New York *Herald Tribune*, September 13, 1950.

VETERAN OF THE 5TH REGIMENTAL COMBAT TEAM *after forty-three days on the front line.*

two stronger regiments usually in the more critical terrain of the division front, but, nevertheless, the 24th Regiment constituted a weak link in the division line that might break at any time and bring disaster to the division and possibly to the army. Eighth Army and the 25th Division assigned officers of an unusually high caliber to the 24th Infantry to give it strong leadership, but this did not solve the problem.

After the enemy breakthrough in the 24th Infantry sector on 1–5 September, General Kean decided he had to seek a solution. On 9 September he recommended to General Walker the immediate removal of the 24th Infantry Regiment from combat, and that the troops of the regiment be transferred as replacements on a percentage basis to other U.S. Army units in Korea. In making these recommendations General Kean said in part, "It is my considered opinion that the 24th Infantry has demonstrated in combat that it is untrustworthy and incapable of carrying out missions expected of an infantry regiment." Nearly all officers serving in the regiment agreed with General Kean, and so did many of the Negro noncommissioned officers and enlisted men themselves. General Walker did not act on General Kean's recommendation since many considerations seemed to make such action impossible at the time.[74]

Coinciding with this heavy fighting at the Pusan Perimeter in the south a new and disturbing element appeared far to the north. In Tokyo and Washington, American military leaders studied reports they received indicating that Chinese Communist troops were moving north through China and concentrating along the Yalu River opposite Korea. An incident at this time added to the build-up of threatening storm clouds to the north. On 4 September, a twin-engine bomber wearing a red star passed over a screening ship of a U.N. naval task force operating in the Yellow Sea off the west coast of Korea, approximately at the 38th Parallel. The bomber continued on toward the center of the naval formation and opened fire on a U.N. fighter plane patrol which re-

[74] Ltr, Kean to CG, Eighth Army, 9 Sep 50, in EUSAK IG Rpt. The 24th Regiment continued to serve in Eighth Army as an all-Negro unit for another year. Its troops were then transferred as replacements to other infantry units of the army, integrated usually in a proportion of about 12 percent.

turned its fire and shot it down. A destroyer of the task force recovered the body of one of the bomber crew members—he was an officer of the Armed Forces of the Soviet Union.[75]

At mid-September the Eighth Army and the ROK Army were still engaged with North Korean forces at nearly all points of the Pusan Perimeter. After two weeks of the heaviest fighting of the war they had just barely turned back the great North Korean offensive on the main axes of the attack: in the east around P'ohang-dong and the Kyongju corridor, in the center at the approaches to Taegu, and in the south around Yongsan and the approaches to Masan. The battles of the Perimeter would go on, that was certain, for the issue there had not been concluded.

But overriding all other factors, favorable and unfavorable, comforting or disquieting, bearing on the Korean War at mid-September was the knowledge—now become widespread among U.N. forces in Korea—that an amphibious landing behind the enemy's lines was imminent. The date set for it was 15 September.

---

[75] New York *Times*, September 5, 1950, gives the State Department note announcing this incident. The *Times* of 7 September gives a summary of the Russian version, and the claim for damages for the bomber and three Russian crewman, which U.S. Ambassador Alan G. Kirk refused to accept.

On 31 August, Ambassador Warren Austin told the U.N. Security Council that a U.S. F-51 fighter plane on 27 August may have strafed the An-tung Airfield in Manchuria, five miles from the Korean border, and thereby have unintentionally violated the territory of Communist China.

## CHAPTER XXV

# The Landing at Inch'on

> The history of war proves that nine out of ten times an army has been destroyed because its supply lines have been cut off. . . . We shall land at Inch'on, and I shall crush them [the North Koreans].
>
> DOUGLAS MACARTHUR

It was natural and predictable that General MacArthur should think in terms of an amphibious landing in the rear of the enemy to win the Korean War. His campaigns in the Southwest Pacific in World War II—after Bataan—all began as amphibious operations. From Australia to Luzon his forces often advanced around enemy-held islands, one after another. Control of the seas gives mobility to military power. Mobility and war of maneuver have always brought the greatest prizes and the quickest decisions to their practitioners. A water-borne sweep around the enemy's flank and an attack in his rear against lines of supply and communications appealed to MacArthur's sense of grand tactics. He never wavered from this concept, although repeatedly the fortunes of war compelled him to postpone its execution.

### *MacArthur's Early Plans*

During the first week of July, with the Korean War little more than a week old, General MacArthur told his chief of staff, General Almond, to begin considering plans for an amphibious operation designed to strike the enemy center of communications at Seoul, and to study the location for a landing to accomplish this. At a Far East Command headquarters meeting on 4 July, attended by Army, Navy, and Air Force representatives, Generals MacArthur and Almond discussed the idea of an amphibious landing in the enemy's rear and proposed that the 1st Cavalry Division be used for that purpose. Col. Edward H. Forney of the Marine Corps, an expert on amphibious operations, was selected to work with the 1st Cavalry Division on plans for the operation.[1]

The early plan for the amphibious operation received the code name BLUEHEARTS and called for driving the North Koreans back across the 38th Parallel. The approximate date proposed for it

[1] Interv, author with Almond, 13 Dec 51; MS review comments, Almond for author, 23 Oct 53; Hq X Corps, Opn CHROMITE, G-3 Sec, 15 Aug-30 Sep 50, p. 1; Lynn Montross and Capt. Nicholas A. Canzona, USMC, *U.S. Marine Operations in Korea, 1950-1953*, vol. II, *The Inchon-Seoul Operation* (Washington: Historical Branch, G-3, Headquarters, U.S. Marine Corps, 1954), pp. 4-7.

was 22 July, but the operation was abandoned by 10 July because of the inability of the U.S. and ROK forces in Korea to halt the southward drive of the enemy.[2]

Meanwhile the planning for an amphibious operation went ahead in the Far East Command despite the cancellation of BLUEHEARTS. These plans were undertaken by the Joint Strategic Plans and Operations Group (JSPOG), Far East Command, which General Wright headed in addition to his duties as Assistant Chief of Staff, G–3. One of Wright's deputies, Col. Donald H. Galloway, was directly in charge of JSPOG. This unusually able group of planners developed various plans in considerable detail for amphibious operations in Korea.

On 23 July, General Wright upon MacArthur's instructions circulated to the GHQ staff sections the outline of Operation CHROMITE. CHROMITE called for an amphibious operation in September and postulated three plans: (1) Plan 100-B, landing at Inch'on on the west coast; (2) Plan 100-C, landing at Kunsan on the west coast; (3) Plan 100-D, landing near Chumunjin-up on the east coast. Plan 100-B, calling for a landing at Inch'on with a simultaneous attack by Eighth Army, was favored.[3]

This same day, 23 July, General MacArthur informed the Department of the Army that he had scheduled for mid-September an amphibious landing of the 5th Marines and the 2d Infantry Division behind the enemy's lines in co-ordination with an attack by Eighth Army.[4]

The North Korean successes upset MacArthur's plans as fast as he made them. He admitted this to the Joint Chiefs in a message on 29 July, saying, "In Korea the hopes that I had entertained to hold out the 1st Marine Division [Brigade] and the 2d Infantry Division for the enveloping counter blow have not been fulfilled and it will be necessary to commit these units to Korea on the south line rather than . . . their subsequent commitment along a separate axis in mid-September. . . . I now plan to commit my sole reserve in Japan, the 7th Infantry Division, as soon as it can be brought to an approximate combat strength." [5]

### *X Corps Troops Assembled*

By 20 July General MacArthur had settled rather definitely on the concept of the Inch'on operation and he spoke of the matter at some length with General Almond and with General Wright, his operations officer. On 12 August, MacArthur issued CINCFE Operation Plan 100-B and specifically named the Inch'on-Seoul area as the target that a special invasion force would seize by amphibious assault.[6]

On 15 August General MacArthur es-

[2] Schnabel, FEC, GHQ Support and Participation in the Korean War, ch. V, pp. 1–18.

[3] *Ibid.*, ch. 5, pp. 12–13; Interv with Wright, 7 Jan 54. The landing at Kunsan called for a drive inland to Taejon; that at Chumunjin-up included a ROK division and called for an advance down the coastal road to Kangnung and then west to Wonju.

[4] GHQ FEC, Ann Narr Hist Rpt, Jan–Oct 50, p. 11.

[5] Schnabel, FEC, GHQ Support and Participation in the Korean War, ch. V, p. 25, quoting Rad C58993, CINCFE to JCS, 29 Jul 50.

[6] Diary of CG X Corps, Opn CHROMITE; Interv, author with Wright, 7 Jan 54; Interv, author with Maj Gen Clark L. Ruffner, 27 Aug 51.

tablished the headquarters group of the Special Planning Staff to take charge of the projected amphibious operation. For purposes of secrecy the new group, selected from the GHQ FEC staff, was designated, Special Planning Staff, GHQ, and the forces to be placed under its control, GHQ Reserve. On 21 August, MacArthur requested the Department of the Army by radio for authority to activate Headquarters, X Corps, and, upon receiving approval, he issued GHQ FEC General Order 24 on 26 August activating the corps. All units in Japan or en route there that had been designated GHQ Reserve were assigned to it.[7]

It appears that General MacArthur about the middle of August had made up his mind on the person he would select to command the invasion force. One day as he was talking with General Almond about the forthcoming landing, the latter suggested that it was time to appoint a commander for it. MacArthur turned to him and replied, "It is you." MacArthur told Almond that he was also to retain his position as Chief of Staff, Far East Command. His view was that Almond would command X Corps for the Inch'on invasion and the capture of Seoul, that the war would end soon thereafter, and Almond would then return to his old position in Tokyo. In effect, the Far East Command would lend Almond and most of the key staff members of the corps for the landing operation. General Almond has stated that MacArthur's decision to place him in command of X Corps surprised him, as he had expected to remain in Tokyo in his capacity as Chief of Staff, FEC. General MacArthur officially assigned General Almond to command X Corps on 26 August.[8]

General Almond, fifty-eight years old when he assumed command of X Corps, was a graduate of Virginia Military Institute. In World War I he had commanded a machine gun battalion and had been wounded and decorated for bravery. In World War II he had commanded the 92d Infantry Division in Italy. Almond went to the Far East Command in June 1946, and served as deputy chief of staff to MacArthur from November 1946 to February 1949. On 18 February 1949 he became Chief of Staff, Far East Command, and, on 24 July 1950, Chief of Staff, United Nations Command, as well.

General Almond was a man both feared and obeyed throughout the Far East Command. Possessed of a driving energy and a consuming impatience with incompetence, he expected from others the same degree of devotion to duty and hard work that he exacted from himself. No one who ever saw him would be likely to forget the lightning that flashed from his blue eyes. To his commander, General MacArthur, he was wholly loyal. He never hesitated before difficulties. Topped by iron-gray hair, Almond's alert, mobile face with its ruddy complexion made him an arresting figure despite his medium stature and the slight stoop of his shoulders.

The corps' chief of staff was Maj. Gen. Clark L. Ruffner, who had arrived from the United States on 6 August and had

[7] Schnabel, Theater Command, ch. VIII. This volume will treat in detail the planning of the Inch'on landing and the policy debate on it. Hq X Corps, Opn CHROMITE.

[8] Interv, author with Almond, 13 Dec 51; Hq X Corps, Opn CHROMITE; Almond biographical sketch.

started working with the planning group two days later. He was an energetic and diplomatic officer with long experience and a distinguished record in staff work. During World War II he had been Chief of Staff, U.S. Army Forces, Pacific Ocean Areas, in Hawaii. The X Corps staff was an able one, many of its members hand-picked from among the Far East Command staff.

The major ground units of X Corps were the 1st Marine Division and the 7th Infantry Division. In the summer of 1950 it was no easy matter for the United States to assemble in the Far East a Marine division at full strength. On 25 July, Maj. Gen. Oliver P. Smith assumed command and on that day the Commandant of the Marine Corps issued an order to him to bring the division to war strength, less one regiment, and to sail for the Far East between 10 and 15 August. This meant the activation of another regiment, the 1st Marines, and the assembly, organization, and equipment of approximately 15,000 officers and enlisted men within the next two weeks. On 10 August, the Joint Chiefs of Staff decided to add the third regiment to the division, and the 7th Marines was activated. It was scheduled to sail for the Far East by 1 September. The difficulty of obtaining troops to fill the division was so great that a battalion of marines on duty with the Sixth Fleet in the Mediterranean was ordered to join the division in the Far East.[9]

General Smith and most of the staff officers of the 1st Marine Division arrived in Japan from the United States on 22 August. The division troops, the 1st Marines, and the staff of the 7th Marines arrived in Japan between 28 August and 6 September. A battalion of marines in two vessels, the *Bexar* and the *Montague*, departed Suda Bay, Crete, in the Mediterranean on 16 August, and sailing by way of Suez arrived at Pusan on 9 September to join the 7th Marines as its 3d Battalion. The remainder of the 7th Marines arrived at Kobe on 17 September. The 5th Marines, in Korea, received a warning order on 30 August to prepare for movement to Pusan to join the division.[10]

Bringing the 7th Infantry Division up to war strength posed an even more difficult problem. During July, FEC had taken 140 officers and 1,500 noncommissioned officers and enlisted men from the division to augment the strength of the 24th and 25th Infantry and the 1st Cavalry Divisions as they in turn had mounted out for Korea. At the end of July the division was at less than half-strength, but in noncommissioned officer weapons leaders and critical specialists the shortage was far greater than that proportion. On 27 July, the 7th Infantry Division was 9,117 men understrength—290 officers, 126 warrant officers, and 8,701 enlisted men. The day before, FEC had relieved it of all occupation duties and ordered it to prepare for movement to Korea.[11]

---

[9] 1st Mar Div SAR, Sep 50; Lt Gen Oliver P. Smith, MS review comments with ltr to Maj Gen Albert C. Smith, Chief Mil Hist, 25 Feb 54; Karig, *et al., Battle Report, The War in Korea*, pp. 123, 172.

[10] 1st Mar Div SAR, vol. III, pp. 1–2; Smith, MS review comments.

[11] EUSAK WD, 31 Jul 50, Memo for CofS, Strategic Status of 7th Inf Div; Schnabel, FEC, GHQ Support and Participation in the Korean War, ch. V, p. 5, citing Ltr, CINCFE to CG Eighth Army, 4 Aug 50; Maj Gen David G. Barr (CG 7th Inf Div), Notes, 1, 6, 31 Jul 50 (copies furnished author by Barr).

From 23 August to 3 September the Far East Command allotted to the 7th Division the entire infantry replacement stream reaching FEC, and from 23 August through 8 September the entire artillery replacement stream. By 4 September the division had received 390 officers and 5,400 enlisted replacements. General MacArthur obtained service units for the X Corps in the same way—by diverting them from scheduled assignments for Eighth Army. The Far East Command justified this on the ground that, while Eighth Army needed them badly, X Corps' need was imperative.[12]

In response to General MacArthur's instructions to General Walker on 11 and 13 August to send South Koreans to augment the 7th Infantry Division, 8,637 of them arrived in Japan before the division embarked for Inch'on. Their clothing on arrival ranged from business suits to shirts and shorts, or shorts only. The majority wore sandals or cloth shoes. They were civilians—stunned, confused, and exhausted. Only a few could speak English. Approximately 100 of the South Korean recruits were assigned to each rifle company and artillery battery; the buddy system was used for training and control.[13]

The quality of the artillery and infantry crew-served weapons troops received from the United States and assigned to the 7th Division during August and early September was high. The superior training provided by the old infantry and artillery noncommissioned officers who arrived from the Fort Benning Infantry and the Fort Sill Artillery Schools brought the 7th Division to a better condition as the invasion date approached than could have been reasonably expected a month earlier. The 7th Division strength on embarkation, including the attached South Koreans, was 24,845.[14]

### *The Landing Controversy*

All through July and August 1950 the Joint Chiefs of Staff gave implied or expressed approval of MacArthur's proposal for an amphibious landing behind the enemy's battle lines. But while it was known that MacArthur favored Inch'on as the landing site, the Joint Chiefs had never committed themselves to it. From the beginning, there had been some opposition to and many reservations about the Inch'on proposal on the part of General Collins, U.S. Army Chief of Staff; the Navy; and the Marine Corps. The FEC senior planning and staff officers—such as Generals Almond and Hickey, Chief of Staff and Deputy Chief of Staff; General Wright, the G–3 and head of JSPOG; and Brig. Gen. George L. Eberle, the G–4—supported the plan.[15]

The Navy's opposition to the Inch'on site centered largely on the difficult tidal

[12] Schnabel, FEC, GHQ Support and Participation in the Korean War, ch. V, pp. 31–32; GHQ FEC, Ann Narr Hist Rpt, 1 Jan–31 Oct 50, p. 45; 7th Div WD, Aug–Sep 50; Barr, Notes, 4 Sep 50.

[13] 7th Inf Div WD, 1 Sep 50; Diary of CG X Corps, Opn CHROMITE, 1 Sep 50; Barr, MS review comments, 1957.

[14] Interv, author with Barr, 1 Feb 54; Barr, Notes, 4 Sep 50.

[15] Interv, author with Wright, 7 Jan 54; Interv, author with Eberle, 12 Jan 54; Ltr, Wright to author, 22 Mar 45; Almond, MS review comments for author, 23 Oct 53; Schabel, FEC, GHQ Support and Participation in the Korean War, ch. V, p. 23; Interv, author with Lutes (FEC Planning Sec), 7 Oct 51.

conditions there, and since this opposition continued, the Joint Chiefs of Staff decided to send two of its members to Tokyo to discuss the matter with MacArthur and his staff. A decision had to be reached. On 20 July General Collins and Admiral Forrest P. Sherman, Chief of Naval Operations, left Washington for their conference with MacArthur. Upon arrival in Japan, Collins and Sherman engaged in private conversations with MacArthur and key members of his staff, including senior naval officers in the Far East. Then, on the afternoon of 23 July, a full briefing on the subject was scheduled in General MacArthur's conference room in the Dai Ichi Building.[16]

The conference began at 1730 in the afternoon. Among those present in addition to General MacArthur were General Collins, Admiral Sherman, Vice Admirals Joy and Struble, Generals Almond, Hickey, and Wright, some members of the latter's JSPOG group, and Rear Adm. James H. Doyle and some members of his staff who were to present the naval problems involved in a landing at Inch'on.

After a short introduction by General MacArthur, General Wright briefed the group on the basic plan. Admiral Doyle then presented the naval considerations. His general tone was pessimistic, and he concluded with the remark, "The operation is not impossible, but I do not recommend it." The naval part of the briefings lasted more than an hour.

During the naval presentation MacArthur, who had heard the main arguments many times before, sat quietly smoking his pipe, asking only an occasional question. When the presentation ended, MacArthur began to speak. He talked as though delivering a soliloquy for forty-five minutes, dwelling in a conversational tone on the reasons why the landing should be made at Inch'on. He said that the enemy had neglected his rear and was dangling on a thin logistical rope that could be quickly cut in the Seoul area, that the enemy had committed practically all his forces against Eighth Army in the south and had no trained reserves and little power of recuperation. MacArthur stressed the strategical, political, and psychological reasons for the landing at Inch'on and the quick capture of Seoul, the capital of South Korea. He said it would hold the imagination of Asia and win support for the United Nations. Inch'on, he said, pointing to the big map behind him, would be the anvil on which the hammer of Walker's Eighth Army from the south would crush the North Koreans.

General MacArthur then turned to a consideration of a landing at Kunsan, 100 air miles below Inch'on, which General Collins and Admiral Sherman had favored. MacArthur said the idea was good but the location wrong. He did not think a landing there would result in severing the North Korean supply lines and destroying the North Korean Army. He returned to his emphasis on Inch'on, saying that the amphibious landing was tactically the most powerful military device available to the United Nations Command and that to employ it properly meant to strike deep and hard into enemy-held territory. He dwelt on the bitter Korean winter campaign that would become necessary if Inch'on was not undertaken. He said the North

[16] Schnabel, Theater Command ch. VIII; New York *Times*, August 19 1950.

Koreans considered a landing at Inch'on impossible because of the very great difficulties involved and, because of this, the landing force would achieve surprise. He touched on his operations in the Pacific in World War II and eulogized the Navy for its part in them. He concluded his long talk by declaring unequivocally for Inch'on and saying, "The Navy has never turned me down yet, and I know it will not now."

MacArthur seems to have convinced most of the doubters present. Admiral Sherman was won over to MacArthur's position. General Collins, however, seemed still to have reservations on Inch'on. He subsequently asked General Wright if the Far East Command had firm plans for a Kunsan landing which could be used as an alternate plan if the Inch'on operation either was not carried out or failed. Wright assured him that there were such plans and, moreover, that it was planned to stage a feint at Kunsan.[17]

Among the alternate proposals to Inch'on, in addition to the Kunsan plan favored by the Navy, was one for a landing in the Posung-myon area thirty miles south of Inch'on and opposite Osan. On the 23d, Admiral Doyle had proposed a landing there with the purpose of striking inland to Osan and there severing the communications south of Seoul. On the 24th, Lt. Gen. Lemuel C. Shepherd, Jr. (USMC), called on General MacArthur and asked him to change the landing site to this area—all to no avail. MacArthur remained resolute on Inch'on.

Upon their return to Washington, Collins and Sherman went over the whole matter of the Inch'on landing with the other members of the Joint Chiefs of Staff. On 28 August the Joint Chiefs sent a message to MacArthur which seemingly concurred in the Inch'on plans yet attached conditions. Their message said in part: "We concur in making preparations for and executing a turning movement by amphibious forces on the west coast of Korea, either at Inch'on in the event the enemy defenses in the vicinity of Inch'on prove ineffective, or at a favorable beach south of Inch'on if one can be located. We further concur in preparations, if desired by CINCFE, for an envelopment by amphibious forces in the vicinity of Kunsan. We understand that alternative plans are being prepared in order to best exploit the situation as it develops.[18]

MacArthur pressed ahead unswervingly toward the Inch'on landing. On 30 August he issued his United Nations Command operation order for it. Meanwhile, the Joint Chiefs in Washington expected to receive from MacArthur further details of the pending operation and, failing to receive them, sent a message to him on 5 September requesting this information. MacArthur replied the next day that his plans remained unchanged. On 7 September, the Joint Chiefs sent another message to MacArthur requesting a reconsideration of

[17] The account of the 23 July conference is based on the following sources: Ltr, Wright to author, 22 Mar 54; Ltr, Joy to author, 12 Dec 52; Ltr, Almond to author, 2 Dec 52; Smith, MS review comments; Montross and Canzona, *The Inchon-Seoul Operation*, pp. 40–47; Karig, *et al.*, *Battle Report, the War in Korea*, p. 169. General MacArthur's MS review comments show no comment on this section.

[18] Schnabel, FEC, GHQ Support and Participation in the Korean War, ch. V, p. 6, citing Msg JCS 89960, JCS to CINCFE, 28 Aug 50.

the whole question and an estimate of the chances for favorable outcome. The energy and strength displayed by the North Koreans in their early September massive offensive had evidently raised doubts in the minds of the Joint Chiefs that General Walker's Eighth Army could go over successfully to the attack or that X Corps could quickly overcome the Seoul defenses. In the meantime, General MacArthur on 6 September in a letter to all his major commanders confirmed previous verbal orders and announced 15 September as D-day for the Inch'on landing.[19]

In response to the Joint Chiefs' request for a reconsideration and an estimate of the chances for a favorable landing at Inch'on, General MacArthur on 8 September sent to Washington a final eloquent message on the subject. His message said in part:

> There is no question in my mind as to the feasibility of the operation and I regard its chance of success as excellent. I go further and believe that it represents the only hope of wresting the initiative from the enemy and thereby presenting an opportunity for a decisive blow. To do otherwise is to commit us to a war of indefinite duration, of gradual attrition, and of doubtful results. . . . There is no slightest possibility . . . of our force being ejected from the Pusan beachhead. The envelopment from the north will instantly relieve the pressure on the south perimeter and, indeed, is the only way that this can be accomplished. . . . The success of the enveloping movement from the north does not depend upon the rapid juncture of the X Corps and the Eighth Army. The seizure of the heart of the enemy distributing system in the Seoul area will completely dislocate the logistical supply of his forces now operating in South Korea and therefore will ultimately result in their disintegration. This, indeed, is the primary purpose of the movement. Caught between our northern and southern forces, both of which are completely self-sustaining because of our absolute air and naval supremacy, the enemy cannot fail to be ultimately shattered through disruption of his logistical support and our combined combat activities. . . . For the reasons stated, there are no material changes under contemplation in the operation as planned and reported to you. The embarkation of the troops and the preliminary air and naval preparations are proceeding according to schedule.

The next day the Joint Chiefs, referring to this message, replied tersely to MacArthur, "We approve your plan and President has been so informed." [20] It appears that in Secretary of Defense Johnson, MacArthur had in Washington a powerful ally during the Inch'on landing controversy, for Johnson supported the Far East commander.[21] Thus on 8 September Washington time and 9 September Tokyo time the debate on the projected Inch'on landing ended.

A co-ordinate part of MacArthur's Inch'on plan was an attack by the Eighth Army north from its Pusan Per-

[19] Schnabel, FEC, GHQ Support and Participation in the Korean War, ch. VIII; Smith MS review comments, 25 Feb 54.

[20] Rad C62423, CINCFE to JCS, 8 Sep 50, and Rad 90958, JCS to CINCFE, 8 Sep 50.

[21] In the course of the MacArthur hearings the next year, Secretary Johnson, in response to an inquiry from Senator Alexander Wiley, said, "I had been carrying along with General MacArthur the responsibility for Inch'on. General Collins—maybe the censor will want to strike this out—did not favor Inch'on and went over to try to argue General MacArthur out of it.

"General MacArthur stood pat. I backed MacArthur, and the President has always, had before backed me on it." See Senate Committees on Armed Services and Foreign Relations, 82d Cong., 1st sess., June, 1951, Hearings on *Military Situation in the Far East and the Relief of General MacArthur*, pt. 4, p. 2618.

imeter beachhead simultaneously with the X Corps landing. This action was intended to tie down all enemy forces committed against Eighth Army and prevent withdrawal from the south of major reinforcements for the North Korean units opposing X Corps in its landing area. The plan called for the Eighth Army to break out of the Perimeter, drive northward, and join forces with X Corps.

On 30 August, General Smith had sent a dispatch to X Corps requesting that the 1st Provisional Marine Brigade in Korea be released from Eighth Army on 1 September to prepare for mounting out for Inch'on. MacArthur ordered that the Marine brigade be available on 4 September for that purpose. But no sooner was this order issued than it was rescinded on 1 September because of the crisis that faced Eighth Army after the great North Korean attack had rolled up the southern front during the night.[22]

Eighth Army's use of the 1st Provisional Marine Brigade in the battle near Yongsan threatened to disrupt the Inch'on landing according to Marine and Navy opinion. A tug of war now ensued between General Smith, supported by the U.S. Naval Forces, Far East, on the one hand and General Walker on the other for control of the 5th Marines. The Marine commander insisted he must have the 5th Marines if he were to make the Inch'on landing. General Walker in a telephone conversation with General Almond said in effect, "If I lose the 5th Marine Regiment I will not be responsible for the safety of the front." Almond sided with Walker despite the fact that he was to be commander of the Inch'on landing force, taking the view that the X Corps could succeed in its plan without the regiment. He suggested that the 32d Infantry Regiment of the 7th Division be attached to the 1st Marine Division as its second assault regiment. General Smith and NAVFE remained adamant. The issue came to a head on 3 September when Admirals Joy, Struble, and Doyle accompanied General Smith to the Dai Ichi Building for a showdown conference with Generals Almond, Ruffner, and Wright.

When it became clear that the group could not reach an agreement, General Almond went into General MacArthur's private office and told MacArthur that things had reached an impasse—that Smith and the Navy would not go in at Inch'on without the 5th Marines. Hearing this, MacArthur told Almond, "Tell Walker he will have to give up the 5th Marine Regiment." Almond returned to the waiting group and told them of MacArthur's decision.[23]

The next day, 4 September, General MacArthur sent General Wright to Taegu to tell General Walker that the 1st Provisional Marine Brigade would have to be released not later than the night of 5–6 September and moved at once to Pusan. At Taegu Wright informed Walker of MacArthur's instructions and told him that the Far East Command was loading the 17th Regiment of the 7th Infantry Division for movement to Pusan, where it would be held in floating reserve and be available

[22] Smith, MS review comments, 25 Feb 54.

[23] Interv, author with Almond, 13 Dec 51; Smith, MS review comments, 25 Feb 54; Diary of CG X Corps, Opn CHROMITE, 2 Sep 50; Schnabel, FEC, GHQ Support and Participation in the Korean War, ch. V, pp. 26–27.

for use by Eighth Army if necessary. (It sailed from Yokohama for Korea on 6 September.) He also said that MacArthur intended to divert to Pusan for assignment to Eighth Army the first regiment (65th Infantry) of the 3d Infantry Division arriving in the Far East, the expected date of arrival being 18–20 September. General Walker, in discussing his part in the projected combined operation set for 15 September, requested that the Eighth Army attack be deferred to D plus 1, 16 September. Wright agreed with this timing and said he would recommend it to MacArthur, who subsequently approved it.[24]

### *Naval Plans*

In making ready its part of the operation, the Commander, NAVFE outlined the tasks the Navy would have to perform. These included the following: maintain a naval blockade of the west coast of Korea south of latitude 39° 35′ north; conduct pre-D-day naval operations as the situation might require; on D-day seize by amphibious assault, occupy, and defend a beachhead in the Inch'on area; transport, land, and support follow-up and strategic reserve troops, if directed, to the Inch'on area; and provide cover and support as required. Joint Task Force Seven was formed to accomplish these objectives with Admiral Struble, Commander, Seventh Fleet, as the task force commander. On 25 August, Admiral Struble left his flagship, USS *Rochester*, at Sasebo and proceeded by air to Tokyo to direct final planning.[25]

On 3 September, Admiral Struble issued JTF 7 Operational Plan 9–50. Marine aircraft from two escort carriers, naval aircraft from the U.S. carrier *Boxer*, and British aircraft from a light British carrier would provide as much support aircraft as could be concentrated in and over the landing area, and would be controlled from the amphibious force flagship (AGC) *Mt. McKinley*. An arc extending inland thirty miles from the landing site described the task force objective area.[26] In order to carry out its various missions, Joint Task Force Seven organized its subordinate parts as follows:

| | |
|---|---|
| TF 90 | Attack Force, Rear Adm. James H. Doyle, USN |
| TF 92 | X Corps, Maj. Gen. Edward M. Almond, USA |
| TF 99 | Patrol & Reconnaissance Force, Rear Adm. G. R. Henderson, USN |
| TF 91 | Blockade & Covering Force, Rear Adm. W. G. Andrews, R.N. |
| TF 77 | Fast Carrier Force, Rear Adm. E. C. Ewen, USN |
| TF 79 | Logistic Support Force, Capt. B. L. Austin, USN |
| TF 70.1 | Flagship Group, Capt. E. L. Woodyard, USN |

For the naval phases, the command post of Admiral Struble was on the *Rochester;* that of Rear Admiral Doyle, second in command, was on the *Mt. McKinley.*

[24] GHQ FEC, G–3 Sec, Wright, Memo for Record, 041930K Sep 50, reporting on his discussions with Walker and subsequent report to General Almond; Barr, Notes, 6 Sep 50.

[25] Commander, Joint Task Force Seven and Seventh Fleet, Inch'on Report, September 1950, I–B–1 (hereafter cited as JTF 7, Inch'on Rpt).

[26] JTF 7, Inch'on Rpt, p. 1; Ltr, Wright to author, 22 Mar 54.

More than 230 ships were assigned to the operation. Surface vessels of JTF 7 were not to operate within twelve miles of Soviet or Chinese territory nor aircraft within twenty miles of such territory.[27]

MacArthur had selected Inch'on as the landing site for one paramount reason: it was the port for the capital city of Seoul, eighteen miles inland, and was the closest possible landing area to that city and the hub of communications centering there.

Inch'on is situated on the estuary of the Yom-ha River and possesses a protected, ice-free port with a tidal basin. The shore line there is a low-lying, partially submerged coastal plain subject to very high tides. There are no beaches in the landing area—only wide mud flats at low tide and stone walls at high tide. Because of the mud flats, the landing force would have to use the harbor and wharfage facilities in the port area. The main approach by sea is from the south through two channels 50 miles long and only 6 to 10 fathoms deep (36–60 feet). Flying Fish Channel is the channel ordinarily used by large ships. It is narrow and twisting.

The Inch'on harbor divides into an outer and an inner one, the latter separated from the former by a long breakwater and the islands of Wolmi and Sowolmi which join by a causeway. The greater part of the inner harbor becomes a mud flat at low tide leaving only a narrow dredged channel of about 12–13 feet in depth. The only dock facilities for deep draft vessels were in the tidal basin, which was 1,700 feet long, 750 feet wide, and had an average depth of 40 feet, but at mean low tide held only 14 feet of water.[28]

Inch'on promised to be a unique amphibious operation—certainly one very difficult to conduct because of natural conditions. Tides in the restricted waters of the channel and the harbor have a maximum range of more than 31 feet. A few instances of an extreme 33-foot tide have been reported. Some of the World War II landing craft that were to be used in making the landing required 23 feet of tide to clear the mud flats, and the LST's (Landing Ship, Tank) required 29 feet of tide—a favorable condition that prevailed only once a month over a period of three or four days. The narrow, shallow channel necessitated a daylight approach for the larger ships. Accordingly, it was necessary to schedule the main landings for the late afternoon high tide. A night approach, however, by a battalion-sized attack group was to be made for the purpose of seizing Wolmi-do during the early morning high tide, a necessary preliminary, the planners thought, to the main landing at Inch'on itself.[29]

Low seas at Inch'on are most frequent from May through August, high seas from October through March. Although September is a period of transition, it was considered suitable for landing operations. MacArthur and his planners had selected 15 September for D-day because there would then be a high tide giving maximum water depth over the Inch'on mud flats. Tidal range for 15 September

---

[27] JTF 7, Inch'on Rpt, p. 4, I–D–3, and ans. I and K.

[28] JTF 7, Inch'on Rpt, an. E, p. 6; Mossman and Middleton, Logistical Problems and Their Solutions. The Navy's operation plan underestimated the size of the basin.

[29] JTF 7, Inch'on Rpt. I–C–1 E–6.

reached 31.2 feet at high and minus .5 feet at low water. Only on this day did the tide reach this extreme range. No other date after this would permit landing until 27 September when a high tide would reach 27 feet. On 11–13 October there would be a tide of 30 feet. Morning high tide on 15 September came at 0659, forty-five minutes after sunrise; evening high tide came at 1919, twenty-seven minutes after sunset. The Navy set 23 feet of tide as the critical point needed for landing craft to clear the mud flat and reach the landing sites.[30]

Another consideration was the sea walls that fronted the Inch'on landing sites. Built to turn back unusually high tides, they were 16 feet in height above the mud flats. They presented a scaling problem except at extreme high tide. Since the landing would be made somewhat short of extreme high tide in order to use the last hour or two of daylight, ladders would be needed. Some aluminum scaling ladders were made in Kobe and there were others of wood. Grappling hooks, lines, and cargo nets were readied for use in holding the boats against the sea wall.

The initial objective of the landing force was to gain a beachhead at Inch'on, a city of 250,000 population. The 3d Battalion, 5th Marines, was to land on Wolmi-do on the early morning high tide at 0630, 15 September (D-day, L-hour). With Wolmi-do in friendly hands, the main landing would be made that afternoon at the next high tide, about 1730 (D-day, H-hour), by the 1st and 5th Marines.

[30] 1st Mar Div SAR, Inch'on-Seoul, 18 Sep–7 Oct 50, p. 12, and G–3 Sec, an. C.

Three landing beaches were selected—Green Beach on Wolmi-do for the preliminary early morning battalion landing, and Red Beach in the sea wall dock area of Inch'on and Blue Beach in the mud flat semiopen area at the south edge of the city for the two-regimental-size force that would make the main landing in the evening. Later, 7th Infantry Division troops would land at Inch'on over what was called Yellow Beach.

The 5th Marines, less the 3d Battalion, was to land over Red Beach in the heart of Inch'on, north of the causeway which joined Wolmi-do with Inch'on, and drive rapidly inland 1,000 yards to seize Observatory Hill. On the left of the landing area was Cemetery Hill, 130 feet high, on which three dual-purpose guns reportedly were located. On the right, a group of buildings dominated the landing area. The 5th Marines considered Cemetery and Observatory Hills as the important ground to be secured in its zone.

Simultaneously with the 5th Marines' landing, the 1st Marines was to land over Blue Beach at the base of the Inch'on Peninsula just south of the city. This landing area had such extensive mud flats that heavy equipment could not be brought ashore over it. It lay just below the tidal basin of the inner harbor and an adjacent wide expanse of salt evaporators. Its principal advantage derived from the fact that the railroad and main highway to Seoul from Inch'on lay only a little more than a mile inland from it. A successful landing there could quickly cut these avenues of escape or access at the rear of Inch'on.[31]

[31] JTF 7, Inch'on Rpt, B–2 Opn Plan and an. B; 1st Mar Div SAR, vol. III, p. 4, and vol. I, p. 13.

An early objective of the 1st Marine Division after securing the beachhead was Kimpo Airfield, sixteen road miles northeast of Inch'on. Then would follow the crossing of the Han River and the drive on Seoul.

As diversions, the battleship *Missouri* was to shell east coast areas on the opposite side of the Korean peninsula, including the rail center and port of Samch'ok, and a small force was to make a feint at Kunsan on the west coast, 100 air miles south of Inch'on.

### *Intelligence Estimate*

General MacArthur's view at the end of August that the North Koreans had concentrated nearly all their combat resources against Eighth Army in the Pusan Perimeter coincided with the official G–2 estimate. On 28 August the X Corps G–2 Section estimated the enemy strength in Seoul as approximately 5,000 troops, in Inch'on as 1,000, and at Kimpo Airfield as 500, for a total of 6,500 soldiers in the Inch'on-Seoul area. On 4 September the estimate remained about the same except that the enemy force in the Inch'on landing area was placed at 1,800–2,500 troops because of an anticipated build-up there. This estimate remained relatively unchanged four days later, and thereafter held constant until the landing.[32]

American intelligence considered the enemy's ability to reinforce quickly the Inch'on-Seoul area as inconsequential. It held the view that only small rear area garrisons, line of communications units, and newly formed, poorly trained groups were scattered throughout Korea back of the combat zone around the Pusan Perimeter. Aerial reconnaissance reported heavy movement of enemy southbound traffic from the Manchurian border, but it was not clear whether this was of supplies or troops, or both. Although reports showed that the Chinese Communist Forces had increased in strength along the Manchurian border, there was no confirmation of rumors that some of them had moved into North Korea.[33]

The Far East Command considered the possibility that the enemy might reinforce the Inch'on-Seoul area from forces committed against Eighth Army in the south. If this were attempted, it appeared that the North Korean *3d*, *13th*, and *10th Divisions*, deployed on either side of the main Seoul-Taejon-Taegu highway, could most rapidly reach the Inch'on area.

North Korean air and naval elements were considered incapable of interfering with the landing. On 28 August the Far East Command estimated there were only nineteen obsolescent Soviet-manufactured aircraft available to the North Korean Air Force. The U.N. air elements, nevertheless, had orders to render unusable any known or suspected enemy air facilities, and particularly to give attention to new construction at Kimpo, Suwon, and Taejon. North Korean naval elements were almost nonexistent at this time. Five divisions of small patrol-type vessels comprised the North Korean Navy; one was on the west coast at Chinnamp'o, the others at Wonsan on the east coast. At both places

[32] Hq X Corps, Opn CHROMITE, p. 5; X Corps WD, G–2 Sec, Hist Rpt, 15 Aug–30 Sep 50, p. 1; 1st Mar Div SAR, vol. III, p. 5.

[33] X Corps WD, G–2 Sec, Hist Rpt, 15 Aug–30 Sep 50; JTF 7, Inch'on Rpt, II, E–2.

they were bottled up and rendered impotent. On the morning of 7 September a ROK patrol vessel (PC boat) north of Inch'on discovered and sank a small craft engaged in mine laying; thus it appeared that some mines were to be expected.[34]

As a final means of checking on conditions in Inch'on harbor, the Navy on 31 August sent Lt. Eugene F. Clark to Yonghung-do, an island at the mouth of the ship channel ten sea miles from Inch'on. There, Clark used friendly natives to gather the information needed. He sent them on several trips to Inch'on to measure water depths, check on the mud flats, and to observe enemy strength and fortifications. He transmitted their reports by radio to friendly vessels in Korean waters. Clark was still in the outer harbor when the invasion fleet entered it.[35]

### *The Ships Load Out*

At the end of August the ports of Kobe, Sasebo, and Yokohama in Japan and Pusan in Korea had become centers of intense activity as preparations for mounting the invasion force entered the final stage. The 1st Marine Division, less the 5th Marines, was to outload at Kobe, the 5th Marines at Pusan, and the 7th Infantry Division at Yokohama. Most of the escorting vessels, the Gunfire Support Group, and the command ships assembled at Sasebo.

The ships to carry the troops, equipment, and supplies began arriving at the predesignated loading points during the last days of August. In order to reach Inch'on by morning of 15 September, the LST's had to leave Kobe on 10 September and the transports (AP's) and cargo ships (AK's) on 12 September. Only the assault elements were combat-loaded. Japanese crews manned thirty-seven of the forty-seven LST's in the Marine convoy.[36]

The loading of the 1st Marine Division at Kobe was in full swing on 2 September when word came that the next morning a typhoon would strike the port, where more than fifty vessels were assembled. All unloading and loading stopped for thirty-six hours. At 0600 on 3 September, Typhoon *Jane* screeched in from the east. Wind velocity reached 110 miles an hour at noon. Waves forty feet high crashed against the waterfront and breakers rolled two feet high across the piers where loose cargo lay. Seven American ships broke their lines and one of the giant 200-ton cranes broke loose. Steel lines two and a half inches thick snapped. Only by exhausting and dangerous work did port troops and the marines fight off disaster. By 1530 in the afternoon the typhoon began to blow out to sea. An hour later relative calm descended on the port and the cleanup work began. A few vessels had to go into drydock for repairs, some vehicles were flooded out, and a large quantity of clothing had to be cleaned, dried, and repackaged.[37]

---

[34] JTF 7, Inch'on Rpt, an. E; Schnabel, FEC, GHQ Support and Participation in the Korean War, ch. V, pp. 36–37; Karig, *et al., Battle Report, The War in Korea,* p. 195.

[35] Karig, *et al., Battle Report, The War in Korea,* pp. 176–91, relates the Clark mission in detail.

[36] 1st Mar Div SAR, 15 Sep–7 Oct 50, an. D, p. 4; 7th Inf Div WD, Sep 50.

[37] JTF 7, Inch'on Rpt, I–D–3; 1st Mar Div SAR, 15 Sep–7 Oct 50, an. D, p. 6; SFC William J. K. Griffen, "Typhoon at Kobe," *Marine Corps Gazette* (September, 1951); "Operation Load-up," *The Quartermaster Review* (November–December, 1950), p. 40.

Despite the delay and damage caused by *Jane,* the port of Kobe and the 1st Marine Division met the deadline of outloading by 11 September. On the 10th and the 11th, sixty-six cargo vessels cleared Kobe for Inch'on. They sailed just ahead of another approaching typhoon. This second typhoon had been under observation by long-range reconnaissance planes since 7 September. Named *Kezia,* it was plotted moving from the southwest at a speed that would put it over the Korean Straits on 12–13 September.

On the 11th, the 1st Marine Division sailed from Kobe and the 7th Infantry Division from Yokohama. The next day the 5th Marines departed Pusan to rendezvous at sea. The flagship *Rochester* with Admiral Struble aboard got under way from Sasebo for Inch'on at 1530, 12 September. That afternoon a party of dignitaries, including Generals MacArthur, Almond, Wright, Maj. Gen. Alonzo P. Fox, Maj. Gen. Courtney Whitney, and General Shepherd of the Marine Corps, flew from Tokyo to Itazuke Air Base and proceeded from there by automobile to Sasebo, arriving at 2120. Originally, the MacArthur party had planned to fly from Tokyo on the 13th and embark on the *Mt. McKinley* at Kokura that evening. But Typhoon *Kezia's* sudden change of direction caused the revision of plans to assure that the party would be embarked in time. The *Mt. McKinley,* sailing from Kobe with Admiral Doyle and General Smith aboard, had not yet arrived at Sasebo when MacArthur's party drove up. It finally pulled in at midnight, and departed for the invasion area half an hour later after taking MacArthur's party aboard.[38]

Part of the invasion fleet encountered very rough seas off the southern tip of Kyushu early on 13 September. Winds reached sixty miles an hour and green water broke over ships' bows. In some cases, equipment shifted in the holds, and in other instances deck-loaded equipment was damaged. During the day the course of *Kezia* shifted to the northeast and by afternoon the seas traversed by the invasion fleet began to calm. The aircraft carrier *Boxer,* steaming at forced speed from the California coast with 110 planes aboard, fought the typhoon all night in approaching Japan. At dusk on the 14th, it quickly departed Sasebo and at full speed cut through the seas for Inch'on.[39]

### *Preliminary Bombardment*

Air attacks intended to isolate the invasion area began on 4 September and continued until the landing. On the 10th, Marine air elements struck Wolmi-do in a series of napalm attacks. Altogether, sixty-five sorties hit Inch'on during the day.[40]

The main task of neutralizing enemy batteries on Wolmi-do guarding the Inch'on inner harbor was the mission of Rear Adm. J. M. Higgins' Gunfire Support Group. This group, composed of 2

---

[38] JTF 7, Inch'on Rpt, II, 1; 1st Mar Div SAR, vol. I, pp. 15–18; Diary of CG X Corps, Opn CHROMITE, 12 Sep 50; Ltr, Wright to author, 22 Mar 54; Barr, Notes, 11 Sep 50.

[39] Diary of CG X Corps, Opn CHROMITE, 13 Sep 50; Karig, *et al., Battle Report, The War in Korea,* p. 197.

[40] GHQ FEC, G–3 Opn Rpt 79, 11 Sep 50; Ernest H. Giusti, "Marine Air Over Inchon-Seoul," *Marine Corps Gazette* (June, 1952), p. 19.

United States heavy cruisers, 2 British light cruisers, and 6 U.S. destroyers, entered the approaches to Inch'on harbor at 1010, 13 September. Just before noon the group in Flying Fish Channel sighted an enemy mine field, exposed at low water. It destroyed some of the mines with automatic fire. At 1220, the 4 cruisers anchored from seven to ten miles offshore, while 5 destroyers—the *Mansfield, DeHaven, Swenson, Collett,* and *Gurke*—proceeded on to anchorages close to Wolmi-do under cover of air strikes by planes from Fast Carrier Task Force 77. The destroyers began the bombardment of Wolmi-do at 1230.[41]

Five enemy heavily revetted 75-mm. guns returned the fire. In the intense ship-shore duel, the *Collett* received nine hits and sustained considerable damage. Enemy shells hit the *Gurke* three times, but caused no serious damage. The *Swenson* took a near miss which caused two casualties: one was Lt. (jg.) David H. Swenson, the only American killed during the bombardment. The destroyers withdrew at 1347.

At 1352 the cruisers, anchored out of range of the Wolmi-do batteries, began an hour and a half bombardment. Planes of Task Force 77 then came in for a heavy strike against the island. After the air strike terminated, the cruisers resumed their bombardment at 1610 for another half hour. Then at 1645 the Gunfire Support Group got under way and withdrew back down the channel.[42]

[41] JTF 7, Inch'on Rpt, I-E-1, Recon in Force; EUSAK WD, 24 Oct 50, G-2 Sec, ADVATIS 1225, Interrog of Sr Lt Cho Chun Hyon.

[42] JTF 7, Inch'on Rpt, I-E-1, and II-1; GHQ FEC Sitrep, 14 Sep 50.

The next day, D minus 1, the Gunfire Support Group returned. Just before 1100, planes of Task Force 77 again delivered heavy strikes against the island. The heavy cruisers began their second bombardment at 1116, this time also taking under fire targets within Inch'on proper. The destroyers waited about an hour and then moved to their anchorages off Wolmi-do. The cruisers ceased firing while another air strike came in on the island. After it ended, the five destroyers began their bombardment at 1255 and in an hour and fifteen minutes fired 1,732 5-inch shells into Wolmi-do and Inch'on. When they left there was no return fire—the Wolmi-do batteries were silent.[43]

### *Securing the Inch'on Beachhead*

The X Corps expeditionary troops arriving off Inch'on on 15 September numbered nearly 70,000 men.[44] At 0200

[43] JTF 7, Inch'on Rpt, I-E-2; Karig, *et al., Battle Report, The War in Korea,* p. 210.

[44] Hq X Corps, Opn CHROMITE, G-3 Sec Hist Rpt, (gives strength of X Corps as 69,450); 1st Mar Div SAR, vol. I, an. A, 5.

The major units were the 1st Marine Division, the 7th Infantry Division, the 92d and 96th Field Artillery Battalions (both 155-mm. howitzers), the 50th Antiaircraft Artillery (Automatic Weapons) Battalion (SP), the 56th Amphibious Tank and Tractor Battalion, the 19th Engineer Combat Group, and the 2d Engineer Special Brigade. The 1st Marine Division on invasion day had a strength of 25,040 men—19,494 organic to the Marine Corps and the Navy, 2,760 Army troops attached, and 2,786 Korean marines attached. Later, after the 7th Marines arrived, the organic Marine strength increased about 4,000 men. On invasion day the GHQ UNC reserve consisted of the 3d Infantry Division and the 187th Airborne Regimental Combat Team (composed of troops from the 11th Airborne Division). The ROK 17th Regiment was in the act of moving from Eighth Army to join X Corps.

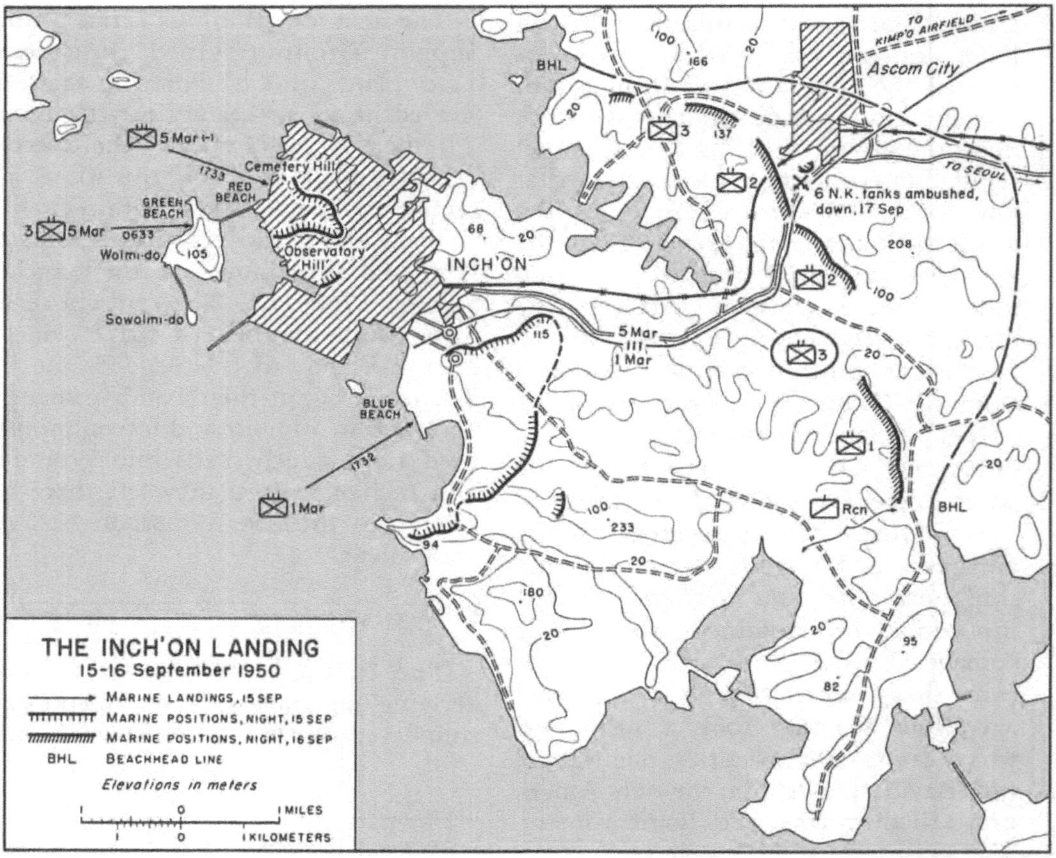


MAP 16

the Advance Attack Group, including the Gunfire Support Group, the rocket ships (LSMR's) and the Battalion Landing Team, began the approach to Inch'on. A special radar-equipped task force, consisting of three high speed transports (APD's) and one Landing Ship Dock (LSD), carried the Battalion Landing Team—Lt. Col. Robert D. Taplett's 3d Battalion, 5th Marines, and a platoon of nine M26 Pershing tanks from A Company, 1st Tank Battalion—toward the transport area off Wolmi-do. Dawn of invasion day came with a high overcast sky and portent of rain.[45]

Wolmi-do, or Moon Tip Island, as it might be translated, is a circular hill (Hill 105) about 1,000 yards across and rising 335 feet above the water. A rocky hill, it was known to be honeycombed with caves, trenches, gun positions, and dugouts. (*Map 16*)

The first action came at 0500. Eight

[45] JTF 7, Inch'on Rpt, I-L-1 and I-F-1; 1st Mar Div SAR, vol. I, p. 13.

WOLMI-DO

Marine Corsairs left their escort carrier for a strike on Wolmi-do. The first two planes caught an armored car crossing the causeway from Inch'on and destroyed it. There was no other sign of life visible on the island as the flight bombed the ridge line. At 0530 the Special Task Force was in its designated position ready to land the assault troops. Twenty minutes later, Taplett's 3d Battalion began loading into 17 landing craft (LCVP's); the 9 tanks loaded into 3 landing ships (LSV's). L-hour was fifty minutes away.

Air strikes and naval gunfire raked Wolmi-do and, after this, three rocket ships moved in close and put down an intense rocket barrage. The landing craft straightened out into lines from their circles and moved toward the line of departure. Just as a voice announced over the ship's loud speaker, "Landing force crossing line of departure," MacArthur came on the bridge of the *Mt. McKinley*. It was 0625. The first major amphibious assault by American troops against an enemy since Easter Sunday, 1 April 1945, at Okinawa was under way. About one mile of water lay between the line of departure and the Wolmi-do beach.[46]

The 3d Battalion moved toward Wolmi-do with G and H Companies in assault and I Company in reserve. Even after the American rocket barrage lifted there was still no enemy fire. The first wave of troops reached the bathing beach on the northern arm of the island unopposed at 0633.

The first troops ashore moved rapidly inland against almost no resistance. Within a few minutes the second wave landed. Then came the LSV's carrying

[46] X Corps WD, Opn CHROMITE, 15 Sep 50; 1st Mar Div SAR, vol. III, an. P, p. 4; Lynn Montross, "The Inchon Landing," *Marine Corps Gazette* (July, 1951), pp. 26ff; Geer, *The New Breed*, pp. 213-25.

the tanks, three of which carried dozer blades for breaking up barbed wire, filling trenches, and sealing caves; three other tanks mounted flame throwers. One group of marines raised the American flag on the high ground of Wolmi-do half an hour after landing. Another force crossed the island and sealed off the causeway leading to Inch'on. The reduction of the island continued systematically and it was secured at 0750.[47]

A little later in the morning, Colonel Taplett sent a squad of marines and three tanks over the causeway to Sowolmi-do where they destroyed an estimated platoon of enemy troops; some surrendered, others swam into the sea, and still others were killed. Taplett's battalion assumed defensive positions and prepared to cover the main Inch'on landing later in the day.

In the capture of Wolmi-do and Sowolmi-do the Battalion Landing Team killed 108 enemy soldiers and captured 136. About 100 more in several caves refused to surrender and were sealed by tank dozers into their caves. Marine casualties were light—seventeen wounded.[48]

The preinvasion intelligence on Wolmi-do proved to be essentially correct. Prisoners indicated that about 400 North Korean soldiers, elements of the *3d Battalion, 226th Independent Marine Regiment,* and some artillery troops of the *918th Artillery Regiment* had defended Wolmi-do.

After the easy capture of Wolmi-do came the anxious period when the tide began to fall, causing further activity to cease until late in the afternoon. The enemy by now was fully alerted. Marine and naval air ranged up and down the roads and over the countryside isolating the port to a depth of twenty-five miles, despite a rain which began to fall in the late afternoon. Naval gunfire covered the closer approaches to Inch'on.

Assault troops of the 5th and 1st Marines began going over the sides of their transports and into the landing craft at 1530. After a naval bombardment, rocket ships moved in close to Red and Blue Beaches and fired 2,000 rockets on the landing areas. Landing craft crossed lines of departure at 1645, and forty-five minutes later neared the beaches. The first wave of the 5th Marines breasted the sea wall on Red Beach at 1733. Most of the A Company men in the fourteen boats of the first three waves climbed over the sea wall with scaling ladders; a few boats put their troops ashore through holes in the wall made by the naval bombardment.[49]

On the left flank of the landing area, the 3d Platoon of A Company encountered enemy troops in trenches and a bunker just beyond the sea wall. There in an intense fight the marines lost eight men killed and twenty-eight wounded. Twenty-two minutes after landing, the company fired a flare signaling that it held Cemetery Hill. On top of Cemetery Hill, North Koreans threw down their arms and surrendered to the 2d Pla-

[47] 3d Bn, 5th Mar, Special Act Rpt, an. P to 5th Mar Special Rpt, in 1st Mar Div SAR; JTF 7, Inch'on Rpt, I-F-1, 15 Sep 50.

[48] 1st Mar Div SAR, vol. I, an. B, App. 1, 1; JTF 7, Inch'on Rpt, I-H-2; X Corps WD, G-2 Sec, Hist Rpt, Intel Estimate 8.

[49] 1st Bn, 5th Mar SAR, p. 3; 1st Mar Div SAR, vol. III; Geer, *The New Breed,* pp. 24-25. Montross and Canzona, *The Inchon-Seoul Operation,* covers the 1st Marine Division part of the Inch'on operation in detail. Much of this fine work is based on extensive interviews with participants.

LANDING CRAFT AND BULLDOZERS *stuck in the Wolmi-do mud after the tide fell.*

toon. Other elements of the battalion by midnight had fought their way against sporadic resistance to the top of Observatory Hill.

The 2d Battalion, 5th Marines, landing on the right side of Red Beach, encountered only spotty resistance and at a cost of only a few casualties gained its objective.

Assault elements of the 1st Marines began landing over Blue Beach at 1732, one minute ahead of the 5th Marines at Red Beach. Most of the men were forced to climb a high sea wall to gain exit from the landing area. One group went astray in the smoke and landed on the sea wall enclosing the salt flats on the left of the beach. The principal obstacle the 1st Marines encountered was the blackness of the night. Lt. Col. Allan Sutter's 2d Battalion lost one man killed and nineteen wounded in advancing to the Inch'on-Seoul highway, one mile inland. The landing force had taken its final D-day objectives by 0130, 16 September.[50]

Following the assault troops, eight specially loaded LST's landed at Red Beach just before high tide, and unloading of equipment to support the forces ashore the next day continued throughout the night. Beaching of the LST's brought tragedy. Just after 1830, after receiving some enemy mortar and machine gun fire, gun crews on three of the LST's began firing wildly with 20-mm. and 40-mm. cannon, and, before they could be stopped, had killed 1 and wounded 23 men of the 2d Battalion, 5th Marines. The Marine landing force casualties on D-day were 20 men killed, 1 missing in action, and 174 wounded.[51]

The U.N. preinvasion estimate of en-

---

[50] 1st Mar Div SAR, Vol. I, an. P, p. 6, and an. C, p. 6; X Corps WD, Opn CHROMITE, 15 Sep 50; Diary of CG X Corps, 15 Sep 50.

[51] Montross and Canzona, *The Inchon-Seoul Operation*, pp. 110–11; Geer, *The New Breed*, p. 128.

MARINES MOPPING UP *on Wolmi-do.*

emy strength at Inch'on was accurate. Prisoners disclosed that about 2,000 men had comprised the Inch'on garrison. Some units of the N.K. *22d Regiment* moved to Inch'on to reinforce the garrison before dawn of the 15th, but they retreated to Seoul after the main landing that evening. To the rank and file of the North Korean soldiers in Seoul the landing came as a surprise.[52]

On the morning of 16 September the two regiments ashore established contact with each other by 0730. Thereafter a solid line existed around Inch'on and escape for any enemy still within the city became unlikely. The ROK Marines now took over mop-up work in Inch'on and went at it with such a will that hardly anyone in the port city, friend or foe, was safe.[53]

Early in the morning of the 16th, Marine aircraft took off from the carriers to aid the advance. One flight of eight Corsairs left the *Sicily* at 0548. Soon it sighted six enemy T34 tanks on the

[52] ATIS Supp Enemy Documents, Issue 2, pp. 114–16, Opn Ord 8–10 Sep 50, CO *226th Unit*, captured 16 Sep 50; ATIS Interrog Rpt (N.K.) Issue 10, p. 7; *Ibid.*, Issue 8, Rpt 1345, Lt Il Chun Son, and Rpt 1346, Lt Lee San Kak; X Corps WD, Opn CHROMITE, p. 26, interrog of Capt Chan Chul, and p. 30, interrog of Lt Col Kim Yonh Mo.

[53] Diary of CG X Corps, 16 Sep 50; Montross, "The Inchon Landing," *op. cit.*

Seoul highway three miles east of Inch'on moving toward the latter place. Ordered to strike at once, the Corsairs hit the tanks with napalm and 500-pound bombs, damaging three of them and scattering the accompanying infantry. The enemy returned the fire, hitting one of the Corsairs. Capt. William F. Simpson's plane crashed and exploded near the burning armor, killing him. A second flight of eight Corsairs continued the attack on the tanks with napalm and bombs and, reportedly, destroyed them all. Later in the morning, however, when the advance platoon of the 1st Marines and accompanying tanks approached the site, three of the T34's began to move, whereupon the Pershings engaged and destroyed them.[54]

Both Marine regiments on the second day advanced rapidly against light resistance and by evening had reached the Beachhead Line, six miles from the landing area. Their casualties for the day were four killed and twenty-one wounded.

Thus, within twenty-four hours of the main landing, the 1st Marine Division had secured the high ground east of Inch'on, occupied an area sufficient to prevent enemy artillery fire on the landing and unloading area, and obtained a base from which to mount the attack to seize Kimpo Airfield. In the evening of 16 September General Smith established his command post east of Inch'on and from there at 1800 notified Admiral Doyle that he was assuming responsibility for operations ashore.[55]

### *Capture of Kimpo Airfield and Advance to the Han River*

During the advance thus far the boundary between the 5th and 1st Marines had followed generally the main Inch'on-Seoul highway, which ran east-west, with the 5th Marines on the north and the 1st Marines astride and on its south side. Just beyond the beachhead line the boundary left the highway and slanted northeast. This turned Colonel Murray's 5th Marines toward Kimpo Airfield, seven miles away, and the Han River just beyond it. Col. Lewis B. Puller's 1st Marines, astride the Inch'on-Seoul highway, headed toward Yongdungp'o, the large industrial suburb of Seoul on the south bank of the Han, ten air miles away.

During the night of 16–17 September, the 2d Battalion, 5th Marines, occupied a forward defensive position commanding the Seoul highway just west of Ascom City. Behind it the 1st Battalion held a high hill. From a forward roadblock position, members of an advanced platoon of D Company, at 0545 on the 17th, saw the dim outlines of six tanks on the road eastward. Infantry accompanied the tanks, some riding on the armor.

The enemy armored force moved past the hidden outpost of D Company. At 0600, at a range of seventy-five yards, rockets fired from a bazooka set one of

---

[54] X Corps WD, G–3 Sec, Msgs J–2, 4, 6, 7 from 160705 to 160825, Sep 50; 1st Mar Div SAR, vol. II, an. OO, p. 15; Geer, *The New Breed*, p. 128; ATIS Enemy Documents, Issue 10, Rpt 1529, Hang Yong Sun, and Rpt 1534, Lt Lee Song Yol.

[55] 1st Mar Div SAR, vol. I, p. 22, and an. C, p. 7; Smith, MS review comments, 25 Feb 54.

DESTROYED ENEMY TANKS *on the road to Seoul.*

the tanks on fire. Pershing tanks now opened fire on the T34's. The recoilless rifles joined in. Within five minutes combined fire destroyed all six enemy tanks and killed 200 of an estimated 250 enemy infantry. Only one man in the 2d Battalion was wounded.[56]

Early that morning, General MacArthur, accompanied by Admiral Struble, and Generals Almond, Wright, Fox, Whitney, and others came ashore and proceeded to General Smith's command post, and from there went on to the position of the 2d Battalion, 5th Marines, where they saw the numerous enemy dead and the still-burning T34 tanks. On the way they had passed the six tanks destroyed the morning before. The sight of twelve destroyed enemy tanks seemed to them a good omen for the future.[57]

The 5th Marines advanced rapidly on the 17th and by 1800 its 2d Battalion was at the edge of Kimpo Airfield. In the next two hours the battalion seized

[56] 5th Mar SAR, pp. 7–8, in 1st Mar Div SAR, vol. III; 1st Mar Div SAR, vol. II, an. OO, p. 16; Montross and Canzona, *The Inchon-Seoul Operation*, pp. 147–51.

[57] Ltr, Wright to author, 22 Mar 54; Diary of CG X Corps, 17 Sep 50.

the southern part of the airfield. The 400–500 enemy soldiers who ineffectively defended it appeared surprised and had not even mined the runway. During the night several small enemy counterattacks hit the perimeter positions at the airfield between 0200 and dawn, 18 September. The marines repulsed these company-sized counterattacks, inflicting heavy casualties on the enemy troops, who finally fled to the northwest E Company and supporting tanks played the leading role in these actions. Kimpo was secured during the morning of 18 September.[58]

The capture on the fourth day of the 6,000-foot-long, 150-foot-wide, hard surfaced Kimpo runway, with a weight capacity of 120,000 pounds, gave the U.N. Command one of its major objectives. It broadened greatly the capability of employing air power in the ensuing phases of the attack on Seoul; and, more important still, it provided the base for air operations seeking to disrupt supply of the North Korean Army.

On the 18th, the 2d Battalion, 5th Marines, sent units on to the Han River beyond the airfield, and the 1st Battalion captured Hill 99 northeast of it and then advanced to the river. At 1409 in the afternoon a Marine Corsair landed at Kimpo and, later in the day, advance elements of Marine Air Group 33 flew in from Japan. The next day more planes came in from Japan, including C–54 cargo planes, and on 20 September land-based Corsairs made the first strikes from Kimpo.[59]

Continuing its sweep along the river, the 1st Battalion, 5th Marines, on the 19th swung right and captured the last high ground (Hills 118, 80, and 85) a mile west of Yongdungp'o. At the same time, the 2d Battalion seized the high ground along the Han River in its sector. At nightfall, 19 September, the 5th Marines held the south bank of the Han River everywhere in its zone and was preparing for a crossing the next morning. (*Map VII*)

Meanwhile, the 2d Engineer Special Brigade relieved the ROK Marines of responsibility for the security of Inch'on, and the ROK's moved up on the 18th and 19th to the Han River near Kimpo. Part of the ROK's Marines extended the left flank of the 5th Marines, and its 2d Battalion joined them for the projected crossing of the Han River the next day.[60]

In this action, the 1st Marines had attacked east toward Yongdungp'o astride the Seoul highway. Its armored spearheads destroyed four enemy tanks early on the morning of the 17th. Then, from positions on high ground (Hills 208, 107, 178), three miles short of Sosa, a village halfway between Inch'on and Yongdungp'o, a regiment of the N.K. *18th Division* checked the advance. At nightfall the Marine regiment dug in for the night a mile from Sosa. At Ascom

[58] 5th Mar SAR, 17 Sep 50, pp. 7–8, in the 1st Mar Div SAR, vol. III; 1st Mar Div SAR, vol. I, p. 18; X Corps WD, G–3 Sec, 18 Sep 50; New York *Herald Tribune*, September 18, 1950, Bigart dispatch.

[59] 1st Mar Div SAR, G–3 an. C, vol. I, p. 10; JTF 7, Inch'on Rpt, 18 Sep 50; USAF Hist Study 71, p. 66; New York *Herald Tribune*, September 19, 1950, Bigart dispatch.

[60] 1st Mar Div SAR, vol. I, G–3 Sec, an. C, pp. 10–13, 18–19 Sep 50; Geer, *The New Breed*, pp. 133–34.

City, just west of Sosa, American troops found 2,000 tons of ammunition for American artillery, mortars, and machine guns, captured there by the North Koreans in June, all still in good condition.[61]

Not all the action that day was on and over land. Just after daylight, at 0550, two enemy YAK planes made bombing runs on the *Rochester* lying in Inch'on harbor. The first drop of four 100-pound bombs missed astern, except for one which ricocheted off the airplane crane without exploding. The second drop missed close to the port bow, causing minor damage to electrical equipment. One of the YAK's strafed H.M.S. *Jamaica,* which shot down the plane but suffered three casualties.[62]

Ashore, the 1st Marines resumed the attack on the morning of the 18th and passed through and around the burning town of Sosa at midmorning. By noon the 3d Battalion had seized Hill 123, a mile east of the town and north of the highway. Enemy artillery fire there caused many casualties in the afternoon, but neither ground nor aerial observers could locate the enemy pieces firing from the southeast. Beyond Sosa the North Koreans had heavily mined the highway and on 19 September the tank spearheads stopped after mines damaged two tanks. Engineers began the slow job of removing the mines and, without tank support, the infantry advance slowed. But at nightfall advanced elements of the regiment had reached Kal-ch'on Creek just west of Yongdungp'o.[63]

Other elements of the X Corps had by now arrived to join in the battle for Seoul. Vessels carrying the 7th Infantry Division arrived in Inch'on harbor on the 16th. General Almond was anxious to get the 7th Division into position to block a possible enemy movement from the south of Seoul, and he arranged with Admiral Doyle to hasten its unloading. The 2d Battalion of the 32d Regiment landed during the morning of the 18th; the rest of the regiment landed later in the day. On the morning of 19 September, the 2d Battalion, 32d Infantry, moved up to relieve the 2d Battalion, 1st Marines, in its position on the right flank south of the Seoul highway. It completed the relief without incident by noon. The total effective strength of the 32d Infantry when it went into the line was 5,114 men—3,241 Americans and 1,873 ROK's. Responsibility for the zone south of the highway passed to the 7th Division at 1800, 19 September. During the day, the 31st Regiment of the 7th Division came ashore at Inch'on.[64]

The Navy had supported the ground action thus far with effective naval gunfire. The *Rochester* and *Toledo* had

[61] 1st Mar Div SAR, vol. I, G-3 Sec, an. C, p. 8 and an. B, app. 2, p. 1; New York *Herald Tribune,* September 18, 1950, Bigart dispatch; CINCFE, Sitrep, 250600-260600 Sep 50.

[62] JTF 7, Inch'on Rpt, I-F-2 and IV-1.

[63] 1st Mar Div SAR, vol. I, G-3 Sec, an. C, pp. 10-13, 18-19 Sep 50; Geer, *The New Breed,* p. 136; Montross and Canzona, *The Inchon-Seoul Operation,* p. 178.

[64] 32d Inf WD, 16-19 Sep 50; Diary of CG X Corps, 18-19 Sep 50; JTF 7, Inch'on Rpt, I-F-2; 1st Mar Div SAR, G-3 Sec, an. C, p. 13, 19 Sep 50; Almond, MS review comments for author, 23 Oct 53; 7th Inf Div WD, 16-19 Sep 50; 31st Inf WD, 19 Sep 50.

KIMPO RUNWAY *after capture by the 5th Marines.*

been firing at ranges up to 30,000 yards in support of the marines and the ROK's on their left flank. Now, on the 19th, the *Missouri* arrived in Inch'on harbor from the east coast of Korea and began delivering naval gunfire support to the 7th Division on the right flank. Despite difficult tide conditions and other restrictive factors in Inch'on harbor, the Navy by the evening of 18 September had unloaded 25,606 persons, 4,547 vehicles, and 14,166 tons of cargo.[65]

The battle for Seoul lay ahead. Mounting indications were that it would be far more severe than had been the action at Inch'on and the advance to the Han. Every day enemy resistance had increased on the road to Yongdungp'o. Aerial observers and fighter pilots reported large bodies of troops moving toward Seoul from the north. The N.K. *18th Division*, on the point of moving from Seoul to the Naktong front when the landing came at Inch'on, was instead ordered to retake Inch'on, and its advanced elements had engaged the 1st Marines in the vicinity of Sosa. On the 17th, enemy engineer units began mining the approaches to the Han River near Seoul. About the same time, the N.K. *70th Regiment* moved from Suwon to join in the battle. As they prepared to cross the Han, the marines estimated that there might be as many as 20,000 enemy troops in Seoul to defend the city. The X Corps intelligence estimate on 19 September, however, undoubtedly expressed the opinion prevailing among American commanders—that the enemy was "capable of offering stubborn resistance in Seoul but unless substantially reinforced, he is not considered

[65] JTF 7, Inch'on Rpt, 18 Sep 50, and II–6.

TOP-LEVEL BRIEFING, *1st Marine Division Headquarters, Inch'on, 17 September. Seated in front row, from left: Generals MacArthur and Smith, Admiral Struble, and Generals Whitney and Wright.*

capable of making a successful defense." [66]

[66] JTF 7, Inch'on Rpt, I–F–2 and an. B, app. 2, p. 2, 18 Sep 50; ATIS Interrog Rpts (N.K.) Issue 8, Rpt 1300, p. 1, Hon Gun Mun, p. 40, Kim So Sung; Rpt 1336, p. 45, Kim Won Yong; Rpt 1365, p. 90, Kan Chun Kil; Rpt 1369, p. 96, Maj Chu Yong Bok; *Ibid.*, Issue 9, p. 29, Kim Te Jon; X Corps PIR 1, 19 Sep 50; Giusti, "Marine Air Over Inchon-Seoul," *op. cit.*, p. 19.

Not until their 18 September communiqué did the North Koreans mention publicly anything connected with the Inch'on landing and then they merely stated that detachments of the coastal defense had brought down two American fighter planes.[67]

[67] New York *Times*, September 19, 1950.

## CHAPTER XXVI

# The Capture of Seoul

By the grace of a merciful Providence our forces fighting under the standard of that greatest hope and inspiration of mankind, the United Nations, have liberated this ancient city of Seoul.

DOUGLAS MACARTHUR (at Seoul, 29 September 1950)

With the crossing of the Han River projected for 20 September, the drive on Seoul was about to begin. During the planning stages for the Inch'on landing, MacArthur had prophesied to Almond, "You will be in Seoul in five days." Almond replied, "I cannot do that but I will have the city in two weeks." [1]

The plan for crossing the Han River called for Colonel Murray's 5th Marines to cross at the ferry site three miles northeast of Kimpo Airfield and eight miles west of Seoul. (*See Map VII.*) A swimming party of fourteen men, mostly from the Reconnaissance Company, stepped into the Han at 2000 on the evening of the 19th, crossed safely to the north side, and found that the crossing site was suitable for LVT's. A 5-man patrol then continued up the slope of Hill 125 but turned back short of the crest. The swimming party gave the signal for the rest of the company to cross. When eight of nine amphibious tractors carrying the Reconnaissance Company were in the water enemy mortar and machine gun fire suddenly struck among them. The tractors turned around and made for the south bank. An hour later the swimming party arrived there with three wounded and one missing. Its plans disrupted, the 5th Marines now began preparing for an assault crossing of the Han after daylight.[2]

After a heavy artillery preparation against Hill 125, I Company, 5th Marines, began the assault crossing at 0645, 20 September. Enemy fire from automatic weapons and small arms on Hill 125 caused rather heavy casualties in I Company but it secured the hill by 0940. Other elements of the 3d Battalion, still riding LVT's, encountered little or no resistance and proceeded a mile inland to cut the Seoul-Kaesong railroad and a road at the village of Nung-dong by 0830. Still riding in LVT's they now turned right and moved southeast along the railroad track toward Seoul.[3]

---

[1] Interv, author with Almond, 13 Dec 51.

[2] 5th Mar SAR in 1st Mar Div SAR, vol. III, p. 9, 19 Sep 50; Diary of CG X Corps, 19 Sep 50; Montross and Canzona, *The Inchon-Seoul Operation*, pp. 191–93.

[3] 1st Mar Div SAR, vol. I, an. B, app. 2, pp. 5, 19; *Ibid.*, an. C, G–3 Sec, p. 14, 20 Sep 50; *Ibid.*, vol. III, p. 10.

The 2d Battalion followed the 3d Battalion across the river at 1000, passed through it, and continued the advance. By nightfall the 5th Marines with twelve tanks, and the 2d Battalion, ROK Marines, were across the river. Engineers had begun constructing a ponton ferry at the crossing site.

On the morning of 21 September, the 5th Marines, after repulsing an enemy company-sized counterattack, advanced southeast astride the rail and road lines paralleling the Han River. Resistance, at first light, steadily increased. The 3d Battalion captured and turned over to Korean marines Hill 104, north of the rail line and five and a half miles from the crossing site, and then turned northeast toward Hill 296 at the northwest edge of Seoul. In the meantime, the 1st Battalion attacked and captured a series of lower hills south of the rail and road lines. That evening the 5th Marines faced a line of hills running generally north-south along the western edge of Seoul. At the southern end of the line, near the village of Sogang, the 1st Battalion was within three miles of the main (Yongsan) railroad station in Seoul.[4]

Although unaware of the fact as they rested on their arms that night at the gate to the city, the men of the 5th Marines were to be held at this line of hills for four days of bloody battle. There the North Koreans had chosen to fight their battle for the western approaches to the city.

After receiving word of the rapid advance of the 5th Marines toward Seoul, General MacArthur returned to Tokyo on 21 September.[5]

### *The Capture of Yongdungp'o*

Advancing on the right (south) of the 5th Marines, the 1st Marines gradually approached Yongdungp'o. Relieved by the 32d Infantry Regiment, 7th Infantry Division, in the early afternoon of 19 September, the 1st Battalion, 1st Marines, was ready to shift northeast to hill positions captured during the day by the 5th Marines at the west edge of Yongdungp'o. Because its transportation was late in arriving, darkness had fallen before the 1st Battalion reached its detrucking point. Company A climbed to the summit of Hill 118 to relieve the occupying force. Later C Company joined it there. Meanwhile, elements of the 1st Battalion, 5th Marines, had departed at 2100 from Hills 80 and 85 nearby, because there was a deadline set for their departure that would enable them to march the six to eight miles to the 5th Marines' Han River crossing site.

After having lost these hills that afternoon to the 5th Marines, the North Koreans counterattacked just before daylight. Their attack groups left Yongdungp'o, crossed the rice paddies and Kal-ch'on Creek, and arrived on Hills 80 and 85 to find them undefended. Part of the enemy force continued on to Hill 118 where A and C Companies of the 1st Marines repulsed them. The unfortunate loss of Hills 80 and 85 at

---

[4] *Ibid.*, vol. I, an. C, G–3 Sec, 21 Sep 50, p. 15; ATIS Interrog Rpts, Issue 8, p. 108, Rpt 1376, Lt Chai Chan Ya; Geer, *The New Breed*, p. 155.

[5] Diary of CG X Corps, 21 Sep 50.

MARINES ON HILL 125 *under heavy enemy fire.*

the edge of Yongdungp'o that night made it necessary for the 1st Battalion, 1st Marines, to assault them again in the morning. It recaptured Hill 85 only after intense close combat in which there were many Marine casualties.[6]

Simultaneously with the North Korean predawn attack against Hills 80, 85, and 118, a battalion-size force led by five T34 tanks left Yongdungp'o on the Inch'on highway to counterattack the 1st Marines. It ran headlong into the Marine position before daylight and in a flaming battle in the darkness the North Koreans were all but annihilated. In this fight, Pfc. Gonegan boldly approached the T34 tanks. He knocked out two of them with his 3.5-inch bazooka and was in the act of firing on a third when he was killed. Dawn disclosed 300 dead North Korean soldiers on the road, in the ditches, and strewn about on the adjacent slopes.

By 0945, the main body of the 1st Marines reached the high ground overlooking Yongdungp'o from the west side of Kal-ch'on Creek, a sizable stream flowing north past the west edge of the city into the Han. General Almond arrived there at 1000 and in a conference with Colonel Puller, the Marine regimental commander, authorized him to shell Yongdungp'o. For the rest of the day the regiment held its place on the west side of the stream while artillery shelled Yongdungp'o and planes bombed it. The

[6] Montross and Canzona, *The Inchon-Seoul Operation*, pp. 216–19; Geer, *The New Breed*, pp. 136–37.

artillery barrage continued into the night.[7]

The arc of high ground west of Yongdungp'o was nearest to the city at its northern and southern extremities. At the center, a wide expanse of rice paddies and dikes, as well as Kal-ch'on Creek, separated the hills from the city. Accordingly, the best points from which to attack seemed to be the two extremities of the arc. The 1st Battalion was on the north, closest to the Han; the 2d Battalion was on the south along the Inch'on highway; the 3d Battalion was in reserve.

At daybreak, 21 September, the artillery resumed its preparation against Yongdungp'o, the big industrial suburb on the south bank of the Han, three miles southwest of Seoul. Then, at 0630, the marines attacked. At the northern end of the arc the 1st Battalion moved off Hills 85 and 80 onto the flat rice paddy ground and across Kal-ch'on Creek into the edge of Yongdungp'o. There enemy fire caused many casualties and slowed its advance. North Koreans held the dikes at the northwest approach to the city.

In its part of the attack, the 2d Battalion at the southern end of the arc had even harder going. Enemy mortar and artillery fire from high ground on that flank took a heavy toll in Marine casualties. By early afternoon, the 2d Battalion had suffered 85 casualties in crossing the rice paddies bordering Kal-ch'on Creek. Here at the edge of Yongdungp'o, large parts of which were now burning, the North Koreans fought the battalion to a standstill. The 3d Battalion, 1st Marines, passed through the hard-hit 2d Battalion late in the afternoon and continued the attack under heavy artillery fire.[8]

The North Koreans had held off the attacks at either end of their line, when in an unexpected manner the key to Yongdungp'o fell into the marines' hands. While the heavy battles were in progress on either flank, A Company left Hill 118 and moved behind low, masking hills to approach the dike system in the rice fields near the center of the line directly west of the main part of Yongdungp'o. The company formed an assault line behind a high dike, crossed it, and then advanced through chest-high rice to the deep mud of Kal-ch'on Creek, crossed the stream, reformed in front of another dike on the far side, and then entered the city streets undiscovered. As A Company moved through the empty heart of the town the men could hear sounds of heavy fighting on their right and left. Capt. Robert Barrow, the company commander, soon found that he and his men were 700 yards inside the town without contact with other friendly units. His reaction was not, as is so often the case when troops are in such a situation, that he and his company were alone, isolated, and surrounded. Instead, he realized that he was in the enemy's rear and proceeded to exploit the situation.[9]

Soon Barrow's advanced platoon on the left saw enemy troops hurrying west-

[7] 1st Mar Div SAR, vol. I, Annex Charlie, G–3 Sec, 20 Sep 50, p. 14; Diary of CG X Corps, 20 Sep 50; X Corps WD, Opn CHROMITE, 20 Sep 50.

[8] 1st Mar Div SAR, vol. I, G–3 Sec, 21 Sep 50, pp. 15, 19; Diary of CG X Corps, 21 Sep 50; Montross and Canzona, *The Inchon-Seoul Operation*, pp. 221–25.

[9] Geer, *The New Breed*, p. 149.

ward along the concrete highway from Seoul—reinforcements for the battle in progress. Its fire surprised and either destroyed or dispersed these troops. The company moved on. Shortly before noon, and after passing almost through the city, Barrow stopped at its eastern side. There he placed his men in a defensive perimeter on both sides of a 30-foot-high dike upon whose crest ran a surfaced road which joined at this point with the Seoul-Inch'on highway.

That afternoon the North Koreans apparently were too busy in the battle at the western edge of the city to give much attention to the unit in their rear, although small groups did make feeble efforts against it. But at dusk five tanks attacked A Company. In the battle between bazooka and tank the bazooka was victorious, the latter knocking out one and damaging two of the tanks. The two undamaged tanks, with machine guns blazing and cannon booming, made five passes along the deeply dug-in infantry at thirty yards' distance from the levee. Then the tanks withdrew into the city. At 2100 an enemy infantry force attacked the 3d Platoon at the northern end of the company perimeter. The platoon repulsed five separate attacks there before midnight. Morning disclosed more than 275 enemy dead in the vicinity of the dike and road intersection and many automatic weapons scattered about on the ground.[10]

On 21 September two developments behind the front occurred that were to affect future tactical operations. First, the third regiment of the 1st Marine Division, the 7th Marines, arrived in Inch'on harbor and began unloading. Second, command of the operation passed from Admiral Struble to General Almond, who at 1800 assumed command of the Seoul operation ashore at the X Corps command post in Inch'on. At this time there were 49,568 persons, 5,356 vehicles, and 22,222 tons of cargo ashore.[11]

The North Koreans, after their failure during the night of 21–22 September to drive Captain Barrow's company from its advanced position at the eastern edge of Yongdungp'o, apparently abandoned the city before daybreak. The 1st Marines occupied the city the next morning. On the left near the river they reached the destroyed railroad and highway bridges over the Han River two miles east of Yongdungp'o.[12]

The *87th Regiment* of the N.K. *9th Division* and elements of the N.K. *18th Division* had defended Yongdungp'o. One battalion of the *87th Regiment* reportedly suffered 80 percent casualties in the fighting there. Prisoners revealed that this regiment had left Kumch'on on 16 September to reinforce the Seoul area, traveling in trains that hid in tunnels during the day, and had arrived in the Yongdungp'o area on 20 September, barely in time to enter the fight there.[13]

On the 22d, the 1st Marine Division issued an operations order setting forth its plan for the seizure of Seoul. The

[10] Montross and Canzona, *The Inchon-Seoul Operation*, pp. 228–31; Geer, *The New Breed*, pp. 151–53.

[11] TF 7, Inch'on Rpt, I–F–3; Diary of CG X Corps, 21 Sep 50.

[12] 1st Mar Div SAR, vol. I, an. C, G–3 Sec, 22–23 Sep 50, pp. 17–19; X Corps WD, G–2 Sec Hist Rpt, PIR 4, 22 Sep 50.

[13] GHQ FEC, History of the North Korean Army (section of *9th Div*), 31 Jul 52; ATIS Res Supp Interrog Rpts, Issue 100, p. 53; X Corps WD, PIR 4, 22 Sep 50.

1st Marines was to cross the Han in the Yongdungp'o area and join the 5th Marines north of the river, forming the division right, while the 7th Marines was to move up from Inch'on and go into the line north of the 5th Marines, which then would form the center of a 3-regiment line. The plan contemplated that the 1st Marine Division, without the help of other ground units, would capture the city. But that same day, General Almond introduced one change in the plan—he indicated that the ROK Marines and the ROK 17th Regiment were also to be committed in securing the city.[14]

### *Securing the Southern Flank*

As the 1st Marines fought its way along the Inch'on-Seoul highway and into Yongdungp'o, the 7th Infantry Division protected its right flank and engaged enemy units moving toward the battle area from the south. An extensive mine field delayed the 32d Regiment on the 20th as it attacked toward Anyang-ni where it was to cut the Seoul-Suwon highway. Exploding mines damaged three tanks of A Company, 73d Tank Battalion, and completely blocked the narrow dirt road the column was following. Colonel Beauchamp, the regimental commander, had a narrow escape. A mine destroyed his jeep, killing the driver and wounding the radio operator a few minutes after he had left it. Engineer troops removed more than 150 mines from this field. The regiment during the day captured T'ongdok Mountain and part of Copper Mine Hill.[15]

On the 21st, the 32d Infantry seized the rest of Copper Mine Hill. It also captured the high ground two miles south of Yongdungp'o and Hill 300, the high ground immediately northeast of Anyang-ni. The 7th Division Reconnaissance Company arrived at Anyang-ni at 1430. When darkness fell, the 3d Battalion, 32d Infantry, held blocking positions astride the Suwon highway two miles south of Anyang-ni, the 1st Battalion held the road east and the high ground northeast of the town, and elements of the regimental combat team had established contact northward at Toksan-ni with the 2d Battalion, where the latter had captured a considerable quantity of ordnance and medical supplies.[16]

After arriving at Anyang-ni with the Reconnaissance Company, Maj. Irwin A. Edwards, Assistant G–2, 7th Division, received radio orders from the division to turn south to Suwon and secure the airfield below the town. Approximately at 1600, 2d Lt. Jesse F. Van Sant, commanding a tank platoon, took the point with his tanks and, followed by the Reconnaissance Company and Major Edwards, started toward Suwon. Naval aircraft bombed Suwon just before they arrived there at 1800 and destroyed a large wooden structure on top of the ancient great stone wall at its East Gate. Debris from this structure blocked the gateway and forced the company to turn

[14] Diary of CG X Corps, 22 Sep 50; 1st Mar Div SAR, vol. I, an. C, G–3 Sec, p. 17, Opn Plan 3–50, 22 Sep 50.

[15] 32d Inf WD, 20 Sep 50; 7th Div POR 2, overlay, 202200 Sep 50; Interv, author with Beauchamp, 1 Aug 52.

[16] Ltr, Lt Col Irwin A. Edwards to author, 5 Aug 53; Interv, author with Barr, 1 Feb 54; Beauchamp, MS review comments for author, 19 Jan 54.

aside to find another entrance into the town. At this point, Lt. Col. Henry Hampton, 7th Division G–3, arrived from Anyang-ni with a platoon of B Company, 18th Engineer Combat Battalion, and joined the group.

Hampton and Edwards with two enlisted men led the column through the streets. Near the center of Suwon the four men surprised two North Korean officers in the act of trying to escape in an American jeep. Edwards shot the driver; the other officer, a major of the N.K. *105th Armored Division,* surrendered. The armored column engaged in some street fighting with scattered groups of enemy soldiers, capturing altogether, thirty-seven North Koreans. Three miles south of Suwon the column went into a perimeter defense astride the highway. Being without maps, it had unwittingly passed the airfield a mile back up the road.[17]

About 2100 a full moon rose and Maj. Gen. David G. Barr, having lost radio contact with the Reconnaissance Company, decided to send an armored force toward Suwon to find it. Colonel Hampton and the platoon of engineers had already loaded into a truck and gone ahead. Task Force Hannum, named after its commander, Lt. Col. Calvin S. Hannum, commanding the 73d Tank Battalion, started from Anyang-ni at 2125. This motorized force—comprised of B Company, 73d Tank Battalion, and the battalion Advance Command Group; K Company, 32d Infantry; C Battery, 48th Field Artillery Battalion; and a medical detachment—hurried south in the moonlight with all possible speed. Lt. Col. John W. Paddock, 7th Division G–2, accompanied it. On the way to Suwon, Colonel Paddock established radio contact with Major Edwards and asked for guides to direct him and his force into the perimeter.[18]

Hannum's armored column reached Suwon near midnight, found the East Gate blocked, and turned aside to enter the town from another point through the ancient stone wall that girds the town on that side. Inside the town an enemy tank hidden behind a building opened fire on the leading American tank, knocking it out with one shot and killing Capt. Harold R. Beavers, the B Company tank commander who was inside it. In the fight that flared in the next few minutes other American tanks destroyed this T34, but a second enemy tank escaped. Hannum's force tried to follow it but became lost at the edge of town. Hannum decided to wait for daylight rather than to risk another enemy tank ambush in the darkness.

Meanwhile, Edwards' party in its perimeter south of Suwon heard the sound of tanks northward. Lieutenant Van Sant thought their clatter sounded like T34's, but the others discounted his comments and hastened preparations to send a party to meet Hannum. Major Edwards put a Korean civilian and eight men from the Reconnaissance Company into two jeeps. Colonel Hampton said he would go along and possibly continue on to rejoin the 7th Division headquarters at Anyang-ni. The party started with Edwards driving the first of four jeeps. A mile northward Ed-

[17] Ltr, Edwards to author, 5 Aug 53; X Corps WD, G–2 Hist Rpt, PIR 4, 22 Sep 50.

[18] Interv, author with Hannum, 21 Jul 53; 7th Div WD, Narr, 21–22 Sep 50; Ltr, Edwards to author, 5 Aug 53.

wards saw four tanks approaching in the moonlight. He flicked his lights in a recognition signal for what he thought was Hannum's lead tank. The tank stopped. Then suddenly its machine guns started firing, and it came on toward the halted vehicles. The men jumped from the jeeps and scrambled into the ditches. Colonel Hampton, however, started toward the tank waving his arms, evidently still thinking them friendly. Machine gun fire cut him down and the oncoming tank crunched into Edwards' jeep. Edwards escaped and rejoined the Reconnaissance Company the next morning.

The North Korean tanks rumbled on south and a few minutes later the first one entered the Reconnaissance Company's perimeter. Just ahead of it, an escapee from the jeep party ran into the perimeter and gave the alarm. The second enemy tank reached the edge of the perimeter. Van Sant gave the order to fire. The American M26 tanks destroyed both T34's at point-blank range of forty yards or less. The other two T34's turned and clattered back toward Suwon.[19]

At daylight Hannum led his armored column south through the town, now deserted. Below it he passed the crushed jeeps and the bodies of Hampton and two or three other men killed there. At midmorning Hannum's armored force joined the Reconnaissance Company at Suwon Airfield where Major Edwards had moved it and Van Sant's tanks at daybreak. Before noon, Col. Richard P. Ovenshine's 31st Infantry Regiment of the 7th Division (less the 3d Battalion in division reserve) arrived at Suwon and relieved Task Force Hannum at the airfield. The Reconnaissance Company then reconnoitered south toward Osan. Task Force Hannum rejoined the 7th Division in the Anyang-ni area.[20]

The big event of 22 September was securing Suwon Airfield and opening it to United Nations air traffic. This field, 21 miles south of Seoul, could accommodate the large C–54 transport planes with its 5,200-foot runway.

Meanwhile, 7 miles northeast of Anyang-ni, enemy forces succeeded in ambushing the lead platoon of B Company, 32d Infantry, and badly disorganized it. Lt. Col. Don C. Faith, Jr., the 1st Battalion commander, withdrew B Company 2 miles, to the vicinity of Kwanmun-dong, closely pursued by the enemy. There the battalion checked the North Koreans. During the day, Lt. Col. Charles M. Mount's 2d Battalion, 32d Infantry, seized the series of hills from 1 to 2 miles south of the rail and highway bridges that crossed the Han into Seoul.

On 23 September, the 1st Battalion captured its objective, Hill 290. This hill, 3 miles below the Han River and 7 miles southeast of Yongdungp'o, dominates the approaches to the Han River and Seoul from that direction.[21]

On the morning of 24 September,

---

[19] Interv, author with Hannum, 21 Jul 53; Ltrs, Edwards to author, 5 and 13 Aug 53; Ltr, Beauchamp to author, 15 Jul 53; Beauchamp, MS review comments for author, 19 Jan 54; 7th Div WD, 21 Sep 50; 32d Inf WD, 21 Sep 50; New York *Herald Tribune,* September 23, 1950, Bigart dispatch.

[20] Interv, Hannum with author, 21 Jul 53; Ltrs, Edwards to author, 5 and 13 Aug 53; 7th Inf Div WD, 22 Sep 50; 31st Inf Narr Rpt, 22 Sep 50; X Corps WD, 21–22 Sep 50.

[21] 32d Inf WD, 22 Sep 50; 7th Div WD, 22 Sep 50; 7th Div POR's 4 and 5 and accompanying overlays, 22–23 Sep 50; Beauchamp, MS review comments for author, 19 Jan 54; Ltr, Mount to author, 12 Mar 56.

Mount's battalion in a predawn attack caught North Koreans asleep in their positions and overran them. In this surprise action the battalion captured a regimental headquarters and much equipment, and broke the remaining enemy strength close to the south bank of the river opposite Seoul. During the day the battalion cleared the south bank of the Han in the fold of the river southeast of the city. This made possible an important action the next morning.[22]

### *Seoul's Western Rampart*

While the 7th Division was securing X Corps' southern flank, the heaviest fighting in the battle for Seoul began at the city's western edge on 22 September and lasted four days.

The North Korean defense line at the western edge of Seoul was anchored at the north on Hill 296 just south of the Kaesong highway and west of Seoul's Sodaemun Prison. From the crest of Hill 296 the North Korean line curved in a gentle half-moon eastward and southward down spur ridges two and a half miles to the Han River, the concave side facing west toward the United Nations troops. The greater part of this uneven ridge line was dominated by three hills, each 105 meters high, and accordingly known as Hills 105. Hills 105 North and 105 Center lay north of the rail and highway lines running into Seoul along the northern bank of the Han River; Hill 105 South lay between the rail and road lines and the river. Hills 105 Center and 105 South completely dominated the Pusan-to-Manchuria Kyonggi main rail line and the road that passed through the saddle between them to enter the city. These hills had been a training area for Japanese troops during the period of Japanese domination and since then of both South and North Korean soldiers. The area was well covered with various types of field fortifications and susceptible to quick organization for defense. The main railroad station and Government House lay in the center of Seoul two miles east of these positions.

The principal enemy unit manning this line was the N.K. *25th Brigade*. Newly formed a month earlier at Ch'orwon, it had started moving by train from that place to Seoul on the day of the Inch'on landing, most of it arriving there four days later on 19 September. Maj. Gen. Wol Ki Chan, forty-five years of age and formerly a student in Russia, commanded the brigade. Most of the brigade's officers and noncommissioned officers had had previous combat experience with the Chinese Communist Forces. The brigade numbered about 2,500 men, and apparently was composed of two infantry battalions, four heavy machine gun battalions, an engineer battalion, a 76-mm. artillery battalion, a 120-mm. mortar battalion, and miscellaneous service troops. It and the *78th Independent Regiment* defended both the military and topographic crests. Foxholes, undercut into the slopes, gave protection from overhead shell air bursts. Concrete caves held supplies. More than fifty heavy machine guns with interlocking fields of fire dotted this defensive position.[23]

---

[22] 7th Div WD, 24 Sep 50.

[23] 1st Mar Div SAR, vol. I, an. B, p. 17, and vol. III, 5th Mar SAR, an. B, app. 1, pp. 5–6, and an. O, p. 9.

On the morning of 22 September the 5th Marines set out to capture these last hills in front of Seoul. On the north flank the 3d Battalion's objective was Hill 296. In the center, the objective of the 2d Battalion, ROK Marines, was Hill 105 Center, but the battalion had to take two knobs called Hills 66 and 88 before reaching the main hill behind them. On the south flank across the railroad track the 1st Battalion, 5th Marines, objective was Hill 105 South. The attack began at 0700. Two hours later the 3d Battalion on the north reported it had captured its objective against only moderate resistance, but this report was misleading because the battalion did not have control of the southern slopes and ridges of Hill 296 where the North Korean strength was concentrated. On the southern flank heavy enemy fire stopped the 1st Battalion for a while, but late in the day it took Hill 105 South after a smashing artillery and mortar preparation. But enemy artillery scored too, some of its fire landing in the 1st Battalion rear areas and inflicting thirty-nine casualties there during the day, including six killed.[24]

In the center, enemy fire decimated the ROK Marine battalion in its attack against Hills 66 and 88. The fighting was heavy there all day long. Marine air strikes tried in vain to destroy the enemy positions. Later, North Korean prisoners said that the *25th Brigade* had 40 percent casualties that day. The next morning, 23 September, the Korean marines resumed the battle in the center and suffered continuing heavy casualties, while accomplishing little. At midafternoon the 2d Battalion, 5th Marines, on orders from Colonel Murray, took over the attack in the center. After sustaining many casualties with little gain, the lead company (D) dug in for the night, short of the enemy-held ridge. In another furious fight, one platoon of F Company suffered so many casualties it had only seven men left for duty at nightfall. Meanwhile, the rest of the regiment had held in place on the flanks and repelled counterattacks during the day.[25]

At noon on 23 September General Smith had ordered the 7th Marines, which had begun unloading at Inch'on on the 21st, to cross the Han and come up behind the 5th Marines. During the day X Corps headquarters moved from Inch'on to Ascom City, about halfway to Seoul on the main Inch'on-Seoul highway.

After daylight on the 24th, elements of the 3d Battalion, 5th Marines, started down the finger ridge from Hill 296, hoping to outflank the enemy in front of the 2d Battalion in the center. Simultaneously, D Company moved out in assault against the ridge line. A heavy morning mist shrouded the company as it crossed the low ground and reached the base of Hill 66. Unexpectedly, the lead elements came upon enemy troops in their trenches. Neither side saw the other because of the fog and smoke until they were at close quarters. A grenade battle started immediately. One

---

[24] 1st Mar Div SAR, vol. III, p. 9, 22 Sep 50, and vol. I, an. C, G-3 Sec, p. 17, 22 Sep 50, and an. B, app. 2, p. 8; Geer, *The New Breed,* pp. 155-57; New York *Herald Tribune,* September 23, 1950, Bigart dispatch.

[25] 1st Mar Div SAR, vol. I, an. C, G-3 Sec, pp. 17-19, 22-23 Sep 50, and an. B, app. 2, p. 10; *Ibid.,* vol. III, an. B, app. 1, 1st Mar Div Opn Ord 9-50, 231200 Sep 50, p. 6; X Corps WD, 23 Sep 50; Giusti, "Marine Air Over Inchon-Seoul," *Marine Corps Gazette* (June 1952).

squad of twelve marines in trying to maneuver around the southern tip of the ridge was wiped out except for three wounded who escaped.

In an effort to break the deadlock, Marine air strikes came in repeatedly. In the course of two such attacks, North Korean antiaircraft fire damaged five of ten planes. Enemy automatic and mortar fire became intense after the fog lifted.

In the early afternoon the 30 remaining effectives in D Company's rifle platoons and 14 other men assembled from the Weapons Platoon, ammunition carriers, and company headquarters prepared for a desperate assault against the ridge line of Hill 66. Thirty-three men were to make the assault up the last 150 yards of the slope—11 others were to follow with machine guns and ammunition. Corsairs came over for final strafing runs, bombing, and napalming of the enemy position. This done, the 33 men, at a prearranged signal of a Corsair's second dry run over the enemy, jumped from their holes and charged forward in a 100-yard-long line. The D Company commander, 1st Lt. H. J. Smith, was killed in front of his men. The others kept going and 26 of them reached the top. The headlong charge surprised the North Koreans; in a sudden panic many ran down the back slope, others feigned death, and some fought back. Enemy dead were stacked up everywhere—in foxholes, in bunkers—and many were strewn about over the ground. When all of D Company's men reached the top there were 56 men to defend it, 26 of them wounded but refusing evacuation. They held the hill against a counterattack. During this day, D Company suffered 176 casualties among its 206 men—36 killed, 116 wounded and evacuated, and 26 more wounded but present for duty.[26]

Events were to prove that D Company's capture of Hill 66 on the afternoon of the 24th was the decisive action in the battle at the western gate to Seoul. The 2d Battalion on the morning of the 25th resumed the attack toward Hill 105 Center. Artillery and fighter bombers pounded the enemy-held hill line all morning. From recently captured Hill 66, D Company advanced northward slowly during the morning up the shank of the fishhook ridge line that slanted southwest from Hill 296, and then turned southeast to capture Hill 88 at the point of the hook just after noon. By midafternoon other elements of the 2d Battalion had captured Hill 105 Center, and the 3d Battalion had gained control of Hill 105 North after very heavy fighting. According to prisoners, three enemy battalions lost 500 men during the day trying to hold the northern hill. The western defenses of Seoul had fallen. More than 1,200 dead enemy soldiers lay on their stubbornly defended positions. Marine estimates placed the total number of enemy killed there by all arms at 1,750.[27]

When the enemy defenses at the western edge of Seoul fell on 25 September, the 1st Marine Division had all its regi-

---

[26] 1st Mar Div SAR, vol. I, an. C, G–3 Sec, p. 21, 24 Sep 50; *Ibid.*, vol. III, 5th Mar SAR, p. 11, 24 Sep 50; X Corps WD, 24 Sep 50; Montross and Canzona, *The Inchon-Seoul Operation*, pp. 245–50; Geer, *The New Breed*, pp. 161–62; Giusti, "Marine Air Over Inchon-Seoul," *op. cit.*

[27] 1st Mar Div SAR, vol. III, an. B, app. 1, p. 6 and an. P, 5th Mar SAR, p. 9; *Ibid.*, vol. I, an. C, G–3 Sec, p. 24, 25 Sep 50; X Corps WD, G–2 Hist Rpt, PIR 7, 25 Sep 50. Montross and Canzona, *The Inchon-Seoul Operation*, pages 239–59, has a detailed account of this battle.

AMERICAN TROOPS MOVE ON SEOUL *along the north bank of the Han River.*

ments together north of the Han River. At 2200 on the night of 23 September, the division had issued an operations order, confirming earlier verbal orders, directing the 1st Marines to cross the Han River early the next morning.[28]

During the morning of 24 September the 1st Marines began crossing the Han from Yongdungp'o in the shadow of Hill 105 South, where the 1st Battalion, 5th Marines, protected the crossing site. Before dark the regiment had crossed to the north side and the 1st and 2d Battalions had taken over from the 1st Battalion, 5th Marines, the southern flank of the Marine line at the western edge of Seoul. By now the 7th Marines had moved up on the left flank of the 5th Marines, with the mission of cutting across the northern edge of Seoul and blocking escape routes there. The 3d Battalion of the 187th Airborne Regiment airlifted from Ashiya, Japan, to Kimpo Airfield on the 24th and upon arrival there assumed responsibility for the airfield.[29]

On the morning of 25 September, two platoons of tanks from B Company, 1st Tank Battalion, including two dozer tanks and a section of flame-throwing tanks, accompanied by a platoon of engineers and another of infantry set out to join the 1st Marines in Seoul. Near the base of Hill 105 South, an enemy force with several antitank guns ambushed the column. The fighting was heavy and the outcome in doubt for several minutes until a flame-thrower tank reached a point from which it

[28] 1st Mar Div Opn Ord 9–50, 231200 and Opn Ord 10–50, 232200 Sep 50.

[29] 1st Mar Div SAR, vol. I, an. C, G–3 Sec, p. 21, 24 Sep 50, and an. B, G–2 Sec, p. 11, 24 Sep 50; 187th RCT Act Rpt, 22 Sep–2 Oct 50.

spurted flame into the enemy trenches. Many North Korean soldiers broke from cover. Machine gun fire from other tanks cut them down as they ran. Several North Koreans came out of a previously undiscovered cave and surrendered. When a large group inside the cave saw these men unhurt they too surrendered. Of the nearly 300 North Koreans that attacked this armored column, approximately 150 were killed, 131 captured. The tank column joined Colonel Puller's 1st Marines in Seoul at noon.[30]

During the morning the 3d Battalion, 1st Marines, made a 90-degree turn northward to change the regimental direction of attack from eastward to northward toward the heart of the city. While it was doing this, the 1st Battalion on its right held a blocking position at the southern edge of Seoul. Once the 3d Battalion had turned northward, the 1st Battalion pivoted to orient its attack northward abreast of and on the right of it. Street fighting now began in Seoul, 25 September, in the zone of the 1st Marines just as the 5th Marines completed its capture of the North Korean defensive hill line at the western edge of the city.[31]

### *The 32d Infantry Enters Seoul*

By this time an important change had taken place in the plan to capture Seoul. The original operations plan required the 1st Marine Division to clear the city. But the expected capture of Seoul by the marines was moving behind schedule. The stubborn enemy defense had denied the Marine division any important advance for three days. General Almond, the corps commander, had been growing increasingly impatient. Seoul was a symbol in the Korean War, just as Paris, Rome, and Berlin had been in World War II. It was a political and psychological as well as a military target. General MacArthur desired to capture the city as soon as possible and restore the Korean capital to its people.[32]

Dissatisfied with the marines' progress, General Almond on 23 September told General Smith that he could continue his frontal assaults but that he strongly urged him to use the space south of the Han River for an envelopment maneuver by the 1st Marines. Smith was unwilling to act on Almond's suggestion because he wanted to unite the 1st and 5th Marines on the north side of the Han instead of having them on opposite sides of the river. Almond told Smith that he would give him twenty-four hours longer to make headway. If Smith could not, Almond said, he would change division boundaries and bring the 7th Infantry Division and its 32d Regiment into the battle for the envelopment of the enemy defenses in Seoul.[33]

On the morning of 24 September the North Koreans still held the marines at

[30] 1st Mar Div SAR, vol. II, an. OO, p. 8, Rpt 1st Tk Bn, 25 Sep 50; Geer, *The New Breed*, p. 165; Montross and Canzona, *The Inchon-Seoul Operation*, pp. 259–61.

[31] 1st Mar Div SAR, vol. I, an. C, G–3 Sec, p. 23, 25 Sep 50, and situation overlay 251800 Sep.

[32] Most of the responsible Marine officers felt that X Corps hurried them too much in the Seoul operation. Col John H. Chiles (G–3 X Corps Sep 50), MS review comments for author, 15 Dec 53; Lt Gen Oliver P. Smith, MS review comments, 25 Feb 54; Almond, MS review comments for author, 15 Dec 53; Interv, author with Lt Col Earle W. Williamson (G–3 Sec, X Corps), 28 Aug 51.

[33] Interv, author with Almond, 13 Dec 51; Smith, MS review comments, 25 Feb 54.

the west edge of Seoul. About 0930 General Almond arrived at 7th Division headquarters and conferred with General Barr, the 7th Division commander, Brig. Gen. Henry I. Hodes, assistant 7th Division commander, and Col. Louis T. Heath, the division chief of staff. Almond told Barr he had tentatively decided that the 7th Division would attack across the Han River into Seoul the next morning. Almond then returned to his command post and there told Colonel Paik, commander of the ROK 17th Regiment, that he expected to attach his regiment to the 32d Infantry for the attack on Seoul.[34]

His mind now made up, Almond called a commanders' conference to meet with him at 1400 at Yongdungp'o Circle. Present besides Almond were Generals Smith, Barr, and Hodes, Colonels Forney and Beauchamp, and Col. John H. Chiles. In this open-air meeting, Almond quickly told the assembled commanders that he was changing the boundary between the 1st Marine Division and the 7th Infantry Division, and that the 32d Regiment, with the ROK 17th Regiment attached, would attack across the Han River into Seoul at 0600 the next morning. The meeting was brief. At its conclusion the officers dispersed at once to make their respective plans.[35]

In the afternoon and evening, X Corps attached the ROK 17th Regiment, the Marine 1st Amphibious Tractor Battalion (less one company), and two platoons of A Company, 56th Amphibious Tank and Tractor Battalion, to the 7th Division to support the crossing.[36]

The crossing was to be at the Sinsa-ri ferry, three miles east of the main rail and highway bridges over the Han River. On the opposite (north) bank, South Mountain (Nam-san) extended from the river northwest two miles into the heart of Seoul, culminating in a peak 900 feet high, the highest point in the city, about one mile east of the main Seoul rail station. A long, ridgelike, shallow saddle connected this peak with a slightly lower one. On a western finger ridge of the main peak, near the 350-foot elevation and only half a mile from the rail station, was a large shrine and a formally landscaped park. From the western base of South Mountain a long series of steps led up to this shrine and park. Viewing Seoul on a north-south axis, the peak of South Mountain was halfway into the city. Government House, at the northern edge of the city, lay two miles away. The main highway and rail line running east out of the city passed about a mile beyond the northern base of South Mountain. On this mountain nearly three months before, a company of ROK soldiers had conducted the last action in the defense of Seoul, dying, it has been said, to the last man.

The 32d Infantry's mission was first to seize and secure South Mountain, then to secure Hill 120 situated two miles eastward at the southeast edge of Seoul, and finally to seize and secure

[34] 7th Div WD, 24 Sep 50; Diary of CG X Corps, 24 Sep 50; Hist, Off CofS, X Corps, 24 Sep 50.

[35] Interv, author with Almond, 13 Dec 51; Ltr, Beauchamp to author, 15 Jul 53; Diary of CG X Corps, 24 Sep 50; 7th Div WD, 24 Sep 50; 32d Inf WD, 24 Sep 50. Forney, a Marine officer, was X Corps Deputy CofS.

[36] 32d Inf WD, 24 Sep 50; 7th Div WD, 24 Sep 50; Diary of CG X Corps, 24 Sep 50; Hist, Off CofS, X Corps, 24 Sep 50.

SEOUL AS SEEN FROM THE AIR. *Government House is in the center.*

Hill 348, a large, high hill mass five miles east of Seoul and dominating the highway and rail line entering the city from that side. The regiment had a strength of 4,912 men as it prepared for the crossing—3,110 Americans and 1,802 ROK's.[37]

Before daybreak of the 25th, General Hodes established an advanced division command post near the river from which he was to direct the crossing operation. At 0400, General Almond, Admiral Struble, and members of the corps staff departed the X Corps headquarters at Ascom City to watch the crossing of the 32d Regiment. General Barr went forward at 0430 to the 32d Infantry's command post and an hour later he and Colonel Beauchamp left for an observation post near the river. At 0600, the 48th Field Artillery Battalion began firing a 30-minute artillery preparation, and the heavy mortars joined in to

[37] 7th Div WD, 25 Sep 50; 32d Inf WD, 25 Sep 50.

pound the cliffs lining the opposite side beyond the river bank.[38]

Colonel Mount's 2d Battalion, selected to make the assault crossing, loaded into amphibious tractors in its assembly area and at 0630 F Company started across the Han. A ground fog obscured the river area. The entire 2d Battalion reached the north bank without loss of personnel or equipment. The 2d Battalion hurried across the narrow river beach, scaled the 30- to 60-foot cliffs, and moved rapidly to the slopes of South Mountain. An hour after the first troops had crossed the river the bright morning sun dispersed the ground fog. Air strikes then came in on South Mountain and Hill 120. Apparently this crossing surprised the North Koreans. Their works on South Mountain were only lightly manned.

The 1st Battalion, commanded by Colonel Faith, followed the 2d across the Han and at 0830 started to move east along the river bank toward Hill 120. Just after noon the 3d Battalion crossed the river, followed the 1st Battalion eastward, and passed through it to occupy Hill 120. The 1st Battalion then took a position between the 3d and 2d Battalions. The ROK 17th Regiment crossed the Han immediately behind the 3d Battalion and moved to the extreme right flank of the 32d Infantry line where, at 2150, it began an all-night attack toward Hill 348.[39]

While the rest of the regiment crossed the Han behind it and moved eastward, the 2d Battalion climbed the slopes of South Mountain, reaching and clearing the summit against moderate resistance by 1500. Once there, it immediately began digging in on a tight perimeter.

The North Koreans did not counterattack South Mountain as quickly as expected. The night passed tensely but quietly for the waiting 2d Battalion. Finally, at 0430 on the morning of the 26th, the soldiers heard tanks moving about and the sound of automatic weapons fire to their front. In semidarkness half an hour later a large enemy force, estimated to number approximately 1,000 men, violently counterattacked the 2d Battalion perimeter on top of South Mountain. On the higher western knob of the mountain, G Company held its position against this attack, but on the lower eastern knob North Koreans overran F Company. Using all its reserves, Colonel Mount's battalion finally restored its positions at 0700 after two hours of battle and drove the surviving enemy down the slopes. Mount's men counted 110 enemy dead within its perimeter and 284 more outside for a total of 394 enemy killed. They took 174 prisoners.[40]

E Company mopped up enemy troops on the rear slopes of the mountain and in the area at its base near the river. Later in the morning, elements of the 1st Battalion had a sharp engagement in the streets immediately north of South Mountain, capturing there some eighty enemy soldiers, apparently a remnant

[38] 32d Inf WD, 25 Sep 50; Diary of CG X Corps, 25 Sep 50.

[39] 32d Inf WD, 25 Sep 50; 7th Div WD, 25 Sep 50; Diary of CG X Corps, 25 Sep 50; Beauchamp, MS review comments for author, 19 Jan 54. Where time for events differs in the various levels of command records, the author has followed that of the lower unit, the one closest to the action, unless other evidence supports another choice.

[40] 32d Inf WD, 26 Sep 50; Ltr, Beauchamp to author, 15 Jul 53.

of the force that had counterattacked South Mountain.

To the east, the 1st Battalion on the morning of the 26th engaged in a heavy fire fight while the 3d Battalion, commanded by Lt. Col. Heinrich G. Schumann, advanced from Hill 120 toward Hill 348, four miles farther east. In this advance, L Company saw a large column of enemy troops on the highway leaving Seoul. The company commander, 1st Lt. Harry J. McCaffrey, Jr., seized the opportunity for surprise and immediately ordered his men to attack. His initiative paid off. In the ensuing action, L Company killed about 500 North Korean soldiers, destroyed 5 tanks, destroyed or captured more than 40 vehicles, 3 artillery pieces, 7 machine guns, 2 ammunition dumps, much clothing and POL products, and overran and captured a large headquarters of corps size, which may have been the principal enemy headquarters in the defense of Seoul.[41]

By midafternoon (26 September) the ROK 17th Regiment had captured Hills 348 and 292 dominating the highway four miles east of Seoul. That evening the 32d Infantry and the ROK 17th Regiment cleared their zone of the enemy, and E Company established contact with the marines on the regimental left at the western base of South Mountain.

### *Battle of the Barricades*

On 25 September, while the 32d Infantry crossed the Han and seized South Mountain, the 1st Marine Division entered Seoul proper. When the 1st Marine Regiment turned north that day, ahead of it lay the main Seoul railroad station. the French, American, and Russian consulates, the City Hall, the Duk Soo Palace of ancient rulers of Korea and Museum of Art, and the main business and hotel area. The 5th Marines, on the other hand, was just entering the city in the northwest quarter, pointed generally eastward toward Government House two miles away. Its course would take it past big Sodaemun Prison. Beyond Government House lay Changdok Palace and the Royal Gardens. That evening the 1st and 5th Marines made plans for a co-ordinated attack the next morning.

Just before dusk an air report claimed that enemy columns were streaming north out of the city. General Almond at X Corps headquarters immediately sent a message to the Far East Air Forces requesting a flare mission to illuminate the roads so that Marine night fighters could attack the enemy troops. A B–29 dropped flares for several hours and two long columns of enemy soldiers came under air attack. Corps artillery placed interdiction fire on the closer portions of the escape route.[42]

At 2040 that evening a X Corps flash message from General Almond came over the teletype to the 1st Marine Division saying, "X Corps TACAir Commander reports enemy fleeing city of Seoul on road north. . . . He is conducting heavy air attack and will con-

[41] Ltr, Beauchamp to author, 15 Jul 53, and MS review comments, 19 Jan 54; 32d Inf WD, 26 Sep 50; 7th Div WD, Narr, 26 Sep 50, and Narr, 7th Div POR 8, 26 Sep 50. Headquarters, 7th Infantry Division, General Order 69, 13 October 1950, awarded the Silver Star to Lieutenant McCaffrey.

[42] Hist, Off CofS, X Corps, 25 Sep 50; USAF Hist Study 71, p. 76.

tinue same. You will push attack *now* to the limit of your objective in order to insure maximum destruction of enemy forces." [43] Col. Alpha L. Bowser, 1st Marine Division operations officer, doubted that the enemy was fleeing the city. He telephoned the X Corps operations officer and questioned the order to "attack now," but that officer told him to attack as ordered. Bowser then gave the message to General Smith who in turn telephoned General Ruffner, X Corps chief of staff, objecting to the order. Smith did not believe the enemy was leaving Seoul and he did not want to attack through the city at night. Ruffner told him that Almond personally had dictated the order and that it was to be executed without delay. General Smith then telephoned Colonels Puller and Murray at 2200 and transmitted the order. He told them to concentrate their advance along avenues that could be identified easily at night, and ordered the three Marine regiments to establish contact with each other.[44]

Within a few minutes after the 5th Marines received the order for the night attack, an enemy force of approximately 200 men struck its 3d Battalion. Fighting continued until 0445 when the battalion repulsed the North Koreans. Meanwhile, regimental patrols sent south and southeast to establish contact with the 1st Marines failed to do so. Likewise, patrols of the 7th Marines from the north failed to establish contact with the 5th Marines. Except for its patrols, the 5th Marines did not move forward during the night from its evening positions.[45]

After receiving the order for the night attack, Colonel Puller ordered the 1st Marines to prepare to attack at 0145 after a 15-minute artillery preparation. A patrol from the 3d Battalion moved out at the conclusion of this preparation and a short distance away encountered a large enemy force preparing to counterattack. Some members of the patrol escaped and gave the alarm. The battalion at 0153 sent a flash message to Colonel Puller that an enemy tank-led force was on the point of attacking it. Puller thereupon ordered a second 15-minute artillery barrage to be fired directly in front of the 3d Battalion, the Marine attack to follow that.

This second barrage, together with mortar, tank, and automatic fire caught an estimated force of 700 North Koreans, supported by twelve tanks, two self-propelled guns, and 120-mm. mortar fire, attacking straight down the main boulevard, and decimated it. The lead enemy tank struck a mine at a Marine street block and bazookas destroyed others. Burning buildings illuminated the street scene in front of the 3d Battalion. The enemy attack continued until daylight although it became progressively weaker. After daylight the marines captured 83 prisoners, counted 250 enemy dead, and saw four tanks and two self-propelled guns knocked out in front of them.[46]

Because of the enemy counterattack

[43] 1st Mar Div SAR, vol. I, an. C, G–3 Sec, p. 25, 25 Sep 50.

[44] *Ibid.*

[45] *Ibid.*, p. 26, 26 Sep 50.

[46] *Ibid.*, vol. I, an. C and an. B, app. 2, p. 14, 26 Sep 50; *Ibid.*, vol. II, an. PP, p. 9; X Corps WD, 26 Sep 50; Diary of CG X Corps, 26 Sep 50; Geer, *The New Breed*, pp. 168–70.

TANKS ENTERING SEOUL *through cleared barricade.*

during the night and the lack of contact between its regiments, the Marine division did not launch a night attack as ordered. Its lines after daylight, 26 September, were substantially the same as they had been the evening before.

It appears that after the seizure of South Mountain by the 32d Infantry and the reduction of the hill defenses at the western edge of the city by the 5th Marines during the 25th, the North Korean commander in Seoul decided the city was doomed and began the withdrawal of certain units that evening while leaving others to fight desperate delaying actions. The aerial reports of enemy columns fleeing the city and General Almond's conclusion that the North Koreans were evacuating Seoul were therefore not without foundation. The major enemy unit withdrawing at this time was the N.K. *18th Division* which had fought south of the Han River in the Yongdungp'o area. On 24 September it had assembled in Seoul, and the next evening the approximately 5,000 men remaining in the division retreated northward on the Uijongbu road, headed for Ch'orwon. At the same time, to cover this withdrawal, the North Korean commander struck with desperate counterattacks at every point of American advance into the city. Against the 2d Battalion, 32d Infantry, on South Mountain he struck with one battalion and at other elements of the regiment eastward with another battalion. The heavy counterattack against the 3d Battalion, 1st Marines, he launched in reinforced battalion strength; while against the 3d Battalion, 5th Marines, he sent a reinforced company.[47]

After the enemy attacks died away with the coming of daylight on the 26th, the marines launched their attack. In a day-long effort down Ma-Po Boulevard the 2d Battalion, 1st Marines, gained less than a mile, and very little at all after 1400. Snipers fired from houses along the way and enemy soldiers manned barricades, making of each one a small battlefield. The 5th Marines had even stronger opposition in trying to advance from a spur of Hill 296 into the city and made only slight gains, but it did establish contact with the 1st Marines.

Just after noon, the Marine division brought Col. Homer L. Litzenberg's 7th Marines into the fight for Seoul proper, directing it to seize the mountain pass north of the city, and to cut the highway running northeast out of Seoul for Uijongbu and Ch'orwon at a point one mile northeast of Government House. The regiment's D Company turned down the Kaesong-Seoul highway toward the city, seeking to establish contact with the 5th Marines, but came under heavy enemy fire at 0830 opposite Sodaemun Prison at the northwest corner of Seoul. The company suffered many casualties there and, unable to advance farther, withdrew to a road cut between Hills 296 and 338 where it established a perimeter defense. That afternoon two planes dropped ammunition and medical supplies to it. Enemy fire hit both planes, and one crash-landed at Kimpo. Friendly tanks succeeded in reaching D Company's perimeter and carried out the wounded. At dusk on Tuesday, 26 September, X Corps troops held approximately half the city.[48]

About twenty hours earlier, just before midnight of the 25th, General Almond had announced the liberation of Seoul, three months to the day after the North Koreans began the invasion. Almond apparently based his announcement on air reports of North Korean evacuation of the city and the seizure of South Mountain during the day. On the 26th, General MacArthur signed and released United Nations Command Communiqué 9 at 1410 announcing the fall of Seoul. The communiqué said in part, "Seoul, the capital of the Republic of Korea, is again in friendly hands. United Nations forces, including the 17th Regiment of the ROK Army and elements of the U.S. 7th and 1st Marine Divisions, have completed the envelopment and seizure of the city." [49] In subsequent communiqués MacArthur made no mention of further fighting in Seoul, confining comment to combat operations in the Suwon area south of the city.

But on 27 September, the battle of the barricades in Seoul continued. In the middle part of Seoul the barricades stretched across the streets from side to side and were usually placed at intersections. Mostly they were chest-high and made of rice and fiber bags filled with earth. From behind them and at their

[47] FEC, Inf Order of Battle, N.K. Army, Supp, Gen Hist of N.K. Army Units, Chart 14 (N.K. *18th Div*), 16 Sep 51; ATIS Interrog Rpts, Issue 10, p. 78, Rpt 1534, Lt Lee Song Yol.

[48] 1st Mar Div SAR, vol. I, an. C, G–3 Sec, pp. 26–28, 26 Sep 50; *Ibid.*, vol. III, 5th Mar SAR, p. 13, 27 Sep 50; X Corps WD, 26 Sep 50; Geer, *The New Breed*, p. 176; New York *Herald Tribune*, September 27, 1950, Bigart dispatch from Seoul.

[49] UNC Communiqué 9, 26 Sep 50; GHQ UNC G–3 Opn Rpt, 26 Sep 50; X Corps WD, 26 Sep 50; New York *Times*, September 26, 1950, Lawrence dispatch.

sides enemy soldiers fired antitank guns and swept the streets with machine gun fire. Other soldiers were posted in adjacent buildings. Antitank mines belted the streets in front of the barricades.

The Marine attack settled into a routine for reducing the barricades. Navy and Marine planes would rocket and strafe them, mortarmen and infantry would set up a base of fire covering the engineers while they exploded the mines, two or three Pershing tanks would advance to the barricade, take it under fire destroying antitank guns and automatic weapons, and breach it. Occasionally, flame-throwing tanks rumbled up to stubbornly held positions and helped reduce the barrier. Infantry accompanying and following the tanks gave them protection, destroyed snipers, and cleared the area. A single barricade might hold up a battalion advance as much as an hour.

On the 27th, the 2d Battalion, 1st Marines, drove down Ma-Po Boulevard into the heart of the city, capturing the French Embassy and raising the American flag over it just before 1100. Richard J. H. Johnston, a correspondent for the New York *Times* who had lived in Seoul for four years, guided Capt. Charles D. Frederick and his E Company men in the afternoon to the Soviet Embassy, where they pulled down the Red flag and raised the American at 1530. They then crossed over to the adjacent American Embassy and raised the American flag over it seven minutes later. North Korean machine gunners at the gate of the American Embassy surrendered without firing. The 1st Battalion meanwhile captured the railroad station in the morning in fairly heavy action and then encountered a series of strongly defended barricades along the main thoroughfare leading northward into the center of the city. ROK marines followed the 1st Battalion, mopping up behind it.[50]

In the northwestern part of the city, the 5th Marines advanced on the 27th against relatively light resistance. Overnight the opposition of the previous day had largely vanished. On the regimental north flank, E Company without opposition entered Sodaemun Prison. Earlier a tank-led force from the 7th Marines had relieved D Company in its perimeter just beyond the prison. A Korean civilian informed troops of E Company, 5th Marines, that about five days earlier approximately 400 American prisoners held in the prison had been removed and taken northward. At midafternoon, the 5th Marines established contact with the 7th Marines in the northwest corner of Seoul.

The main axis of attack of the 5th Marine Regiment, however, was farther south. There the 3d Battalion secured the Seoul Middle School and Hill 79 just to the north of it by 1015, and reorganized for the attack toward Government House, its major objective. From the Seoul Middle School the battalion advanced due east to Kwang Who Moon Circle, where a memorial shrine stood. From behind a barricade at this intersection the North Koreans put up their last organized resistance in the heart of the city. A flame-throwing tank clanked across the large circular plaza and ended this resistance.

---

[50] 1st Mar Div SAR, vol. I, an. C, p. 29; *Ibid.*, vol. II, 1st Mar Regt SAR, an. PP, p. 10, 27 Sep 50; X Corps WD, G-3 Sec, Msg J-17, 271145 Sep 50; New York *Times*, September 28, and Johnston dispatch from Seoul, September 27, 1950.

From Kwang Who Moon Circle a broad and, in peacetime, imposingly beautiful, modern boulevard also named Kwang Who Moon runs north one-half mile, terminating in front of the modern and handsome Government House. After the tanks had reduced the barricade at Kwang Who Moon Circle, G Company, 5th Marines, advanced without opposition down the boulevard to Government House "as fast as they could walk," as one who was present has written. The company had possession of the building at 1508, and immediately thereafter struck the North Korean flags flying from the flag poles on either side of the Court of Lions in front of the building and raised in their place the American flag.

The breakthrough to Government House apparently caught some North Korean officials or stragglers there by surprise, compelling a hasty getaway, for upon entering the troops found hot food ready for eating. The battalion cleared the area of snipers and stragglers during the afternoon, and that night the 3d Battalion established its command post in the building.[51]

During that morning, the 1st Battalion, 5th Marines, following behind the 3d Battalion, turned off to the left after reaching the Seoul Middle School and attacked north to Hill 338, a key terrain feature a mile northwest of Government House. At 1900 it secured the hill which dominated the Seoul-Pyongyang highway at the northwest corner of the city. Except for scattered snipers and stragglers, the last defenders of Seoul withdrew from the city that night.

The next day, 28 September, although its 1st Battalion had to contend with many mines, the 1st Marines swept through the northeast corner of Seoul against only light resistance to complete its occupation. By evening the regiment had taken Hills 132 and 133, at the northeast edge of Seoul, dominating the Seoul-Uijongbu-Ch'orwon highway. A mile farther north, the enemy held the 7th Marines in check short of its objective, Hill 224, the key terrain feature on the other, west, side of the highway.[52]

Enemy resistance in Seoul had ended —the North Korean forces were withdrawing northward in the direction of Uijongbu just ninety days after they had victoriously entered the city in their bid for conquest of South Korea.

## *MacArthur Re-establishes Syngman Rhee in Seoul*

After his return to Tokyo, General MacArthur on 23 September sent a message to the Joint Chiefs of Staff in Washington saying in part, "I plan to return President Rhee, his cabinet, senior members of the legislature, the U.N. Commission and perhaps others of similar official category to domicile in Seoul as soon as conditions there are sufficiently stable to permit reasonable security." [53] The combat situation in Seoul did not permit final plans for the ceremony until 27 and 28 September.

General MacArthur and his party ar-

[51] 5th Mar Regt SAR in 1st Mar Div SAR, vol. III, an. P, 10–13, 27 Sep 50; 1st Mar Div SAR, vol. I, an. C, G–3 Sec, p. 30; New York *Herald Tribune*, September 28, 1950, Bigart dispatch from Seoul, 27 September.

[52] 1st Mar Div SAR, vol. I, p. 19, an. C, G–3 Sec, pp. 30–31; and an. B, app. 2, p. 16, 28 Sep 50; X Corps WD, G–2 Sec Hist Rpt, PIR 10, 28 Sep 50.

[53] Msg C64159, MacArthur to JCS, 23 Sep 50, CM–IN 14748.

rived at Kimpo Airfield from Tokyo at 1000, 29 September. General Almond and other high-ranking officers met the party and proceeded with it to Seoul. During the night bulldozers had worked to clear the main streets of barricades and the litter of battle. Wildly cheering throngs of South Koreans assembled and lined the streets in the shell-torn, burning city. The 3d Battalion, 1st Marines, provided security along the route from the Han River ponton bridge; the 3d Battalion, 5th Marines, provided security around Government House.[54]

The National Assembly Hall in Government House was packed with selected South Korean officials and citizens and representatives of the combat units that had liberated Seoul. At 1200, General MacArthur came into the chamber with President Rhee and proceeded to the dais where important officers and officials were seated, including General Walker and a few Eighth Army officers who had flown to Seoul on MacArthur's invitation. The Austrian-born wife of President Rhee took her place beside her husband. MacArthur began forthwith to deliver his short address in sonorous voice and unhurried words:

> Mr. President: By the grace of a merciful Providence our forces fighting under the standard of that greatest hope and inspiration of mankind, the United Nations, have liberated this ancient capital city of Korea.

He spoke of the ravages of war that had visited the land, of the righteous wrath and indignation that had caused fifty-three nations to pledge their aid to the Republic of Korea, and of the spiritual revulsion against Communism. Then, turning to President Rhee, he continued:

> In behalf of the United Nations Command I am happy to restore to you, Mr. President, the seat of your government that from it you may better fulfill your constitutional responsibilities.

The assemblage then joined MacArthur in reciting the Lord's Prayer. While MacArthur spoke slivers of glass fell from the partially shattered glass-paneled roof, but he gave them no heed.[55]

Aging but indomitable, Syngman Rhee rose to express the gratitude of the Republic of Korea for the liberation of its capital city. All but overcome with emotion, he departed from his prepared text and stretching out his hands, clenching and unclenching them, spoke to that part of the audience made up of American soldiers, "How can I ever explain to you my own undying gratitude and that of the Korean people?"

The short ceremony over, General MacArthur left at once for Kimpo Airfield and at 1335 departed for Tokyo.

The capture of Seoul led to a series of exchanges between officials of the United States Government and MacArthur. President Truman sent a message which said in part, "Few operations in military history can match either the delaying action where you traded space for time in which to build up your forces, or the brilliant maneuver which has now resulted in the liberation of Seoul." From the Joint Chiefs of Staff

---

[54] Diary of CG X Corps, 29 Sep 50; New York *Herald Tribune*, September 30, 1950; 1st Mar Div SAR, vol. I, an. C, G–3 Sec, 28–29 Sep 50.

[55] DA Public Info Div (PID) file, Miscellaneous Public Statements, Letters, and Documents relating to the Korean War, September 1950; New York *Herald Tribune*, September 30, 1950.

MacArthur received a message which read in part, "Your transition from defensive to offensive operations was magnificently planned, timed, and executed. . . . We remain completely confident that the great task entrusted to you by the United Nations will be carried to a successful conclusion." [56]

MacArthur sent his thanks to the President and the Joint Chiefs of Staff, saying he would publish their messages to all elements of his command.

### *The Blocking Force South of Seoul*

While the greater part of X Corps concentrated its strength before Seoul and was preoccupied with its capture, the blocking force of the 31st Infantry Regiment, 7th Division, thirty miles below the city, was not without action. On 23 September, when Colonel Ovenshine's 31st Infantry Regiment assumed responsibility for Suwon and Suwon Airfield, its mission was to clear the enemy from Suwon and to seize and hold the high ground south of the airfield. Prisoners captured in Suwon by the Reconnaissance Company reported that a regiment of the *105th Armored Division* was in Choch'iwon on the 18th, only fifty air miles to the south, and on its way north to help the Seoul garrison. If this was true, it had to be assumed that this armored force must be approaching the 31st Infantry position. Accordingly, the regiment kept the area south of Suwon under close observation.[57]

During the night of 24 September, the 2d Battalion, 31st Infantry, on high ground (Hill 142) two miles south of the Suwon Airfield came under attack an hour before midnight, and enemy armor struck the battalion's left flank resting on the Suwon-Osan highway. The battalion, with the supporting artillery fire of the 57th Field Artillery Battalion and of B Battery, 15th Field Artillery Battalion, repulsed the attack and knocked out four T34 tanks. The next day the 92d Field Artillery Battalion moved to Suwon to strengthen the forces there. Aerial reconnaissance on the 25th and 26th reported enemy entrenchments in the hills dominating both sides of the highway and rail line just north of Osan, eight to ten miles south of Suwon and two to three miles south of the American position.[58]

On the 26th, Colonel Ovenshine ordered the 2d Battalion to attack and seize the high ground held by the North Koreans near Osan. Interestingly enough, this included the positions where Task Force Smith had met and delayed the North Koreans briefly on 5 July in the first American ground action of the war. The 3d Battalion, less I Company, stood ready to reinforce the 2d Battalion. Ovenshine started the 2d Battalion Task Force—composed of E, F, and part of G Companies and two platoons of tanks—on a wide, flanking movement southeastwardly toward Osan to attack the enemy positions from the rear. At the same time, he formed another attack force composed of elements of G and H Companies, and A

[56] DA PID file, Miscellaneous Public Statements, Letters, and Documents relating to the Korean War, September 1950.

[57] 31st Inf WD, 23 Sep 50; 7th Div WD, 23 Sep 50; X Corps WD, G–2 Hist Rpt, PIR 6, 24 Sep 50, Interrog of Maj Lee Ki J'un.

[58] 31st Inf WD, overlay to POR 10, 24 Sep 50; 7th Div WD, 24–25 Sep 50.

Company, 73d Tank Battalion, to attack south along the highway.

By daylight of the 27th, the flanking force arrived at Osan. A bazooka team destroyed an enemy tank that fired on the column. The force then moved through Osan and engaged the enemy on the hills northward. Attacking down the road simultaneously from the north, the second force was stopped by strong enemy tank, antitank gun, mortar, and small arms fire. Fighting continued throughout the clear, warm autumn day, with the 31st Infantry making only small gains. Prisoners captured in the action said that the enemy force was indeed from the *105th Armored Division*. While its ground gains were slight, the 31st Infantry claimed the destruction or immobilization of 14 tanks, 6 antitank guns, and several mortars, and the infliction of 300 North Korean casualties. The 31st Infantry's two attack forces dug in that night around Hill 113 where the main enemy force was concentrated. Maj. Lester K. Olson, the regimental S-3, and Lt. Col. Robert R. Summers, the 2d Battalion commander, were both seriously wounded during the day.[59]

On the morning of the 28th, the American infantry withdrew at 0830 almost a mile westward from their overnight positions to make sure that they would not suffer casualties from air strikes scheduled to come in against Hills 113 and 92. Beginning at noon and continuing for fifty minutes, seven Navy planes attacked both hills and the railroad tunnel area just east of Hill 92, using napalm extensively. When the air strikes ended, the 57th and 92d Field Artillery Battalions pounded the hills for thirty minutes, the Heavy Mortar Company joining in the preparation. When it ended, K and L Companies attacked eastward against Hill 113. By 1515 they had secured the hill against only light resistance, and from there L Company attacked across the saddle to Hill 92, 600 yards away, supported by K Company fire from Hill 113. An hour later the 31st Infantry held both hills—taken without a single casualty to itself. Surviving enemy troops withdrew eastward. The road between Suwon and Osan was open. The next day the 31st Infantry buried more than a hundred enemy dead on the captured positions.[60]

While the 31st Infantry was clearing the Osan highway, the 2d Battalion, 17th Infantry, 7th Division, fought its first battle of the war on 29 September in a heavy fire fight against an enemy force at the southeast side of Seoul. In this action, which continued after dark, the battalion suffered seventy-nine casualties. That night the 49th Field Artillery Battalion laid down a barrage which effectively broke up an attempted enemy counterattack. The North Koreans reportedly suffered more than 400 men killed.[61]

On 30 September the 1st Marine Di-

[59] 31st Inf WD, 26–28 Sep 50; 7th Div WD, 26–27 Sep 50; X Corps WD, G–3 Sec, Msg J–52, 2315 27 Sep 50; Interv, author with Barr, 1 Feb 54.

[60] 31st Inf WD, and PIR 9, an. 2, 28 Sep 50; 7th Div WD, 28–29 Sep 50. General Almond was dissatisfied with Colonel Ovenshine's handling of the 31st Infantry in the action below Suwon and relieved him of command on 5 October. Intervs, author with Almond, 13 Dec 51, and Barr, 1 Feb 54.

[61] 7th Div WD, 29 Sep 50; 17th Inf WD, 28–29 Sep 50; Barr, Notes, 29 Sep 50, and MS review comments, Nov 57. The third regiment of the 7th Division, the 17th Infantry, arrived at Inch'on on the 24th from Pusan, where it had been held in floating reserve for Eighth Army, and began unloading the next day.

vision assumed responsibility for the 32d Infantry zone in Seoul and that unit then crossed back to the south side of the Han River.

### *The X Corps Situation*

After the capture of Seoul, the 1st Marine Division cleared enemy troops from the northern environs of the city. On 1 October, elements of the 5th Marines patrolled the P'yongyang highway as far as Munsan-ni and the Imjin River. They encountered only scattered individual enemy riflemen except in the vicinity of Munsan-ni.

The 7th Marines, at the same time, advanced up the Uijongbu road north of the city against almost no resistance, hampered only by mines. But the next day, 2 October, the regiment made virtually no gains. Three battalions of the *31st Regiment*, N.K. *31st Division*, well dug-in on either side of the highway, stopped the regiment in hard fighting three miles south of Uijongbu in the vicinity of Nuwon-ni. There high mountains closing in on either side of the highway created a natural fortress.[62]

During the night, the enemy blocking force withdrew northward, and on 3 October tanks led the 2d Battalion into Uijongbu in the afternoon. Marine and Navy air strikes had completely destroyed the town. The 7th Marines occupied the high ground just north of Uijongbu and consolidated its position around the town for the night. The fighting of 2–3 October in front of Uijongbu was the last organized resistance the 1st Marine Division encountered in the Inch'on-Seoul operation.[63]

Before being driven from Seoul, the North Koreans had taken ghastly revenge on men, women, and children in the families of South Korean soldiers, policemen, and guerrillas.

The Seoul operation disclosed that the preinvasion estimate of 5,000 organized troops in the city was low and that, instead, there were approximately 8,000 such troops in Seoul and 5,000 more in the Yongdungp'o area. Reinforcements after the landing at Inch'on brought the total enemy troops in Seoul to at least 20,000. And there were at least 10,000 enemy soldiers between the Han River and Suwon. Below Suwon in the Osan area there were from 2,000 to 3,000 more. It appears that altogether somewhat more than 30,000 North Korean troops entered battle in the Inch'on-Suwon-Seoul area, and that there were perhaps 10,000 more miscellaneous soldiers in the vicinity, uncommitted or who arrived too late to be used. The X Corps reported 7,000 North Korean prisoners taken in the fighting and estimated enemy troops killed at 14,000.[64]

The 1st Marine Division did not lose a single tank to enemy tank action in the Seoul operation but lost several to enemy infantry action. An accurate

[62] 1st Mar Div SAR, vol. I, an. C, G–3 Sec, pp. 36–37, 1–2 Oct 50; *Ibid.*, vol. III, an. RR, p. 24.

[63] 1st Mar Div SAR, an. C, G–3 Sec, p. 38, 3 Oct 50; Interv, author with Col Harold K. Johnson, 4 Jan 52.

[64] 1st Mar Div SAR, vol. I, p. 21, and an. B, app. 2, p. 16, 28 Sep, and p. 24, 5 Oct 50; X Corps WD, 30 Sep 50; 32d Inf WD, 30 Sep 50. The 1st Marine Division captured 4,792 prisoners and claimed to have inflicted 13,666 enemy casualties. In the 7th Division the 32d Infantry Regiment captured 1,203 prisoners and estimated it had killed 3,000 enemy troops; the 31st and 17th Infantry Regiments each inflicted several hundred casualties. Estimates of enemy casualties inflicted by ROK units are not available.

count of the enemy tanks destroyed in the X Corps operation is hard to make, but it appears that approximately 45 to 50 were destroyed in the Inch'on-Yongdungp'o-Seoul area and about 10 to 15 more in the Suwon-Osan area, or about 60 altogether. The North Koreans lost a great amount of other military equipment in the Seoul operation. The 1st Marine Division alone reported that it had destroyed or captured 23 120-mm. mortars, 19 45-mm. antitank guns, 56 heavy machine guns, 337 light and sub-machine guns, 59 14.5 antitank rifles, and 7,543 rifles.[65]

The Inch'on-Seoul victory cost the United Nations forces approximately 3,500 casualties. The 7th Infantry Division suffered 572 battle casualties, including 106 killed, 409 wounded, and 57 missing in action. Of the total, 166 were ROK soldiers integrated into the division. Within the division, the 32d Regiment lost 66 killed, 272 wounded, and 47 missing. The heaviest losses in X Corps occurred in the 1st Marine Division which suffered total casualties of 2,383 men—364 killed, 53 who died of wounds, 1,961 wounded, and 5 missing. Marine losses were heaviest for the six days from 21 to 27 September. During that time it suffered 1,482 battle casualties, the greatest single day's loss being 285 on 24 September.[66]

[65] 1st Mar Div SAR, vol. II, an. OO, p. 42, 7 Oct 50, and an. PP, p. 14; *Ibid.*, vol. I, an. B, app. 2, p. 24, 5 Oct 50; 7th Div WD, 18–30 Sep 50.

[66] X Corps WD, 30 Sep 50; 7th Div WD, Narr, 30 Sep 50; 32d Inf WD, 30 Sep 50; 1st Mar Div SAR, vol. I, p. 20, and an. A, G–1 Sec, p. 6; Smith, MS review comments, 25 Feb 54. In a letter to the author, 13 February 1954, General Smith claimed the 1st Marine Division suffered 2,430 battle casualties in the Inch'on-Seoul operation. Montross and Canzona, *The Inchon-Seoul Operation*, page 297, gives Marine casualities as 366 KIA, 49 DOW, 6 MIA, and 2,029 WIA, for a total of 2,450. The figure 2,383 is that given in Marine records cited.

# CHAPTER XXVII

# Breaking the Cordon

Tactics are based on weapon-power . . . strategy is based on movement . . . movement depends on supply.

J. F. C. Fuller, *The Generalship of Ulysses S. Grant*

The Inch'on landing put the United States X Corps in the enemy's rear. Concurrently, Eighth Army was to launch a general attack all along its front to fix and hold the enemy's main combat strength and prevent movement of units from the Pusan Perimeter to reinforce the threatened area in his rear. This attack would also strive to break the enemy cordon that had for six weeks held Eighth Army within a shrinking Pusan Perimeter. If Eighth Army succeeded in breaking the cordon it was to drive north to effect a juncture with X Corps in the Seoul area. The battle line in the south was 180 air miles at its closest point from the landing area in the enemy's rear, and much farther by the winding mountain roads. This was the distance that at first separated the anvil from the hammer which was to pound to bits the enemy caught between them.

Most Eighth Army staff officers were none too hopeful that the army could break out with the forces available. And to increase their concern, in September critical shortages began to appear in Eighth Army's supplies, including artillery ammunition. Even for the breakout effort Eighth Army had to establish a limit of fifty rounds a day for primary attack and twenty-five rounds for secondary attack. Fortunately, the *Aripa* arrived in the Far East with a cargo of 105-mm. howitzer shells in time for their use in the offensive. But, despite some misgivings, General Walker and his chief of staff, General Allen, believed that if the Inch'on landing succeeded Eighth Army could assume the offensive and break through the enemy forces encircling it.[1]

## *The Eighth Army Plan*

The Eighth Army published its attack plan on 6 September and the next day General Allen sent it to Tokyo for approval. Eighth Army revised the plan on 11 September, and on the 16th made it an operations directive. It set the hour for attack by United Nations and ROK forces in the Perimeter at 0900, 16 September, one day after the Inch'on landing. The U.S. Eighth and the ROK Armies were to attack "from present

[1] Interv, author with Lt Col Paul F. Smith, 7 Oct 52; Ltr, Maj Gen Leven C. Allen to author, 10 Jan 54; Brig Gen Edwin K. Wright, FEC, Memo for Record, 4 Sep 50; Interv, author with Stebbins (EUSAK G–4 Sep 50), 4 Dec 53; Interv, author with Maj Gen George L. Eberle, 12 Jan 54.

bridgehead with main effort directed along the Taegu-Kumch'on-Taejon-Suwon axis," to destroy the enemy forces "on line of advance," and to effect a "junction with X Corps." [2]

The operations directive required the newly formed United States I Corps in the center of the Perimeter line to strive for the main breakthrough. The following reasons dictated this concept: (1) the distance to the link-up area with X Corps was shorter than that from elsewhere around the Perimeter, (2) the road net was better and had easier grades, (3) the road net offered the armor better opportunity to exploit a breakthrough, and (4) supply to advancing columns would be easier. The plan called for the 5th Regimental Combat Team and the 1st Cavalry Division to seize a bridgehead over the Naktong River near Waegwan. The 24th Division would then cross the river and drive on Kumch'on-Taejon, followed by the 1st Cavalry Division which would patrol its rear and lines of communications. While this breakthrough attempt was in progress, the 25th and 2d Infantry Divisions in the south on the army left flank and the ROK II and I Corps on the east and right flank were to attack and fix the enemy troops in their zones and to exploit any local breakthrough. The ROK 17th Regiment was to move to Pusan for water movement to Inch'on to join X Corps.

Supplementing the 5th Regimental Combat Team's mission of establishing a bridgehead across the Naktong, the U.S. 2d and 24th Divisions were to strive for crossings of the river below Waegwan and the ROK 1st Division above it. Execution of this plan was certain to run into difficulties because the Engineer troops and bridging equipment available to General Walker were not adequate for several quick crossings. Eighth Army had equipment for only two ponton treadway bridges across the Naktong.

To help replace the Marine Air squadrons taken from the Eighth Army front for the X Corps operation at Inch'on, General Stratemeyer obtained the transfer from the 20th Air Force on Okinawa to Itazuke, Japan, of the 51st Fighter-Interceptor Wing and the 16th and 25th Fighter-Interceptor Squadrons.

The situation at the Pusan Perimeter did not afford General Walker an opportunity to concentrate a large force for the breakout effort in the center. The enemy held the initiative and his attacks pinned down all divisions under Eighth Army command except one, the U.S. 24th Infantry Division, which Walker was able to move piecemeal from the east to the center only on the eve of the projected attack. The problem was to change suddenly from a precarious defense to the offensive without reinforcement or opportunity to create a striking force. In theater perspective, Eighth Army would make a holding attack while the X Corps made the envelopment. A prompt link-up with the X Corps along the Taejon-Suwon axis was a prerequisite for cutting off a large force of North Koreans in the southwestern part of the peninsula.

Eighth Army anticipated that the news of the Inch'on landing would have a demoralizing effect on the North Koreans in front of it and an opposite effect on the spirit of its own troops. For this rea-

[2] EUSAK Opn Plan 10, 6 Sep 50, and Revision, 11 Sep 50; EUSAK WD, G-3 Plans Sec, 7 Sep 50; *Ibid.*, 15-16 Sep 50; I Corps WD, G-3 Sec, 2 Aug-30 Sep 50.

son, General Walker had requested that the Eighth Army attack not begin until the day after the Inch'on landing. While the news of the successful landing spread to Eighth Army troops at once on the 15th, apparently it was not allowed to reach the enemy troops in front of Eighth Army until several days later.

The corridors of advance in the event of a breakout from the Perimeter necessarily would be the same that the North Korean Army had used in its drive south. Enemy forces blocked every road leading out of the Perimeter. The axis of the main effort required the use of the highway from the Naktong opposite Waegwan to Kumch'on and across the Sobaek Range to Taejon. A second corridor, the valley of the Naktong northward to the Sangju area, could be used if events warranted it. The Taegu–Tabudong–Sangju road traversed this corridor, with crossings of the Naktong River possible at Sonsan and Naktong-ni. From Sangju the line of advance could turn west toward the Kum River above Taejon or bypass Taejon for a more direct route to the Suwon-Seoul area.

Eastward in the mountainous central sector, the ROK's would find the best route of advance by way of Andong and Wonju. On the east coast they had no alternative to a drive straight up the coastal road toward Yongdok and Wonsan.

An important step taken by the Far East Command in preparation for the offensive was the establishment of corps organization within Eighth Army. Up to this time Eighth Army had controlled directly the four infantry divisions and other attached ground forces of regimental and brigade size. Beginning in August, preparations were made to provide Eighth Army with two corps.

On 2 August, I Corps was activated at Fort Bragg, N.C., with General Coulter in command. Eleven days later General Coulter and a command group arrived in Korea and began studies preparatory to a breakout effort from the Perimeter. The main body of the corps staff arrived in Korea on 6 September, but it still had no troops assigned to it.[3]

The IX Corps was activated on 10 August at Fort Sheridan, Ill., with Maj. Gen. Frank W. Milburn in command. General Milburn and a small group of staff officers departed Fort Sheridan on 5 September by air for Korea. The main body of the corps staff, however, did not reach Korea until the end of September and the first part of October. Both I and IX Corps had previously been part of Eighth Army in Japan, the I Corps with the 24th and 25th Divisions with headquarters in Kyoto, and the IX Corps with the 1st Cavalry and the 7th Divisions with headquarters in Sendai.[4]

General Walker had decided to group the main breakout forces under I Corps. He gave long and serious thought to the question of a commander for the corps. Walker eventually shifted General Milburn on 11 September from IX Corps to I Corps and General Coulter from I Corps to IX Corps. Milburn assumed command of I Corps that day at Taegu and Coulter assumed command of IX Corps the next day at Miryang. I Corps

[3] I Corps WD, Hist Narr, 2 Aug–30 Sep 50; EUSAK WD, Aug 50 Summ.

[4] IX Corps WD, Hist Narr, 23–30 Sep 50. It is interesting to note that I and IX Corps had been deactivated in Japan only a few months before, in the early part of 1950, in line with maintaining the framework of four divisions and remaining within reduced army personnel ceilings.

became operational at 1200, 13 September, with the U.S. 1st Cavalry Division, the 5th Regimental Combat Team (–), and the ROK 1st Division attached. On 15 and 16 September the 5th Regimental Combat Team and the 24th Division moved to the Taegu area, and by the evening of 16 September I Corps comprised the U.S. 24th and 1st Cavalry Divisions, the 5th Regimental Combat Team, the British 27th Infantry Brigade, the ROK 1st Division, and supporting troops.[5]

During the first week of the Eighth Army offensive the IX Corps was not operational. It became so at 1400, 23 September, on Eighth Army orders which attached to it the U.S. 25th and 2d Infantry Divisions and their supporting units. Until 23 September, therefore, these two divisions operated directly under Eighth Army command.[6]

IX Corps was not made operational at the same time as I Corps principally because of a critical lack of communications personnel and equipment. The Signal battalion and the communications equipment intended for this corps had been diverted to X Corps. Even after IX Corps became operational the lack of proper communications facilities hampered its operations.[7]

[5] Landrum, Notes for author, recd 8 Mar 54; EUSAK Special Ord 49, 11 Sep 50; EUSAK WD, G–3 Sec, 13 and 16 Sep 50; I Corps WD, G–3 Sec, 12–19 Sep 50; EUSAK POR 195, 15 Sep 50; 24th Div WD, 15–16 Sep 50. The 3d Battalion of the 19th Regiment, 24th Division, remained at Samnangjin on the lower Naktong as a left flank guard force.

[6] IX Corps WD, Hist Summ, 23–30 Sep 50; EUSAK WD, POR 217, 22 Sep 50 (POR erroneously dated 200001).

[7] Landrum, Notes for author, recd 8 Mar 54.

### *The Enemy Strength*

On the eve of Eighth Army's attack, the intelligence annex to the army order presented an elaborate estimate of the enemy's strength, order of battle, and capabilities. This gave the North Koreans 13 infantry divisions on line supported by 1 armored division and 2 armored brigades, with the N.K. *I Corps* on the southern half of the front having 6 infantry divisions with armored support—a strength of 47,417 men, and the *II Corps* on the northern and eastern half of the front having 7 infantry divisions with armored support—a strength of 54,000 men. This made a total of 101,417 enemy soldiers around the Perimeter. Eighth Army intelligence estimated enemy organizations at an average of 75 percent strength in troops and equipment.[8]

The Eighth Army estimate credited the enemy with sufficient strength to be able to divert three divisions from the Pusan Perimeter to the Seoul area without endangering his ability to defend effectively his positions around the Perimeter. The estimate stated, "Currently the enemy is on the offensive and retains this capability in all general sectors of the Perimeter. It is not expected that this capability will decline in the immediate future."

With respect to both enemy troop strength and equipment the Eighth Army estimate was far too high. Although it is not possible to state precisely the strength of the North Korean units facing Eighth Army in mid-September and the state of their equipment, an examination of prisoner of war inter-

[8] EUSAK WD, 16 Sep 50, app. 1 to an. A (Intel) to Opn Plan 10 (as of 10 Sep 50).

rogations and captured documents reveals that it was far less than Eighth Army thought it was. The Chief of Staff, N.K. *13th Division,* Col. Lee Hak Ku, gave the strength of that divison as 2,300 men (not counting 2,000 untrained and unarmed replacements not considered as a part of the division) instead of the 8,000 carried in the Eighth Army estimate. The N.K. *15th Division,* practically annihilated by this time, numbered no more than a few hundred scattered and disorganized men instead of the 7,000 men in the Eighth Army estimate. Also, the N.K. *5th Division* was down to about 5,000 men instead of 6,500, and the N.K. *7th Division* was down to about 4,000 men instead of the 7,600 accorded it by the Eighth Army estimate. The N.K. *1st, 2d,* and *3d Divisions* almost certainly did not begin to approach the strength of 7,000–8,000 men each in mid-September accorded to them in the estimate.[9]

Enemy losses were exceedingly heavy in the first half of September. No one can accurately say just what they were. Perhaps the condition of the North Korean Army can best be glimpsed from a captured enemy daily battle report, dated 14 September, and apparently for a battalion of the N.K. *7th Division.* The report shows that the enemy battalion on 14 September had 6 officers, 34 noncommissioned officers, and 111 privates for a total of 151 men. There were 82 individual weapons in the unit: 3 pistols, 9 carbines, 57 rifles, and 13 automatic rifles. There was an average of somewhat more than 1 grenade for every 2 men—a total of 92 grenades. The unit still had 6 light machine guns but less than 300 rounds of ammunition for each.[10]

A fair estimate of enemy strength facing Eighth Army at the Perimeter in mid-September would be about 70,000 men. Enemy equipment, far below the Eighth Army 75 percent estimate of a few days earlier, particularly in heavy weapons and tanks, was probably no more than 50 percent of the original equipment.

Morale in the North Korean Army was at a low point. No more than 30 percent of the original troops of the divisions remained. These veterans tried to impose discipline on the recruits, most of whom were from South Korea and had no desire to fight for the North Koreans. It was common practice in the North Korean Army at this time for the veterans to shoot anyone who showed reluctance to go forward when ordered or who tried to desert. Food was scarce, and undernourishment was the most frequently mentioned cause of low morale by prisoners. Even so, there had been few desertions up to this time because the men were afraid the U.N. forces would kill them if they surrendered and that their own officers would shoot them if they made the attempt.[11]

[9] ATIS Interrog Rpts, Issue 9 (N.K. Forces), Rpt 1468, pp. 158ff, Col Lee Hak Ku; *Ibid.,* Issue 7, Rpt 1253, p. 112, Sr Lt Lee Kwan Hyon; ATIS Res Supp Interrog Rpts, Issue 3 (N.K. *15 Div*), p. 44; *Ibid.,* Issue 96 (N.K. *5th Div*), pp. 43–44; *Ibid.,* Issue 99 (N.K. *7th Div*), p. 38; *Ibid.,* Issue 4 (N.K. *105th Armored Div*), p. 39; *Ibid.,* Issue 100 (N.K. *9th Div*), p. 52; ATIS Interrog Rpts, Issue 22 (N.K. Forces), p. 4; EUSAK WD, 30 Sep 50, G–2 Sec, interrog of Maj Lee Yon Gun, Asst Regt CO, *45th Regt, 15th Div.*

[10] 35th Inf WD, PW Interrrog Team Rpt by Lt Herada, 151500 Sep 50.

[11] U.N. forces had captured and interned at the Eighth Army enclosure 3,380 N.K. prisoners by 15 September. The ROK Army had captured 2,254 of them; Eighth Army, 1,126. See EUSAK WD, Incl 16, Provost Marshal Rpt, 15 Sep 50.

### *United Nations' Perimeter Strength*

Standing opposite approximately 70,000 North Korean soldiers at the Pusan Perimeter in mid-September were 140,000 men in the combat units of the U.S. Eighth and ROK Armies. These comprised four U.S. divisions with an average of 15,000 men each for a total of more than 60,000 men, to which more than 9,000 attached South Korean recruits must be added, and six ROK divisions averaging about 10,000 men each with a total of approximately 60,000 men. The three corps headquarters added at least another 10,000 men, and if the two army headquarters were counted the total would be more than 150,000 men. The major U.N. units had an assigned strength at this time as follows: [12]

| | | |
|---|---|---|
| U.S. Eighth Army | | 84,478 |
| U.S. I Corps (plus attached Koreans, 1,110) | 7,475 | |
| U.S. 1st Cavalry Division (plus attached Koreans, 2,338) | 13,904 | |
| U.S. 24th Division (plus attached Koreans, 2,786) | 16,356 | |
| U.S. 2d Division (plus attached Koreans, 1,821) | 15,191 | |
| U.S. 25th Division (plus attached Koreans, 2,447) | 15,334 | |
| British 27th Infantry Brigade | 1,693 | |
| ROK Army | | 72,730 |

Since it marked a turning point in the Korean War, the middle of September 1950 is a good time to sum up the cost in American casualties thus far. From the beginning of the war to 15 September 1950, American battle casualties totaled 19,165 men. Of this number, 4,280 men were killed in action, 12,377 were wounded, of whom 319 died of wounds, 401 were reported captured, and 2,107 were reported missing in action. The first fifteen days of September brought higher casualties than any other 15-day period in the war, before or afterward, indicating the severity of the fighting at that time.[13]

The assigned strength of the U.S. divisions belied the number of men in the rifle companies, the men who actually did the fighting. Some of the rifle companies at this time were down to fifty or fewer effectives—little more than 25 percent strength. The Korean augmentation recruits, virtually untrained and not yet satisfactorily integrated were of little combat value at this time.

While perhaps 60,000 of the 70,000 ROK Army soldiers were in the line, most of the ROK divisions, like those of the North Korean Army, had sunk to a low level of combat effectiveness because of the high casualty rate among the trained commissioned and noncommissioned officers and the large percentage of recruits among the rank and file. After taking these factors into account, however, any realistic analysis of the strength of the two opposing forces must

[12] GHQ FEC Sitrep, 16 Sep 50; GHQ FEC G–3 Opn Rpt, 16 Sep 50. U.S. Air Force strength in Korea was 4,726 men. The total U.N. supported strength in Korea was 221,469 men, of which about 120,000 were in the ROK Army, 83,000 assigned, and 30,000-odd in training. See EUSAK POR 65, G–4 Sec, 15 Sep 50. NAVFE strength was 52,011 men.

[13] Battle Casualties of the Army, 31 May 52, DA TAGO.

give a considerable numerical superiority to the United Nations Command.[14]

In the matter of supporting armor, artillery, and heavy weapons and the availability of ammunition for these weapons, the United Nations Command had an even greater superiority than in troops, despite the rationing of ammunition for most artillery and heavy weapons. Weapon fire power superiority was probably about six to one over the North Koreans. In the air the Far East Air Forces had no rival over the battleground, and on the flanks at sea the United Nations naval forces held unchallenged control.[15]

### *The 38th Infantry Crosses the Naktong*

The morning of 16 September dawned over southern Korea with murky skies and heavy rain. The weather was so bad the Air Force canceled a B–29 saturation bombing scheduled against the enemy positions in the Waegwan area.

The general attack set for 0900 did not swing into motion everywhere around the Perimeter at the appointed hour for the simple reason that at many places the North Koreans were attacking and United Nations troops defending. In most sectors an observer would have found the morning of 16 September little different from that of the 15th or the 14th or the 13th. It was the same old Perimeter situation—attack and counterattack. The battle for the hills had merely gone on into another day. Only in a few places were significant gains made on the first day of the offensive. (*Map 17*) The 15th Regiment of the ROK 1st Division advanced to the right of the North Korean strongpoint at the Walled City north of Taegu in a penetration of the enemy line. Southward, the U.S. 2d Division after hard fighting broke through five miles to the hills overlooking the Naktong River.[16]

The most spectacular success of the first day occurred in the 2d Division zone. There, west of Yongsan and Changnyong, the 2d Division launched a 3-regiment attack with the 9th Infantry on the left, the 23d Infantry in the center, and the 38th Infantry on the right. Its first mission was to drive the enemy *4th, 9th,* and *2d Divisions* back across the Naktong. The attack on the left failed as the enemy continued to hold Hill 201 against all attacks of the 9th Infantry. In the center, a vicious enemy predawn attack penetrated the perimeter of C Company, 23d Infantry, and caused twenty-five casualties, which included all company officers and the platoon leader of the attached heavy weapons platoon.

On the 15th, the 3d Battalion had returned to regimental control from attachment to the 1st Cavalry Division, and because it had not been involved in the preceding two weeks of heavy fighting, Colonel Freeman assigned it the main attack effort in the 23d Infantry zone. After the early morning attack on

[14] Lt. Gen. Chung Il Kwon commanded the ROK Army. The six ROK divisions were the following: 1st Division—11th, 12th, 15th Regiments; 3d Division—22d, 23d, 26th Regiments; 6th Division—2d, 7th, 19th Regiments; 7th Division—3d, 5th, 8th Regiments; 8th Division—10th, 16th, 21st Regiments; and Capital Division—1st, 17th, 18th Regiments. See EUSAK WD, Br for CG, 12 Sep 50.

[15] EUSAK WD, Arty Rpt, 11 Sep 50.

[16] EUSAK POR 198, 16 Sep 50; I Corps WD, 16 Sep 50; 2d Inf WD, Sep 50; 38th Inf WD, 16 Sep 50.

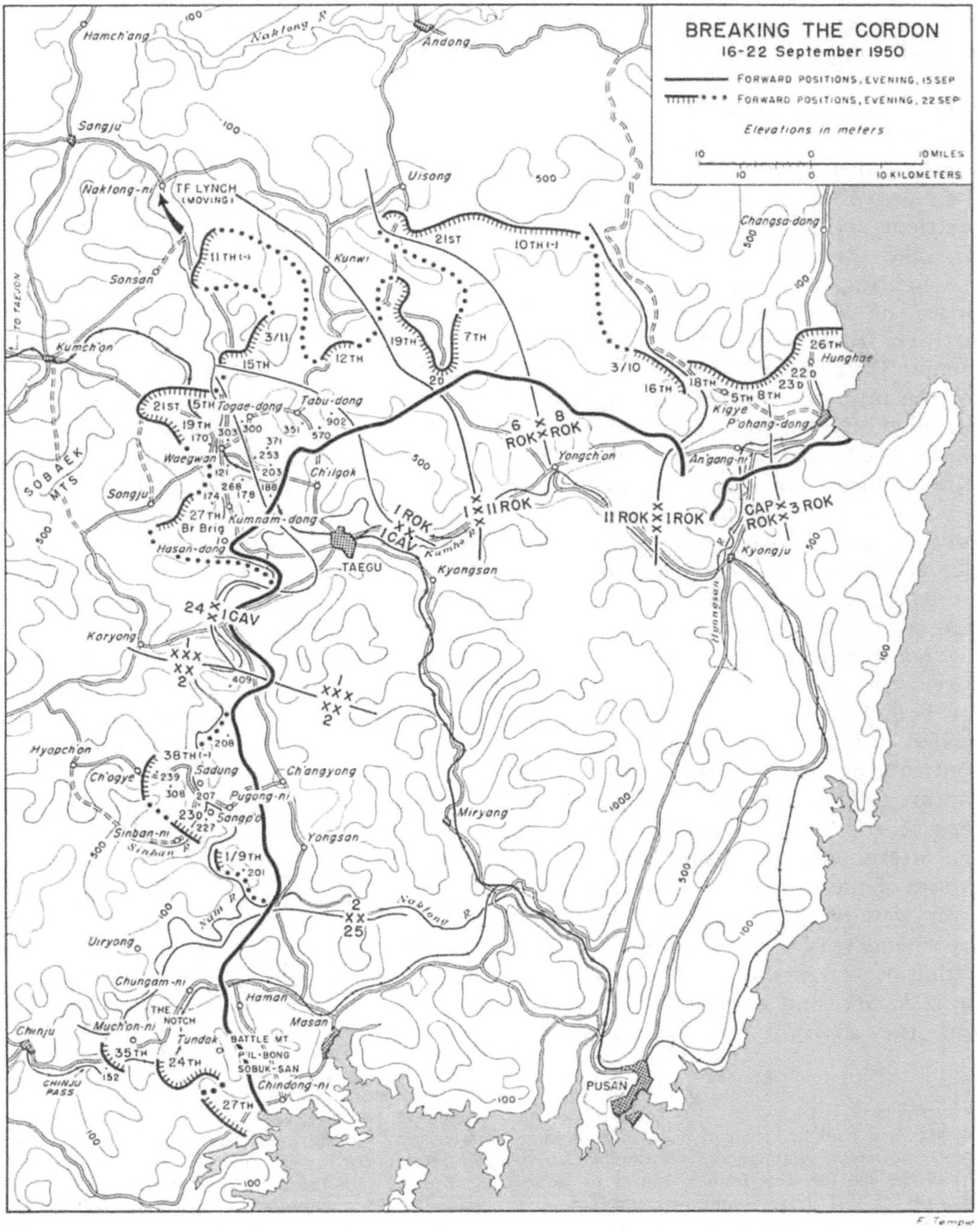

BREAKING THE CORDON
16-22 September 1950
FORWARD POSITIONS, EVENING, 15 SEP
FORWARD POSITIONS, EVENING, 22 SEP
Elevations in meters
10 0 10 MILES
10 0 10 KILOMETERS
Hamch'ang
Naktong R
Andong
Sangju
Uisong
Naktong-ni
TF LYNCH (MOVING)
Changsa-dong
Kunwi
Sonsan
Kumch'on
Hunghae
Kigye
P'ohang-dong
Togae-dong
Tabu-dong
Waegwan
Yongch'on
An'gang-ni
Ch'ilgok
SOBAEK MTS
Songju
Kumnam-dong
Br Brig
Hasan-dong
TAEGU
Kumho R
Kyongsan
Kyongju
Koryong
Hyopch'on
Ch'ogye
Sadung
Ch'angyong
Pugong-ni
Sangp'o
Sinban-ni
Sinban R
Miryang
Yongsan
Nam R
Uiryong
Chungam-ni
THE NOTCH
Haman
Masan
Chinju
Much'on-ni
Tundok
BATTLE MT
P'IL-BONG
SOBUK-SAN
Chindong-ni
CHINJU PASS
PUSAN
6 ROK
8 ROK
1 ROK
1 CAV
II ROK
CAP ROK
3 ROK
24
2
25
F. Temple


MAP 17

the 16th was repulsed, Lt. Col. R. G. Sherrard ordered his 3d Battalion to move out at 1000 in attack, with C Company of the 72d Tank Company in support. Enemy resistance was stubborn and effective until about midafternoon when the North Koreans began to vacate their positions and flee toward the Naktong. To take advantage of such a break in the fighting, a special task force comprised of B Battery, 82d Antiaircraft Artillery Battalion, and the 23d Regimental Tank Company had been formed for the purpose of advancing rapidly to cut off the North Korean soldiers. From about 1600 until dark this task force with its heavy volume of automatic fire cut down large numbers of fleeing enemy along the river. The weather had cleared in the afternoon and numerous air strikes added to the near annihilation of part of the routed army.[17]

The 38th Infantry on the right kept pace with the 23d Infantry in the center. Four F–51's napalmed, rocketed, and strafed just ahead of the 38th Infantry, contributing heavily to the 2d Battalion's capture of Hill 208 overlooking the Naktong River. Fighter planes operating in the afternoon caught and strafed large groups of enemy withdrawing toward the river west of Changnyong. That night the enemy's *2d Division* command post withdrew across the river, followed by the *4th, 6th,* and *17th Rifle Regiments* and the division artillery regiment. Their crossings continued into the next day.[18]

On the 17th, air attacks took a heavy toll of enemy soldiers trying to escape across the Naktong in front of the 2d Division. During the day, fighter planes dropped 260 110-gallon tanks of napalm on the enemy in this sector and strafed many groups west of Changnyong. The fleeing enemy troops abandoned large quantities of equipment and weapons. In pursuit the 23d Infantry captured 13 artillery pieces, 6 antitank guns, and 4 mortars; the 38th Infantry captured 6 artillery pieces, 12 antitank guns, 1 SP gun, and 9 mortars. General Allen, Eighth Army chief of staff, in a telephone conversation with General Hickey in Tokyo that evening said, "Things down here [Pusan Perimeter] are ripe for something to break. We have not had a single counterattack all day.[19]

During the morning of 18 September patrols of the 2d and 3d Battalions, 38th Infantry, crossed the Naktong near Pugong-ni, due west of Changnyong, and

[17] Maj Gen Paul L. Freeman, Jr., MS review comments, 30 Oct 57; Highlights of the Combat Activities of the 23d Inf Regt from 5 Aug 50 to 30 Sep 50; 23d Inf WD, Narr Summ, Sep 50; *Ibid.*, G–3 Jnl, entry 151, 160207 Sep 50. Task Force Haynes, which had defended the Changnyong area since 1 September, was dissolved on 15 September.

[18] 2d Div WD, G–3 Sec, Sep 50; 23d Inf WD, Narr Summ, Sep 50; *Ibid.*, G–3 Jnl, entry 151, 160207 Sep 50; EUSAK PIR 66, 16 Sep 50; 38th Inf Comd Rpt, Sep–Oct 50; FEAF Opns Hist, vol. I, 25 Jun–31 Oct 50, p. 168; ATIS Res Supp Interrog Rpts, Issue 94 (N.K. *2d Div*), p. 38; ATIS Interrog Rpts, Issue 7 (N.K. Forces), Rpts 1208, 1233, 1242, pp. 19, 69, 82 and 131. The senior medical officer of the *17th Regiment*, captured on the 17th, estimated that each of the three regiments of the *2d Division* had only approximately 700 men left. See EUSAK WD, 21 Sep 50, ADVATIS Interrog Rpt of Sr Lt Lee Kwan Hyon.

[19] FEAF Opns Hist, vol. I, 25 Jun–31 Oct 50, p. 170; "Air War in Korea," *Air University Quarterly Review* IV, (Spring, 1951), No. 3, 70; 23d Inf Comd Rpt, Jnl entry 175, 17 Sep 50, and Pers Rpt 13, 17 Sep 50; 38th Inf Comd Rpt, Sep–Oct 50, p. 8; ATIS Res Supp Interrog Rpts, Issue 106 (N.K. Arty), p. 51; Fonecon, Allen with Hickey, GHQ FEC CofS, 17 Sep 50; EUSAK WD, G–3 Air, 17 Sep 50.

found the high ground on the west side of the river clear of enemy troops. Colonel Peploe, regimental commander, thereupon ordered Lt. Col. James H. Skeldon, 2d Battalion commander, to send two squads across the river in two-man rubber boats, with a platoon to follow, to secure a bridgehead. Peploe requested authority to cross the river in force at once. At 1320 Col. Gerald G. Epley, 2d Division chief of staff, authorized him to move one battalion across the river.

Before 1600, E and F Companies and part of G Company had crossed the 100-yard-wide and 12-foot-deep current. Two hours later the leading elements secured Hill 308 a mile west of the Naktong, dominating the Ch'ogye road, against only light resistance. This quick crossing clearly had surprised the enemy. From Hill 308 the troops observed an estimated enemy battalion 1,000 yards farther west. That evening Colonel Skeldon requested air cover over the bridgehead area half an hour after first light the next morning.

During the day, the 38th Infantry captured 132 prisoners; 32 of them were female nurses, 8 were officers—1 a major. Near the crossing site on the east bank buried in the sand and hidden in culverts, it found large quantities of supplies and equipment, including more than 125 tons of ammunition, and new rifles still packed in cosmoline.[20]

The 38th Infantry's crossing of the Naktong by the 2d Battalion on 18 September was the first permanent crossing of the river by any unit of Eighth Army in the breakout, and it was the most important event of the day. The crossing was two days ahead of division schedule.

On the 19th the 3d Battalion, 38th Infantry, crossed the river, together with some tanks, artillery, and heavy mortars. The 3d Battalion was to protect the bridgehead while the 2d Battalion pushed forward against the enemy. In order to support the two battalions now west of the river it was necessary to get vehicles and heavy equipment across to that side. The two destroyed spans of the Changnyong-Ch'ogye highway bridge across the Naktong could not be repaired quickly, so the 2d Engineer Combat Battalion prepared to construct a floating bridge downstream from the crossing site.

By the end of the third day of the attack, 18 September, the U.S. 2d Division had regained control of the ground in its sector east of the Naktong River except the Hill 201 area in the south and Hill 409 along its northern boundary. Elements of the N.K. *9th Division* had successfully defended Hill 201 against repeated air strikes, artillery barrages, and attacks of the 9th Infantry. At its northern boundary Eighth Army, for the moment, made no effort to capture massive Hill 409. There, air strikes, artillery barrages, and patrol action of the 1st Battalion, 38th Infantry, merely attempted to contain and neutralize this enemy force of the *10th Division*. Behind the 2d Division lines there were many enemy groups, totaling several hundred soldiers, cut off and operating as far as twenty miles east of the river. During the 18th, a 22-man patrol of the 23d Infantry came to grief in trying to

[20] Interv, author with Peploe, 12 Aug 51; 2d Div POR 132, 18 Sep 50; 2d Div WD, 18 Sep 50 and CofS Log entries 121 and 124, 18 Sep 50; 38th Inf Comd Rpt, Sep–Oct 50, pp. 9–10; 2d Div WD, G–3 Jnl, Msg 74, 181625 Sep 50; EUSAK WD, 18 Sep 50 and G–3 Jnl, Msg 181745; 2d Div WD, G–2 Jnl, entry 2107, 181525 Sep 50, and PIR 25, 18 Sep 50.

THE BATTLE FOR HILL 201. *This photograph, taken on 18 September, shows 9th Infantry soldiers helping a wounded man to the rear.*

cross the Naktong, partly because of the river's depth. Enemy fire from the west bank killed three, wounded another, and drove the rest of the patrol back to the east side.[21]

### *The 5th Regimental Combat Team Captures Waegwan*

The 5th Regimental Combat Team was attached to the 1st Cavalry Division on 14 September. It went into an assembly area west of Taegu along the east bank of the Naktong River six miles below Waegwan and prepared for action. On 16 September it moved out from its assembly area to begin an operation that was to prove of great importance to the Eighth Army breakout. Numbering 2,599 men, the regiment was 1,194 short of full strength. The three battalions were nearly equal, varying between 586 and 595 men in strength. On the 16th only the 2d Battalion engaged the enemy as it attacked north along the Naktong River road toward Waegwan. But by the end of the second day the 3d Battalion had joined in the battle and the 1st Battalion was deployed to enter it.[22]

[21] 2d Div POR 132, 18 Sep 50; 2d Div WD, G–3 Sec, 18 Sep 50; 2d Div PIR 24, 17 Dec 50; see Capt. Russell A. Gugeler, Small Unit Actions in Korea, ch. V, "Patrol Crossing of the Naktong, I&R Platoon, 23d Infantry, 18 September 1950," gives details of this incident. MS in OCMH. For an account of a typical rear area action, see Capt. Edward C. Williamson, Attack of the 38th Ordnance Medical Maintenance Company by a Guerrilla Band, 20 September 1950. MS in OCMH.

[22] 24th Div WD, G–1 Hist Rpt, 26 Aug–28 Sep 50; 1st Cav Div WD, 14–16 Sep 50. The earliest 5th RCT document found in the official records is the Personnel Report for 17 September and is included in the 24th Division records.

The next day, 19 September, as the 38th Infantry crossed the Naktong, the 5th Regimental Combat Team began its full regimental attack against Hill 268, southeast of Waegwan.

An estimated 1,200 soldiers of the N.K. *3d Division,* supported by tanks, defended this southern approach to Waegwan. The hills there constituted the left flank of the enemy *II Corps.* If the North Koreans lost this ground their advanced positions in the 5th Cavalry zone eastward along the Taegu highway would become untenable. The tactical importance of Hill 268 and related positions was made the greater by reason of the gap in the enemy line to the south. At the lower side of this gap the British 27th Infantry Brigade held vital blocking positions just above strong forces of the N.K. *10th Division.*

In hard fighting all day the 5th Regimental Combat Team gained Hill 268, except for its northeast slope. By night the 3d Battalion was on the hill, the 1st Battalion had turned northwest from it toward another enemy position, and the 2d Battalion had captured Hill 121, only a mile south of Waegwan along the river road. Air strikes, destructive and demoralizing to the enemy, had paced the regimental advance all the way. In this important action along the east bank of the Naktong, the 5th Cavalry and part of the 7th Cavalry protected the 5th Regimental Combat Team's right flank and fought very heavy battles co-ordinated with the combat team on the adjoining hills east of Waegwan.[23] At 1800 that evening, 18 September, the 5th Regimental Combat Team and the 6th Medium Tank Battalion reverted to 24th Division control.

The next morning the battle for Hill 268 continued. More than 200 enemy soldiers in log-covered bunkers still fought the 3d Battalion. Three flights of F–51's napalmed, rocketed, and strafed these positions just before noon. This strike enabled the infantry to overrun the enemy bunkers. Among the North Korean dead was a regimental commander. About 250 enemy soldiers died on the hill. Westward to the river, other enemy troops bitterly resisted the 2d and 1st Battalions, losing about 300 men in this battle. But Colonel Throckmorton's troops pressed forward. The 2d Battalion entered Waegwan at 1415. Fifteen minutes later it joined forces there with the 1st Battalion. After surprising an enemy group laying a mine field in front of it, the 2d Battalion penetrated deeper into Waegwan and had passed through the town by 1530.[24]

On 19 September the N.K. *3d Division* defenses around Waegwan broke apart and the division began a panic-stricken retreat across the river. At 0900 aerial observers reported an estimated 1,500 enemy troops crossing to the west side of the Naktong just north of Waegwan, and in the afternoon they reported roads north of Waegwan jammed with enemy groups of sizes varying from 10 to 300 men pouring out of the town. By midafternoon observers reported enemy soldiers in every draw and pass north of Waegwan. During the day the 5th Regimental Combat Team captured 22 45-mm. antitank guns, 10 82-mm. mor-

[23] 5th RCT WD, 18 Sep 50.

[24] 5th RCT WD, 19 Sep 50; *Ibid.*, Unit Rpt 38, 19 Sep 50; 24th Div Opn Instr 44, 17 Sep 50; 24th Div WD, G–1 Sec, 19 Sep 50; Throckmorton, MS review comments, recd 16 Apr 54.

tars, 6 heavy machine guns, and approximately 250 rifles and burp guns.[25]

On 20 September the 5th Regimental Combat Team captured the last of its objectives east of the Naktong River when its 2d Battalion in the afternoon seized important Hill 303 north of Waegwan. In securing its objectives, the 5th Regimental Combat Team suffered numerous casualties during the day—18 men killed, 111 wounded, and 3 missing in action. At 1945 that evening the 1st Battalion started crossing the river a mile above the Waegwan railroad bridge. By midnight it had completed the crossing and advanced a mile westward. The 2d Battalion followed the 1st Battalion across the river and dug in on the west side before midnight. During the day the 3d Battalion captured Hill 300, four miles north of Waegan. The following afternoon, 21 September, after the 2d Battalion, 5th Cavalry Regiment, relieved it on position, the 3d Battalion crossed the Naktong. The 5th Regimental Combat Team found large stores of enemy ammunition and rifles on the west side of the river.[26]

The 5th Regimental Combat Team in five days had crushed the entire right flank and part of the center of the N.K. *3d Division*. This rendered untenable the enemy division's advanced positions on the road to Taegu where it was locked in heavy fighting with the 5th Cavalry Regiment.

From 18 to 21 September, close air support reached its highest peak in the Korean campaign. Fighters and bombers returned several times a day from Japanese bases to napalm, bomb, rocket, and strafe enemy strongpoints of resistance and to cut down fleeing enemy troops caught in the open.[27]

### *The 24th Division Deploys West of the Naktong*

The Eighth Army and I Corps plans for the breakout from the Pusan Perimeter called for the 24th Division to make the first crossings of the Naktong River. Accordingly, General Church on 17 September received orders to force a crossing in the vicinity of the Hasandong ferry due west of Taegu. The 5th Regimental Combat Team had just cleared the ground northward and secured the crossing site against enemy action from the east side of the river. The 21st Infantry was to cross the river after dark on 18 September in 3d Engineer Combat Battalion assault boats. Once landed on the other side, the regiment was to attack north along the west bank of the Naktong to a point opposite Waegwan where it would strike the main highway to Kumch'on. The 24th Reconnaissance Company and the 19th Infantry Regiment were to cross at the same time a little farther south and block the roads leading from Songju, an enemy concentration point, some six miles west of the river. The unexpected crossing of the Naktong during the day by the 2d Battalion, 38th Infantry, farther south

[25] 24th Div WD, 19 Sep 50; 5th RCT WD, 19 Sep 50 and Unit Rpt 38, 19 Sep 50; EUSAK PIR 69, 19 Sep 50; ATIS Interrog Rpts, Issue 13 (N.K. Forces), Rpt 1880, p. 189. MSgt Son Tok Hui, *105th Armored Division*.

[26] 5th RCT WD, 20–21 Sep 50; EUSAK WD, Br for CG, 20 Sep 50; Throckmorton, MS review comments, recd 16 Apr 54.

[27] EUSAK WD, Sep 50 Summ, p. 30; "Air War in Korea," *Air University Quarterly Review*, IV, No. 2 (Fall, 1950), 19–39.

did not alter the Eighth Army plan for the breakout.[28]

In moving up to the Naktong, the 24th Division had to cross one of its tributaries, the Kumho River, that arched around Taegu. On the morning of the 18th, Colonel Stephens, the 21st Infantry regimental commander, discovered that the I Corps engineers had not bridged the Kumho as planned. The division thereupon hurried its own Engineer troops to the stream and they began sandbagging the underwater bridge that the 5th Regimental Combat Team had already used so that large vehicles could cross. A makeshift ferry constructed from assault boats moved jeeps across the Kumho. Constant repair work on the underwater sandbag bridge was necessary to keep it usable. By nightfall there was a line of vehicles backed up for five miles east of the Kumho, making it clear that the regiment would not be in position to cross the Naktong that evening after dark as planned. As midnight came and the hours passed, General Church began to fear that daylight would arrive before the regiment could start crossing and the troops consequently would be exposed to possibly heavy casualties. He repeatedly urged on Stephens the necessity of crossing the Naktong before daylight. During the night supporting artillery fired two preparations against the opposing terrain.[29]

Despite nightlong efforts to break the traffic jam and get the assault boats, troops, and equipment across the Kumho and up to the crossing site, it was 0530, 19 September, before the first wave of assault boats pushed off into the Naktong. Six miles below Waegwan and just south of the village of Kumnan-dong on the west side, Hill 174 and its long southern finger ridge dominated the crossing site. In the murky fog of dawn there was no indication of the enemy on the opposite bank. The first wave landed and started inland. Almost at once enemy machine gun fire from both flanks caught the troops in a crossfire. And now enemy mortar and artillery fire began falling on both sides of the river. The heaviest fire, as expected, came from Hill 174 and its long southern finger ridge.

For a while it was doubtful that the crossing would succeed. The 1st Battalion, continuing its crossing under fire, suffered approximately 120 casualties in getting across the river. At 0700 an air strike hit Hill 174. On the west side the 1st Battalion reorganized and, supported by air napalm and strafing strikes, attacked and captured Hill 174 by noon. That afternoon the 3d Battalion crossed the river and captured the next hill northward. During the night and the following morning the 2d Battalion crossed the Naktong. The 1st Battalion on 20 September advanced north to Hill 170, the high ground on the west side of the river opposite Waegwan, while the 3d Battalion occupied the higher hill a mile northwestward.[30]

Meanwhile, two miles south of the 21st Infantry crossing site, the 2d Battalion,

---

[28] 24th Div WD, 17 Sep 50 and an. B, overlay accompanying 24th Div Opn Instr 44, 17 Sep 50.

[29] 24th Div WD, 18–19 Sep 50; 21st Inf WD, 18–19 Sep 50, and Summ, 26 Aug–28 Sep 50; Throckmorton, MS review comments, recd 16 Apr 54; Col Emerson C. Itschner, "The Naktong Crossings in Korea," *The Military Engineer*, XLIII, No. 292 (March–April, 1951), 96ff; Interv, author with Alkire (21st Inf), 1 Aug 51.

[30] 24th Div WD, 19 Sep 50; 21st Inf Unit Rpts 73–74, 18–20 Sep 50; 3d Engr C Bn WD, Narr Summ, Sep 50.

19th Infantry, began crossing the Naktong at 1600 on the afternoon of the 19th and was on the west side by evening. Enemy mortar and artillery fire inflicted about fifty casualties while the battalion was still east of the river. Beach operations were hazardous. Once across the river, however, the battalion encountered only light enemy resistance.

In the 24th Division crossing operation the engineers' role was a difficult and dangerous one, as their casualties show. The 3d Engineer Combat Battalion lost 10 Americans and 5 attached Koreans killed, 37 Americans and 10 Koreans wounded, and 5 Koreans missing in action.[31]

On 20 September the 19th Infantry consolidated its hold on the high ground west of the river along the Songju road. The 24th Reconnaissance Company, having crossed the river during the night, passed through the 19th Infantry and started westward on the Songju road. During the day I Corps attached the British 27th Infantry Brigade to the 24th Division and it prepared to cross the Naktong and take part in the division attack. Relieved in its position by the 2d Battalion, 7th Cavalry Regiment, the British 27th Brigade moved north to the 19th Infantry crossing site and shortly after noon started crossing single file over a rickety footbridge that Engineer troops had thrown across the river. An enemy gun shelled the crossing site sporadically but accurately all day, causing some British casualties and hampering the ferrying of supplies for the 19th Regiment. Despite special efforts, observers could not locate this gun because it remained silent while aircraft were overhead.[32]

Thus, on 20 September, all three regiments of the 24th Division and the attached British 27th Brigade were across the Naktong River. The 5th Regimental Combat Team held the high ground north of the Waegwan-Kumch'on highway, the 21st Infantry that to the south of it, the 19th was below the 21st ready to move up behind and support it, and the 24th Reconnaissance Company was probing the Songju road west of the Naktong with the British brigade preparing to advance west on that axis. The division was ready to attack west along the main Taegu-Kumch'on-Taejon-Seoul highway.

With the 24th Division combat elements west of the river, it was necessary to get the division transport, artillery, tanks, and service units across to support the advance. The permanent bridges at Waegwan, destroyed in early August by the 1st Cavalry Division, had not been repaired by the North Koreans except for ladders at the fallen spans to permit foot traffic across the river. A bridge capable of carrying heavy equipment had to be thrown across the Naktong at once. Starting on 20 September and working continuously for thirty-six hours, the 11th Engineer Combat Battalion and the 55th Engineer Treadway Bridge Company completed at 1000, 22 September, an M2 ponton float treadway bridge

---

[31] 24th Div WD, 19 Sep 50; 3d Engr C Bn WD, Narr Summ, Sep 50.

[32] 21st Inf Unit Rpts, 19–20 Sep 50; Ltr, Gay to author, 30 Sep 53; 2d Bn, 7th Cav Jnl, 20 Sep 50; 24th Div WD, 21–22 Sep 50; EUSAK WD, 22 Sep 50; 19th Inf WD, 19–21 Sep 50 and Summ, 19–21 Sep; Maj. Gen. B. A. Coad, "The Land Campaign in Korea," *op. cit.*, p. 4; Eric Linklater, *Our Men in Korea, The Commonwealth Part in the Campaign, First Official Account* (London: Her Majesty's Stationery Office, 1952), p. 18.

CROSSING THE KUMHO RIVER *via underwater bridge and makeshift ferry.*

across the 700-foot-wide and 8-foot-deep stream at Waegwan. Traffic began moving across it immediately. Most 24th Division vehicles were on the west side of the Naktong by midnight. Many carried signs with slogans such as "One side, Bud—Seoul Bound," and "We Remember Taejon." [33]

In the action of 20–21 September near Waegwan the North Koreans lost heavily in tanks, as well as in other equipment and troops on both sides of the Naktong. In these two days the 24th Division counted 29 destroyed enemy tanks, but many of them undoubtedly had been destroyed earlier in August and September. According to enemy sources, the *203d Regiment* of the *105th Armored Division* retreated to the west side of the Naktong with only 9 tanks, and the *107th Regiment* with only 14. Nevertheless, the enemy covered his retreat toward Kumch'on with tanks, self-propelled guns, antitank guns, and small groups of supporting riflemen.[34]

Except for the muddle in bridging the Kumho River and the resulting delayed crossing of the Naktong by the 21st Infantry Regiment, the 5-day operation of the 24th Division beginning on 18 Sep-

[33] 24th Div WD, 21–22 Sep 50; 61st FA Bn WD, 20 Sep 50; I Corps WD, Engr Sec, 22 Sep 50; 3d Engr C Bn WD, Narr Summ, Sep 50; Itschner, "The Naktong River Crossings in Korea," *op. cit.*

[34] 24th Div WD, 20–22 Sep 50; 5th RCT WD, 21 Sep 50; EUSAK WD, G–3 Air, 22 Sep 50; GHQ FEC Sitrep, 22 Sep 50; ATIS Res Supp Interrog Rpts, Issue 4 (*105th Armored Div*), pp. 39–40; *Ibid.*, Issue 14 (N.K. Forces), p. 4, Rpt 1901, Lt Lee Kim Chun.

PONTON TREADWAY BRIDGE *across the Naktong, built in thirty-six hours.*

tember left little to be desired. On the 22d the division was concentrated and poised west of the river ready to follow up its success. Its immediate objective was to drive twenty miles northwest to Kumch'on, headquarters of the North Korean field forces.

### *The Indianhead Division Attacks West*

Below the 24th Division, the 2d Division waited for the 9th Infantry to capture Hill 201. On the 19th, the 1st and 2d Battalions, 23d Infantry, were put into the fight to help reduce the enemy stronghold. While the 1st Battalion helped the 9th Infantry at Hill 201, the 2d Battalion attacked across the 9th Infantry zone against Hill 174, a related enemy defense position. In this action Sgt. George E. Vonton led a platoon of tanks from the regimental tank company to the very top of Hill 201 in an outstanding feat which was an important factor in driving the enemy from the heights. That evening this stubbornly held enemy hill on the 2d Division left flank was in 9th Infantry hands and the way was open for the 2d Division crossing of the Naktong.

In predawn darkness, 20 September, the 3d Battalion, 23d Infantry, without opposition slipped across the river in assault boats at the Sangp'o ferry site, just south of where the Sinban River enters the Naktong from the west. The battalion achieved a surprise so complete that its leading element, L Company, captured a North Korean lieutenant colonel and his staff asleep. From a map captured at this time, American troops learned the locations of the N.K. *2d, 4th,* and *9th Divisions* in the Sinban-ni area. By noon the 3d Battalion had captured Hill 227, the critical terrain dominating the crossing site on the west side.[35]

In the afternoon, the 1st Battalion, 23d Infantry, crossed the river. Its objective was Hill 207, a mile upstream from the crossing site and dominating the road which crossed the Naktong there. In moving toward this objective, the lead company soon encountered the Sinban River which, strangely enough, no one in the company knew was there. After several hours of delay in attempting to find a method of crossing it, the troops finally crossed in Dukw's and, in a night attack, moved up the hill which they found undefended.[36]

Meanwhile, the 3d Battalion had dug in on Hill 227. That night it rained hard and, under cover of the storm, a company of North Koreans crept up near the crest. The next morning (21 September) while L Company men were eating breakfast the enemy soldiers charged over the hill shooting and throwing grenades. They drove one platoon from its position and inflicted twenty-six casualties. Counterattacks regained the position by noon.[37]

While this action was taking place on the hill south of it, the 1st Battalion, 23d Infantry, with a platoon of tanks from the 72d Tank Battalion, attacked up the road toward Sinban-ni, a known enemy headquarters command post five miles west of the river. The advance against strong enemy opposition was weakened by ineffective co-ordination between tanks and infantry. The great volume of fire from supporting twin-40 and quad-50 self-propelled AA gun vehicles was of greatest help, however, in enabling the troops to make a two-and-a-half mile advance which bypassed several enemy groups.

The next morning an enemy dawn attack drove B Company from its position and inflicted many casualties. Capt. Art Stelle, the company commander, was killed. During the day an estimated two battalions of enemy troops in heavy fighting held the 23d Infantry in check in front of Sinban-ni. The 2d Battalion of the regiment crossed the Naktong and moved up to join the 1st Battalion in the battle north of the road. South of it the 3d Battalion faced lighter resistance. The next day, 23 September, the 23d Regiment gained Sinban-ni, and was ready then to join the 38th Infantry in a converging movement on Hyopch'on.[38]

On the next road northward above the 23d Infantry, six miles away, the 38th Infantry had hard fighting against strong enemy delaying forces as it attacked toward Ch'ogye and Hyopch'on. An air strike with napalm and fragmentation bombs helped its 2d Battalion on 21 September break North Korean resistance

---

[35] 23d Inf Comd Rpt, Narr, Sep 50, p. 12; *Ibid.,* Jnl entry 183, 19 Sep 50, and entries 192 and 197, 20 Sep 50; 2d Div WD, G-3 Sec, Sep-Oct 50, p. 18 and PIR 27, 20 Sep 50; EUSAK WD POR 206, 19 Sep 50; Freeman, MS review comments, 30 Oct 57; Combat Activities of the 23d Infantry.

[36] 23d Inf WD, entries 199-202, 20 Sep 50; EUSAK WD, G-3 Sec, 20 Sep 50; Glasgow, Platoon Leader in Korea, pp. 207-24.

[37] Interv, author with Radow (M Co, 23d Inf, Sep 50), 16 Aug 50; 23d Inf WD, entries 206-10, 20 Sep 50.

[38] 23d Inf WD, 20-22 Sep 50, entries 206-210, 214-222, and 226; 23d Inf POR 26, 20-21 Sep 50; and an. 1, Overlay; Glasgow, Platoon Leader in Korea, pp. 234-62. Glasgow, a platoon leader in B Company and critically wounded in the fight, indicates that part of the company behaved poorly in the enemy dawn attack.

ADVANCING TO THE CREST OF HILL 201

on Hill 239, the critical terrain overlooking Ch'ogye. The next day the battalion entered the town in the early afternoon. Before midnight the 1st Battalion turned over its task of containing elements of the N.K. *10th Division* on Hill 409 east of the Naktong to the 2d Battalion, 9th Infantry, and started across the river to join its regiment.[39]

Although it had only 276 feet of bridging material, the 2d Division, by resorting to various expedients, completed a bridge in the afternoon of 22 September across the 400-foot-wide stream at the Sadung ferry site, and was ready to start moving supplies to the west side of the river in support of its advanced units.

### *Encirclement Above Taegu*

In the arc above Taegu and on the right of the 5th Regimental Combat Team, the 1st Cavalry Division and the ROK 1st Division had dueled for days

[39] 38th Inf Comd Rpt, Sep–Oct 50, pp. 10–12; 2d Div PIR 18, 21 Sep 50; 2d Div WD, G–3 Sec, Sep–Oct 50, p. 20; *Ibid.*, G–4 Sec; EUSAK WD, G–3 Jnl, entry 1456, 22 Sep 50. An enemy sketch captured on the 21st near Ch'ogye showed accurately every position the 1st Battalion, 38th Infantry, had occupied east of the Naktong.

VIEW FROM CREST OF HILL 201

with the N.K. *3d, 1st,* and *13th Divisions* in attack and counterattack. The intensity of the fighting there in relation to other parts of the Perimeter is apparent in the casualties. Of 373 casualties evacuated to Pusan on 16 September, for instance, nearly 200 came from the Taegu area. The fighting centered, as it had for the past month, on two corridors of approach to Taegu: (1) the Waegwan-Taegu highway and railroad, where the 5th Cavalry Regiment blocked the advanced elements of the N.K. *3d Division* five miles southeast of Waegwan and eight miles northwest of Taegu, and (2) the Tabu-dong road through the mountains north of Taegu where other elements of the 1st Cavalry Division and the ROK 1st Division had been striving to hold off the N.K *13th* and *1st Divisions* for nearly a month. There the enemy was still on hills overlooking the Taegu bowl and only six miles north of the city.

General Gay's plan for the 1st Cavalry Division in the Eighth Army breakout effort was (1) to protect the right flank of the 5th Regimental Combat Team as it drove on Waegwan by having the 5th Cavalry Regiment attack and hold the enemy troops in its zone east of the Waegwan-Taegu highway; (2) to maintain pressure by the 8th Cavalry Regiment on the enemy in the Ch'ilgok area north of Taegu, and be prepared on order to make a maximum effort to drive north to Tabu-dong; and (3) the 7th Cavalry Regiment on order to shift, by successive battalion movements, from the division right flank to the left flank and make a rapid encirclement of the enemy over a trail and secondary road between Waegwan and Tabu-dong. If the plan worked, the 7th and 8th Cavalry Regiments would meet at Tabu-dong and enclose a large number of enemy troops in the Waegwan–Taegu–Tabu-dong triangle. General Gay started shifting forces from right to left on 16 September by moving the 2d Battalion, 7th Cavalry, to Hill 188 in the 5th Cavalry area.[40]

North of Taegu on the Tabu-dong road enemy units of the N.K. *13th Division* fought the 8th Cavalry Regiment to a standstill during the first three days of the Eighth Army offensive. Neither side was able to improve its position materially. The enemy attacked the 2d Battalion, 8th Cavalry, repeatedly on Hill 570, the dominating height east of the mountain corridor, ten miles north of Taegu. West of the road, the 3d Battalion made limited gains in high hills closer to Taegu. The North Koreans on either side of the Tabu-dong road had some formidable defenses, with a large number of mortars and small field pieces dug in on the forward slopes of the hills. Until unit commanders could dispose their forces so that they could combine fire and movement, they had to go slow or sacrifice the lives of their men.

General Walker was displeased at the slow progress of the 8th Cavalry Regiment. On the 18th he expressed himself on this matter to General Gay, as did also General Milburn, commander of I Corps. Both men believed the regiment was not pushing hard. The next day the division attached the 3d Battalion, 7th Cavalry, to the 8th Cavalry Regiment, and Colonel Holmes, the division chief of staff, told Colonel Palmer that he must take Tabu-dong during the day. But the enemy *13th Division* frustrated the 8th Cavalry's attempt to reach Tabu-dong. Enemy artillery, mortar, and automatic weapons crossfire from the Walled City area of Ka-san east of the road and the high ground of Hill 351 west of it turned back the regiment with heavy casualties. On 20 September the 70th Tank Battalion lost seven tanks in this fight.[41]

But on the right of the 1st Cavalry Division, the ROK 1st Division made impressive gains. General Paik's righthand regiment, the 12th, found a gap in the enemy's positions in the high mountains and, plunging through it, reached a point on the Tabu-dong–Kunwi road ten miles northeast of Tabu-dong, and approximately thirteen miles beyond the most advanced units of the 1st Cavalry Division. There the ROK troops were in the

[40] Ltr, Gay to author, 30 Sep 53; 2d Bn, 7th Cav, Unit Jnl, 16 Sep 50.

[41] 8th Cav Regt WD, 16–20 Sep 50; 1st Cav Div WD, 16–20 Sep 50; *Ibid.*, PIR 181, 20 Sep 50; I Corps WD, 16–20 Sep 50; EUSAK WD, G–3 Sec, 18 Sep 50, situation overlay 0630, 18 Sep 50; Ltr, Gay to author, 30 Sep 53; Col Harold K. Johnson, MS review comments for author, Aug 54. Department of the Army General Order 38, 16 April 1952, awarded the Distinguished Unit Citation to the 2d Battalion, 8th Cavalry Regiment, and attached units for defense of Hill 570.

rear of the main body of the N.K. *1st* and *13th Divisions* and in a position to cut off one of their main lines of retreat. The U.S. 10th AAA Group accompanied the ROK 12th Regiment in its penetration and the artillerymen spoke glowingly of "the wonderful protection" given them, saying, "The 10th AAA Group was never safer than when it had a company of the 12th Regiment acting as its bodyguard. Everywhere the Group moved, Company 10 of the 12th Regiment moved too." This penetration caused the N.K. *1st Division* on 19 September to withdraw its *2d* and *14th Regiments* from the southern slopes of Ka-san (Hill 902) to defend against the new threat. That day also a ROK company penetrated to the south edge of the Walled City.[42]

Along the Waegwan-Taegu road at the beginning of the U.N. offensive on 16 September, the 5th Cavalry Regiment attacked North Korean positions, centering on Hills 203 and 174 north of the road and Hill 188 opposite and south of it. Approximately 1,000 soldiers of the *8th Regiment, 3d Division,* held these key positions. The 1st Battalion, 5th Cavalry, began the attack on the 16th. The next day the 2d Battalion, 7th Cavalry, joined in, moving against Hill 253 farther west. There North Koreans engaged F and G Companies of the 7th Cavalry in heavy combat. When it became imperative to withdraw from the hill, G Company's Capt. Fred P. DePalina, although wounded, remained behind to cover the withdrawal of his men. Ambushed subsequently by enemy soldiers, DePalina killed six of them before he himself died. The two companies were forced back south of the road.[43]

For three days the North Koreans on Hill 203 repulsed every attempt to storm it. "Get Hill 203" was on every tongue. In the fighting, A Company of the 70th Tank Battalion lost nine tanks and one tank dozer to enemy action on 17 and 18 September, six of them to mines, two to enemy tank fire, and two to enemy antitank fire. In one tank action on the 18th, American tank fire knocked out two of three dug-in enemy tanks. Finally, on 18th September, Hill 203 fell to the 1st Battalion, 5th Cavalry, but the North Koreans continued to resist from the hills northwest of it, their strongest forces being on Hill 253. In this battle the three rifle companies of the 2d Battalion, 7th Cavalry, were reduced to a combined strength of 165 effective men—F Company was down to forty-five effectives. The enemy's skillful use of mortars had caused most of the casualties. At the close of 18 September the enemy *3d Division* still held the hill mass three miles east of Waegwan, centering on Hills 253 and 371.[44]

---

[42] EUSAK WD, G–3 Sec, 18 Sep 50, situation overlay 0630 18 Sep; I Corps WD, Narr Hist, 19 Sep 50; ATIS Interrog Rpts, Issue 9 (N.K. Forces), Interrog Rpt 1468, pp. 158–74, Col Lee Hak Ku, CofS *13th Div*, captured 21 Sep 50; Capt. Arthur C. Brooks, Jr., "From Pusan to Unsan with the 10th AAA Group," *Antiaircraft Journal,* XCIV, No. 1 (January–February, 1951), 14.

[43] 2d Bn, 7th Cav, Unit Jnl, 17–18 Sep 50; 1st Cav Div WD, 17 Sep 50. General Order 182, 30 March 1951, awarded the Distinguished Service Cross posthumously to Captain DePalina. EUSAK. (The order is apparently in error on the date of his death, giving it as 19 September.)

[44] 1st Cav Div WD, 16–18 Sep 50; 1st Cav Div Arty Unit Hist, 17 Sep 50; 5th Cav Regt WD, 16–18 Sep 50; *Ibid.*, Narr Rpt; 7th Cav Regt WD, 17–18 Sep 50; 1st Cav Div, G–2 Hist, Sep 50; Summ of Act, A Co, 70th Tk Bn, 17–24 Sep 50; I Corps WD, 18 Sep 50; USAF Hist Study 71, p. 66.

40-MM. ANTIAIRCRAFT BATTERY *attached to the ROK 1st Division fighting north of Tabu-dong.*

On 18 September forty-two B–29 bombers of the 92d and 98th Groups bombed west and northwest of Waegwan across the Naktong but apparently without damage to the enemy.

The battle on the hills east of Waegwan reached a climax on the 19th when the 1st Battalion, 5th Cavalry, and the 2d Battalion, 7th Cavalry, engaged in very heavy fighting with fanatical, die-in-place North Koreans on Hills 300 and 253. Elements of the 1st Battalion, 5th Cavalry, gained the crest of Hill 300. On that hill the 1st Battalion suffered 207 battle casualties—28 American soldiers killed, 147 wounded, and 4 missing in action, for a total of 179, with 28 additional casualties among the attached South Koreans. At noon, F Company reported 66 men present for duty; E and G Companies between them had 75 men. That afternoon the battalion reported it was only 30 percent combat effective. The 5th Cavalry's seizure of the 300 and 253 hill mass dominating the Taegu road three miles southeast of Waegwan unquestionably helped the 5th Regimental Combat Team to capture Waegwan that day. But one mile to the north of these hills, the enemy on Hill 371 in a stubborn holding action turned back for the moment all efforts of the 5th Cavalry to capture that height.[45]

[45] 2d Bn, 7th Cav, Unit Jnl, 19 Sep 50; 7th Cav Regt WD, 19 Sep 50; 1st Cav Div WD, 19–20 Sep 50. There were 205 counted enemy dead on Hill 300.

In its subsequent withdrawal from the Waegwan area to Sangju the N.K. *3d Division* fell from a strength of approximately 5,000 to about 1,800 men. Entire units gave way to panic. Combined U.N. ground and air action inflicted tremendous casualties. In the area around Waegwan where the 5th Cavalry Regiment reoccupied the old Waegwan pocket a count showed 28 enemy tanks—27 T34's and one American M4 refitted by the North Koreans—as destroyed or captured.[46]

During the 19th General Gay started maneuvering his forces for the encirclement movement, now that the hard fighting east of Waegwan had at last made it possible. Colonel Clainos led his 1st Battalion, 7th Cavalry, from the division right to the left flank, taking position in front of the 2d Battalion, 7th Cavalry, to start the movement toward Tabu-dong. Gay ordered the 3d Battalion, 7th Cavalry, to shift the next morning from the right flank to the left, and prepare to follow the 1st Battalion in its dash for Tabu-dong. On the morning of 20 September the 3d Battalion entrucked north of Taegu and rolled northwest on the road toward Waegwan. The regimental commander, apparently fearing that enemy mortar and artillery fire would interdict the road, detrucked his troops short of their destination. Their foot march tired the troops and made them late in reaching their assembly area. This overcaution angered General Gay because the same thing had happened when the 2d Battalion of the same regiment had moved to the left flank four days earlier.[47]

In the meantime, during the morning the 1st Battalion, 7th Cavalry, led off down the road toward Waegwan past Hill 300. Two miles short of Waegwan the lead elements at 0900 turned off the main highway onto a poor secondary road which cut across country to a point three miles east of Waegwan, where it met the Waegwan–Tabu-dong road. This latter road curved northeast, winding along a narrow valley floor hemmed in on both sides by high mountains all the way to Tabu-dong, eight miles away.

Even though an armored spearhead from C Company, 70th Tank Battalion, led the way, roadblocks and enemy fire from the surrounding hills held the battalion to a slow advance. By midafternoon it had gained only two miles, and was only halfway on the cutoff road that led into the Waegwan–Tabu-dong road. The column stopped completely when a tank struck a mine. General Gay showed his irritation over the slow progress by ordering the regimental commander to have the 1st Battalion bypass enemy on the hills and "high-tail it" for Tabu-dong.[48]

Acting on General Gay's orders, the 1st Battalion pushed ahead, reached the Tabu-dong road, and turned northeast on it toward the town eight miles away. This road presented a picture of devastation—dead oxen, disabled T34 tanks, wrecked artillery pieces, piles of abandoned ammunition, and other military

[46] ATIS Res Supp Interrog Rpts, Issue 96 (N.K. *3d Div*), pp. 34–35; I Corps WD, Narr Hist, 20 Sep 50 and G–2 Sec, 22 Sep 50; 5th Cav Regt WD, 22 Sep 50; EUSAK WD, 22 Sep 50, and POR 216, 22 Sep 50.

[47] 1st Bn, 7th Cav, Unit Jnl, 19 Sep 50; 3d Bn, 7th Cav, Unit Jnl, Msg 5, 191920, and Msg 6, 201400 Sep 50; Ltr, Gay to author, 30 Sep 53.

[48] 7th Cav Regt WD, 20 Sep 50 (entries for 19 and 20 Sep are run together with no date entry for the 20th).

equipment and supplies littered its course. As the battalion halted for the night, an exploding mine injured Colonel Clainos. He refused evacuation, but the next day was evacuated on orders of the regimental commander. That evening the 1st Battalion, with the 3d Battalion following close behind, advanced to the vicinity of Togae-dong, four miles short of Tabu-dong.

The premature detrucking of the 3d Battalion during the day was the final incident that caused General Gay to replace the 7th Cavalry regimental commander. That evening General Gay put in command of the regiment Colonel Harris, commanding officer of the 77th Field Artillery Battalion, which had been in support of the regiment. Harris assumed command just before midnight.[49]

Colonel Harris issued orders about midnight to assembled battalion and unit commanders that the 7th Cavalry would capture Tabu-dong on the morrow, and that the element which reached the village first was to turn south to contact the 8th Cavalry Regiment and at the same time establish defensive positions to secure the road.

The next morning, 21 September, the 1st Battalion resumed the attack and arrived at the edge of Tabu-dong at 1255. There it encountered enemy resistance, but in a pincer movement from southwest and northwest cleared the village by 1635. An hour later the battalion moved out of Tabu-dong down the Taegu road in attack southward toward the 8th Cavalry Regiment.

Late that afternoon, General Gay was accompanying the 1st Battalion, 8th Cavalry, advancing northward toward Tabu-dong. He and Colonel Kane, the battalion commander, were standing close to a tank when a voice came over its radio saying, "Scrappy, this is Skirmish Red, don't fire." A few minutes later a sergeant, commanding the lead platoon of C Company, 7th Cavalry, came into the position and received the personal congratulations of the division commander upon completing the encircling movement.[50]

Meanwhile, the 3d Battalion, 7th Cavalry, arrived at Tabu-dong and turned north to deploy its troops in defensive positions on both sides of the road. By this time, elements of the ROK 1st Division had cut the Sangju road above Tabu-dong and were attacking south toward the village. The ROK 12th Regiment, farthest advanced, had a roadblock eight miles to the northeast below Kunwi. It appeared certain that the operations of the 1st Cavalry Division and the ROK 1st Division had cut off large numbers of the N.K. *3d, 13th,* and *1st Divisions* in the mountains north of Taegu. The next day, 22 September, the 11th Regiment of the ROK 1st Division and units of the ROK National Police captured the Walled City of Ka-san, and elements of the ROK 15th Regiment reached Tabu-dong from the north to link up with the 1st Cavalry Division.[51]

---

[49] 7th Cav Regt WD, 20 Sep 50; 1st Bn, 7th Cav, Unit Jnl, 20 Sep 50; 3d Bn, 7th Cav, Unit Jnl, Msgs 10, 12, 14, 20 Sep 50; 77th FA Bn WD, 20 Sep 50; Ltr, Gay to author, 30 Sep 53.

[50] 7th Cav WD, 20–21 Sep 50; 3d Bn, 7th Cav, Unit Jnl, Msg 15, 20 Sep 50; 1st Bn, 7th Cav, Unit Jnl, 21 Sep 50; I Corps WD, 21 Sep 50; 1st Cav Div WD, 21 Sep 50; Ltr, Gay to author, 30 Sep 53; Interv, author with Harris, 30 Apr 54; Interv, author with Clainos, 30 Apr 54.

[51] 3d Bn, 7th Cav, Unit Jnl, Msgs 22 and 26, 211500 and 211700 Sep 50; 61st FA Bn WD, 21 Sep 50; EUSAK WD, 21 Sep 50.

## *The Right Flank*

In the mountainous area of the ROK II Corps the enemy *8th Division* was exhausted and the *15th* practically destroyed. The ROK divisions were near exhaustion, too, but their strength was greater than the enemy's and they began to move slowly north again. The ROK 6th Division attacked against the N.K. *8th Division,* which it had held without gain for two weeks, and in a 4-day battle destroyed the division as a combat force. According to enemy sources, the N.K. *8th Division* suffered about 4,000 casualties at this time. The survivors fled north toward Yech'on in disorder. By 21 September the ROK 6th Division was advancing north of Uihung with little opposition.[52]

Eastward, the ROK 8th Division, once it had gathered itself together and begun to move northward, found little resistance because the opposing enemy *15th Division* had been practically annihilated.

In the battle-scarred Kigye–An'gang-ni–Kyongju area of the ROK I Corps sector, units of the Capital Division fought their way through the streets of An'gang-ni on 16 September, the day the U.N. offensive got under way. Beyond it, the ROK 3d Division had moved up to the north bank of the Hyongsan-gang just below P'ohang-dong. The next day a battalion of the ROK 7th Division, advancing from the west, established contact with elements of the Capital Division and closed the 2-week-old gap between the ROK II and I Corps.

Retiring northward into the mountains, the N.K. *12th Division* fought stubborn delaying actions and did not give up Kigye to the Capital Division until 22 September. It then continued its withdrawal toward Andong. This once formidable organization, originally composed largely of Korean veterans of the Chinese Communist Army, was all but destroyed—its strength stood at approximately 2,000 men. The North Korean and ROK divisions on the eastern flank now resembled exhausted wrestlers, each too weak to press against the other. The ROK divisions, however, had numerical superiority, better supply, daily close air support and, in the P'ohang-dong area, naval gunfire.[53]

On the 16th, naval support was particularly effective when Admiral Charles C. Hartman's Task Group, including the battleship USS *Missouri,* appeared off P'ohang-dong. The big battleship pounded the enemy positions below the town, along the dike north of the Hyongsan-gang, with 2,000-pound shells from its 16-inch guns. Two days later the battleship again shelled these dike positions under observed radio fire direction by Colonel Emmerich, KMAG adviser to the ROK 3d Division. ROK troops then assaulted across the bridge, but enemy machine gunners cut them down. The number killed is unknown, but 144 were wounded in trying to cross the bridge. In a final desperate step, thirty-one ROK

[52] EUSAK WD, G–3 Jnl, entry 1050, 21 Sep 50; *Ibid.*, G–3 Sec, 19 Sep 50; ATIS Res Supp Interrog Rpts, Issue 4 (N.K. *8th Div*), p. 25; ATIS Interrog Rpts, Issue 10 (N.K. Forces), Rpt 1517, p. 44, Lt Choe Yun Ju.

[53] ATIS Res Supp Interrog Rpts, Issue 99 (N.K. *12th Div*), pp. 48–49; EUSAK WD, G–2 Stf Sec Rpt, 14 Sep 50, and Summ, 19 Sep 50, p. 31; *Ibid.*, POR 217, 22 Sep 50.

soldiers volunteered to die if necessary in trying to cross the bridge. Fighter planes helped their effort by making dummy strafing passes against the enemy dike positions. Of the thirty-one who charged, nineteen fell on the bridge. Other ROK soldiers quickly reinforced the handful of men who gained a foothold north of the river. There they found dead enemy machine gunners tied to their dike positions.[54]

As a preliminary move in the U.N. offensive in the east, naval vessels on the night of 14–15 September had transported the ROK Miryang Guerrilla Battalion, specially trained and armed with Russian-type weapons, to Changsa-dong, ten miles above P'ohang-dong, where the battalion landed two and a half hours after midnight in the rear of the N.K. *5th Division*. Its mission was to harass the enemy rear while the ROK 3d Division attacked frontally below P'ohang-dong. That evening the enemy division sent a battalion from its *12th Regiment* to the coastal hills where the Miryang Battalion had taken a position and there engaged it. The ROK guerrilla battalion's effort turned into a complete fiasco. The U.S. Navy had to rush to its assistance and place a ring of naval gunfire around it on the beach, where enemy fire had driven the battalion. This saved it from total destruction. Finally, on 18 September, with great difficulty, the Navy evacuated 725 of the ROK's, 110 of them wounded, by LST. Thirty-nine dead were left behind, as well as 32 others who refused to try to reach the evacuating ships.[55]

Although this effort to harass the enemy rear came to nothing and gave the ROK 3d Division little help, elements of that division had combat patrols at the edge of P'ohang-dong on the evening of 19 September. The next morning at 1015 the division captured the destroyed fishing and harbor village. One regiment drove on through the town to the high ground north of it. And in the succeeding days of 21 and 22 September the ROK 3d Division continued strong attacks northward, supported by naval gunfire and fighter planes, capturing Hunghae, and driving the N.K. *5th Division* back on Yongdok in disorder.[56]

### *The Left Flank—The Enemy Withdraws From Sobuk-san*

At the other end of the U.N. line, the left flank in the Masan area, H-hour on 16 September found the 25th Division in an embarrassing situation. Instead of being able to attack, the division was still fighting enemy forces behind its lines, and the enemy appeared stronger than ever on the heights of Battle Mountain, P'il-bong, and Sobuk-san.

General Kean and his staff felt that the division could advance along the roads toward Chinju only when the mountainous center of the division front

---

[54] Interv, author with Emmerich, 5 Dec 51; EUSAK WD, 18 Sep 50; GHQ FEC, G–3 Opn Rpt, 17 Sep 50; Karig, *et al.*, *Battle Report, The War in Korea*, pp. 254–55.

[55] EUSAK WD, G–3 Sec, 14–15 Sep 50; *Ibid.*, Br for CG, 15 Sep 50; GHQ FEC, History of the N.K. Army, pp. 51, 61; GHQ FEC Sitrep, 19–20 Sep 50; Karig, *et al.*, *Battle Report, The War in Korea*, pp. 243–55.

[56] Interv, author with Emmerich, 5 Dec 51; EUSAK WD, Summ, 20 Sep 50, p. 32; *Ibid.*, Br for CG, and G–3 Sec, 20 Sep 50; EUSAK WD, 21–22 Sep 50; USAF Hist Study 71, p. 67; ATIS Res Supp Interrog Rpts, Issue 96 (N.K. *5th Div*), p. 44.

ENEMY-HELD AREA, *showing high ground north of P'ohang-dong.*

was clear of the enemy. The experience of Task Force Kean in early August, when the enemy had closed in behind it from the mountains, was still fresh in their minds. They therefore believed that the key to the advance of the 25th Division lay in its center where the enemy held the heights and kept the 24th Infantry Regiment under daily attack. The 27th Infantry on the left and the 35th Infantry on the right, astride the roads between Chinju and Masan, could do little more than mark time until the situation in front of the 24th Infantry improved.

To carry out his plan, General Kean on 16 September organized a composite battalion-sized task force under command of Maj. Robert L. Woolfolk, commanding officer of the 3d Battalion, 35th Infantry, and ordered it to attack the enemy-held heights of Battle Mountain and P'il-bong the next day, with the mission of restoring the 24th Infantry positions there. On the 17th and 18th the task force repeatedly attacked these heights, heavily supported by artillery fire from the 8th and 90th Field Artillery Battalions and by numerous air strikes, but enemy automatic fire from the heights drove back the assaulting troops every time with heavy casualties. Within twenty-four hours, A Company, 27th Infantry, alone suffered fifty-seven casualties. Woolfolk's force abandoned its effort to drive the enemy from the peaks after its failure on the 18th, and the task group was dissolved the next day.[57]

During the morning of 19 September it was discovered that the enemy had abandoned the crest of Battle Mountain during the night, and the 1st Battalion, 24th Infantry, moved up and occupied it. On the right, the 35th Infantry began moving forward. There was only light resistance until it reached the high ground in front of Chungam-ni where cleverly hidden enemy soldiers in spider holes shot at 1st Battalion soldiers from the rear. The next day the 1st Battalion captured Chungam-ni, and the 2d Battalion captured the long ridge line running northwest from it to the Nam River. Meanwhile, the enemy still held strongly against the division left where the 27th Infantry had heavy fighting in trying to move forward.[58]

On 21 September the 35th Infantry Regiment captured the well-known Notch, three miles southwest of Chungam-ni, and then swept westward eight air miles without resistance, past the Much'on-ni road fork, to the high ground at the Chinju pass. There at 2230 the lead battalion halted for the night. At the same time, the 24th and 27th Regiments in the center and on the division left advanced, slowed only by the rugged terrain they had to traverse. They passed abandoned position after position from which the North Koreans previously had fought to the death, and saw that enemy automatic positions had honeycombed the hills.[59]

The events of the past three days made it clear that the enemy in front of the 25th Division in the center and on the right had started his withdrawal the night of 18–19 September. The N.K. *7th Division* withdrew from south of the Nam River while the *6th Division* sideslipped elements to cover the entire front. Covered by the *6th Division,* the *7th* had crossed to the north side of the Nam River by the morning of the 19th. Then the N.K. *6th Division* had withdrawn from its positions on Sobuk-san.[60]

Although the North Korean withdrawal had been general in front of the 25th Division, there were still delaying groups and stragglers in the mountains. Below Tundok on the morning of 22 September some North Koreans slipped into the bivouac area of A Company, 24th Infantry. One platoon leader awoke to find an enemy soldier standing over him. He grabbed the enemy's bayonet and struggled with the North Korean until someone else shot the man. Nearby another enemy dropped a grenade into a foxhole on three sleeping men, killing

[57] EUSAK WD, 16 Sep 50; 24th Inf WD, 16 Sep 50; 27th Inf WD, 17 Sep 50; 1st Bn, 27th Inf Unit Rpt, Sep 50; 25th Div WD, Narr Rpt, Sep 50, p. 31; Barth MS, p. 33. Woolfolk's task group: Hq, 3d Bn, 35th Inf; I Co, 35th Inf; A Co, 27th Inf; B Co and 1 plat, C Co, 65th Engr C Bn. The 25th Reconnaissance Company and the Heavy Weapons Company, 24th Infantry, gave support.

[58] 24th Inf WD, 19 Sep 50; 35th Inf WD, 19–20 Sep 50; 1st Bn, 35th Inf WD, 19–20 Sep 50; 27th Inf WD, Act Rpt, Sep 50, p. 3; 25th Div WD, Narr, Sep 50, p. 25; EUSAK WD, G–3 Jnl, entry 1610, 20 Sep 50.

[59] 35th Inf Unit Rpt, 21 Sep 50; 1st Bn, 35th Inf, Unit Rpt, 21 Sep 50; 27th Inf Act Rpt, Sep 50, p. 3; 2d Bn, 24th Inf, WD, 21 Sep 50; Barth MS, p. 35.

[60] 25th Div WD, 19 Sep 50, and Narr Rpt, Sep 50, p. 31.

two and wounding the third. A little later mortar fire fell on a company commanders' meeting at 1st Battalion headquarters and inflicted seven casualties, including the commanding officer of Headquarters Company killed, and the battalion executive officer, the S–1, and the S–2 wounded.[61]

Up ahead of the division advance, elements of the N.K. *6th Division* at the Chinju pass blocked the 35th Infantry all day on 22 September, covering the withdrawal of the main body across the Nam River and through Chinju, six miles westward. The assault companies of the 1st Battalion, 35th Infantry, got within 200 yards of the top of Hill 152 at the pass but could go no farther.[62]

Just before the Eighth Army breakout MacArthur had revived a much debated proposal. On 19 September General Wright sent a message from Inch'on to General Hickey, Acting Chief of Staff, FEC, in Tokyo, saying General MacArthur directed that Plan 100-C, which provided for a landing at Kunsan, be readied for execution. He indicated that MacArthur wanted two U.S. divisions and one ROK division prepared to make the landing on 15 October. This proposal indicates quite clearly that on the 19th General MacArthur entertained serious doubts about the Eighth Army's ability to break out of the Pusan Perimeter. In truth it did not look very promising. General Walker, when informed of this plan, opposed taking any units out of the Eighth Army line in the south. By the 22d the situation had brightened considerably for a breakout there, and after discussing the matter with Walker that day General MacArthur gave up the idea of a Kunsan landing; General Hickey penciled on the plan, "File." [63]

Aerial observers' reports on 22 September gave no clear indication of enemy intentions. While there were reports of large enemy movements northward there were also large ones seen going south. Eighth Army intelligence on that day estimated the situation to be one in which, "although the enemy is apparently falling back in all sectors, there are no indications of an over-all planned disengagement and withdrawal." [64] This estimate of enemy intentions was wrong. Everywhere, even though it was not yet apparent to Eighth Army, the enemy units were withdrawing, covering their withdrawal by strong blocking and delaying actions wherever possible.

In any analysis of Eighth Army's unanticipated favorable position at this time it is imperative to calculate the effect of the Inch'on landing on the North Koreans fighting in the south. There can be little doubt that when this news reached them it was demoralizing in the extreme and was perhaps the greatest single factor in their rapid deterioration. The evidence seems to show that news of the Inch'on landing was kept from most of the North Korean officers as well as nearly all the troops at the Pusan Perimeter for nearly a week. It would appear that the North Korean

[61] Interv, author with 1st Lt Robert J. Tews (Plat Ldr A Co, 24th Inf), 21 Sep 51; 24th Inf WD, 22 Sep 50; 25th Div WD, Narr, p. 27, Sep 50.

[62] 35th Inf WD, 22–23 Sep 50; 25th Div WD, 22 Sep 50 and Narr, p. 40.

[63] Msg 063180, CINCUNC to CINCFE (Wright to Hickey), 19 Sep 50, FEC CofS file.

[64] EUSAK PIR 72, 22 Sep 50.

High Command did not decide on a withdrawal from the Perimeter and a regrouping somewhere farther north until three or four days after the landing when it became evident that Seoul was in imminent danger. The pattern of fighting and enemy action at the Perimeter reflects this fact.

Nowhere on 16 September, when Eighth Army began its offensive, did it score material gains except in certain parts of the 2d Division zone where the 38th and 23d Infantry Regiments broke through decimated enemy forces to reach the Naktong River. Until 19 September there was everywhere the stoutest enemy resistance and no indication of voluntary withdrawal, and, generally, U.N. advances were minor and bought only at the cost of heavy fighting and numerous casualties. Then during the night of 18–19 September the enemy *7th* and *6th Divisions* began withdrawing in the southern part of the line where the enemy forces were farthest from North Korean soil. The *6th Division* left behind well organized and effective delaying parties to cover the withdrawals.

On 19 September Waegwan fell to the U.S. 5th Regimental Combat Team, and the ROK 1st Division in the mountains north of Taegu penetrated to points behind the enemy *1st* and *13th Divisions'* lines. These divisions then started their withdrawals. The next day the ROK 3d Division on the east coast recaptured P'ohang-dong and in the ensuing days the *5th Division* troops in front of it fell back rapidly northward on Yongdok. And at the same time the ROK Army made sweeping advances in the mountains throughout the eastern half of the front. The 1st Cavalry Division was unable to make significant gains until 20 and 21 September. On the 21st it finally recaptured Tabu-dong. West of the Naktong the U.S. 2d Division fought stubborn enemy delaying forces on 21 and 22 September.

The effect of the Inch'on landing and the battles around Seoul on enemy action at the Pusan Perimeter from 19 September onward was clearly apparent. By that date the North Korean High Command began to withdraw its main forces committed in the south and start them moving northward. By 23 September this North Korean retrograde movement was in full swing everywhere around the Perimeter. This in itself is proof of the theater-wide military effectiveness of the Inch'on landing. The Inch'on landing will stand as MacArthur's masterpiece.

By 23 September the enemy cordon around the Pusan Perimeter was no more—Eighth Army's general attack combined with the effect of the Inch'on landing had rent it asunder. The Eighth Army and the ROK Army stood on the eve of pursuit and exploitation, a long-awaited revenge for the bitter weeks of defeat and death.

## CHAPTER XXVIII

# Pursuit and Exploitation

> Once the enemy has taken flight, they can be chased with no better weapons than air-filled bladders . . . attack, push, and pursue without cease. All maneuvers are good then; it is only precautions that are worthless. MARSHAL MAURICE DE SAXE, *Reveries on the Art of War*

By 23 September the North Korean Army was everywhere in retreat from the Pusan Perimeter. Eighth Army, motorized and led by armored spearheads, was ready to sweep forward along the main axes of advance.

The Eighth Army decision to launch the pursuit phase of the breakout operation came suddenly. On 20 September General Allen, in a telephone conversation with General Hickey in Tokyo, reported, "We have not had any definite break yet. They [the North Koreans] are softening but still no definite indication of any break which we could turn into a pursuit." [1] The next day General Allen thought the break had come, and on 22 September General Walker issued his order for the pursuit.

The Eighth Army order stated:

> Enemy resistance has deteriorated along the Eighth Army front permitting the assumption of a general offensive from present positions. In view of this situation it is mandatory that all efforts be directed toward the destruction of the enemy by effecting deep penetrations, fully exploiting enemy weaknesses, and through the conduct of enveloping or encircling maneuver get astride enemy lines of withdrawal to cut his attempted retreat and destroy him.[2]

The order directed a full-scale offensive: I Corps to continue to make the main effort along the Taegu-Kumch'on-Taejon-Suwon axis and to effect a juncture with X Corps; the 2d Infantry Division to launch an unlimited objective attack along the Hyopch'on-Koch'ang-Anui-Chonju-Kanggyong axis; the 25th Division on the army's southern flank to seize Chinju and be ready to attack west or northwest on army order; and the ROK Army in the east to destroy the enemy in its zone by deep penetrations and enveloping maneuvers. An important section of the Eighth Army order and a key to the contemplated operation stated, "Commanders will advance where necessary without regard to lateral security."

Later in the day, Eighth Army issued radio orders making IX Corps, under General Coulter, operational at 1400, 23

---

[1] Fonecon, Gen Allen with Gen Hickey, 1145 20 Sep 50, CofS GHQ files.

[2] Eighth Army Opn Ord 101, 22 Sep 50, copy in EUSAK WD, G-3 Sec, 22 Sep 50.

September, and attaching the U.S. 2d and 25th Infantry Divisions to it. This order charged IX Corps with the responsibility of carrying out the missions previously assigned to the 2d and 25th Divisions. In preparing for the pursuit, Eighth Army moved its headquarters from Pusan back to Taegu, reopening there at 1400, 23 September.[3]

The U.N. forces around the Pusan Perimeter at this time numbered almost 160,000 men, of whom more than 76,000 were in Eighth Army and about 75,000 in the ROK Army. United Nations reinforcements had begun arriving in Korea by this time. On 19 September the Philippine 10th Infantry Battalion Combat Team began unloading at Pusan, and on 22 September the 65th Regimental Combat Team started unloading there, its principal unit being the 65th Puerto Rican Infantry Regiment. The next day Swedish Red Cross Field Hospital personnel arrived at Pusan. On the 19th, the Far East Command deactivated the Pusan Logistical Command and reconstituted it as the 2d Logistical Command—its mission of logistical support unchanged.[4]

Since pursuit of the enemy along any one of the several corridors leading away from the Pusan Perimeter did not materially affect that in any other, except perhaps in the case of the Taejon and Poun-Ch'ongju central roadways, the pursuit phase of the breakout operation will be described by major corridors of advance. The story will move from south to north and northeast around the Perimeter. It must be remembered that the various movements were going on simultaneously all around the Perimeter.

### *The 25th Division Crosses Southwest Korea*

On the day he assumed command of operational IX Corps, 23 September, General Coulter in a meeting with General Walker at the 25th Division command post requested authority to change the division's axis of attack from southwest to west and northwest. He thought this would permit better co-ordination with the 2d Division to the north. Walker told Coulter he could alter the division boundaries within IX Corps so long as he did not change the corps boundaries.[5]

The change chiefly concerned the 27th Regiment which now had to move from the 25th Division's south flank to its north flank. General Kean formed a special task force under Capt. Charles J. Torman, commanding officer of the 25th Reconnaissance Company, which moved through the 27th Infantry on the southern coastal road at Paedun-ni the evening of the 23d. The 27th Regiment then began its move from that place to the division's north flank at Chungam-ni. The 27th Infantry was to establish a bridgehead across the Nam River and attack through Uiryong to Chinju.[6] (*Map VIII*)

On the morning of 24 September Task

[3] EUSAK WD, G-3 Sec, 22–23 Sep 50.

[4] GHQ FEC, G-3 Opn Rpt, 22 Sep 50. The assigned strength as of 1800, 21 September 1950, was: Eighth Army, 76,837; British 27th Infantry Brigade, 1,679; Air Force in Korea, 4,794; Philippine 10th Battalion Combat Team, 1,200; ROK Army, 74,987. EUSAK WD, 19–23 Sep 50; 2d Log Comd Act Rpt, Sep 50.

[5] Intervs, author with Coulter, 20 Jul 51 and 2 Apr 53; IX Corps WD, Sep 50, Personal Recollections of Coulter.

[6] 27th Inf Hist Rpt, Sep 50, p. 10; 25th Div WD, 23 Sep 50; EUSAK WD, 23 Sep 50; 25th Recon Co WD, 23 Sep 50.

Force Torman attacked along the coastal road toward Chinju. North of Sach'on the task force engaged and dispersed about 200 enemy soldiers of the *3d Battalion, 104th Security Regiment.* By evening it had seized the high ground at the road juncture three miles south of Chinju. The next morning the task force moved up to the Nam River bridge which crossed into Chinju. In doing so one of the tanks hit a mine and fragments from the explosion seriously wounded Captain Torman, who had to be evacuated.[7]

Meanwhile, on the main inland road to Chinju the N.K. *6th Division* delayed the 35th Infantry at the Chinju pass until the evening of 23 September, when enemy covering units withdrew. The next day the 35th Infantry consolidated its position at the pass. That night a patrol reported that enemy demolitions had rendered the highway bridge over the Nam at Chinju unusable.

On the strength of this information the 35th Regiment made plans to cross the Nam downstream from the bridge. Under cover of darkness at 0200, 25 September, the 2d Battalion crossed the river two and a half miles southeast of Chinju. It then attacked and seized Chinju, supported by tank fire from Task Force Torman across the river. About 300 enemy troops, using mortar and artillery fire, served as a delaying force in defending the town. The 3d and 1st Battalions crossed the river into Chinju in the afternoon, and that evening Task Force Torman crossed on an underwater sandbag ford that the 65th Engineer Combat Battalion built 200 yards east of the damaged highway bridge. Working all night, the engineers repaired the highway bridge so that vehicular traffic began crossing it at noon the next day, 26 September.[8]

Sixteen air miles downstream from Chinju, near the blown bridges leading to Uiryong, engineer troops and more than 1,000 Korean refugees worked all day on the 25th constructing a sandbag ford across the Nam River. Enemy mortars fired sporadically on the workers until silenced by counterbattery fire of the 8th Field Artillery Battalion. Before dawn of the 26th the 1st Battalion, 27th Infantry, crossed the Nam. Once on the north bank, elements of the regiment attacked toward Uiryong, three miles to the northwest, and secured the town just before noon after overcoming an enemy force that defended it with small arms and mortar fire. The regiment pressed on to Chinju against negligible resistance on 28 September.[9]

On 24 September Eighth Army had altered its earlier operational order and directed IX Corps to execute unlimited objective attacks to seize Chonju and Kanggyong. To carry out his part of the order, General Kean organized two main task forces with armored support centered about the 24th and 35th Infantry Regiments. The leading elements of these two task forces were known respectively as Task Force Matthews and Task Force Dolvin. Both forces were to start their drives from Chinju. Task Force

[7] 25th Recon Co WD, G-3 Sec, Sep 50; 25th Div WD, 23-24 Sep 50; ATIS Interrog Rpts, Issue 12, p. 193, Rpt 1769, Lt Col Pak Chong Song, CO, *3d Bn, 104th Security Regt.*

[8] 35th Inf Unit Rpt, 25-26 Sep 50; 25th Recon Co WD, Sep 50; 25th Div WD, Hist Narr, p. 44; IX Corps WD, 25 Sep 50; 65th Engr C Bn WD, 25-26 Sep 50.

[9] 1st Bn, 27th Inf, Unit Rpt, 26 Sep 50; 27th Inf Hist Rpt, 25-26 Sep 50, p. 11; 25th Div WD, Sep 50, Narr Hist, p. 46.

Matthews, the lefthand column, was to proceed west toward Hadong and there turn northwest to Kurye, Namwon, Sunch'ang, Kumje, Iri, and Kunsan on the Kum River estuary. Taking off at the same time, Task Force Dolvin, the righthand column, was to drive north out of Chinju toward Hamyang, there turn west to Namwon, and proceed northwest to Chonju, Iri, and Kanggyong on the Kum River.[10]

Three blown bridges west of Chinju delayed the departure of Task Force Matthews (formerly Task Force Torman) until 1000, 27 Septebmer. Capt. Charles M. Matthews, commanding officer of A Company, 79th Tank Battalion, replaced Torman in command of the latter's task force after Torman had been wounded and evacuated. He led the advance out of Chinju with the 25th Reconnaissance Company and A Company, 79th Tank Battalion. The 3d Battalion, 24th Infantry, followed Task Force Matthews, and the rest of the regiment came behind it. Matthews reached Hadong at 1730.[11]

In a sense, the advance of Task Force Matthews became a chase to rescue a group of U.S. prisoners that the North Koreans moved just ahead of the pursuers. Korean civilians and bypassed enemy soldiers kept telling of them being four hours ahead, two hours ahead—but always ahead. At Hadong the column learned that some of the prisoners were only thirty minutes ahead. From Hadong, in bright moonlight, the attack turned northwest toward Kurye. About ten miles above Hadong at the little village of Komdu the advanced elements of the task force liberated eleven American prisoners. They had belonged to the 3d Battalion, 29th Infantry Regiment. Most of them were unable to walk and some had open wounds.[12]

Just short of Namwon about noon the next day, 28 September, several vehicles at the head of the task force became stuck in the river crossing below the town after Sgt. Raymond N. Reifers in the lead tank of the 25th Reconnaissance Company had crossed ahead of them. While the rest of the column halted behind the stuck vehicles Reifers continued on into Namwon.

Entering the town, Reifers found it full of enemy soldiers. Apparently the North Koreans' attention had been centered on two F-84 jet planes that could be seen sweeping in wide circles, rocketing and strafing the town, and they were unaware that pursuing ground elements were so close. Surprised by the sudden appearance of the American tank, the North Koreans in wild disorder jumped over fences, scurried across roof tops, and dashed madly up and down the streets.

---

[10] Eighth Army, Ltr of Inst to CG, I and IX Corps, 24 Sep 50; EUSAK WD, 23 Sep 50; 25th Div WD, 24–25 Sep 50.

[11] The name Task Force Torman is used frequently in the official records when TF Matthews is meant. TF Matthews consisted of the 25th Rcn Co; A Co, 79th Tk Bn; a platoon of B Co, 67th Engr C Bn; an air TAC Party; and the medical section of the 27th Inf Regt.

There were three separate task forces in this movement: (1) TF Matthews which formed the point; (2) TF Blair, named after Maj Melvin R. Blair, CO 3d Bn, 24th Inf, which followed close behind Matthews; and (3) TF Corley, named for Col John T. Corley, CO 24th Inf Regt, which included the rest of the RCT, following TF Blair. See 24th Inf WD, 26–30 Sep 50; 3d Bn, 24th Inf Unit Rpt, 26–30 Sep 50.

[12] Sawyer, Notes for author, 1 Oct 52 (Sawyer, then a 1st lieutenant, commanded the 3d Plat, 25th Recon Co, in TF Matthews); 25th Div WD, Med Co Unit Rpt (24th Inf), 28 Sep 50; 3d Bn, 24th Inf WD, 27 Sep 50; Corley, MS review comments, Oct 57.

Reifers said later that the scene would have appeared ludicrous if his own plight had not been so precarious. Suddenly he heard American voices calling out, "Don't shoot! Americans! GI's here!" A second later a gateway leading into a large courtyard burst open and the prisoners—shouting, laughing, and crying—poured out into the street.

Back at the head of the stuck column, 1st Lt. Robert K. Sawyer over his tank radio heard Reifer's voice calling out, "Somebody get up here! I'm all alone in this town! It is full of enemy soldiers and there are American prisoners here." Some of the tanks and vehicles now pushed ahead across the stream. When Sawyer's tank turned into the main street he saw ahead of him, gathered about vehicles, "a large group of bearded, haggard Americans. Most were bare-footed and in tatters, and all were obviously half starved. We had caught up with the American prisoners," he said, "there were eighty-six of them." [13]

Task Forces Matthews and Blair cleared Namwon of enemy soldiers. In midafternoon Task Force Dolvin arrived there from the east. Task Force Matthews remained overnight in Namwon, but Task Force Blair continued on toward Chongup, which was secured at noon the next day, 29 September. That evening Blair's force secured Iri. There, with the bridge across the river destroyed, Blair stopped for the night and Task Force Matthews joined it. Kunsan, the port city on the Kum River estuary, fell to the 1st Battalion, 24th Infantry, without opposition at 1300, 30 September.[14]

Eastward of and generally parallel to the course of Task Force Matthews and the 24th Infantry, Task Force Dolvin and the 35th Infantry moved around the eastern and northern sides of the all but impenetrable Chiri-san area, just as the 24th Infantry had passed around its southern and western sides. This almost trackless waste of 750 square miles of 6,000 to 7,000-foot-high forested mountains forms a rough rectangle northwest of Chinju about thirty by twenty-five miles in dimension, with Chinju, Hadong, Namwon, and Hamyang at its four corners. This inaccessible area had long been a hideout for Communist agents and guerrillas in South Korea. Now, as the North Korean forces retreated from southwest Korea, many enemy stragglers and some organized units with as many as 200 to 400 men went into the Chiri Mountain fastnesses. There they planned to carry on guerrilla activities.[15]

[13] Intervs, author with Sawyer at numerous times in the fall of 1951 and in 1952; Sawyer, Notes for author, 1 Oct 52; 24th Inf WD, 28 Sep 50; 25th Div WD, Med Co Unit Rpt (24th Inf), 28 Sep 50; 3d Bn, 24th Inf WD, 28 Sep 50. Sawyer is the source for details of the Reifers incident; his information was based on personal experience and numerous extended conversations held with Reifers at the time of and shortly after the event. Reifers was later killed near Unsan in North Korea on 27 November 1950 in the CCF offensive.

Most of the official Army records erroneously give credit to TF Dolvin for liberating these prisoners (EUSAK WD, G-3 Jnl, 28 Sep 50; IX Corps PIR, Msg 281645 Sep 50; 25th Div WD, Narr, Sep 50, p. 49). The 3d Bn, 24th Inf Unit Rpt, the Med Co Rpt, and Sawyer's Notes, however, leave no doubt that TF Matthews effected the liberation. TF Dolvin did not reach the outskirts of Namwon until midafternoon, about 1515, nearly three hours after TF Matthews had entered the town. See 89th Med Tk Bn WD, Sep 50.

[14] 3d Bn, 24th Inf, Unit Rpt, 28–30 Sep 50; 24th Inf WD, 30 Sep 50.

[15] ATIS Interrog Rpts, Issue 12 (N.K. Forces), Rpt 1728, p. 86, Jr Lt Ku Sung Son, *110th Security Regt;* Rpt 1764, p. 175, Jr Lt Kang Myong Ho, *110th Security Regt;* Rpt 1711, p. 36, Jr Lt Kim Tok Ho, *102d Security Regt.*

Lt. Col. Welborn G. Dolvin, commanding officer of the 89th Tank Battalion, led Task Force Dolvin out of Chinju at 0600, 26 September, on the road northwest toward Hamyang, the retreat route taken by the main body of the N.K. *6th Division*. The tank-infantry task force included as its main elements A and B Companies, 89th Medium Tank Battalion, and B and C Companies, 35th Infantry. It had two teams, A and B, each formed of an infantry company and a tank company. The infantry rode the rear decks of the tanks. The tank company commanders commanded the teams.[16]

Three miles out of Chinju the lead M26 tank struck a mine. While the column waited, engineers removed eleven more from the road. Half a mile farther on, a second tank was damaged in another mine field. Still farther along the road a third mine field, covered by an enemy platoon, stopped the column again. After the task force dispersed the enemy soldiers and cleared the road of mines, it found 6 antitank guns, 9 vehicles, and an estimated 7 truckloads of ammunition in the vicinity abandoned by the enemy. At dusk, the enemy blew a bridge three miles north of Hajon-ni just half an hour before the task force reached it. During the night the task force constructed a bypass.[17]

The next morning, 27 September, a mine explosion damaged and stopped the lead tank. Enemy mortar and small arms fire from the ridges near the road struck the advanced tank-infantry team. Tank fire cleared the left side of the road, but an infantry attack on the right failed. The column halted, and radioed for an air strike. Sixteen F-51 fighter-bombers came in strafing and striking the enemy-held high ground with napalm, fragmentation bombs, and rockets. General Kean, who had come forward, watched the strike and then ordered the task force to press the attack and break through the enemy positions. The task force broke through on the road, bypassing an estimated 600 enemy soldiers. Another blown bridge halted the column for the night while engineers constructed a bypass.

Continuing its advance at first light on the 28th, Task Force Dolvin an hour before noon met elements of the 23d Infantry, U.S. 2d Division, advancing from the east, at the road junction just east of Hamyang. There it halted three hours while engineers and 280 Korean laborers constructed a bypass around another blown bridge. Ever since leaving Chinju, Task Force Dolvin had encountered mine fields and blown bridges, the principal delaying efforts of the retreating N.K. *6th Division*.

When it was approaching Hamyang the task force received a liaison plane report that enemy forces were preparing to blow a bridge in the town. On Colonel Dolvin's orders the lead tanks sped ahead, machine-gunned enemy troops who were placing demolition charges, and seized the bridge intact. This success upset the enemy's delaying plans. The rest of the afternoon the task force dashed ahead at a speed of twenty miles

---

[16] Interv, author with 1st Lt Francis G. Nordstrom (89th Med Tk Bn elements of TF), 31 Aug 51; 89th Med Tk Bn Unit Rpt, 25–26 Sep 50; 25th Div WD, 24 Sep 50; Joseph M. Quinn, "Catching the Enemy Off Guard," *Armor*, vol. 60 (July–August, 1951), pp. 47–48.

TF Dolvin also included the 1st Plat, A Co, 65th Engr C Bn; 2d Plat, Hv Mort Co, 35th Regt; Medical Det, 89th Med Tk Bn; and TF trains.

[17] 89th Med Tk Bn WD, 26 Sep 50.

an hour. It caught up with numerous enemy groups, killing some of the soldiers, capturing others, and dispersing the rest. At midafternoon Task Force Dolvin entered Namwon to find that Task Force Matthews and elements of the 24th Infantry were already there.

Refueling in Namwon, Task Force Dolvin just after midnight continued northward and in the morning reached Chonju, already occupied by elements of the 38th Infantry Regiment, and continued on through Iri to the Kum River. The next day at 1500, 30 September, its mission accomplished, Task Force Dolvin was dissolved. It had captured or destroyed 16 antitank guns, 19 vehicles, 65 tons of ammunition, 250 mines, captured 750 enemy soldiers, and killed an estimated 350 more. It lost 3 tanks disabled by mines and 1 officer and 45 enlisted men were wounded in action.[18]

In crossing southwest Korea from Chinju to the Kum River, Task Force Matthews had traveled 220 miles and Task Force Dolvin, 138 miles. In the wake of Task Force Dolvin the 27th Regiment moved north from Chinju to Hamyang and Namwon on 29 September and maintained security on the supply road. This same day, 29 September, ROK marines captured Yosu on the south coast.

### *The 2d Division Pushes West*

Opposite the old Naktong Bulge area, the N.K. *9th, 4th,* and *2d Divisions* retreated westward. At Sinban-ni the *4th Division* turned toward Hyopch'on. The *9th* withdrew on Hyopch'on, and the *2d,* after passing through Ch'ogye, continued on to the same place. Apparently the *9th Division,* in the lead, had passed through Hyopch'on before elements of the U.S. 2d Division closed in on the place.[19]

On 23 September the 38th Infantry of the U.S. 2d Division had hard fighting in the hills around Ch'ogye before overcoming enemy delaying forces. The next day the 23d Infantry from the southeast and the 38th Infantry from the northeast closed on Hyopch'on in a double envelopment movement. Elements of the 38th Infantry established a roadblock on the north-south Chinju-Kumch'on road running northeast out of Hyopch'on and cut off an estimated two enemy battalions still in the town. During the day the 3d Battalion, 23d Infantry, entered Hyopch'on after a rapid advance of eight miles from the southeast. As the North Koreans fled Hyopch'on in the afternoon, 38th Infantry fire killed an estimated 300 of them at the regiment's roadblock northeast of the town. Two flights of F-51 fighter planes caught the rest in the open and continued their destruction. The surviving remnant fled in utter disorder for the hills. The country around Hyopch'on was alive with hard-pressed, fleeing North Koreans on 24 September, and the Air Force, flying fifty-three sorties in the area, wrought havoc among them. That night elements of the 1st Battalion, 38th

[18] 89th Med Tk Bn WD, 28–30 Sep 50; A Co, 79th Tk Bn, Unit Rpt, Sep 50; IX Corps WD, Hist Narr, 23–30 Sep 50; 25th Div WD, Narr, Sep 50, pp. 49, 58; EUSAK WD, Recommendation for Distinguished Unit Citation, 89th Med Tk Bn, Armor Sec, 11 Mar 51.

[19] ATIS Res Supp Interrog Rpts, Issue 100 (N.K. *9th Div*), p. 53; *Ibid.*, Issue 94 (N.K. *4th Div*), p. 50; *Ibid.*, Issue 94 (N.K. *2d Div*), p. 38; EUSAK WD, G–2 Sec, 3 Oct 50.

Infantry, entered Hyopch'on from the north.[20]

At daylight on the 25th, the 38th Infantry started northwest from Hyopch'on for Koch'ang. The road soon became impassable for vehicles and the men had to detruck and press forward on foot.

In retreating ahead of the 38th Infantry on 25 September the N.K. *2d Division,* according to prisoners, abandoned all its remaining vehicles and heavy equipment between Hyopch'on and Koch'ang. This apparently was true, for in its advance from Hyopch'on to Koch'ang the 38th Infantry captured 17 trucks, 10 motorcycles, 14 antitank guns, 4 artillery pieces, 9 mortars, more than 300 tons of ammunition, and 450 enemy soldiers, and killed an estimated 260 more. Division remnants, numbering no more than 2,500 men, together with their commander, Maj. Gen. Choe Hyon, who was ill, scattered into the mountains.

Up ahead of the ground troops, the Air Force in the late afternoon bombed, napalmed, rocketed, and strafed Koch'ang, leaving it virtually destroyed. After advancing approximately thirty miles during the day, the 38th Infantry stopped at 2030 that night only a few miles from the town.[21]

Elements of the 38th Infantry entered Koch'ang at 0830 the next morning, 26 September, capturing there a North Korean field hospital containing forty-five enemy wounded. Prisoners disclosed that elements of the N.K. *2d, 4th, 9th,* and *10th Divisions* were to have assembled at Koch'ang, but the swift advance of the U.S. 2d Division had frustrated the plan.[22]

The 23d Infantry was supposed to parallel the 38th Infantry on a road to the south, in the pursuit to Koch'ang, but aerial and road reconnaissance disclosed that this road was either impassable or did not exist. General Keiser then directed Colonel Freeman to take a road to the north of the 38th Infantry. Mounted on organic transportation, the regiment, less its 1st Battalion, started at 1600 on the 25th and made a night advance to Koch'ang, fighting three skirmishes and rebuilding four small bridges on the way. It arrived at Koch'ang soon after the 38th Infantry, in daylight on 26 September.

That evening the 23d Infantry continued the advance to Anui, fourteen miles away, which it reached at 1930 without enemy opposition. Except for the small town itself, the area was a maze of flooded paddies. The regimental vehicles could find no place to move off the roads except into the village streets where they were dispersed as well as possible. At least one enemy group remained in the vicinity of Anui. At 0400 the next morning, 27 September, a heavy enemy artillery and mortar barrage struck in the town. The second round

[20] 38th Inf Comd Rpt, 24 Sep 50; 2d Div WD, G-3 Sec, 24 Sep 50; EUSAK WD, Br for CG, 24 Sep 50, and PIR 74, 24 Sep 50; Interv, author with Peploe, 12 Aug 51. The Air Force claimed it destroyed on 24 September in the Hyopch'on area 3 tanks, 5 artillery pieces, 1 antitank gun, an ammunition dump, a supply dump, a POL dump, and an estimated 1,400 enemy soldiers. See EUSAK WD, G-3 Air Br Rpt, 24 Sep 50.

[21] ATIS Interrog Rpts, Issue 12 (N.K. Forces), Rpt 1741, p. 118, Lt Ko Kon Su, Aide to CG, N.K. *2d Div;* EUSAK WD, 25 Sep 50; 38th Inf Comd Rpt, 25 Sep 50; 2d Div PIR 32, 25 Sep 50; ATIS Res Supp Interrog Rpts, Issue 94 (N.K. *2d Div*), p. 38.

[22] 38th Inf Comd Rpt, Sep–Oct 50, p. 13; 23d Inf Comd Rpt, 26 Sep 50; EUSAK WD, 26 Sep 50; IX Corps PIR, Msg 332, 261700 Sep 50.

hit the 3d Battalion command post, killing the battalion executive officer, the S–2, the assistant S–3, the motor officer, the artillery liaison officer, and an antiaircraft officer. Lt. Col. R. G. Sherrard, the battalion commander, was severely wounded; also wounded were twenty-five enlisted men of the Regimental and Headquarters Companies.[23]

At least passing notice should be taken of another event on 27 September. The last organized unit of the North Korean forces east of the Naktong River, elements of the N.K. *10th Division,* withdrew from notorious Hill 409 near Hyongp'ung and crossed to the west side of the river before daylight. Patrols of the 9th Infantry Regiment entered Hyongp'ung in the afternoon, and two companies of the 2d Battalion occupied Hill 409 without opposition. On 28 September the 2d Battalion, 9th Infantry, crossed the Naktong to join the 2d Division after the newly arrived 65th Regimental Combat Team of the U.S. 3d Division relieved it on Hill 409.[24]

At 0400, 28 September, Colonel Peploe started the 38th Infantry, with the 2d Battalion leading, from Koch'ang in a motorized advance toward Chonju, an important town in the west coastal plain seventy-three miles away across the mountains. The 25th Division also was approaching Chonju through Namwon. Meeting only light and scattered resistance, the 2d Battalion, 38th Infantry, entered Chonju at 1315, having covered the seventy-three miles in nine and a half hours. At Chonju the battalion had to overcome about 300 enemy soldiers of the *102d* and *104th Security Regiments,* killing about 100 of them and taking 170 prisoners.

There the 38th Infantry ran out of fuel for its vehicles. Fortunately, a 2d Division liaison plane flew over the town and the pilot learned the situation. He reported it to the 2d Division and IX Corps which rushed gasoline forward. At 1530 on 29 September, after refueling, the 3d Battalion departed Chonju for Nonsan and continued to Kanggyong on the Kum River, arriving there without incident at 0300 the morning of 30 September.

The IX Corps had only two and a half truck companies with which to transport supplies to the 25th and 2d Divisions in their long penetrations, and the distance of front-line units from the railhead increased hourly. When the 2d Division reached Nonsan on the 29th the supply line ran back more than 200 road miles, much of it over mountainous terrain and often on one-way roads, to the railhead at Miryang. The average time for one trip was forty-eight hours. In one 105-hour period, Quartermaster truck drivers supporting the 2d Division got only about thirteen hours' sleep.[25]

At the end of September the 2d Division was scattered from the Kum River southward, with the 38th Infantry in

[23] 23d Inf Comd Rpt, 26–27 Sep 50; 2d Div WD, G–3 Sec, 26–27 Sep 50; Freeman, MS review comments, 30 Oct 57; Combat Activities of the 23d Infantry.

[24] 2d Div PIR 34, 27 Sep 50; EUSAK POR 232 and PIR 77, 27 Sep 50; EUSAK WD, Br for CG, 27–28 Sep 50; 2d Div Hist, vol. II, p. 31.

[25] Intervs, author with Coulter, 20 Jul and 2 Apr 53; Interv, author with Peploe, 12 Aug 51; 2d Div Comd Rpt Hist, vol. II, 28 Sep 50; 2d Div WD, G–4 Sec, Sep–Oct 50; 38th Inf Comd Rpt, Sep–Oct 50, p. 14; Capt Perry Davis, The 2d Infantry Division in Korea, July–September 1950, MS, copy in OCMH; Coulter, MS review comments, 22 Nov 57.

the Chonju-Kanggyong area, the 23d Infantry in the Anui area, and the 9th Infantry in the Koryong-Samga area.

### *Taejon Regained*

On the right flank of the U.S. 2d Division, the British 27th Infantry Brigade, attached to the U.S. 24th Division for the pursuit, was to move against Songju while the 24th Division simultaneously attacked parallel to and north of it on the main highway toward Kumch'on. After passing through Songju, the British brigade was to strike the main highway halfway between the Naktong River and Kumch'on. Its path took it along the main retreat route of the N.K. *10th Division*. The brigade was across the Naktong and ready to attack before daylight on 22 September.

At dawn the 1st Battalion, Middlesex Regiment, seized a small hill, called by the men Plum Pudding Hill, on the right of the road three miles short of Songju. The battalion then attacked the higher ground immediately to the northeast, known to the British as Point 325 or Middlesex Hill. Supported by American tank fire and their own mortar and machine gun fire, the Middlesex Battalion took the hill from dug-in enemy soldiers before dark.[26]

While the Middlesex Battalion attacked Hill 325, the Scottish Highlander Argyll Battalion moved up to attack neighboring Hill 282 on the left of the road. Starting before dawn on 23 September, B and C Companies after an hour's climb seized the crest of Hill 282, surprising there a North Korean force at breakfast. Across a saddle, and nearly a mile away to the southwest, higher Hill 388 dominated the one they had just occupied. C Company started toward it.

But enemy troops occupying this hill already were moving to attack the one just taken by the British. The North Koreans supported their attack with artillery and mortar fire, which began falling on the British. The action continued throughout the morning with enemy fire increasing in intensity. Shortly before noon, with American artillery fire inexplicably withdrawn and the five supporting U.S. tanks unable to bring the enemy under fire because of terrain obstacles, the Argylls called for an air strike on enemy-held Hill 388.[27]

Just after noon the Argylls heard the sound of approaching planes. Three F-51 Mustangs circled Hill 282 where the British displayed their white recognition panels. The enemy on Hill 388 also displayed white panels. To his dismay, Captain Radcliff of the tactical air control party was unable to establish radio contact with the flight of F-51's. Suddenly, at 1215, the Mustangs attacked the wrong hill; they came in napalming and machine-gunning the Argyll position.

The terrible tragedy was over in two minutes and left the hilltop a sea of orange flame. Survivors plunged fifty feet down the slope to escape the burning napalm. Maj. Kenneth Muir, second

[26] 24th Recon Co (24th Div) WD, Summ, 22 Sep 50; Linklater, *Our Men in Korea*, p. 19. Linklater was with the British brigade in Korea and used its records in preparing this small book. Middlesex Hill is shown on revised maps as Hill 341.

[27] Linklater, *Our Men in Korea;* Malcolm, *The Argylls in Korea*, pp. 17–22.

in command of the Argylls, who had led an ammunition resupply and litter-bearing party to the crest before noon, watching the flames on the crest die down, noticed that a few wounded men still held a small area on top. Acting quickly, he assembled about thirty men and led them back up the hill before approaching North Koreans reached the top. There, two bursts of enemy automatic fire mortally wounded him as he and Maj. A. I. Gordon-Ingram, B Company commander, fired a 2-inch mortar. Muir's last words as he was carried from the hilltop were that the enemy "will never get the Argylls off this ridge." But the situation was hopeless. Gordon-Ingram counted only ten men with him able to fight, and some of them were wounded. His three Bren guns were nearly out of ammunition. At 1500 the survivors were down at the foot of the hill.

The next day a count showed 2 officers and 11 men killed, 4 officers and 70 men wounded, and 2 men missing for a total of 89 casualties; of this number, the mistaken air attack caused approximately 60.[28]

That night, after the Argyll tragedy, the 1st Battalion, 19th Infantry, attacked south from Pusang-dong on the Waegwan-Kumch'on highway and captured Songju at 0200, 24 September. From there it moved to link up with the British 27th Brigade below the town. That day and the next the 19th Infantry and the British brigade mopped up in the Songju area. On the afternoon of 25 September the British brigade, released from attachment to the U.S. 24th Division, reverted to I Corps control.

The N.K. *10th Division* which had been fighting in the Songju area, its ammunition nearly gone and its vehicles out of fuel, withdrew on the 24th and 25th after burying its artillery. A captured division surgeon estimated the *10th Division* had about 25 percent of its original strength at this time. The N.K. *I Corps,* about 25 September, ordered all its units south of Waegwan to retreat northward.[29]

On 23 September, the day that disaster struck the British from the air near Songju, the U.S. 24th Division started its attack northwest along the Taejon-Seoul highway. General Church had echeloned his three regiments in depth so that a fresh regiment would take the lead at short intervals and thus maintain impetus in the attack. Leading off for the division, the 21st Infantry headed for Kumch'on, the N.K. headquarters. Elements of the N.K. *105th Armored Division* blocked the way with dug-in camouflaged tanks, antitank guns, and extensive mine fields.

[28] Malcolm, *The Argylls in Korea,* pp. 22–23; Linklater, *Our Men in Korea,* pp. 20–21; I Corps WD, G–3 Sec, 2 Aug–30 Sep 50, Rpt of Lt Col Thomas C. Gillis; I Corps WD, 23 Sep 50; FEAF Opn Hist, I, 25 Jun–31 Oct 50, p. 184 (23 Sep). The British Government awarded Major Muir posthumously the Victoria Cross, England's highest military award for valor. Text of citation in Brig. C. N. Barclay, *The First Commonwealth Division* (Aldershot: Gale and Polden, 1954), app. D, p. 218.

[29] 24th Div WD, 23 Sep 50; 19th Inf WD, 23–26 Sep 50; EUSAK WD, British Forces, 23–24 Sep 50; *Ibid.,* POR 228, 26 Sep 50; *Ibid.,* G–2 Sec, 5 Oct 50, Interrog ADVATIS 1039, Pak In Hyok, Surgeon *10th Div* Hosp; GHQ FEC, History of the N.K. Army, pp. 41 and 69; ATIS Res Supp Interrog Rpts, Issue 106 (N.K. Arty), pp. 68 and 76.

KUMCH'ON FROM THE AIR

In the afternoon a tank battle developed in which D Company, 6th Medium Tank Battalion (Patton M46), lost four tanks to enemy tank and antitank fire. During a slow advance, American tanks and air strikes in turn destroyed three enemy tanks.[30]

Just as this main Eighth Army drive started it was threatened with a supply breakdown. Accurate enemy artillery fire during the night of the 22d destroyed the only raft at the Naktong River ferry and cut the footbridge three times. The ferrying of vehicles and supplies during the day practically stopped, but at night local Koreans carried across the river the supplies and ammunition needed the next day.

Shortly after midnight, 23–24 September, the 5th Regimental Combat Team passed through the 21st Infantry to take the lead. Enemy troops in positions on Hill 140, north of the highway, stopped the regiment about three miles east of Kumch'on. There the North Koreans fought a major delaying action to permit large numbers of their retreating units to escape. The North Korean command diverted its *9th Division,* retreating from the lower Naktong toward Taejon, to Kumch'on to block the rapid Eighth Army advance. Remaining tanks of two regiments of

[30] 21st Inf WD, 23 Sep 50; 6th Med Tk Bn WD, 22–24 Sep 50; EUSAK WD, G–3 Jnl, 1715, 23 Sep 50; 24th Div WD, 23 Sep 50.

ON THE OUTSKIRTS OF KUMCH'ON *infantrymen take a brief rest.*

the N.K. *105th Armored Division* and the *849th Independent Anti-Tank Regiment,* the latter recently arrived at Kumch'on from the north, also joined in the defense of the town.

In the battle that followed in front of Kumch'on, the 24th Division lost 6 Patton tanks to enemy mines and antitank fire, while the North Koreans lost 8 tanks, 5 to air attack and 3 to ground fire. In this action the enemy *849th Regiment* was practically destroyed. The 5th Regimental Combat Team and supporting units lost approximately 100 men killed or wounded, most of them to tank and mortar fire. Smaller actions flared simultaneously at several points on the road back to Waegwan as bypassed enemy units struck at elements of the 19th Infantry bringing up the rear of the 24th Division advance.[31]

As a result of the battle in front of Kumch'on on 24 September, the 21st Infantry swung to the north of the highway and joined the 5th Regimental Combat Team that night in a pincer attack on the town. The 3d Battalion of the 5th Regimental Combat Team entered Kumch'on the next morning, and by 1445 that afternoon the town, a mass of rubble from bombing and artillery barrages, was cleared of the enemy.

[31] I Corps WD, Narr Hist, 24 Sep 50; EUSAK WD, G–2 Sec, 30 Sep 50, interrog Lt Kim Chong Song (N.K. *9th Div*); 24th Div WD, 24 Sep 50, and G–1 Hist Rpt, 25 Sep; 5th RCT Unit Rpt 43, 24 Sep 50; 6th Med Tk Bn WD, 24–25 Sep 50; New York *Times*, September 26, 1950.

That evening the 21st Infantry continued the attack westward. The 24th Division was interested only in the highway. If it was clear, the column went ahead. With the fall of Kumch'on on the 25th, enemy resistance melted away and it was clear that the North Koreans were intent only on escaping.[32]

On 26 September the 19th Infantry took the division lead and its 2d Battalion entered Yongdong without resistance. In the town jail the troops found and liberated three American prisoners. The regiment continued on and reached Okch'on, ten miles east of Taejon, at 0200, 27 September. There it halted briefly to refuel the tanks and give the men a little rest.

At 0530 the regiment resumed the advance—but not for long. Just outside Okch'on the lead tank hit a mine and enemy antitank fire then destroyed it. The 1st Battalion deployed and attacked astride the road but advanced only a short distance. The North Koreans held the heights west of Okch'on in force and, as at Kumch'on three days earlier, were intent on a major delaying operation. This time it was to permit thousands of their retreating fellow soldiers to escape from Taejon. An American tank gunner moving up to join the fight in front of Taejon sang, "The last time I saw Taejon, it was not bright or gay. Today I'm going to Taejon and blow the place away." [33]

This fight in front of Taejon on 27 September disclosed that the city, as expected, was an assembly point for retreating North Korean units south and west of Waegwan. The 300 prisoners taken during the day included men from seven North Korean divisions. The reports of enemy tanks destroyed in the Taejon area during the day are confusing, conflicting, and, taken together, certainly exaggerated. The ground forces reported destroying 13 tanks on the approaches to the city, 3 of them by A Company, 19th Infantry, bazooka teams. The Air Force claimed a total of 20 tanks destroyed during the day, 13 of them in the Taejon area, and another 8 damaged.[34]

On the morning of the 28th an air strike at 0700 hit the enemy blocking position. When the 2d Battalion advanced cautiously up the slopes, it was unopposed. It then became clear that the North Koreans had withdrawn during the night. Aerial reconnaissance at the time of the air strike disclosed approximately 800 North Korean troops moving out of Taejon on the road past the airstrip. At noon aerial observers saw more enemy troops assembling at the railroad station and another concentration of them a few miles west of Taejon turning toward Choch'iwon. The Air Force napalmed and strafed still another force of 1,000 enemy soldiers west of the city.

Scouts of the 2d Battalion, 19th Infantry, and engineers of C Company, 3d Engineer Combat Battalion, entered the outskirts of Taejon at 1630. An hour later the 19th Infantry secured the city

[32] 21st Inf WD, 25 Sep 50; 24th Div WD, 25 Sep 50; I Corps WD, Narr Hist, 25 Sep 50; EUSAK POR 226, 25 Sep 50; ATIS Res Supp Interrog Rpts, Issue 4 (N.K. *105th Armd Div*), pp. 39–40; Interv, author with Alkire, 1 Aug 51.

[33] 24th Div WD, 26–28 Sep 50 and G–1 Summ, 26 Sep; 19th Inf WD, Opn Summ, 26 Sep 50.

[34] 19th Inf WD, 27 Sep 50; 24th Div WD, 27 Sep 50; EUSAK WD, G–3 Sec and G–3 Air Br Rpt, 27 Sep 50; GHQ FEC Sitrep, 271603 Sep 50.

after engineers had cleared mines ahead of tanks leading the main column. At 1800 a 24th Division artillery liaison plane landed at the Taejon airstrip.[35]

On 28 September, the day it entered Taejon, the 19th Infantry captured so many North Korean stragglers that it was unable to keep an accurate count of them. The capture of large numbers of prisoners continued during the last two days of the month; on the 30th the 24th Division took 447 of them. At Taejon the division captured much enemy equipment, including four U.S. howitzers lost earlier and fifty new North Korean heavy machine guns still packed in cosmoline. At Choch'iwon the North Koreans were destroying equipment to prevent its capture.[36] Already other U.S. forces had passed Taejon and Choch'iwon on the east to cut the main highway farther north at Ch'onan and Osan.

With the capture of Taejon, the 24th Division accomplished its mission in the pursuit. And sweet revenge it was for the Taro Leaf Division to re-enter this now half-destroyed town where it had suffered a disastrous defeat nine weeks earlier. Fittingly enough, it was the 19th Infantry Regiment and engineers of the 3d Engineer Combat Battalion, among the last to leave the burning city on that earlier occasion, who led the way back in. But there was bitterness too, for within the city American troops soon discovered that the North Koreans had perpetrated there one of the greatest mass killings of the entire Korean War. American soldiers were among the victims.

While this is not the place to tell in detail the story of the North Korean atrocities perpetrated on South Korean civilians and soldiers and some captured American soldiers, an account of the breakout and pursuit would not be complete without at least a brief description of the grisly evidence that came to light at that time. Everywhere the advancing columns found evidence of atrocities as the North Koreans hurried to liquidate political and military prisoners held in jails before they themselves retreated in the face of the U.N. advance. At Sach'on the North Koreans burned the jail, causing some 280 South Korean police, government officials, and landowners held in it to perish. At Anui, at Mokp'o, at Kongju, at Hamyang, at Chonju, mass burial trenches containing the bodies of hundreds of victims, including some women and children, were found, and near the Taejon airstrip the bodies of about 500 ROK soldiers, hands tied behind backs, lay in evidence of mass killing and burial.

Between 28 September and 4 October a frightful series of killings and burials were uncovered in and around the city. Several thousand South Korean civilians, estimated to number between 5,000 and 7,000, 17 ROK Army soldiers, and at least 40 American soldiers had been killed. After Taejon fell to the North Koreans on 20 July civilian prisoners had been packed into the Taejon city jail and still others into the Catholic Mission. Beginning on 23 September, after the first U.S. troops had crossed the Naktong, the North Koreans began executing these people. They were

[35] 24th Div WD, 28 Sep 50; EUSAK WD, G-3 Air Br Rpt, PIR 78, G-3 Jnl 1715 and 1920, 28 Sep 50; 3d Engr C Bn Unit Rpt, Narr Summ, Sep 50.

[36] 24th Div WD, 28-30 Sep 50; EUSAK WD, G-2 Sec, 10 Oct 50, PW Rpt, ADVATIS 1028, Lt Pak Kyu, *I Corps Rear Command.*

taken out in groups of 100 and 200, bound to each other and hands tied behind them, led to previously dug trenches, and shot. By 26 September American forces had approached so close to Taejon that the N.K. Security Police knew they had to hurry. The executions were speeded up and the last of them took place just before the city fell.

Of the thousands of victims only six survived—two American soldiers, one ROK soldier, and three South Korean civilians. Wounded and feigning death, they had been buried alive. The two wounded Americans had only a thin layer of loose soil over them, enabling them to breathe sufficiently to stay alive until they could punch holes to the surface, one of them with a lead pencil. Still wired to their dead comrades beneath the soil and partially buried themselves, they were rescued when the city fell to the 24th Division. Hundreds of American soldiers, including General Milburn, the I Corps commander, and General Church, the 24th Division commander, saw these ghastly burial trenches and the pathetic bodies of the victims.[37]

On 29 September the 24th Division command post moved to Taejon. From there the division had the task of protecting the army line of communications back to the Naktong River. Its units were strung out for nearly 100 miles: the 19th Infantry held the Taejon area up to the Kum River, the 21st Infantry extended from Taejon southeast to Yongdong, the 5th Regimental Combat Team was in the Kumch'on area, and the 24th Reconnaissance Company secured the Waegwan bridges.

### *From Tabu-dong to Osan—Eighth Army Link-up With X Corps*

The Eighth Army breakout plans initially required the 1st Cavalry Division to cross the Naktong River at Waegwan and follow the 24th Division toward Kumch'on and Taejon. As the breakout action progressed, however, I Corps changed the plan so that the 1st Cavalry Division would cross the river at some point above Waegwan, pursue a course east of and generally parallel to that of the 24th Division, and seize Sangju. General Milburn left to General Gay the decision as to where he would cross. General Gay, the 1st Cavalry Division commander, and others, including Colonel Holmes, his chief of staff, and Colonel Holley of the 8th Engineer Combat Battalion, had proposed a crossing at Naktong-ni where a North Korean underwater bridge was known to exist. General Walker rejected this proposal. He himself flew in a light plane along the Naktong above Waegwan and selected the ferry site at Sonsan as the place the division should cross.[38]

In front of the 1st Cavalry Division

[37] The documentation on these atrocities is voluminous in the official records of the Army. Only a few citations will be given here: 25th Div WD, 5 Oct 50; 24th Div WD, Stf Secs, 29 Sep–31 Oct and G–1 Hist Rpt, 4 Oct 50; 2d Div WD, JA Stf Sec Rpt, Sep–Oct 50; Interim Hist Rpt, War Crimes Div, JA Sec, Korean Communications Zone (cumulative to 30 Jun 53); Lt Gen Frank W. Milburn, MS review comments, Nov 57; New York *Times*, October 3, 1950.

[38] Ltrs, Gay to author, 30 Sep and 21 Oct 53; Interv, author with Holmes, 27 Oct 53; Interv, author with Holley, 20 Feb 52; 1st Cav Div WD, 21 Sep 50; 7th Cav Regt Opn Plan 18; EUSAK WD, Br for CG, 21 Sep 50; Milburn, MS review comments, Nov 57.

COL. LEE HAK KU, *Chief of Staff, N.K. 13th Division, being transported to 1st Cavalry Division headquarters for questioning.*

two enemy divisions were retreating on Sangju. The N.K. *3d Division* reportedly had only 1,800 men when its survivors arrived there. The other division, the *13th,* was in complete disorder in the vicinity of Tabu-dong and northward along the road to Sangju when the 1st Cavalry Division prepared to engage in the pursuit.[39]

Shortly before noon, 21 September, General Walker telephoned from Taegu to General Hickey in Tokyo. He had important news—the chief of staff of the N.K. *13th Division* had surrendered that morning. Walker told Hickey that, based on the prisoner's testimony, the N.K. *II Corps* had ordered its divisions on 17 September to go on the defensive and that the *13th Division* knew nothing of the Inch'on landing.[40]

The *13th Division*'s chief of staff had indeed surrendered that morning. Shortly after daylight Sr. Col. Lee Hak Ku gently shook two sleeping American soldiers of the 8th Cavalry Regiment on the roadside near the village of Samsan-dong, four miles south of Tabu-dong. Once they were awake, the 30-year-old North Korean surrendered to them.

[39] ATIS Res Supp Interrog Rpts, Issue 96 (N.K. *3d Div*), p. 35; *Ibid.*, Issue 106 (N.K. Arty), p. 76; GHQ FEC, History of the N.K. Army, p. 57.

[40] Fonecon, Walker with Hickey, 1145, 21 Sep 50, GHQ CofS files.

Colonel Lee had slipped away from his companions during the night and approached the American lines alone. He was the ranking North Korean prisoner at the time and remained so throughout the war. Before he became chief of staff of the *13th Division,* Lee had been operations officer (G–3) of the N.K. *II Corps.* Later he was to become notorious as the leader of the Communist prisoners of the Compound 76 riots on Koje Island in 1952.

Now, however, on the day of his voluntary surrender, Colonel Lee was most co-operative. He gave a full report on the deployment of the *13th Division* troops in the vicinity of Tabu-dong, the location of the division command post and the remaining artillery, the status of supply, and the morale of the troops. He gave the strength of the division on 21 September as approximately 1,500 men. The division, he said, was no longer an effective fighting unit, it held no line, and its survivors were fleeing from the Tabu-dong area toward Sangju. The regiments had lost communication with the division and each, acting on its own impulse and according to necessity, was dispersed in confusion. Many other *13th Division* prisoners captured subsequently confirmed the situation described by Colonel Lee.

Colonel Lee said the *19th Regiment* had about 200 men, the *21st Regiment* about 330, the *23d* about 300; that from 70 to 80 percent of the troops were South Korean conscripts and this condition had existed for a month; that the officers and noncommissioned officers were North Korean; that all tanks attached to the division had been destroyed and only 2 of 16 self-propelled guns remained; that there were still 9 122-mm. howitzers and 5 120-mm. mortars operational; that only 30 out of 300 trucks remained; that rations were down one-half; and that supply came by rail from Ch'orwon via Seoul to Andong.[41]

At the time of Colonel Lee's surrender, General Gay had already directed Lt. Col. William A. Harris, Commanding Officer, 7th Cavalry Regiment, to lead the pursuit movement for the 1st Cavalry Division. Colonel Harris, now with a 2-battalion regiment (the 2d Battalion had relieved the British 27th Brigade on the Naktong), organized Task Force 777 for the effort. Each digit of the number represented one of the three principal elements of the force: the 7th Cavalry Regiment, the 77th Field Artillery Battalion, and the 70th Tank Battalion. Harris assigned Lt. Col. James H. Lynch's 3d Battalion as the lead unit, and this force in turn was called Task Force Lynch. In addition to the 3d Battalion, 7th Cavalry, it included B Company, 8th Engineer Combat Battalion; two platoons of C Company, 70th Tank Battalion (7 M4 tanks); the 77th Field

[41] The first extensive interrogation report on Colonel Lee is in EUSAK WD, 21 Sep 50. Others more complete include ATIS Interrog Rpts, Issue 9 (N.K. Forces), Interrog Rpt 1468, pp. 158–74. See also ATIS Interrog Rpts, Issue 14 (N.K. Forces), Rpt 1915, Col Mun Che Won, CO *23d Regt, 13th Div,* p. 43; *Ibid.,* Issue 27 (N.K. Forces), p. 82, Rpt 2978, Lt Col Yun Bong Hun, CO *19th Regt, 13th Div; Ibid.,* Issue 10 (N. K. Forces), p. 38, Interrog Rpt 1516, Maj Yu Pong Sun, Medical Off of *19th Regt; Ibid.,* p. 52, Lt Kim Pyon Jon. There is a very large number of prisoner interrogation reports from the N.K. *13th Division* available in the ATIS documents not cited here which, collectively, give a detailed picture of the nearly complete destruction of this division by 22 September 1950. Capt Jones N. Epps, Reduction of Compound 76, UN POW Camp, Koje-do, Korea, Student MS, Advanced Inf Course, Ft. Benning, Class 1, 1952–53, is of interest relative to Colonel Lee in the prisoner riots in 1952.

Artillery Battalion (less one battery); the 3d Platoon, Heavy Mortar Company; the regimental I&R Platoon; and a tactical air control party.[42]

After helping to repel an attack by a large force of North Koreans cut off below Tabu-dong and seeking to escape northward, Task Force Lynch started to move at 0800, 22 September from a point just west of Tabu-dong. Brushing aside small scattered enemy groups, Colonel Lynch put tanks in the lead and the column moved forward. Up ahead flights of planes coursed up and down the road attacking fleeing groups of enemy soldiers.

Near Naksong-dong, where the road curved over the crest of a hill, enemy antitank fire suddenly hit and stopped the lead tank. No one could see the enemy guns. General Gay, who was with the column, sent the remaining four tanks in the advance group over the crest of the hill at full speed firing all weapons. In this dash they overran two enemy antitank guns. Farther along, the column halted while men in the point eliminated a group of North Koreans in a culvert in a 10-minute grenade battle.[43]

After the task force had turned into the river road at the village of Kumgok but was still short of its initial objective, the Sonsan ferry, a liaison plane flew over and dropped a message ordering it to continue north to Naktong-ni for the river crossing. The column reached the Sonsan ferry at 1545. There, before he turned back to the division command post in Taegu, General Gay approved Lynch's decision to stop pending confirmation of the order not to cross the river there but to proceed to Naktong-ni. At 1800 Lynch received confirmation and repetition of the order, and an hour later he led his task force onto the road, heading north for Naktong-ni, ten miles away.

A bright three-quarter moon lit the way as the task force hastened forward. Five miles up the river road it began to pass through burning villages, and then suddenly it came upon the rear elements of retreating North Koreans who surrendered without resistance.

An hour and a half before midnight the lead tanks halted on the bluff overlooking the Naktong River crossing at Naktong-ni. Peering ahead, men in the lead tank saw an antitank gun and fired on it. The round struck a concealed enemy ammunition truck. Shells in the truck exploded and a great conflagration burst forth. The illumination caused by the chance hit lighted the surrounding area and revealed a fascinating and eerie sight. Abandoned enemy tanks, trucks, and other vehicles littered the scene, while below at the underwater bridge several hundred enemy soldiers were in the water trying to escape across the river. The armor and other elements of the task force fired into them, killing an estimated 200 in the water.[44]

Task Force Lynch captured a large

[42] Ltrs, Harris to author, 25 Nov 52 and 23 Dec 53; Ltr, Col James H. Lynch to author, 11 Dec 53; 1st Cav Div WD, 21 Sep 50.

[43] 3d Bn, 7th Cav Regt, G-3 Jnl, Msg 4, 220800 Sep 50; 1st Cav Div, G-3 Jnl, 22 Sep 50; Lt. Col. James H. Lynch, "Task Force Penetration," *Combat Forces Journal* (January, 1951), pp. 11-16; Lynch, "Tie-in in Korea," *Armor*, vol. 59, No. 6 (November-December 1950), pp. 34-36; Ltr, Gay to author, 30 Sep 53.

[44] Ltr, Gay to author, 30 Sep 53; Ltr, Lynch to author, 11 Dec 53; 3d Bn, 7th Cav Regt, Unit Jnl, 22 Sep 50; 1st Cav Div WD, 22 Sep 50; C Co, 70th Tk Bn, Summ of Action, 17-24 Sep 50; EUSAK WD, Br for CG, 22 Sep 50.

amount of enemy equipment at the Naktong-ni crossing site, including 2 abandoned and operable T34 tanks; 50 trucks, some of them still carrying U.S. division markings; and approximately 10 artillery pieces. According to prisoners taken at the time, this enemy force consisted principally of units of the N.K. *3d Division,* but it included also some men from the *1st* and *13th Divisions.*

Reconnaissance parties reported the ford crossable in waist-deep water and the far bank free of enemy troops. Colonel Lynch then ordered the infantry to cross to the north bank. At 0430, 23 September, I and K Companies stepped into the cold water of the Naktong and began wading the river. The crossing continued to the accompaniment of an exploding enemy ammunition dump at the other end of the underwater bridge. At 0530 the two companies secured the far bank. Altogether, in the twenty-two hours since leaving Tabu-dong, Task Force Lynch had advanced thirty-six miles, captured 5 tanks, 50 trucks, 6 motorcycles, 20 artillery pieces, secured a Naktong River crossing site, and had killed or captured an estimated 500 enemy soldiers.[45]

During the 23d, Maj. William O. Witherspoon, Jr., led his 1st Battalion across the river and continued on ten miles northwest to Sangju, which he found abandoned by the enemy. Meanwhile, Engineer troops put into operation at Naktong-ni a ferry and raft capable of transporting trucks and tanks across the river, and on the 24th they employed 400 Korean laborers to improve the old North Korean underwater bridge. Tanks were across the river before noon that day and immediately moved forward to join the task force at Sangju.

As soon as the tanks arrived, Colonel Harris sent Capt. John R. Flynn with K Company, 7th Cavalry, and a platoon of tanks thirty miles farther up the road to Poun, which they entered before dark. Colonel Harris had authority only to concentrate the regiment at Poun; he was not to go any farther.

On the 24th also, General Gay sent a tank-infantry team down the road from Sangju toward Kumch'on where the 24th Division was engaged in a hard fight on the main Waegwan-Taejon-Seoul highway. Since this took the force outside the 1st Cavalry Division zone of action, I Corps ordered it to withdraw, although it had succeeded in contacting elements of the 24th Division.[46]

On 24–25 September General Gay concentrated the 1st Cavalry Division in the Sangju–Naktong-ni area while his advanced regiment, the 7th Cavalry, stayed at Poun. About dark on the 25th he received a radio message from I Corps forbidding him to advance his division farther. Gay wanted to protest this message but was unable to establish radio communication with the corps. He was able, however, to send a message to Eighth Army headquarters by liaison plane asking for clarification of what he

[45] 3d Bn, 7th Cav Regt, S–3 Jnl, 222230 Sep 50; 1st Cav Div WD, 23 Sep 50, EUSAK WD, Br for CG, 22 Sep 50; Lynch, articles cited n. 44.

[46] 1st Cav Div WD, 23–24 Sep 50; C Co, 70th Tk Bn Unit Rpt, 17–24 Sep 50; EUSAK WD, G–3 Jnl, 0201 and 1345 24 Sep 50; Ltr, Gay to author, 30 Sep 52; Ltr, Harris to author, 23 Dec 53; Interv, author with Holley, 20 Feb 52; Itschner (CO I Corps Engrs Sep 50), "The Naktong River Crossings in Korea," *op. cit.*, pp. 96ff; Lynch, articles cited n. 44; 70th Tk Bn WD, 23–27 Sep 50.

thought was a confusion of General Walker's orders, and requesting authority to continue the breakthrough and join X Corps in the vicinity of Suwon. During the evening, field telephone lines were installed at Gay's forward echelon division headquarters at the crossing site, and there, just before midnight, General Gay received a telephone call from Col. Edgar T. Conley, Jr., Eighth Army G-1, who said General Walker had granted authority for him to go all the way to the link-up with X Corps if he could do so.[47]

Acting quickly on this authority, General Gay called a commanders' conference in a Sangju schoolhouse the next morning, 26 September, and issued oral orders that at twelve noon the division would start moving day and night until it joined the X Corps near Suwon. The 7th Cavalry Regiment was to lead the advance by way of Poun, Ch'ongju, Ch'onan, and Osan. Division headquarters and the artillery would follow. The 8th Cavalry Regiment was to move on Ansong via Koesan. At noon the 5th Cavalry Regiment, to be relieved by elements of the ROK 1st Division, was to break off its attack toward Hamch'ang and form the division rear guard; upon reaching Choch'iwon and Ch'onan it was to halt, block enemy movement from the south and west, and await further orders.[48]

On the right of the 1st Cavalry Division the ROK 1st Division, as part of the U.S. I Corps and the only ROK unit operating as a part of Eighth Army, had passed through Tabu-dong from the north on 22 September and headed for the Sonsan ferry of the Naktong. It crossed the river there on the 25th, and moved north on the army right flank to relieve elements of the 1st Cavalry Division, and particularly the 5th Cavalry Regiment, in the Hamch'ang-Poun area above Sangju. The 1st Cavalry Division was now free to employ all its units in the pursuit.[49]

Upon receiving General Gay's orders at the commanders' conference in Sangju, Colonel Harris in turn ordered Colonel Lynch at Poun to lead northwest with his task force as rapidly as possible to effect a juncture with 7th Division troops of the X Corps somewhere in the vicinity of Suwon. This task force was the same as in the movement from Tabu-dong on the 22d, except that now the artillery contingent comprised only C Battery of the 77th Field Artillery Battalion.

The regimental I&R Platoon and 1st Lt. Robert W. Baker's 3d Platoon of tanks, 70th Tank Battalion, led Task Force Lynch out of Poun at 1130, 26 September. Baker had orders from Lynch to move at maximum tank speed and not to fire unless fired upon. For mile after mile they encountered no enemy opposition—only cheers from South Korean villagers watching the column go past. Baker found Ch'ongju deserted except for a few civilians when he entered it at midafternoon.

Approximately at 1800, after traveling sixty-four miles, Baker's tanks ran out of gasoline and the advance stopped at

[47] Ltrs, Gay to author, 30 Sep 52 and 31 Dec 53; Ltr, Harris to author, 23 Dec 53; Ltr, Brig Gen Edgar T. Conley, Jr., to author, 26 Mar 56; I Corps WD, 22 and 24 Sep 50.

[48] Ltrs, Gay to author, 30 Sep 52 and 31 Dec 53; Ltrs, Harris to author, 8 and 23 Dec 53.

[49] Ltr, Gay to author, 30 Sep 53; Itschner, "The Naktong River Crossings in Korea," *op. cit.;* I Corps WD, Narr Hist, 25 Sep 50; EUSAK WD, G-3 Sec, and Br for CG, 29 Sep 50.

Ipchang-ni. For some reason the refuel truck had not joined the tank-led column. Three of the six tanks refueled from gasoline cans collected in the column.

Just after these three tanks had refueled, members of the I&R Platoon on security post down the road ran up and said a North Korean tank was approaching. Instead, it proved to be three North Korean trucks which approached quite close in the near dark before their drivers realized that they had come upon an American column. The drivers immediately abandoned their vehicles, and one of the trucks crashed into an I&R jeep. On the trucks was enough gasoline to refuel the other three tanks. About 2000 the column was at last ready to proceed.[50]

Colonel Harris ordered Colonel Lynch, at the latter's discretion, to drive on in the gathering darkness with vehicular lights on. This time Baker's platoon of tanks, rather than the I&R Platoon, was to lead the column. The other platoon of three tanks was to bring up the rear. At his request, Colonel Lynch gave Baker authority to shoot at North Korean soldiers if he thought it necessary. Shortly after resuming the advance at 2030 the task force entered the main Seoul highway just south of Ch'onan.

It soon became apparent that the task force was catching up with enemy soldiers. Ch'onan was full of them. Not knowing which way to turn at a street intersection, Baker stopped, pointed, and asked a North Korean soldier on guard, "Osan?" He received a nod just as the soldier recognized him as an American and began to run away. The rest of the task force followed through Ch'onan without opposition. Groups of enemy soldiers just stood around and watched the column go through. Beyond Ch'onan, Baker's tanks caught up with an estimated company of enemy soldiers marching north and fired on them with tank machine guns. Frequently they passed enemy vehicles on the road, enemy soldiers on guard at bridges, and other small groups.

Soon the three lead tanks began to outdistance the rest of the column, and Colonel Lynch was unable to reach them by radio to slow them. In this situation, he formed a second point with a platoon of infantry and a 3.5-inch bazooka team riding trucks, the first truck carrying a .50-caliber ring-mounted machine gun. Actions against small enemy groups began to flare and increase in number. When they were ten miles south of Osan men in the task force heard from up ahead the sound of tank and artillery fire. Lynch ordered the column to turn off its lights.[51]

Separated from the rest of Task Force Lynch, and several miles in front of it by now, Baker's three tanks rumbled into Osan at full speed. After passing through the town, Baker stopped just north of it and thought he could hear vehicles of the task force on the road behind him, although he knew he was out of radio communication with it. T34 tank tracks in the road indicated that enemy armor might be near.

[50] Ltrs, Harris to author, 25 Nov 52 and 23 Dec 53; Ltr, Lynch to author, 11 Dec 53; 70th Tk Bn WD, Act Rpt of 1st Lt Robert W. Baker, C Co, Oct 50.

[51] Baker Rpt, C Co, Oct 50; Lynch, articles cited n. 44; 3d Bn, 7th Cav Regt, S-3 Jnl, 262000 Sep 50; 61st FA Bn WD, 26 Sep 50; EUSAK WD, G-3 Jnl, 0350 27 Sep 50.

Starting up again, Baker encountered enemy fire about three or four miles north of Osan. His tanks ran through it and then Baker saw American M26 tank tracks. At this point fire against his tanks increased. Antitank fire sheared off the mount of the .50-caliber machine gun on the third tank and decapitated one of its crew members. Baker's tanks, now approaching the lines of the U.S. 31st Infantry, X Corps, were receiving American small arms and 75-mm. recoilless rifle fire. American tanks on the line held their fire because the excessive speed of the approaching tanks, the sound of their motors, and their headlights caused the tankers to doubt that they were enemy. One tank commander let the first of Baker's tanks go through, intending to fire on the second, when a white phosphorus grenade lit up the white star on one of the tanks and identified them in time to avoid a tragedy. Baker stopped his tanks inside the 31st Infantry lines. He had established contact with elements of X Corps. The time was 2226, 26 September; the distance, 106.4 miles from the starting point at Poun at 1130 that morning.[52]

That Baker ever got through was a matter of great good luck for, unknown to him, he had run through a strong enemy tank force south of Osan which apparently thought his tanks were some of its own, then through the North Korean lines north of Osan, and finally into the 31st Infantry position just beyond the enemy. Fortuitously, American antitank and antipersonnel mines on the road in front of the American position had just been removed before Baker's tanks arrived, because the 31st Infantry was preparing to launch an attack.

Baker's tanks may have escaped destruction from American weapons because of a warning given to X Corps. Shortly after noon of 26 September MacArthur's headquarters in Tokyo had radioed a message to X Corps, to NAVFE, and to the Far East Air Forces saying that elements of Eighth Army might appear at any time in the X Corps zone of action and for the corps to take every precaution to prevent bombing, strafing, or firing on these troops. A little later, at midafternoon, Generals Walker and Partridge, flying from Taegu unannounced, landed at Suwon Airfield and conferred with members of the 31st Infantry staff for about an hour. Walker said that elements of the 1st Cavalry Division attacking from the south would probably arrive in the Osan area and meet the 7th Division within thirty-six hours.[53]

After his miraculous escape, Baker and the 31st Infantry tank crews at the front line tried unsuccessfully to reach Task Force Lynch by radio.

Instead of being right behind Baker at Osan, the rest of Task Force Lynch was at least an hour behind him. After turning out vehicular lights approximately ten miles south of Osan, Task Force Lynch continued in blackout. Just south of the village of Habong-ni, Colonel Lynch, about midnight, noticed a T34 tank some twenty yards off the road and commented to Captain Webel, the regimental S-3 who accompanied the task force, that the Air Force must have destroyed it. Many men in the column

[52] Baker Rpt, C Co, Oct 50; 70th Tk Bn WD, Oct 50, Opn Rpt, 23–27 Sep.

[53] X Corps WD, G–3 Sec, Msgs J–5 at 0200, J–33 at 1400, and J–39 at 1738; 31st Inf WD, 26 Sep 50; 7th Div WD, Narr, 26 Sep 50.

saw the tank. Suddenly it opened fire with cannon and machine gun. A second enemy tank, unnoticed up to that time, joined in the fire. Task Force Lynch's vehicular column immediately pulled over and the men hit the ditch.

Lt. John G. Hill, Jr., went ahead to the point to bring back its rocket launcher team. This bazooka team destroyed one of the T34's, but the second one moved down the road firing into vehicles and running over several of them. It finally turned off the road into a rice paddy where it continued to fire on the vehicles. A 75-mm. recoilless rifle shell immobilized the tank, but it still kept on firing. Captain Webel had followed this tank and at one point was just on the verge of climbing on it to drop a grenade down its periscope hole when it jerked loose from a vehicle it had crashed into and almost caught him under its tracks. Now, with the tank immobilized in the rice paddy, a 3.5-inch bazooka team moved up to destroy it but the weapon would not fire. Webel pulled a 5-gallon can of gasoline off one of the vehicles and hurried to the side of the tank. He climbed on it and poured the gasoline directly on the back and into the engine hatch. A few spurts of flame were followed by an explosion which blew Webel about twenty feet to the rear of the tank. He landed on his side but scrambled to his feet and ran to the road. He had minor burns on face and hands and two ribs broken. The burning tank illuminated the entire surrounding area.

Up at the head of the halted column, Colonel Lynch heard to the north the sound of other tank motors. He wondered if Baker's three tanks were returning. Watching, he saw two tanks come over a hill 800 yards away. Fully aware that they might be enemy tanks, Lynch quickly ordered his driver, Cpl. Billie Howard, to place the lead truck across the road to block it. The first tank was within 100 yards of him before Howard got the truck across the road and jumped from it. The two tanks halted a few yards away and from the first one a voice called out in Korean, "What the hell goes on here?" A hail of small arms fire replied to this shout.

The two tanks immediately closed hatches and opened fire with cannon and machine guns. The truck blocking the road burst into flames and burned.

The three tanks still with Task Force Lynch came up from the rear of the column and engaged the enemy tanks. Eight more T34's quickly arrived and joined in the fight. The American tanks destroyed one T34, but two of them in turn were destroyed by the North Korean tanks. Webel, in running forward toward the erupting tank battle, came upon a group of soldiers who had a 3.5-inch bazooka and ammunition for it which they had just pulled from one of the smashed American trucks. No one in the group knew how to operate it. Webel took the bazooka, got into position, and hit two tanks, immobilizing both. As enemy soldiers evacuated the tanks, he stood up and fired on them with a Thompson submachine gun.

Sgt. Willard H. Hopkins distinguished himself in this tank-infantry melee by mounting an enemy tank and dropping grenades down an open hatch, silencing the crew. He then organized a bazooka team and led it into action against other tanks. In the tank-infantry battle that raged during an hour or more, this bazooka team was credited by some sources with destroying or helping to destroy 4 of the enemy tanks. Pfc. John R. Muhoberac was an outstanding member of this team. One of the enemy tanks ran all the way through the task force position shooting up vehicles and smashing into them as it went. At the southern end of the column a 105-mm. howitzer had been set up and there, at a point-blank range of twenty-five yards, it destroyed this tank. Unfortunately, heroic Sergeant Hopkins was killed in this exchange of crossfire as he was in the act of personally attacking this tank. Combined fire from many weapons destroyed another tank. Of the 10 tanks in the attacking column, 7 had been destroyed. The 3 remaining T34's withdrew northward. In this night battle Task Force Lynch lost 2 men killed, 28 wounded, and 2 tanks and 15 other vehicles destroyed.[54]

After the last of the enemy tanks had rumbled away to the north, Colonel Harris decided to wait for daylight before going farther. At 0700 the next morning, 27 September, the task force started forward again. The men were on foot and prepared for action. Within a few minutes the point ran into an enemy tank which a 3.5-inch bazooka team destroyed. An enemy machine gun crew opened fire on the column but was quickly overrun and the gunners killed in a headlong charge by Lt. William W. Woodside and two enlisted men. A little later the column came upon two abandoned enemy tanks and blew them up. The head of Task Force Lynch reached Osan at 0800.

At 0826, 27 September, north of Osan at a small bridge, Platoon Sgt. Edward C. Mancil of L Company, 7th Cavalry, met elements of H Company, 31st Infantry, 7th Division. Task Force 777 sent a message to General Gay which said in part, "Contact between H Company, 31st Infantry Regiment, 7th Division, and forward elements of Task Force 777 established at 0826 hours just north of Osan, Korea." [55]

After the link-up with elements of the 31st Infantry, elements of Task Force 777 did not actually participate in this regiment's attack against the North Koreans on the hills north of Osan. Their communication equipment, including the forward air controllers, and their medical troops, however, did assist the 31st Infantry. General Gay arrived at Osan before noon and, upon seeing the battle in progress on the hills to the north, conferred with a 31st Infantry battalion commander. He offered to use the 8th Cavalry Regiment as an enveloping force and assist in destroying the enemy. He also offered to the 31st Infantry, he said, the use of the 77th and 99th Field Artillery Battalions and one tank company. The battalion com-

[54] 7th Cav Regt WD, 27 Sep 50; Lynch, articles cited n. 44; Ltr, Lynch to author, 11 Dec 53; Ltrs, Harris to author, 8 and 23 Dec 53; Lt Col James B. Webel, MS review comments, 15 Nov 57. Eighth Army General Order 132, 11 March 1952, awarded Colonel Lynch the Oak Leaf Cluster to the Distinguished Service Cross. Department of the Army General Order 35, 8 April 1952, awarded the Distinguished Unit Citation to the 3d Battalion, 7th Cavalry Regiment, and attached units. GHQ FEC General Order 21, 3 February 1951, awarded the Distinguished Service Cross to Captain Webel and the Distinguished Service Cross posthumously to Sergeant Hopkins.

[55] Brig Gen William A. Harris, MS review comments, 29 Oct 57, quoting copy of message in his possession; 7th Cav Regt WD, 27 Sep 50; Lynch, articles cited n. 44. It would appear that some elements of K Company, 7th Cavalry, were also in the contacting group.

mander said he would need concurrence of higher authority. Just what happened within the 31st Infantry after General Gay made this offer has not been learned. But elements of the 1st Cavalry Division stood idly by at Osan while the 31st Infantry fought out the action which it did not win until the next afternoon, 28 September. General Barr, commanding general of the 7th Infantry Division, has said he was never informed of General Gay's offer of assistance.[56]

In this rapid advance to Osan, the 1st Cavalry Division cut off elements of the *105th Armored Division* in the Ansong and P'yongt'aek area and miscellaneous units in the Taejon area. On the 28th, elements of C Company, 70th Tank Battalion, and K Company, 7th Cavalry, with the strong assistance of fighter-bombers, destroyed at least seven of ten T34's in the P'yongt'aek area, five by air strikes. Elements of the 16th Reconnaissance Company barely escaped destruction by these enemy tanks, and did suffer casualties.[57]

As late as 29 September, L Company of the 5th Cavalry Regiment ambushed approximately fifty enemy soldiers in nine Russian-built jeeps driving north from the vicinity of Taejon.

### *The ROK Army Arrives at the 38th Parallel*

Eastward, the ROK Army made advances from Taegu that kept pace with Eighth Army, and in some instances even outdistanced it. This performance is all the more remarkable because the ROK Army, unlike the Eighth Army, was not motorized and its soldiers moved on foot.

In the ROK II Corps, the 6th and 8th Divisions on 24 September gained approximately sixteen miles. The 6th Division advanced on Hamch'ang and entered it the night of 25 September. By the 27th it was advancing across the roughest part of the Sobaek Range, past Mun'gyong in the high passes, on its way to Chungju. On the last day of the month the 6th Division encountered enemy delaying groups as it approached Wonju.

The ROK 8th Division made similarly rapid advances on the right of the 6th Division. Its reconnaissance elements entered Andong before midnight of the 24th. Five spans of the 31-span bridge over the Naktong there were down. Remnants of two enemy divisions, the N.K. *12th* and *8th,* were retreating on and through Andong at this time. The *12th Division* was pretty well through the town, except for rear guard elements, when advanced units of the ROK 8th Division arrived, but the main body of the N.K. *8th Division* had to detour into the mountains because ROK troops arrived there ahead of it. After two days of fighting, during which it encountered extensive enemy mine fields, the ROK 8th Division secured Andong on 26 September. That evening the division's advanced elements entered Yech'on, twenty miles northwest of Andong. The next day some of its troops were at Tanyang preparing to cross the

[56] Ltr, Gay to author, 31 Dec 53; Harris, MS review comments, 29 Oct 57; Interv, author with Barr, 1 Feb 54.

[57] 70th Tk Bn WD, Oct 50, Opn Rpt, C Co, 23–27 Sep 50; Ltr, Capt Charles A. Rogers to author, 1 Feb 54 and his attached MS on 16th Recon Co action in Korea, 1950 (written May 1951 in Korea); Webel, MS review comments, 15 Nov 57.

upper Han River.[58] On the last day of the month the division met strong enemy resistance at Chech'on and bypassed the town in the race northward.

The ROK Capital Division was keeping pace with the others in the pursuit. On the 27th it had entered Ch'unyang, about thirty-one miles east of the ROK 8th Division, and was continuing northward through high mountains.

On the night of 1–2 October, shortly after midnight, an organized North Korean force of from 1,000 to 2,000 soldiers, which had been bypassed some place in the mountains, struck with savage fury as it broke out in its attempt to escape northward. Directly in its path was Wonju where the ROK II Corps headquarters was then located. This force overran the corps headquarters and killed many of its men, including five American officers who were attached to the corps or who had come to Wonju on liaison missions. The North Koreans ran amok in Wonju until morning, killing an estimated 1,000 to 2,000 civilians.[59]

Along the east coast the ROK 3d Division, with heavy U.S. naval gunfire support, captured Yongdok on 25 September. A huge cloud of black smoke hung overhead from the burning city. The fall of the town apparently caught the N.K. *5th Division* by surprise. Some Russian-built trucks were found with motors running, and artillery pieces were still in position with ammunition at hand. Horse-drawn North Korean signal carts were found with ponies hitched and tied to trees. After the fall of Yongdok it appears that remnants of the *5th Division*, totaling now no more than a regiment, turned inland for escape into the mountains. One North Korean regimental commander divided his three remaining truckloads of ammunition and food among his men and told them to split into guerrilla bands.

In the pursuit up the coastal road above Yongdok, Maj. Curtis J. Ivey, a member of KMAG, with the use of twenty-five 2½-ton trucks made available for the purpose through the efforts of Colonel McPhail, KMAG adviser to the ROK I Corps, led the ROK's northward in shuttle relays. When a roadblock was encountered it was Major Ivey who usually directed the action of the point in reducing it.[60]

The impressive gains by the ROK units prompted General Walker to remark on 25 September, "Too little has been said in praise of the South Korean Army which has performed so magnificently in helping turn this war from the defensive to the offensive." [61]

[58] EUSAK WD, Br for CG, 25–26 Sep 50; GHQ FEC Sitrep, 26 Sep 50; EUSAK WD, 5 Oct, G–2 Sec, PW Interrog ADVATIS 1038, and ATIS Interrog Rpts, Issue 12 (N.K. Forces), Rpt 1753, p. 148, Lt Col Kim Chong Ung, Supply Officer, *31st Regt*, N. K. *12th Div.*

[59] EUSAK WD, 2 Oct 50. Eighth Army General Order 35, 21 January 1951, awarded the Distinguished Service Cross posthumously to Capt. Walt W. Bundy and 2d Lt. George E. Mannan, 205th Signal Repair Company, who were killed in Wonju while covering the escape of seventeen enlisted men.

[60] EUSAK WD, G–3 Sec, 24–26 Sep 50; *Ibid.*, POR 229, 26 Sep 50; *Ibid.*, Br for CG, 24–26 Sep 50; *Ibid.*, Engr Hist Rpt, 24 Sep 50; *Ibid.*, G–2 Sec, 30 Sep 50, interrogs of Lt Kim Kun Bong, Arty Maintenance Off, *11th Regt*, N.K. *5th Div*, and of Sr Lt Pak Kwan Hyok, Medical Off, *2d Regt*, N.K. *8th Div;* ATIS Interrog Rpts, Issue 13 (N.K. Forces), Rpt 1865, p. 153, Lt Han Si Hong; GHQ FEC G–3 Opn Rpt, 25 Sep 50; Emmerich, MS review comments, Dec 57.

[61] New York *Times*, September 26, 1950.

On up the coast road raced the ROK 3d Division. It secured Samch'ok on the morning of 29 September, and then continued on toward Kangnung. It moved north as fast as feet and wheels could take it over the coastal road. It led all ROK units, in fact, all units of the United Nations Command, in the dash northward, reaching a point only five miles below the 38th Parallel on the last day of the month.[62]

### *The Invaders Expelled From South Korea*

The last week of September witnessed a drastic change in the pattern of North Korean military activity. Enemy targets were disappearing from the scene. On 24 September some fighter pilots, unable to find targets, returned to their bases without having fired a shot. Survivors of the once victorious North Korea People's Army were in flight or in hiding, and, in either case, they were but disorganized and demoralized remnants. On 1 October there occurred an incident illustrating the state of enemy demoralization. An Air Force Mosquito plane pilot dropped a note to 200 North Korean soldiers northeast of Kunsan ordering them to lay down their arms and assemble on a nearby hill. They complied. The pilot then guided U.N. patrols to the waiting prisoners.

The virtual collapse of the North Korean military force caused General MacArthur on 1 October to order the Air Force to cease further destruction of rail, highway, bridge, and other communication facilities south of the 38th Parallel, except where they were known to be actively supporting an enemy force. Air installations south of the 40th Parallel were not to be attacked, and he halted air action against strategic targets in North Korea.[63]

The extent of his collapse was truly a death blow to the enemy's hopes for continuing the war with North Korean forces alone. Loss of weapons and equipment in the retreat north from the Pusan Perimeter was of a scope equal to or greater than that suffered by the ROK Army in the first week of the war. For the period 23–30 September, the IX Corps alone captured 4 tanks, 4 self-propelled guns, 41 artillery pieces, 22 antitank guns, 42 mortars, and 483 tons of ammunition. In I Corps, the 24th Division on one day, 1 October, captured on the Kumsan road below Taejon 7 operable tanks and 15 artillery pieces together with their tractors and ammunition. On the last day of September the 5th Cavalry Regiment captured three trains complete with locomotives hidden in tunnels. A few miles north of Andong advancing ROK forces found approximately 10 76-mm. guns, 8 120-mm. mortars, 5 trucks, and 4 jeeps, together with dead enemy soldiers, in a tunnel—all had been destroyed earlier by air force napalm attacks at either end of the tunnel. At Uisong, ROK forces captured more than 100 tons of rice, other supplies, and most of the remaining equipment of one North Korean division. The North Koreans had abandoned

[62] EUSAK WD, G–3 Sec, 27–30 Sep 50; *Ibid.*, Summ, p. 43, Sep 50; *Ibid.*, Br for CG, 29–30 Sep 50. Beginning on 25 September, LST's from Pusan helped supply the ROK 3d Division in this advance. Sixty tons a day were required for a ROK division. See 2d Log Comd Activities Rpt Sep 50.

[63] USAF Study 71, pp. 66–67; "Air War in Korea," *Air University Quarterly Review*, IV (Fall, 1950), p. 35.

## Captured Enemy Equipment

*76-mm. howitzer*

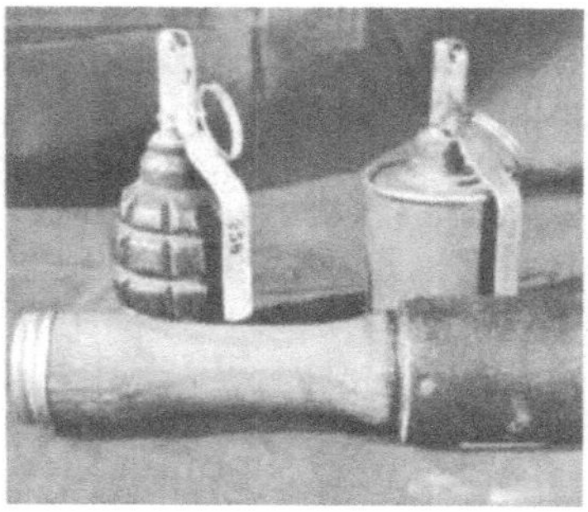

*Chinese grenades*

*Russian-made SU76*

*Small arms (front to rear): Russian 7.62-mm. submachine (burp) gun; Russian 7.62-mm. carbine M1944; Japanese 7.7-mm. rifle; Russian 7.62 Tokarev semiautomatic rifle with flash hider; Russian 14.5-mm. antitank rifle PTRD-1941 (Degtyarev).*

*120-mm. mortar*

*76-mm. gun*

*Coryunov heavy machine gun*

many tanks, guns, vehicles, ammunition, and other equipment because they lacked gasoline to operate their vehicles.[64]

During the approximately three months of the war up to the end of September, all U.N. combat arms had made various claims regarding destroyed enemy equipment, especially tanks and self-propelled guns. Air Force claims for the period, if totaled from daily reports, would be extravagantly high. After the U.N. breakout from the Pusan Perimeter, in the period from 26 September to 21 October 1950, seven survey teams traveled over all major routes of armored movement between the Perimeter line and the 38th Parallel, and also along the Kaesong-Sariwon-P'yongyang highway above the Parallel. This survey disclosed 239 destroyed or abandoned T34 tanks and 74 self-propelled 76-mm. guns. The same survey counted 60 destroyed U.S. tanks.[65]

According to this survey, air action destroyed 102 (43 percent) of the enemy tanks, napalm accounting for 60 of them or one-fourth the total enemy tank casualties; there were 59 abandoned T34's without any visible evidence of damage, also about one-fourth the total; U.N. tank fire accounted for 39 tanks (16 percent); and rocket launchers were credited with 13 tanks (5 percent). The number credited to bazooka fire is in error, for the number certainly is much higher. Very likely air action is credited in this survey with many tanks that originally were knocked out with infantry bazooka fire. There are many known cases where aircraft attacked immobilized tanks after bazooka fire had stopped them. There was an almost complete absence of enemy tanks destroyed by U.S. antitank mines.

No reliable information is available concerning the number of damaged tanks the North Koreans were able to repair and return to action. But the figure of 239 found destroyed or abandoned comes close to being the total number used by the North Korea People's Army in South Korea. Very few escaped from the Pusan Perimeter into North Korea at the end of September.

From July through September 1950 United States tank losses to all causes was 136. A survey showed that mine explosions caused 70 percent of the loss. This high rate of U.S. tank casualties in Korea to mines is all the more surprising since in World War II losses to mines came to only 20 percent of tank losses in all theaters of operations.[66]

In the two weeks beginning with 16 September, the breakout and pursuit period, the U.N. forces in the south placed 9,294 prisoners in the Eighth Army stockade. This brought the total to 12,777. Eighth Army had captured 6,740 of them and the ROK Army 6,037. Beginning with 107 prisoners on 16 September the number had jumped to 435

[64] IX Corps WD, sec. IV, Sep 50; EUSAK PIR 81, 1 Oct 50; GHQ FEC Sitrep, 1 Oct 50; 1st Cav Div WD, 30 Sep 50; 5th Cav Regt Unit Hist, 30 Sep 50; USAF Hist Study 71, p. 40; 24th Div WD, Narr Summ, 29 Sep–31 Oct 50.

[65] ORO, The Employment of Armor in Korea, ORO–R–1 (FEC), vol. II, and app. K, 8 Apr 51. The figures for 239 tanks and 74 SP guns are adjusted from a larger total in overlapping reports from the seven surveys.

As late as July 1951, the author counted 74 enemy T34 tanks and 14 76-mm. SP guns rusting along the main highway running from Waegwan on the Naktong River to Seoul; 30 of the T34's were between the river and Taejon.

[66] ORO, The Employment of Armor in Korea, 8 Apr 51; DA *Intelligence Review*, March 1951, No. 178, p. 53.

on 23 September and passed the 1,000 mark daily with 1,084 on 27 September, 1,239 the next day, and 1,948 on 1 October.[67]

The rapid sweep of the U.N. forces northward from the Pusan Perimeter in the last week of September bypassed thousands of enemy troops in the mountains of South Korea. One of the largest groups, estimated to number about 3,000 and including soldiers from the N.K. *6th* and *7th Divisions* with about 500 civil officials, took refuge initially in the Chiri Mountains of southwest Korea. At the close of the month, of the two major enemy concentrations known to be still behind the U.N. lines, one was south of Kumch'on in the Hamyang area and the other northeast and northwest of Taejon.

Just before midnight, 1 October, a force of approximately sixty North Korean riflemen, using antitank and dummy mines, established and maintained a roadblock for nearly ten hours across the main Seoul highway about fifteen miles northwest of Kumch'on. A prisoner said this roadblock permitted about 2,000 North Korean soldiers and a general officer of the N.K. *6th Division* to escape northward. The *6th* at the time apparently still had its heavy machine guns and 82-mm. mortars but had discarded all heavier weapons in the vicinity of Sanch'ong.[68]

[67] EUSAK WD, Provost Marshal Sec, 16 Sep–1 Oct 50; U.S. Military Academy, Operations in Korea (25 Jun 50–1 Apr 51), p. 12. As contrasted with the number of prisoners in the stockade, the total number reported captured by 20 September, however, was 23,620 (4,305 by 15 September), and by 1 October approximately 30,000.

[68] 25th Div WD, Sep 50; *Ibid.*, G–2 Sec, 8 Oct 50, ADVATIS 1069, Maj Chi Ki Chol, N.K. *6th Div Arty Regt;* EUSAK PIR 79 and 80, 29 and 30 Sep 50; 24th Div WD, Enemy Situation, 29 Sep–31 Oct 50; Throckmorton, Notes for author.

Enemy sources make quite clear the general condition of the North Korean Army at the end of September. The *5th Division* started its withdrawal in good order, but most of its surviving troops scattered into the Chiri Mountain area and elsewhere along the escape route north so that only a part reached North Korea. The *7th Division* commander reportedly was killed in action near Kumch'on as the division retreated northward; remnants assembled in the Inje-Yanggu area above the 38th Parallel in mid-October. In the *2d Division,* Maj. Gen. Choe Hyon, the division commander, had only 200 troops with him north of Poun at the end of September. Other elements of the division had scattered into the hills.

Parts of the *9th* and *10th Divisions* retreated through Taejon, and other parts cut across the Taejon highway below the city in the vicinity of Okch'on when they learned that the city had already fallen. Only a handful of men of the *105th Armored Division* reached North Korea. The commanding general of the N.K. *I Corps* apparently dissolved his headquarters at Choch'iwon during the retreat and then fled with some staff officers northeast into the mountains of the Taebaek Range on or about 27 September. From the central front near Taegu, 1,000 to 1,800 men of the *3d Division* succeeded in reaching P'yonggang in what became known as the Iron Triangle at the beginning of October. The *1st Division,* retreating through Wonju and Inje, assembled approximately 2,000 men at the end of October.

Of all the North Korean divisions fighting in South Korea perhaps no other suffered destruction as complete as the *13th.* Certainly no other yielded so many high-ranking officers as prisoners of war. In August the *13th Division* artillery commander surrendered; on 21 September Col. Lee Hak Ku, the chief of staff, surrendered; three days later the commander of the self-propelled gun battalion surrendered; the division surgeon surrendered on the 27th; and Col. Mun Che Won, a 26-year-old regimental commander, surrendered on 1 October after hiding near Tabu-dong for nearly a week. The commander of the *19th Regiment,* 22-year-old Lt. Col. Yun Bong Hun, led a remnant of his command northward by way of Kunwi, Andong, and Tanyang. Near Tanyang, finding his way blocked by ROK troops, he marched his regiment, then numbering 167 men, into a ROK police station at Subi-myon and surrendered. A few members of the division eventually reached the P'yonggang area in the Iron Triangle.

Remnants of the *8th Division,* numbering perhaps 1,500 men, made their way northeast of P'yonggang and continued on in October to a point near the Yalu River. Some small elements of the *15th Division* escaped northward through Ch'unch'on to Kanggye in North Korea. From Kigye about 2,000 men of the *12th Division* retreated through Andong to Inje, just north of the 38th Parallel, picking up stragglers from other divisions on the way so that the division numbered about 3,000 to 3,500 men upon arrival there. Remnants of the *5th Division* infiltrated northward above Yongdok along and through the east coast mountains in the direction of Wonsan.

The bulk of the enemy troops that escaped from the Pusan Perimeter assembled in the Iron Triangle and the Hwach'on-Inje area of east-central North Korea just above the 38th Parallel. On 2 October an Air Force pilot reported an estimated 5,000 enemy marching in small groups along the edge of the road north of the 38th Parallel between Hwach'on and Kumhwa.

The commanding general of the N.K. *II Corps* and his staff apparently escaped to the Kumhwa area in the Iron Triangle, and the best available evidence indicates that the commanding general and staff of the N.K. *Army Front Headquarters* at Kumch'on also escaped northeast to the Iron Triangle. From there in subsequent months this headquarters directed guerrilla operations on U.N. lines of communications.

It appears that not more than 25,000 to 30,000 disorganized North Korean soldiers reached North Korea from the Pusan Perimeter after the U.N. breakout in late September. For all practical purposes the North Korea People's Army had been destroyed. That was the real measure of the success of the Inch'on landing and Eighth Army's correlated attack—General MacArthur's strategy for winning the war.[69]

---

[69] This summary of the condition of the North Korean Army at the end of September 1950 is based upon an analysis of a large body of enemy materials. The most important of them are GHQ FEC, History of the N.K. Army; ATIS Res Supp Interrog Rpts, Issues 3 (*1st* and *15th Divs*), 94 (*2d* and *4th Divs*), 96 (*3d* and *5th Divs*), 100 (*6th* and *9th Divs*), 99 (*7th* and *12th Divs*), 4 (*8th Div* and *105th Armd Div*), 104 (*10th* and *13th Divs*), and 106 (N.K. Arty); ATIS Interrog Rpts (N.K. Forces), Issues 9, 10, 12, 13, 14, 15, 18, 20, 25; ATIS Enemy Documents, Issue 16, p. 48; ATIS Interrog Rpts (Korean Opns), Issue 27, p. 82; FEC, Order of Battle Information, 16 Sep 51. N.K. Army Supp, Chart 9; EUSAK WD, G–2 Sec, PW Interrog, 16 Sep–31 Oct 50.

TABLE 3—ESTIMATED U.N. STRENGTH AS OF 30 SEPTEMBER 1950[a]

| Organization | U.N. Forces | Attached Koreans |
|---|---|---|
| Total Ground Combat Forces | 198,211 | 22,404 |
| Total U.S. Ground Combat Forces | 113,494 | 22,404 |
| Combat | | |
| Eighth Army | 1,120 | |
| I Corps | 4,141 | 267 |
| 1st Cavalry Division | 13,859 | 2,961 |
| 24th Infantry Division | 15,591 | 3,606 |
| IX Corps | 4,224 | 1,009 |
| 2d Infantry Division | 14,122 | 2,756 |
| 25th Infantry Division | 14,617 | 3,230 |
| X Corps | 8,344 | 600 |
| 7th Infantry Division | 15,865 | 7,975 |
| 1st Marine Division (reinforced) | 21,611 | |
| ROK Army | 81,644 | |
| British Ground Combat Forces | 1,704 | |
| Philippine Ground Combat Forces | 1,369 | |
| FEAF | 36,677 | |
| Other U.N. Air Forces | 330 | |
| NAVFE | 59,438 | 7,045 |
| Total Ground Service Forces | 20,608 | 444 |
| Service | | |
| Eighth Army | 2,820 | |
| I Corps | 1,235 | 305 |
| IX Corps | 187 | 110 |
| X Corps | 2,039 | 29 |
| Pusan Base | 9,792 | |
| Inch'on Base | 4,452 | |
| Seoul Area Command | 83 | |

[a] Figures are adjusted and do not include personnel previously carried as wounded, missing, or injured who had returned to duty by 30 September.

*Source:* EUSAK WD, 30 Sep 50; GHQ FEC Sitrep, 29–30 Sep 50.

The war in Korea by the end of September had cost Eighth Army 24,172 battle casualties—5,145 killed in action; 16,461 wounded in action, of whom 422 died of wounds; 402 reported captured; and 2,164 missing in action. Many of the latter were prisoners of war.[70]

Estimated U.N. strength in Korea as of 30 September 1950 is shown in Table 3.

[70] Battle Casualties of the Army, 31 May 52, DA TAGO.

TABLE 4—POSTWAR TABULATION OF U.N. STRENGTH IN KOREA AS OF 30 SEPTEMBER 1950

| Organization | Strength |
|---|---|
| Total Ground Combat Forces | 229,772 |
| U.S. Army | 103,601 |
| 1st Cavalry Division | 14,472 |
| 2d Infantry Division | 17,656 |
| 7th Infantry Division | 15,620 |
| 24th Infantry Division | 11,876 |
| 25th Infantry Division | 14,824 |
| Nondivisional units | 29,153 |
| U.S. 1st Marine Division (reinforced) | 21,525 |
| ROK Army | 101,573 |
| Korean divisions | 82,786 |
| Attached to U.S. Army combat units | 18,787 |
| British Brigade | 1,704 |
| Philippine Battalion Combat Team | 1,369 |
| Total Ground Service Forces | 119,559 |
| U.S. Army | 28,507 |
| ROK Army | 91,052 |
| Korean Service units | 90,608 |
| Attached to U.S. Army service units | 444 |
| U.S. Far East Air Forces | 36,677 |
| U.S. Naval Forces, Far East | 59,438 |
| Other U.N. Air Forces | 330 |

*Source:* ROK and UN Ground Forces Strength in Korea, 31 July 1950–31 July 1953, DA, COA, 7 Oct 54, and GHQ FEC Sitrep, 29–30 Sep 50.

# CHAPTER XXIX

# The Plan for Complete Victory

It is better to abandon a whole province than to divide an army.

VON SCHLIEFFEN

The question whether U.N. forces should cross the 38th Parallel became a most difficult one as soon as the Inch'on landing succeeded and the Eighth Army broke out from the Pusan Perimeter. As a result of long and detailed consideration at high levels on the future course of action, the government in Washington decided Eighth Army should cross into North Korea.

Pursuant to this decision, the Joint Chiefs of Staff on 27 September sent to General MacArthur a comprehensive directive to govern his future actions. They stated that his first objective was to be the destruction of the North Korean forces. He was to unite all of Korea under Syngman Rhee if possible. But they warned him that he was not to consider the directive final since developments might require its modification. They particularly enjoined him to make special efforts to determine whether Soviet or Chinese intervention appeared likely, and to report any such threat to them at once.

Subject to these injunctions, the directive stated that MacArthur's mission was "the destruction of the North Korean Armed Forces" and authorized him to conduct military operations for that purpose north of the 38th Parallel, provided that at the time there was no major Chinese Communist Forces or Soviet entry into North Korea or announced intention to enter in order to counter U.N. military operations there. The Joint Chiefs added that in no circumstances would any of the U.N. forces cross the Manchurian or Soviet borders of Korea and that non-Korean ground forces, as a matter of policy, should not be used in the area along the Manchurian border or in the northeast provinces bordering the Soviet Union. They instructed MacArthur to submit his plan for operations north of the 38th Parallel to them for approval. Thus the Joint Chiefs of Staff in Washington held in their own hands final approval for any operation north of the 38th Parallel.[1]

Upon receiving this directive MacArthur urged on the JCS removal of the restriction requiring specific approval from the United States Government before his forces crossed the 38th Parallel. He urged that he be allowed to cross the Parallel and seek out and

[1] JCS 92801, 27 Sep 50, Personal for MacArthur, from JCS to CINCUNC.

destroy the remaining parts of the North Korean Army if North Korea did not surrender in accordance with a proclamation he intended to issue.

Two days later Secretary of Defense George C. Marshall sent him a personal message, marked for his eyes only, which stated that he should feel free tactically and strategically to proceed north of the 38th Parallel. President Truman himself had approved this message.[2]

It is clear that on 29 September MacArthur had authority from the United States Government to cross the 38th Parallel. In a communication to the Secretary of Defense on 30 September, MacArthur said, "Unless and until the enemy capitulates, I regard all of Korea open for our military operations." [3]

The next day, 1 October, in order to remove any obscurity that might still exist in Washington regarding his plan, MacArthur sent a message that was crystal clear as to his intentions. He said:

> I plan to issue and make public the following general directive to all elements of the United Nations Command at 1200 hours, Monday, 2 October, unless I receive your instructions to the contrary. "Under the provisions of the United Nations Security Council Resolution of 27 June, the field of our military operations is limited only by military exigencies and the international boundaries of Korea. The so-called 38th Parallel, accordingly, is not a factor in the military employment of our forces. To accomplish the enemy's complete defeat, your troops may cross the border at any time, either in exploratory probing or exploiting local tactical conditions. If the enemy fails to accept the terms of surrender set forth in my message to him of 1 October, our forces, in due process of campaign will seek out and destroy the enemy's armed forces in whatever part of Korea they may be located." [4]

MacArthur stated later that the temporary U.N. halt at the 38th Parallel that occurred in early October was due to logistical difficulties.[5]

From the Communist side certain storm signals appeared. In a speech in Peiping on 1 October, the first anniversary of the Chinese Communist state, Premier Chou En-lai warned that the Chinese people "will not tolerate foreign aggression and will not stand aside should the imperialists wantonly invade the territory of their neighbor." [6] This clearly was a threat to intervene in the Korean War if U.N. forces crossed the 38th Parallel. In the United Nations the Soviet delegate proposed on 2 October a plan which called for a cease fire in Korea and the withdrawal of all foreign troops. The next day Sir Benegal Rau, the Indian delegate, stated his government's view that U.N. forces should not cross the 38th Parallel. The Indian view was undoubtedly influenced by a report to Prime Minister Jawaharlal Nehru from India's Ambassador to Peiping

[2] JCS 92985, 29 Sep 50, Marshall to MacArthur; JCS 92975, JCS to CINCFE, 29 Sep 50, Memo for Secy Def from Bradley, Chmn, JCS, 29 Sep 50; S. Comm. on Armed Services and S. Comm. on Foreign Relations, 82d Cong., 1st Sess., 1951, Joint Hearings, *Military Situation in the Far East*, pt. 1, pp. 245, 339–40, and 488 (hereafter cited as Senate MacArthur Hearings), testimony of Marshall.

[3] Msg C65034, CINCFE to DA for Secy Def, 30 Sep 50, and Msg C65035, CINCFE to CG Eighth Army, 30 Sep 50, quoted in Schnabel, FEC, GHQ Support and Participation in the Korean War, ch. VI, pp. 16–17.

[4] Msg C65118, CINCUNC to DA, 1 Oct 50, quoted in Schnabel, *op. cit.*, p. 17.

[5] Senate MacArthur Hearings, pt. 1, p. 245.

[6] New York *Times*, October 2, 1950, a London dispatch; New York *Herald Tribune*, October 1, 1950.

that China would enter the war if the U.N. forces crossed the Parallel.[7]

Meanwhile, MacArthur on 1 October issued his demand that North Korea surrender. He addressed his message to the Commander in Chief of the North Korean forces. He called upon the North Koreans to lay down their arms and cease hostilities under such military supervision as he might direct in order that the decisions of the United Nations might be carried out with a minimum of further loss of life and destruction of property, and to liberate U.N. prisoners of war and civilian internees. There was no answer from North Korea.

On 9 October General MacArthur issued an ultimatum calling "for the last time" for North Korean surrender. There was no official response from North Korea to this demand, but Kim Il Sung in a radio broadcast in P'yongyang on the morning of 10 October, which was monitored in Tokyo, rejected it.[8]

### *MacArthur's Plan of Operations in North Korea*

As Eighth Army approached a junction with X Corps near Seoul, General Walker became concerned about the future relationship of Eighth Army and X Corps. He and his staff felt that X Corps should become part of Eighth Army and that all U.N. forces in Korea should operate under a unified field command.[9] It is not known with certainty whether General Walker ever discussed with General MacArthur his own ideas about operations north of the 38th Parallel. It appears, however, that he never submitted them to him in writing.

So far as is known, the nearest General Walker ever came to broaching the subject to MacArthur in writing was on 26 September when he sent a discreetly worded message to him suggesting that he would like to be informed of X Corps' progress and plans so that he could plan better for the approaching juncture of the two forces. General MacArthur dashed Walker's hopes in a reply the next day, informing him that X Corps would remain in GHQ Reserve, in occupation of the Inch'on-Seoul area ready to undertake a GHQ-directed operation "of which you will be apprised at an early date." [10]

When General MacArthur flew to Seoul on 29 September to return the South Korean capital to the government of Syngman Rhee he already had formulated in his mind plans for the next phase of Korean operations. On 26 September, General Hickey had sent a check sheet to General Wright stating that General MacArthur wanted plans developed for further operations in North Korea which would employ the X Corps in an amphibious landing at Wonsan. Because the Far East Command's Joint Strategic Plans and Opera-

---

[7] New York *Times,* October 3 and 4, 1950; Msg 031344Z, DA to SCAP, 3 Oct 50, cited in Schnabel, *op. cit.,* ch. VII, p. 13.

[8] Dept of State Pub 4015, app. A, *United Nations Command Seventh Report to the Security Council, United Nations, 1–15 October 1950.* See also New York *Times,* October 9 and 11, 1950.

[9] Interv, author with Maj Gen Leven C Allen, 15 Dec 53.

[10] Msg CX64410, 27 Sep 50, CINCFE to CG Eighth Army, and Msg G25090 KG, CG Eighth Army to CINCFE, 26 Sep, both quoted in Schnabel, *op. cit.,* ch. VI, pp. 9–10; Interv, author with Allen, 15 Dec 53.

tions Group had kept active its studies for amphibious operations in areas other than Inch'on, including one for a corps-size landing in the Wonsan-Hamhung area of the east coast, it was only a matter of a few hours until General Wright had the outline of such a plan in MacArthur's hands.

This plan proposed that the advance into North Korea would consist of a "main effort of Eighth Army on the west in conjunction with an amphibious landing at Wonsan or elsewhere." [11] This was the beginning officially of the Far East Command decision that led quickly to the establishment of two separate field commands in Korea for the next phase of the war, and which almost at once became the subject of controversy.

For a period prior to 26 September, it appears that General MacArthur had intended to place X Corps under Eighth Army command once Seoul had fallen. Generals Hickey and Wright favored this course of action, and Maj. Gen. George L. Eberle, the Far East Command G-4, agreed with them. But apparently they did not actively advocate it to General MacArthur. Eberle held the view that although it would be possible to support X Corps logistically in an amphibious operation on the east coast, it could more easily be supported as part of Eighth Army. But if MacArthur ever had been uncertain on the future role of the X Corps, he had decided the point in his own mind by the last week of September. The reasoning which led General MacArthur to decide on two commands in Korea can best be understood by reference to the terrain map of North Korea and the problem of logistics.[12]

Above the Seoul-Wonsan corridor the northern Taebaek Range rises to rugged heights in the east central part of the peninsula, forming an almost trackless mountainous waste in the direction of the Manchurian border. The principal routes of travel follow the deep mountain valleys in a generally north-south direction. The only reasonably good lateral road from east to west in North Korea lay just north of the 39th Parallel, connecting P'yongyang with Wonsan, on the east coast. A rail line also crossed the peninsula here. Any plan for a military campaign north of the P'yongyang-Wonsan corridor in the interior of North Korea would encounter most difficult logistical and supply problems.

In surveying the logistical problems attending any future military operations in Korea, General MacArthur had to note the condition of transport communications in South Korea. U.N. aerial action, together with enemy demolitions, had destroyed nearly all the rail and highway bridges north of the Pusan Perimeter. Weeks of concentrated work by all available Engineer troops would be required to repair the rail lines from the Pusan Perimeter to the 38th Parallel. Aerial action had also badly shattered the communication and transport system of North Korea. In considering this state of affairs, General MacArthur apparently decided that he could

[11] Interv, author with Maj Gen Edwin K. Wright, 7 Jan 54; Check Sheet, Gen Hickey to JSPOG, 26 Sep 50, sub: Plans for Future Operations, quoted in Schnabel, *op. cit.*, ch. VI, pp. 19-20.

[12] Interv, author with Wright, 7 Jan 54; Lt Gen Doyle O. Hickey, MS review comments, 14 Feb 56; Almond, MS review comments, 4 May 55; Interv, author with Eberle, 12 Jan 54.

not supply both Eighth Army and X Corps from Inch'on for a quick continuation of the pursuit northward. He also wanted to get military forces behind the North Koreans retreating from the Pusan Perimeter through the central mountains and up the east coast. MacArthur reasoned that a landing on the northeast coast might accomplish this. The base for operations in Korea actually was Japan. MacArthur believed that two separate forces co-ordinated from there could operate in Korea without impairing the effectiveness of either.

Involved also in his decision was the idea of encirclement of the North Korean capital. While Eighth Army attacked north from the Seoul area toward P'yongyang, MacArthur's plan called for the X Corps, upon landing at Wonsan, to drive west along the P'yongyang corridor and to take the city from the flank and rear.

The first outline of the operational plan for the projected movement into North Korea set the target date for the Wonsan assault, for planning purposes, at twelve days after Eighth Army passed through the X Corps in the Seoul-Inch'on area. It was thought that Eighth Army could initiate its attack three to seven days before the X Corps amphibious assault on Wonsan. General MacArthur approved this plan on 29 September.[13]

After the GHQ plan to move the X Corps by water to Wonsan became known to Eighth Army, Colonel Dabney, Eighth Army G–3, prepared a message to GHQ setting forth a concept to replace it. This plan would have assigned X Corps to Eighth Army and provided for early movement against P'yongyang and Wonsan overland. The Eighth Army staff felt that the GHQ plan to outload X Corps would unnecessarily delay pursuit of the defeated North Korean Army and would impede the advance of Eighth Army northward. It also believed that the ROK advance on the east coast would capture Wonsan before the X Corps could be landed there. Dabney took the message to General Walker who read it and said that he agreed with the plan, but that it was not to be sent to GHQ. According to Dabney, Walker said he had already made his views known and had received contrary orders. In connection with the possible escape into North Korea of large numbers of enemy soldiers from the Pusan Perimeter, Eighth Army earlier had requested X Corps to block the central mountain route through Wonju and Ch'unch'on with at least a regiment, but X Corps had replied that it could not extend "the anvil" to that point.[14] On 11 October a radio message from General MacArthur shattered any remaining hope Walker may have had of directing future operations in the east. It informed him that MacArthur intended to use Wonsan Airfield for land-based aircraft under X Corps control and that the ROK I Corps in the east, then under Eighth Army control, would come under X Corps command as soon as that corps landed.[15]

[13] JSPOG file, FEC, Opn Plan 9–50, 29 Sep 50, cited in Schnabel, *op. cit.*, ch. 6, pp. 21–22.

[14] Intervs, author with Gens Allen, 15 Dec 53, Wright, 7 Jan 54, and Eberle, 12 Jan 54; Hickey, MS review comments, 14 Feb 56; Dabney, MS review comments, 26 Nov 57.

[15] Msg CX66169, 11 Oct 50, CINCFE to CG Eighth Army, quoted in Schnabel, *op. cit.*, ch. VI, p. 27.

It appears that General Walker believed that X Corps, after the fall of Seoul, should continue the attack north to P'yongyang under Eighth Army command, with Eighth Army moving up behind it. He reasoned that this should save a lot of time as X Corps was already in position for continuing the attack in the west, and Eighth Army was not. In such an attack the corps could continue to be supplied from Inch'on. General Walker and most Eighth Army senior officers felt that not to continue the pursuit at once—to halt for a period of almost two weeks while X Corps loaded out at Inch'on and Eighth Army moved into position below the 38th Parallel—would permit the escape of a large part of the remaining North Korean forces retreating northward which might otherwise be destroyed or captured.

In Walker's view, once the X Corps reached P'yongyang, with or without Eighth Army help as the case might be, Eighth Army could then move laterally along the P'yongyang-Wonsan corridor to the east coast where it would join the ROK I Corps already there and advancing northward. This plan contemplated the X Corps continuing the attack in the west from P'yongyang toward the Yalu. An alternate course would be for the X Corps to cross to the east coast by the P'yongyang-Wonsan corridor, while Eighth Army attacked north from P'yongyang. The operations of both forces would be co-ordinated under Walker's command, and both would be supplied from Inch'on and Pusan and by airlift until Wonsan fell. Then the force operating in the east could be supplied largely by sea through that port and Hungnam farther to the north. Generals Hickey, Wright, and Eberle of MacArthur's staff favored such a plan of operations.[16]

Admiral Joy, Commander, NAVFE, and key members of his staff, had objected to the Wonsan amphibious operation as being unnecessary, holding the view that X Corps could march overland from Seoul to Wonsan much faster than it could be lifted and landed there by water. General Smith, commanding the 1st Marine Division, had many reservations about the proposed operations of his division in northeast Korea.[17]

The prediction of the Eighth Army commander and staff that Wonsan would fall to the ROK I Corps before the X Corps could land there became a fact on 10 October. And their view that the North Korean capital of P'yongyang would also fall to Eighth Army attack before the X Corps could move west from Wonsan also proved to be correct. General MacArthur officially acknowledged this fact by issuing on 17 October a new United Nations Command Operations Order which drew a boundary between Eighth Army and X Corps. This boundary, starting at the 39th Parallel, followed generally the watershed of the high Taebaek Range that extended through the eastern part of Korea up to the Yalu River.[18]

### *Eighth Army Deploys for the Attack*

Based on General MacArthur's United

[16] General MacArthur says these officers never expressed any such views to him at the time. MacArthur MS review comments, 15 Nov 57.

[17] Ltr, Smith to author, 13 Feb 54; Smith, MS review comments, 15 Nov 57; Karig, *et al., Battle Report: The War in Korea,* pp. 298–99.

[18] UNC Opn Ord 4, 17 Oct 50, cited in Schnabel, *op. cit.,* ch. 6, pp. 31–32.

TANK TROOPS *of the 1st Cavalry Division pursuing the enemy fourteen miles north of Kaesong on 13 October.*

Nations Command Operations Order 2, dated 2 October, Eighth Army the next day issued an operations order to implement its part in the plan for the attack into North Korea. The army order called for the U.S. I Corps to seize a line west of the Imjin River with not less than a division, and to concentrate the corps in an assembly area there as rapidly as IX Corps could relieve it. The U.S. I Corps was then to conduct operations northward on army orders, making the main effort with the 1st Cavalry Division leading the attack. The 24th Division and the ROK 1st Division were to protect the corps flanks and form a reserve.[19]

[19] EUSAK Opn Ord 103, 3 Oct 50; EUSAK WD, G–3 Sec, 3 and 4 Oct 50.

In addition to relieving the U.S. I Corps in its zone, the U.S. IX Corps was to protect the line of communications, Seoul-Suwon-Taejon-Taegu-Pusan and, together with ROK police, destroy the remaining enemy forces in South Korea.

The ROK Army was directed to move its II Corps, consisting of the 6th, 7th, and 8th Divisions, to the area between Ch'unch'on and Uijongbu in central Korea, and its I Corps, composed of the Capital and 3d Divisions, to the area between Yongp'o and Chumunjin-up on the east coast, all prepared to attack northward. The ROK Army was also to provide a new division (the 11th) by 5 October to help IX Corps in the rear areas of South Korea.

Pursuant to orders, the 1st Cavalry

Division on 5 October advanced north of Seoul for the purpose of securing the U.S. I Corps assembly area near the 38th Parallel. Led by I Company, the 5th Cavalry Regiment in the evening crossed to the north side of the Imjin River at Munsan-ni.

At noon on the 7th, the 16th Reconnaissance Company entered Kaesong, and that evening elements of the 1st Battalion, 8th Cavalry Regiment, arrived there. By evening of 8 October the 7th and 8th Cavalry Regiments of the 1st Cavalry Division had secured the I Corps assembly area in the vicinity of Kaesong. Some of the troops were within small arms range of the 38th Parallel. Behind the 1st Cavalry Division, the 24th Division concentrated in the Seoul area.[20]

At this juncture a new military organization appeared in Korea, and it also concentrated near Seoul. The 3d Battalion, Royal Australian Regiment, commanded by 30-year-old Lt. Col. Charles H. Green, a veteran of World War II, arrived at Pusan on 28 September. It joined the British 27th Brigade at Kumch'on on 3 October, which was then renamed the 27th British Commonwealth Brigade. Two days later the bulk of the brigade moved by air to Kimpo Airfield as part of the I Corps concentration near the 38th Parallel.[21]

With its I Corps concentrated to the north of Seoul, Eighth Army took over control of the Inch'on-Seoul area from X Corps at 1200, 7 October. The command posts of both Eighth Army and the ROK Army moved from Taegu and opened in Seoul on 12 October.[22]

Earlier, on 4 October, the Far East Air Forces and the Fifth Air Force, acting on a directive of 8 July, had assumed control of the Marine squadrons at Kimpo. This was highly displeasing to X Corps, and particularly so to the marines. But the change in control actually made little difference in air operations since FEAF directed that the 1st Marine Air Wing continue to support X Corps. The Fifth Air Force headquarters moved to Seoul on 15 October. As a result of the September victories, the Japan-based fighters and fighter-bombers of the Fifth Air Force moved to Korean bases. This permitted an increase in their armament load, more time over target and combat area, and lengthened flight ranges into North Korea.[23]

### *The ROK I Corps Captures Wonsan and Hungnam*

Regardless of whether the U.N. forces did or did not cross the 38th Parallel, there was always the strong probability that the ROK troops would. Syngman Rhee had often stated his intention of halting the South Korean Army only at the Yalu. Speaking at a mass meeting at Pusan on 19 September he said, "We have to advance as far as the Manchurian border until not a single enemy soldier is left in our country." He said that he did not expect the U.N. forces to stop at the 38th Parallel, but if they

[20] EUSAK POR 259, 6 Oct, and 265, 8 Oct 50; EUSAK PIR 262, 7 Oct 50; 1st Cav Div WD, 6–7 Oct 50; 5th Cav Regt WD, 6 Oct 50; EUSAK WD, G–3 Sec, 7 Oct 50; 24th Div WD, 4–6 Oct 50.

[21] EUSAK WD, G–3 Sec, 28 Sep and 5 Oct 50; *Ibid.*, Br for CG, 5–6 Oct 50; I Corps WD, 3 Oct 50; GHQ UNC, G–3 Opn Rpt, 3 Oct 50.

[22] EUSAK POR No. 277, 12 Oct 50; EUSAK PLR No. 84, 6 Oct 50; EUSAK WD, 7 Oct 50.

[23] USAF Hist Study 71, 1 Jul 52.

did, he continued, "we will not allow ourselves to stop." [24] And stop the ROK troops did not.

A message dropped by a KMAG G–3 officer from a light plane at Samch'ok and delivered to Colonel Emmerich at Kangnung on the afternoon of 29 September ordered the ROK 3d Division to cross the 38th Parallel and proceed to Wonsan as soon as possible. Advanced patrols of the ROK 3d Division crossed the parallel on 30 September. The next day just before noon two rifle companies crossed the border and came under fire from enemy troops in old fixed positions north of the Parallel. On 2 October the ROK 3d and Capital Divisions established their command posts in Yangyang, eight miles north of the parallel. Although General MacArthur made the first official public announcement of forces under U.N. command crossing the 38th Parallel on 3 October, the American press had reported the incident the day before. Anticipating that ROK forces would cross the Parallel, newspaper correspondents flew to Kangnung, just south of the border on the east coast, to get the news.[25]

Now began a remarkable phase of the pursuit. The ROK 3d Division traveled northward night and day, on foot and by vehicle, out of communication most of the time with higher headquarters, without flank protection to the west, and bypassing many enemy groups which often attacked their supply points in the rear. There were some costly fire fights on the road north. The N.K. *5th Division* with about 2,400 survivors, retreating as best it could ahead of the ROK's, kept the pursuing advanced elements under mortar and 76-mm. antitank fire. The road was heavily mined and lead vehicles had many casualties. From fortified positions, including connecting trenches, caves, and dug-in gun positions, North Koreans tried to stop or slow the ROK advance. The 3d Division averaged about fifteen miles a day. Many of its men had no shoes and large numbers trudged ever northward on bloody feet.

The Capital Division followed the 3d, and at intervals sent some of its units inland into the Diamond Mountains, the lofty and beautiful Kumgang-san, which crowded close upon the coast line. In happier days these mountains had been the vacation grounds of people from all parts of Korea.

In central Korea, troops of the ROK II Corps crossed into North Korea later than did the troops of the I Corps on the coast. On 6 October the ROK 6th Division crossed the parallel from the vicinity of Ch'unch'on and advanced on Hwach'on. For three days it fought two regiments of the N.K. *9th Division* which stubbornly defended that town. Late on the afternoon of 8 October the division entered Hwach'on, driving two enemy battalions northwest.

The 8th Division crossed the 38th Parallel on 7 October. On its right, the 7th Division crossed a day or two later. Both divisions headed for the Iron Triangle. ROK troops arrived at the Iron Triangle on 10 October. There in the

---

[24] FEC, CofS files, Associated Press dispatch from Pusan, 1450 19 Sep 50.

[25] Ltr, Emmerich to author, 12 Feb 54; Emmerich, MS review comments, 12 Dec 57; EUSAK WD, G–3 Jnl, Msgs at 2000, 2215, 2305 1 Oct, and Msg at 2030 2 Oct 50; EUSAK WD, Summ, 4 Oct 50; CINCFE Korean Release 522, 031545 Oct 50; New York *Times*, October 2, 1950.

Ch'orwon area a large force of North Koreans attacked the ROK 16th Regiment during the day but was repulsed and forced to withdraw. Elements of the 8th Division then entered Ch'orwon.

The Iron Triangle, a place whose name became famous later in the Korean War, was an area of relatively flat terrain, shaped like an equilateral triangle, in the mountains of east central North Korea. It is situated 20 to 30 miles above the 38th Parallel, halfway across the peninsula, and 50 air miles northeast of Seoul. It is bounded at its three corners by the towns of Ch'orwon at its western base, Kumhwa at its eastern base, and P'yonggang at its northern apex. The Iron Triangle is an important North Korean rail and road communication center, linking east and west coastal areas with each other, and in turn connecting them with the communication net leading south through central South Korea.

On 11 October the ROK 8th Division and the 7th Regiment of the 6th Division converged on P'yonggang. On 13 October the 7th Division arrived there by way of Kumhwa.

All the ROK divisions, except the 1st, which was part of the U.S. I Corps and accordingly under direct American command, were across the Parallel before any of the American divisions crossed.[26]

On 9 October, the ROK 3d and Capital Divisions were at the south edge of Wonsan, 110 air miles up the coast above the 38th Parallel. That day the Capital Division on the Wonsan–Iron Triangle road south of the city captured 6 tanks, 4 artillery pieces, 10 82-mm. mortars, 1 120-mm. mortar, 30 heavy machine guns, 500 submachine guns, 5,000 Russian rifles, 1 boxcar of medical supplies, and another of miscellaneous supplies. The bulk of the ROK 3d Division arrived in front of Wonsan by the coastal road. The N.K. *24th Mechanized Artillery Brigade*, the *945th Regiment* (naval amphibious troops), and other units subordinate to the naval headquarters at Wonsan defended the city. Enemy artillery pieces emplaced behind dikes just south of it delivered direct fire against the ROK's.[27]

Troops of both the ROK 3d and Capital Divisions entered Wonsan on 10 October, with the 3d Division on the coastal road making the greater effort. About two miles long and of irregular, narrow width, the city is shaped by the 450-foot-high hills that rise abruptly from the narrow coastal strip. In order to settle rival claims as to which division entered the city first, the corps commander, Brig. Gen. Kim Baik Yil decreed that both divisions got there simultaneously at 0600 and that both secured it at 1000. But the city was not secured then. Colonel Emmerich, KMAG senior adviser with the 3d Division, entered the city with the front line troops of the ROK 23d Regiment just after noon. The North Koreans had maintained a heavy artillery fire from the city until almost noon. Then, after withdrawing most of their guns from Wonsan, they fired into the city all after-

[26] EUSAK WD, Summ, Oct 50; *Ibid.*, Daily News Bulletin, 4 Oct 50; EUSAK POR's, 7–9 Oct 50; EUSAK WD, G–3 Sec, 4 Oct and 7 Oct 50; *Ibid.*, Br for CG, 4, 5, 9, 10, 11–14 Oct 50; *Ibid.*, PIR's 86–88, 6–8 Oct, and 91, 11 Oct 50.

[27] EUSAK WD, Br for CG, 090001 Oct 50; GHQ FEC, History of the N.K. Army, pp. 81–82.

3D ROK DIVISION OFFICERS AND KMAG ADVISERS

noon from its northwest sector and the hills behind it. That afternoon the 3d Division captured the heavily mined airfield on the peninsula east of the city. At nightfall both ROK divisions were still engaged in street fighting within the city. During the night an enemy armored task force, including about ten 76-mm. self-propelled antitank guns, returned to the airfield and did a good job of shooting it up, burning out most of the buildings and hangars.[28]

The next day, 11 October, the ROK 3d Division fought through Wonsan against enemy artillery, mortar, and small arms fire. It secured the city, and by evening had troops one mile north of it. The Capital Division helped clear the city and occupied the airfield. Generals Walker and Partridge flew into the Wonsan Airfield on the 11th. Finding it in good condition, General Partridge had twenty-two planes of the Combat Cargo Command fly in 131 tons of supplies for the ROK troops the next day.[29]

In the week after the capture of Wonsan the ROK 3d Division remained in the vicinity, securing the area for the expected landing of X Corps. The Capital Division meanwhile moved on

[28] Ltr, Emmerich to author, 12 Feb 54; EUSAK WD, Br for CG, 100001–110800 Oct 50.

[29] EUSAK WD, 11 Oct 50, Aide de Camp Diary; Schnabel, *op. cit.;* USAF Hist Study 71, p. 73.

ROK TROOPS *marching past the Diamond Mountains.*

north fifty air miles up the coast, and, against light resistance, secured both Hamhung and its port, Hungnam, on 17 October.[30]

During its great success in advancing northward into North Korea the ROK Army expanded and reorganized. On 8 October it reactivated the 5th Division at Taegu and once again counted eight divisions, the same number that it had when the war began. Simultaneously, the ROK Army activated the 1st Guerrilla Group of five battalions (1st, 2d, 3d, 5th, 6th). Eight days later, on 16 October, it activated the ROK III Corps. This new corps, to which the 5th and 11th Divisions were attached, was to assume responsibility for the ROK Army zone south of the Seoul-Ch'unch'on-Inje-Yangyang axis, and destroy remaining enemy troops and guerrillas in that part of Korea.[31]

### *The X Corps Prepares To Move Amphibiously to Northeast Korea*

About the time the ROK I Corps crossed the 38th Parallel and started north toward Wonsan, General Almond and Admiral Struble received on 1 October preliminary instructions from GHQ, Far East Command, for the projected landing at Wonsan. Joint Task Force 7 had been re-established to land the X Corps at Wonsan, and Admiral Struble had been named to command it. He received from Admiral Joy the mis-

[30] Interv, author with Emmerich, 5 Dec 51; EUSAK WD, G–3 Sec, 12, 15, and 16 Oct 50; *Ibid.*, Br for CG, 14, 17 Oct 50.

[31] EUSAK POR 300, 20 Oct 50; EUSAK WD, G–4 Sec, 5 Oct 50; GHQ UNC, G–3 Opn Rpt, 20 Oct 50; IX Corps WD, bk. 1, 13 Oct 50. The 5th Division was supposed to move to the Andong area in the east, but it actually stayed in northwest Korea. The 11th Division established headquarters at Namwon in southwest Korea.

sion of JTF 7. This was (1) to maintain a naval blockade of the east coast of Korea south from Ch'ongjin; (2) to load and transport X Corps to the Wonsan area and provide cover and support en route; (3) to conduct pre-D-day naval operations as required; (4) on D-day to seize by amphibious assault, occupy, and defend a beachhead in the Wonsan area; (5) to provide naval gunfire, air, and initial logistic support to X Corps in the Wonsan area.[32]

General MacArthur on 2 October formalized in his United Nations Operation Order 2 instructions for U.N. military operations north of the 38th Parallel, and set forth therein the plan of movement and the mission of X Corps. The X Corps was to revert to GHQ Reserve when Eighth Army passed through it in the Seoul area. The 1st Marine Division and X Corps headquarters were to load at Inch'on while the 7th Infantry Division and most of the X Corps troops moved to Pusan for loading. The problem of outloading X Corps at Inch'on in adverse tidal conditions with the limited amount of amphibious craft available, concurrently with the expected partial use of the port by Eighth Army, was so complicated and difficult that MacArthur decided that part of the force would have to outload at Pusan if the entire corps was to be loaded within two weeks.[33]

The selection of Wonsan as the site of the projected X Corps landing in northeast Korea had been based on a number of factors. Situated at the southwest side of a large bay which bulges inland from the Japan Sea, Wonsan is the principal port on the east coast of Korea; it is the eastern terminus of the easiest route across North Korea; and it is a road and rail communications center. In 1950 when the war started the city had a population of approximately 150,000. The Japanese had developed Wonsan as a naval base, and the North Koreans had continued to use it for the same purpose. It was the principal port of entry for Russian supplies and military equipment received by sea from the Vladivostok area, and it was a key point on the rail line running southwest into Korea from the Soviet Vladivostok base. It was the petroleum refining capital of Korea. From Wonsan a military force could move inland and west across the peninsula to P'yongyang, or north to the Hamhung-Hungnam area, fifty air miles away, the most important industrial area of all Korea.[34]

On 30 September General Smith, commander of the 1st Marine Division, was first informed of the projected X Corps landing at Wonsan. The next day he was requested to submit loading plans by 3 October with a proposed D-day at Wonsan of 15 October. Since ships for the lift had not yet been designated it was impossible to meet these dates. On 7 October, Admiral Doyle, in command of the Attack Force, recommended 20 October as the earliest

[32] Act Rpt, JTF 7, Wonsan Opn, I-A-1; *Ibid.*, I-B-1, COMNAVFE Opn Plan 113-50; X Corps WD, Oct 50, CofS Sec, p. 10.

[33] Act Rpt, JTF 7, Wonsan Opn, I-C-1; X Corps WD, Oct 50, CofS Sec, p. 10; Schnabel, *op. cit.*, ch. VI, pp. 23-24; 2d Log Comd, Oct 50 Act Rpt, G-4 Sec.

[34] Joint Intel Study Pub Board, JANIS 75, ch. VIII (Korea—Cities and Towns), pp. 52-53; GHQ FEC, Terrain Study 6, Northern Korea, sec. XIV, pp. 26-27 and Map 760, Wonsan City Plan, Plate 12; X Corps WD, Oct 50, Opns, pp. 18-19; Diary of CG X Corps, 24 Oct 50.

D-day which the amphibious assault forces could meet. Admirals Struble and Joy concurred and forwarded this recommendation to General MacArthur. He accepted it as a tentative D-day but indicated that every effort should be made to achieve an earlier one. Two days later Admiral Struble published his operation plan outlining the task force organization. JTF 7 was organized as follows:

90 Attack Force, Rear Adm. James H. Doyle
95 Advance Force, Rear Adm. Allan E. Smith
95.2 Covering and Support Group, Rear Adm. Charles C. Hartman
95.6 Minesweeping Group, Capt. Richard T. Spofford
92 X Corps, Maj. Gen. Edward M. Almond
96.2 Patrol and Reconnaissance Group, Rear Adm. George R. Henderson
96.8 Escort Carrier Group, Rear Adm. Richard W. Ruble
77 Fast Carrier Force, Rear Adm. Edward C. Ewen
70.1 Flagship Group (USS *Missouri*), Capt. Irving T. Duke
79 Logistics Support, Capt. Bernard L. Austin

On 10 October General MacArthur ordered U.N. Operations Plan 2 put into effect, thereby canceling all other tentative plans.[35]

General MacArthur's Operation Plan 9-50 of 29 September assigned priority of outloading at Inch'on to the 1st Marine Division, the amphibious assault element of X Corps. On 3 October, X Corps ordered the Marine division to initiate movement to an assembly area in Inch'on. On 4 October, General Almond issued a corps order for the projected operations at Wonsan. The 1st Marine Division had the mission of seizing a corps base of operations while the 7th Infantry Division was to start an attack west to join with Eighth Army in front of P'yongyang. By 6 October the 1st, 5th, and 11th Marines had virtually completed their movement to Inch'on; the next day the 7th Marines began its movement from Uijongbu to the Inch'on assembly area. As it assembled at Inch'on for outloading, the 1st Marine Division numbered 23,591 men, with 40 U.S. Army troops and 4,516 Korean marines attached, for a total of 28,147.[36]

At noon on 6 October the 3d Logistical Command assumed responsibility for all unloading at Inch'on. During the day the X Corps requested it to halt all unloading activities not directly concerned with the corps, because otherwise X Corps outloading would be delayed for an estimated six to twenty days. X Corps reverted to GHQ Reserve at noon on 7 October when Eighth Army assumed responsibility for the Inch'on-Seoul area.

[35] Act Rpt, JTF 7, Wonsan Opn, app. D, Opn Ord Comdr 7th Flt, 16-50, 5 Oct 50; *Ibid.*, I-C-3-4; *Ibid.*, II-3; X Corps WD, Oct 50, CofS Notes, 10 Oct 50.

[36] CINCFE Opn Plan 9-50, 29 Sep 50; X Corps Opn Instr 9, 031600 Oct 50; X Corps WD, catalogue of Plans and Orders, p. 44; X Corps WD, Summ of Opns, 3-5 Oct 50, and POR 18, 6 Oct 50; 1st Mar Div SAR, 15 Sep-7 Oct 50, an. C, pp. 39-42; and vol. I, an. A, G-1 Sec; X Corps Opn Ord 4, 4 Oct 50; X Corps WD, Diary of CG X Corps, 4, 7 Oct, and Notes of CofS, 4, 7 Oct 50.

The 31-foot tides and the great mud banks at low tide made the outloading exceedingly difficult and required carefully co-ordinated schedules in moving troops and supplies. There were only seven berths where LST's or landing craft could beach at Inch'on and these could be used only at high tide. Moreover, there was no adequate staging area. There was only one small pier from which vehicles could be loaded into an LCM, and then only at high tide. Vehicles were loaded on the top decks of LST's and ferried out to the ships in the harbor, and there lifted by crane from the LST's to the APA's and AKA's. The tidal basin was used to outload all bulk cargo for ferrying to the ships in the harbor. This unexpectedly developed into a major and difficult task. The 1st Marine Division had been informed that the 1st Logistical Command would bottom load all the shipping dispatched to Inch'on to outload the division with 10 days' level of supply, Classes I, III, and V. But this was not done, and it resulted in the necessity of unloading from ships in the harbor and reloading on others, and also of reloading on X Corps shipping considerable supplies from the dumps ashore that otherwise could have been left for Eighth Army. From Japan by air came 32,000 assault rations and 100,000 C rations to Kimpo Airfield, and from there they were taken to the port for outloading.

Troops began loading at Inch'on on the 9th. The 1st and 3d Battalions, 1st Marines, went aboard their LST's on 10 October, and were in these cramped quarters for sixteen days before they again got ashore. On 11 October the X Corps command post closed ashore and opened aboard the *Mt. McKinley*. Bulk loading of cargo began on 8 October and continued to 16 October when all X Corps loading at Inch'on was completed. Already the U.S. Eighth Army had crossed the Parallel in the west and was fighting its way north.

## CHAPTER XXX

# Eighth Army and X Corps Enter North Korea

Wherever the enemy goes let our troops go also.

ULYSSES S. GRANT, dispatch to Halleck, 1 August 1864

### *Eighth Army Crosses the Parallel—The Kumch'on Pocket*

On 5 October Eighth Army issued its operations order for the movement across the 38th Parallel, but withheld the date for the attack. Two days later General Allen telephoned General Hickey and told him that General Walker wanted to know when A-day (date for crossing the Parallel) was to be given. Hickey replied, "Your A-day will be at such time as you see it ready." Allen replied, "That's fine, because we're on the verge of it now." A message from Tokyo the same day confirmed the call. Eighth Army at once implemented its order of the 5th by radio messages to General Milburn at U.S. I Corps and to the Chief of Staff, ROK Army. The attack on P'yongyang was about to begin.[1]

Eighth Army expected strong enemy resistance at the 38th Parallel and a stubborn defense of P'yongyang. According to ROK Army combat intelligence, the North Koreans had three known lines of defense across the peninsula, each consisting of pillboxes, gun emplacements, trenches, and barbed wire entanglements. The first line was along the 38th Parallel and was about 500 yards in depth; the second line was about three miles behind the first; the third lay farther back and was based on locally situated critical terrain features. All three lines were oriented to defend against southern approaches.[2]

North of the Parallel the U.N. Command expected to meet newly activated divisions that had been training in North Korea or elements of units that had engaged in the fighting around Seoul. Some intelligence sources indicated there might be as many as six divisions totalling 60,000 men in North Korean training centers. Actually, only

[1] Fonecon, Allen with Hickey, 1115 7 Oct 50, and Msg CX65711, CINCUNC to CG Eighth Army, 7 Oct 50, both quoted in Schnabel, FEC, GHQ Support and Participation in the Korean War, ch. VI, pp. 24–25; EUSAK WD, G–3 Sec, 7 Oct 50, Opn Ord 104, 5 Oct 50.

[2] EUSAK PIR's 82, 2 Oct, 89, 9 Oct, 90, 10 Oct, and 93, 13 Oct 50.

the N.K. *19th* and *27th Divisions* defended the Kumch'on-Namch'onjom area north of Kaesong. Both had been brigades activated in the summer and expanded in September to division status. They engaged in combat for the first time when U.N. forces crossed the Parallel. On the right (west) of these divisions, the *74th Regiment* of the *43d Division* defended the Yesong River crossing site west of Kaesong. The *43d Division,* organized in mid-September, had the task of defending the coastal area beyond the Yesong River. Some elements of the N.K. *17th Armored Division* engaged in action just north of the Parallel in the zone of the ROK 1st Division, east of the 1st Cavalry Division.[3]

Ready for the attack, the 1st Cavalry Division was deployed in three regimental combat teams just below the Parallel in the vicinity of Kaesong. In the center, Colonel Palmer's 8th Cavalry Regiment was to attack frontally along the main highway axis from Kaesong to Kumch'on; on his right, Colonel Crombez' 5th Cavalry Regiment was to swing eastward, then west, in a circular flanking movement designed to envelope enemy forces south of Kumch'on, fifteen miles north of the Parallel. Meanwhile, on the division left, Colonel Harris' 7th Cavalry Regiment faced the task of crossing the Yesong River to get on the road running north from Paekch'on to the little town of Hanp'o-ri, six miles north of Kumch'on, where the main P'yongyang road crossed the Yesong River. At Hanp'o-ri the 7th Cavalry was to establish a blocking position to trap the large enemy forces that General Gay expected the 8th and 5th Cavalry Regiments to be driving northward. These were the maneuvers involved in the action of the Kumch'on Pocket. Because the prospects of forcing a crossing of the Yesong River did not appear very promising with the support available, General Gay and the division staff relied principally on the 8th and 5th Cavalry Regiments for initial success in the attack.

The 1st Cavalry Division sent patrols across the Parallel late on the afternoon of the 7th, and others crossed Sunday night, 8 October. Then on Monday, 9 October, at 0900 General Gay issued his orders, and the division moved up to the Parallel and started fighting its way northward. (*Map 18*)

In the division center along the main highway, the advance was very slow. The highway was heavily mined and the armored spearhead repeatedly came to a halt, waiting for Engineer troops to remove the mines. On 12 October, halfway to Kumch'on, an enemy strongpoint defended with tanks, self-propelled guns, and antiaircraft weapons again stopped the regiment. An air strike by sixteen planes and a 155-mm. howitzer barrage failed to dislodge the enemy. In this action, Lt. Col. Robert W. Kane, the 1st Battalion commander, was severely wounded.[4]

---

[3] EUSAK PIR 85, 5 Oct 50, and PIR 89, Incl No. 3, 9 Oct 50; FEC, Ord of Battle Info, N.K. Army, Chart 14 (N.K. *19th Div*) and Chart 13 (N.K. *17th Div*); ATIS Interrog Rpts, Issue 14 (N.K. Forces), p. 138, and Issue 15, pp. 100, 149, 191, 192; ATIS Enemy Documents, Issue 16, p. 45, diary of Sr Col Chang Tong Mu; GHQ FEC, History of the N.K. Army, p. 27.

[4] 8th Cav Regt WD, S-3 Jnl, Msg 091226 Oct 50; Capt. Charles A. Rogers, History of the 16th Reconnaissance Company in Korea, 18 July 1950–April 1951, MS, May 1951, copy in OCMH; 1st Cav Div WD, 11–12 Oct 50.

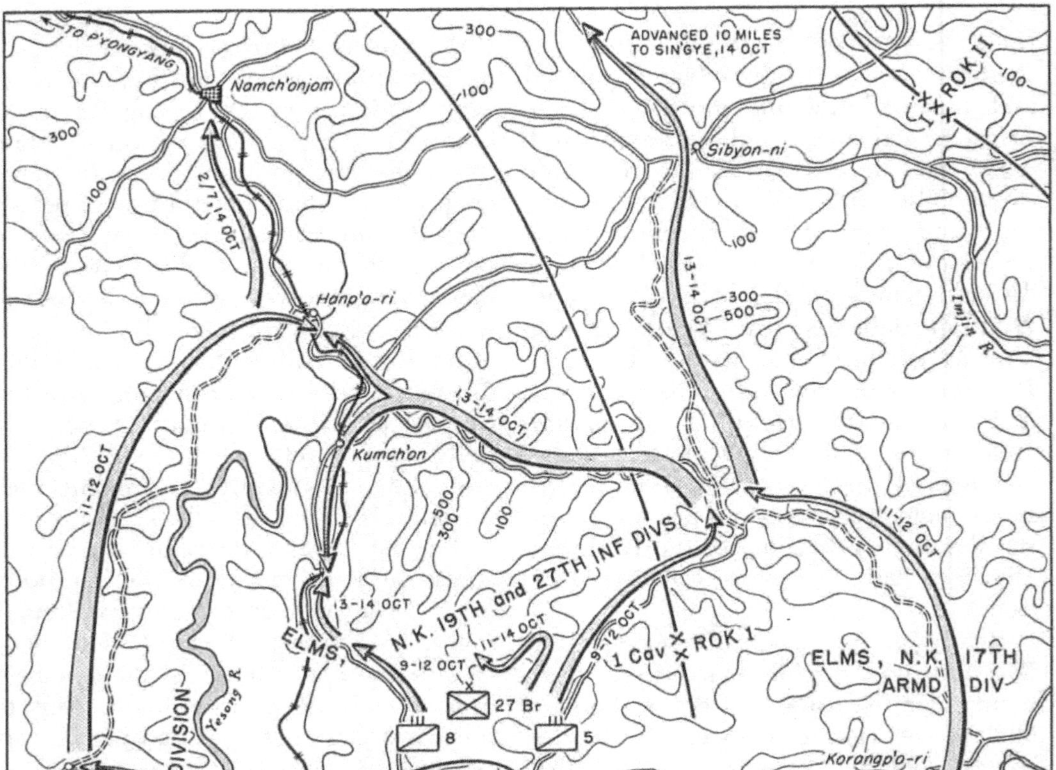
THE KUMCH'ON POCKET
9-14 October 1950
U.S. ASSEMBLY AREAS, EVENING, 8 OCT
AXIS OF U.N. ATTACKS, DATES INDICATED
Elevations in meters
5 0 5 MILES
5 0 5 KILOMETERS
TO P'YONGYANG
Namch'onjom
ADVANCED 10 MILES TO SIN'GYE, 14 OCT
Sibyon-ni
Hanp'o-ri
Kumch'on
ELMS, N.K. 19TH and 27TH INF DIVS
1 Cav
ROK 1
ELMS, N.K. 17TH ARMD DIV
N.K. 43D INFANTRY DIVISION
Paekch'on
Kaesong
Korangp'o-ri
TO HAEJU
1ROK
(FROM SEOUL)
Munsan-ni
Imjin R
Han R
Yesong R
27 Br
D. Holmes, Jr.

MAP 18

On the division right the 5th Cavalry Regiment also had difficulty. It reached the Parallel at 1930 9 October but did not cross until the next morning. In its initial attack it captured the hills flanking and dominating the road on both sides just above the Parallel. Fifteen miles northeast of Kaesong an enemy force held a long ridge with several knobs (Hills 179, 175, 174) dominating a pass. There it stopped the 1st Battalion. The next day, 12 October, the 2d Battalion joined in the battle. The 5th Cavalry drove the North Koreans from the ridge during the afternoon. In the fighting at Hill 174, 1st Lt. Samuel S. Coursen, a platoon leader in C Company, went to the aid of a soldier who had entered an enemy emplacement mistakenly thinking it was empty. The soldier escaped with a wound, but Coursen was later found dead there together with seven enemy soldiers whom he had killed in a desperate hand-to-hand struggle. Several of the North Koreans had crushed skulls from rifle butt blows.[5]

The day before, 11 October, the 27th British Commonwealth Brigade, with tanks of B Company, 6th Medium Tank Battalion, in support, had crossed the Imjin River and followed the 5th Cavalry Regiment northeast out of Kaesong. General Gay's plan was for the brigade to move northwest through the mountains for a close-in envelopment of Kumch'on. His aerial observer, hitherto very reliable, wrongly reported that the roads were as shown on the maps and that the plan was feasible. The road taken by the British, little more than a cart track, dead-ended in the mountains. The Middlesex Battalion got lost on this trail, turned back, and tried another. Despite an arduous effort in the mountains, the British troops never got into the fight for Kumch'on.[6]

While the British were crossing the Imjin, the ROK 1st Division crossed it at Korangp'o-ri at dawn on the 11th, eastward of the 1st Cavalry Division, and attacked northwest on a road that converged with the one taken by the 5th Cavalry Regiment. The 5th Cavalry in the late afternoon of 12 October was engaged in a fire fight with the enemy at the objective crossroads when advance elements of the ROK 1st Division arrived there from the southeast. In a conference on the spot Colonel Crombez and General Paik, the ROK division commander, agreed that the 5th Cavalry would have precedence on the road until Crombez' troops turned west, five miles northward on a lateral road leading into Kumch'on. The ROK 1st Division, following behind the 5th Cavalry, would then continue its attack north to Sibyon-ni where it would veer northwest toward P'yongyang. Tanks of C Company, 6th Medium Tank Battalion, supported the ROK 1st Division.[7]

Of the three regimental attack forces, the 7th Cavalry Regiment on the divi-

---

[5] 5th Cav POR 23, 091800–101800 Oct 50; 5th Cav Reg WD, 9 Oct 50 and Oct Narr Summ; 1st Cav Div WD, 11–12 Oct 50. Department of the Army General Order 57, 2 August 1951, awarded the Medal of Honor posthumously to Lieutenant Coursen.

[6] Ltr, Gay to author, 23 Jan 54; Interv, author with Crombez, 12 Jan 56; 1st Cav Div WD, 12 Oct 50.

[7] Crombez, MS review comments, 12 Jan 56; 6th Med Tk Bn WD, 29 Sep–Oct 50 (with 24th Div records); EUSAK POR 272, 11 Oct 50; 1st Cav Div WD, 12–13 Oct 50.

sion left flank had the most difficult assignment, and in fact General Gay and his staff expected it to accomplish little. The regiment had to cross the wide Yesong River against defending enemy forces before it could turn north as the lefthand column in the Kumch'on Pocket maneuver. Since all of I Corps' bridging troops and equipment were committed to establishing bridges across the Imjin River at Munsan-ni to support the main effort northward, river crossing support could not be supplied for the 7th Cavalry Regiment at the Yesong River.

On 8 October the regiment received orders to move up to the Yesong River, search for crossing sites, and clear enemy troops from the area southwest of Kaesong. The I&R Platoon found that the high, 800-yard-long combination highway and rail bridge over the river on the Kaesong-Paekch'on route was standing, although damaged. It was so weakened, however, that it could support only foot traffic. The I&R Platoon received small arms, automatic, and mortar fire from the enemy on the far side of the river. Colonel Clainos, commander of the 1st Battalion, also personally reconnoitered the area with a platoon of A Company on the afternoon of the 8th and received fire from the west bank of the stream. The I&R Platoon leader told him that enemy forces held the west side of the river from the southern tip of the peninsula to a point one-half mile northeast of the Yesong River bridge. Colonel Harris, the regimental commander, upon receiving the I&R Platoon report that the bridge was usable for foot troops, ordered the platoon to prevent further destruction of the bridge. He then called upon the 1st Battalion to seize the bridge and crossing area.[8]

A full report of the situation was given to the 1st Cavalry Division with the recommendation that the 7th Cavalry Regiment seize this unexpected opportunity for a quick crossing of the river. General Gay feared that the North Koreans had set a trap in leaving the bridge usable for foot troops, and that enemy zeroed-in mortar and artillery fire and automatic weapons would decimate any troops caught on it. The division staff said also that a regimental attack west of the Yesong River northward could not be supported logistically. The untiring efforts of Colonel Harris and his S–3, Captain Webel, however, succeeded in winning from General Gay authority to attempt the crossing on the 9th.

On the afternoon of 9 October the 7th Cavalry Regiment delivered three hours of preparatory artillery fire against enemy positions on the west bank of the river. At 1500, Colonel Clainos ordered a platoon of C Company to cross the bridge under cover of the barrage. In crossing the bridge and seizing the immediate approaches on the far side, the platoon suffered a few casualties from small arms fire. Following this platoon, B Company, 8th Engineer Combat Battalion, went on the bridge and spent all night under fire repairing holes in the roadway. After the first troops reached the far side, Clainos sent the rest of C Company across and it occupied the hill on the right of the

[8] Ltr, Webel (7th Cav S–3 Oct 50) to author, 13 Apr 54; Ltr, Col Peter D. Clainos to author, 24 May 54; Ltrs, Gay to author, 23 Jan and 19 Apr 54; Ltr, Harris to author, 7 Apr 54.

bridge. Next to cross was B Company, which seized the hill just south of the bridge. The artillery and mortar barrage had been unable to silence enemy mortars, and these fired heavy concentrations on the bridge during the 1st Battalion crossing, which took several hours to complete. The overhead steel girders of the bridge gave excellent protection from fire and prevented many casualties. When the supporting artillery barrage had to be lifted from the immediate environs of the bridge, once the 1st Battalion troops crossed to that side, casualties began to increase rapidly from enemy fire. In this crossing attack, the 1st Battalion had 78 casualties; C Company alone had 7 killed and 36 wounded.

After dark the North Koreans launched a counterattack against the 1st Battalion, and Colonel Harris ordered Lt. Col. Gilmon A. Huff to hasten his crossing with the 2d Battalion. Just before midnight Huff's battalion started infiltrating across the bridge which was still under some mortar and small arms fire. On the other side, Huff assembled his battalion on the south flank of the 1st Battalion, approximately 100 yards west of the bridge. He then attacked west along the Paekch'on road in a column of companies with G Company leading. This attack progressed only a short distance when a heavy enemy counterattack from the south struck the flank of G Company. The enemy counterattack threw the 2d Battalion into momentary confusion. In the beginning of the fight enemy small arms fire hit Huff in the shoulder, but he remained with his battalion throughout the night battle. The largest weapons the battalion had at hand were 57-mm. recoilless rifles and 60-mm. mortars. Huff showed superb leadership in this difficult night battle and eventually seized the high ground southeast of the bridge and the road. By dawn it was clear that the battle was all but over and that the 2d Battalion would be able to move forward. Huff then turned over command of the battalion to the executive officer who led it in a continuation of the attack westward. The battalion seized Paekch'on and the high ground north of the town in the afternoon.[9]

The next morning, 11 October, the 3d Battalion, 7th Cavalry, crossed the Yesong River and headed north. Thus, by that morning all three regiments of the 1st Cavalry Division had crossed the 38th Parallel and were driving into North Korea. The 3d Battalion, 7th Cavalry, on the morning of 12 October, seized its objective—the railroad and highway bridges at Hanp'o-ri north of Kumch'on, and the road juncture there. This closed the western escape route of an estimated 1,000 enemy troops in Kumch'on. Friendly fighter-bombers mistakenly strafed and rocketed the 3d Battalion at Hanp'o-ri, wounding several men. That evening the 2d Battalion joined the 3d Battalion at Hanp'o-ri.[10]

During the night at the 3d Battalion roadblock, the pressure from the 8th and 5th Cavalry Regiments on the North Koreans was made evident. MSgt. John H. Smith and his platoon of L Company ambushed 11 enemy trucks running with

---

[9] Ltr, Huff to author, 28 Apr 54; Ltr, Webel to author, 13 Apr 54; Fonecon, Webel with author, 20 May 54; Ltr, Clainos to author, 24 May 54; Ltr, Gay to author, 19 Apr 54; 2d Bn, 7th Cav Unit Jnl, 0355–0700 10 Oct 50; 7th Cav Regt WD, 9–10 Oct 50.

[10] 7th Cav Regt WD, 12 Oct 50; 1st Cav Div WD, 11–12 Oct 50; Webel, MS review comments for author, 13 Apr 54.

their lights on. They destroyed 4 trucks loaded with ammunition, captured 6 others, killed about 50 enemy soldiers, and captured an equal number. Among the latter was a mortally wounded regimental commander who had in his possession a document indicating that two North Korean divisions, the *27th* and *19th,* intended to break out of Kumch'on the night of 14 October. Before he died the officer said part of the enemy force had been ordered to withdraw to Namch'onjom, a fortified area fifteen miles north of Kumch'on.[11]

The drive of the 7th Cavalry Regiment northward to Hanp'o-ri after crossing the Yesong River could not have taken place without one of the most successful logistical supply operations of the Korean War. In the discussions before the 7th Cavalry attack at the Yesong River bridge, the 1st Cavalry Division G–4 had told Colonel Harris and Maj. Lucian Croft, the regimental S–4, that he could not provide the gasoline, rations, and certain types of ammunition for the drive north from Paekch'on even if the river crossing attempt was successful. Colonel Harris and Captain Webel decided to try to obtain the needed logistical support from the 3d Logistical Command at Inch'on by amphibious transport through the Yellow Sea and up the Yesong River. Capt. Arthur Westburg, an assistant regimental S–3 officer, went to Inch'on and presented the matter to Brig. Gen. George C. Stewart, the port commander. General Stewart and his staff loaded 500 tons of supplies on thirteen LCV's. They arrived at the 7th Cavalry crossing site at the Yesong River bridge late in the afternoon of the 10th. Engineer troops from I Corps on the 12th constructed a ponton ferry at the bridge site and it transported the tanks of C Company, 70th Tank Battalion, across the river for support of the regiment.[12]

The 13th of October promised to be a critical day in the efforts of the 1st Cavalry Division to close the Kumch'on Pocket. With the 7th Cavalry blocking the exit road from Kumch'on, the decisive action now rested with the 8th and 5th Cavalry Regiments which were trying to compress the pocket from the south and the east.

After it turned west from the Sibyonni road the 5th Cavalry encountered an almost continuous mine field in its approach to Kumch'on, and it also had to fight and disperse an enemy force estimated to number 300 soldiers, eight miles from the town. Overcoming these difficulties, the regiment pressed ahead and by the evening of the 13th it was approaching Kumch'on.

Strong opposition confronted the 8th Cavalry Regiment on the main highway where the enemy apparently had concentrated most of his available forces and weapons. There, on the morning of the 13th, an artillery preparation employing proximity fuze air bursts blanketed the North Korean positions. Because of the closeness of the American

[11] 7th Cav Regt WD, 12 Oct 50; 7th Cav Regt Opn Ord 28, 141015 Oct 50; EUSAK WD, G–3 Jnl, 1100 13 Oct 50.

[12] Ltr, Webel to author, 13 Apr 54; Ltr, Harris to author, 7 Apr 54; 7th Cav Regt WD, 10–11 Oct 50; 3d Log Comd Hist Opn Rpt, G–4 Sec (Hist Memo: Yesong River Supply and Ferry Mission), 10–12 Oct 50. During the 12th, the 1st Cavalry Division received reinforcements which were attached to it for the drive on P'yongyang. These were B Co, 6th Med Tk Bn; the 89th Tk Bn; the the 13th FA Bn; and the 90th FA Bn.

troops to the enemy, a planned B–26 bomber strike was canceled, but a new flight of fighter planes appeared over the enemy positions every thirty minutes. The North Koreans resisted stubbornly with tanks, artillery, mortars, small arms fire, and counterattacks. In one of the counterattacks, enemy tanks rumbled out of the early morning mist to strike an outpost of B Company, 70th Tank Battalion. Sgt. Marshall D. Drewery said his tank gunner first fired on the lead enemy tank at a range of fifty yards. A second round hit it at a range of twenty yards. Still the T34 came on and rammed into the American tank. Drewery's driver put his tank in reverse, jerked loose, and backed away. At a few yards' range the gunner fired a third round into the enemy tank which now had a split gun muzzle and was burning. Amazingly, the tank rumbled forward and rammed Drewery's tank a second time. The fourth round finally knocked out this stubborn enemy tank. In the day's series of attacks and counterattacks the 8th Cavalry and supporting arms destroyed eight enemy tanks; B Company, 70th Tank Battalion, accounted for seven of them without loss to itself.[13]

While the enemy force south of Kumch'on fought desperately and successfully to prevent the 8th Cavalry from closing in on the town, a large enemy column of trucks and carts with an estimated 1,000 soldiers moved northward out of it on the road toward Namch'onjom. At the Hanp'o-ri bridge it ran into the 7th Cavalry roadblock. In the ensuing action, the 7th Cavalry, aided by air strikes, killed an estimated 500 and captured 201 of this force. Many enemy troops, however, escaped into the hills northeast of the town.

At the same time, elements of the N.K. *43d Division* cut off below Paekch'on were moving around that town and fleeing north. One such group in company strength occupied old North Korean defensive positions just north of the 38th Parallel the night of 12–13 October. The following day it ambushed the tail end of the 1st Battalion, 7th Cavalry, column moving north from Paekch'on. Part of A Battery, 77th Field Artillery Battalion, and B Company, 8th Engineer Combat Battalion, were in the ambushed column. A soldier who escaped raced back into Paekch'on to the 3d Battalion, 21st Infantry, command post. Colonel Stephens, the regimental commander, happened to be there. Upon hearing what had happened, he directed Lt. Col. John A. McConnell, Commanding Officer, 3d Battalion, to send a company to the scene. Colonel McConnell thereupon directed I Company, 21st Infantry, which was on a blocking mission south of the ambush site, to go there. On arrival it engaged and dispersed the enemy force with mortar and small arms fire, and captured 36 North Koreans. In this ambush the North Koreans killed 29 American and 8 South Korean soldiers and wounded 30 Americans and 4 South Koreans. They also destroyed 4 and damaged 14 vehicles. In this episode, as in so many others like it, those caught in the roadblock apparently made little effort to defend themselves. In another ambush on the road that night enemy troops captured the 2d Battalion, 7th Cavalry, supply

[13] 1st Cav Div WD, 13–14 50; 8th Cav Regt WD, Jnl file 0643 to 0800 13 Oct 50; New York *Herald Tribune*, October 14, 1950.

officer and 11 men; subsequently, however, the officer and 5 men escaped.[14]

At midnight of the 13th, the 2d Battalion, 5th Cavalry Regiment, resumed its attack on Kumch'on from the east. After dispersing an enemy force near the town, the battalion then entered and seized the northern part of it. The 3d Battalion following behind seized the southern part. At 0830, 14 October, Colonel Crombez and the regimental command group arrived in Kumch'on. Crombez ordered the 2d Battalion to turn north toward the 7th Cavalry at Hanp'o-ri and the 3d Battalion to turn south to meet the 8th Cavalry on the Kaesong road. The 1st Battalion remained behind to secure the town.

Advancing northwest, the 2d Battalion joined elements of the 7th Cavalry above Hanp'o-ri at noon. An enemy force, estimated to number 2,400 men, which had been attacking the 7th Cavalry roadblock position at Hanp'o-ri, escaped into the hills when the 2d Battalion approached from the south. Meanwhile, attacking south from Kumch'on, the 3d Battalion neared a special task force of the 8th Cavalry Regiment which had attacked north during the morning and already had lost two tanks to enemy action. The two columns, the 3d Battalion, 5th Cavalry, and the special 8th Cavalry task force met just after noon about four miles south of Kumch'on. Even though the 1st Cavalry Division envelopment and capture of Kumch'on had been carried out in five days, a large part of the enemy force in the Kumch on Pocket escaped, mostly to the north and northwest.[15]

The day Kumch'on fell to the 1st Cavalry Division, 14 October, the North Korean Premier and Commander in Chief, Kim Il Sung, issued an order to all troops of the North Korea People's Army explaining the reasons for the army's defeat and outlining harsh measures for future army discipline. Alluding to the recent reverses, Kim Il Sung said, "Some of our officers have been cast into utter confusion by the new situation and have thrown away their weapons and left their positions without orders." He commanded; "Do not retreat one step farther. Now we have no space in which to fall back." He directed that agitators and deserters be executed on the spot, irrespective of their positions in the Army. To carry out this order, he directed that division and unit commanders organize, by the following day, a special group, which he termed the "Supervising Army," its men to be recruited from those who had distinguished themselves in battle.[16]

At the close of 14 October, with U.S. I Corps troops through the principal prepared enemy positions between the 38th Parallel and the North Korean capital, enemy front lines as such had ceased to exist. The North Korean forces were in a state of utter confusion.

In these auspicious circumstances President Truman on 15 October met

[14] 1st Cav Div WD, 13 Oct 50; 7th Cav Regt WD, 13 Oct 50; 24th Div WD, 14 Oct 50; Webel, MS review comments for author, 13 Apr 54; Stephens, MS review comments, Dec 57.

[15] 1st Cav Div WD, 13–14 Oct 50; 7th Cav Regt WD, 13–14 Oct 50; 8th Cav Regt Jnl, 141310–141315 Oct 50; Crombez, MS review comments, 12 Jan 56; Stephens MS review comments, Dec 57.

[16] ATIS Supp, Enemy Docs, Issue 19, pp. 1–4, order dated 14 Oct 50, issued to all NKPA personnel by Kim Il Sung and Pak Hon Yong, Chief of Supreme Political Bureau, NKPA.

KUMCH'ON, NORTH KOREA, *after the battle.*

General MacArthur on Wake Island. A few days earlier, in announcing his intention to make the trip, President Truman had said he would discuss with General MacArthur "the final phase of U.N. action in Korea." [17]

*The X Corps Moves to Northeast Korea*

While the I Corps of Eighth Army was driving into North Korea on the P'yongyang axis and the 1st Marine Division was loading at Inch'on, the 7th Infantry Division was assembling at Pusan to outload there in the X Corps amphibious movement to northeast Korea. On 30 September the division had been relieved of its responsibilities in the Seoul area and its units began to shift south and southeast to the Suwon and Ich'on areas preparatory to the long overland move to Pusan. Ten LST's were reserved at Inch'on for the division's tanks and heavy equipment.

On 4 October Eighth Army indicated the route it wanted the 7th Division to take through its zone, specifying the road through Ch'ungju, Hamch'ang, Kumch'on, Taegu, and Kyongju to Pusan, a road distance of 350 miles from Ich'on. At Taegu the troops were to load on trains for the final part of the journey, whereupon the trucks would return to Suwon and Ich'on for others.[18]

[17] EUSAK WD, 21 Oct 50, has transcript of Truman's address at San Francisco, 17 Oct; New York *Times*, October 11, 1950.

[18] X Corps WD, CofS Notes, 2, 4 Oct 50; 7th Inf Div WD, G–3 Jnl, 2–4 Oct 50, and Msg 17, 5 Oct 50; X Corps WD, Opn Sec, pp. 16–18; Barr, Notes, 30 Sep–2 Oct 50.

The 3d Battalion, 31st Infantry, led the 7th Division movement, passing the initial point at Ich'on at 0350 5 October, with the rest of the regiment following. The command group of the 32d Infantry led the movement of that regiment through Ich'on four hours later. The 17th Regiment remained at Ich'on, holding its blocking position there until relieved on 8 October, and it then began the motor movement to Pusan. Both the 31st and 32d Regiments closed at Pusan on 7 October. On 8 October the 7th Division command post closed at Anyang-ni and opened at Pusan, although most of the headquarters was still on the road.

The movement to Pusan was not without incident. On two occasions enemy forces ambushed convoys in the mountains near Mun'gyong. The first ambush caught the head of the 2d Battalion, 31st Infantry, at 0200, 6 October, and inflicted nine casualties; the second ambush at 0230, 9 October, caught the division headquarters convoy in the pass three miles northwest of Mun'gyong. Enemy machine gun fire killed six men and destroyed several vehicles. Elements of the 1st Battalion, 17th Infantry, succeeded in clearing the pass area that afternoon. This battalion thereafter patrolled the pass above Mun'gyong until it was relieved on 11 October by the 27th Infantry Regiment of the 25th Division.[19]

The division artilllery was the last major unit to leave Ich'on, clearing there at 1700 on 10 October. It and the 1st Battalion, 17th Infantry, arrived at Pusan on 12 October to complete the division movement to the port. About 450 division troops had been airlifted on 11 October from Kimpo Airfield to Pusan. In addition to the 7th Division, the X Corps Medical, Engineer, Ordnance, Transportation, Quartermaster, Chemical, and Signal units moved overland to Pusan for outloading. Altogether, in seven days, approximately 1,460 tons of supplies and equipment and 13,422 troops had moved overland in division vehicles and those of the 52d Truck Battalion.[20]

The loading of the 7th Division vehicles and equipment at Pusan began on 14 October and that of the men two days later. The division was completely loaded on 17 October, the deadline set by X Corps nine days earlier. The loading of corps troops at Pusan began on 19 October.

In its order of 8 October, X Corps had required the 2d Logistical Command to furnish 15 days' supply of all classes for the 25,000 troops outloading at Pusan, 10 days of Class II and IV supplies for the troops outloading at Inch'on, and, for the entire corps, 15 days' resupply to arrive in the Wonsan area on D-day plus 8 (28 October). Except for Class I supply, the 2d Logistical Command had no reliable information as to the requirements of the various units. Providing the 15 days of supply depleted depot stocks in that area, particularly of winter clothing, operational rations, POL, and post exchange comfort items. This resulted in subsequent logistical

[19] 7th Inf Div WD, 6–11 Oct 50; X Corps WD, 6–11 Oct 50; X Corps POR 19, 7 Oct 50; Barr Notes, 9 Oct 50; X Corps WD, G–3 Jnl, Msg 46, 2215 9 Oct 50.

[20] X Corps WD, Oct 50, Log Sec, p. 22.

difficulties for Eighth Army. Much of the 15 days' resupply for X Corps had to be requisitioned on the Japan Logistical Command.[21]

The difficult logistical and outloading problem given the 2d Logistical Command on such short notice was worked out successfully only by the constant mutual effort and co-operation of the staffs of the logistical command and of the 7th Infantry Division. The outloading was completed in time. It was an outstanding performance. On 16 October the 7th Division advance command post opened aboard the USS *Eldorado.* But because mine fields in Wonsan Harbor now delayed sailing of the convoys for nearly two weeks, the hectic work at the port to meet the loading deadline was largely in vain.

### *Mines at Wonsan Harbor*

After the Inch'on landing and Eighth Army's successful breakout from the Pusan Perimeter, evidence began to mount that the North Koreans were mining the coastal waters of North Korea. Three U.S. ships, the *Brush, Mansfield,* and *Magpie,* struck mines and suffered heavy damage. Although intelligence sources indicated enemy mines were being laid in coastal waters, little was known about the location and extent of these mine fields. North Korean interests certainly dictated, however, that the sea approaches to Wonsan should be mined.

In a series of conferences from 2 to 4 October, Admiral Struble and his staff decided to form the Advance Force JTF 7, which would proceed to the objective area and begin minesweeping at the earliest possible date. All minesweepers available were to be concentrated for the task. The group comprised 21 vessels, including 10 American and 8 Japanese minesweepers, and 1 South Korean vessel used as a minesweeper. Minesweeping operations at Wonsan began on 10 October. A search by helicopter over the harbor channel showed it to be heavily mined inside the 30-fathom curve. The plan to sweep this channel was canceled and another substituted—to sweep from the 100-fathom curve down the Russian Hydropac Channel passing between Yo-do and Hwangt'o-do Islands. By 12 October this channel had been swept a distance of twenty-four miles from the 100-fathom curve. Ten miles remained to the inner harbor.[22]

At this point the novel idea was advanced of exploding mines along a narrow passageway by aerial bombing which would permit the lead sweeps to pass. On 12 October thirty-nine planes from the carriers *Philippine Sea* and *Leyte* flew down the Russian channel dropping 1,000-pound bombs.

Three minesweepers, the *Pirate, Pledge,* and *Incredible,* then entered the bombed channel to resume minesweeping. Northwest of Yo-do Island the *Pirate* struck a mine at 1209; the *Pledge* hit one six minutes later. Both vessels sank. As the *Incredible,* third in line, maneuvered into safe water, enemy shore

[21] 7th Inf Div WD, 14 Oct 50; 2d Log Comd Monthly Act Rpt, G–4 Sec, Oct 50; Barr Notes, 14 Oct 50; Interv, author with Col A. C. Morgan, CofS 2d Log Comd, 21 Jul 51; Interv, author with Lt Col Robert J. Fuller, 2d Log Comd, G–4 Sec, 21 Jul 51.

[22] Act Rpt, JTF 7, Wonsan Opn, I–C–2 and 3, and VI–D–1.

batteries opened fire. Twelve men went down with the two sunken ships. One enlisted man died later from wounds. At least 33 others were wounded and injured in varying degrees; some sources place the number of wounded as high as 99. The *Incredible* itself rescued 27 survivors.[23]

The menace of shore batteries was removed on 17 October when ground forces of the ROK I Corps, which had already captured Wonsan, gained control of the peninsulas and islands commanding the harbor approaches.

Casualties from mines continued. On 18 October two ROK Navy vessels struck mines in the Wonsan area; one was disabled at the entrance to the harbor, and the other, a minesweeper, was sunk. The next day a Japanese minesweeper struck a mine and sank.

The risk of sending transports with troops to the beaches was still great. The presence of ground mines in the shallow water made necessary a thorough magnetic sweep of the close-in approaches to the landing beaches. Because troops of the ROK I Corps were now well past Wonsan, the military situation did not warrant an unnecessary risk in unloading the Marine units. Admiral Struble, therefore, recommended that they not be unloaded on 20 October as planned, but that D-day be deferred until the minesweeping could be completed. Admiral Joy and General MacArthur concurred.

A report from the minesweeper group on 23 October indicated that a channel free of mines had been swept to Blue-Yellow Beach, but that sweeping of the beach area itself was being continued. At a conference on board the *Missouri* the next day, Admiral Struble decided that landings could start on the 25th; actually they did not begin until the morning of the 26th. The conference on the 24th also decided that the minesweepers should clear the Wonsan inner harbor. Then they were to sweep the approaches to Hungnam to clear that port. General Almond had urged this so that logistical support could be centered there for the X Corps operations in northeast Korea. Not until 4 November did the minesweepers complete their work in the Wonsan inner harbor. Ships of the task force then stood into the harbor and pulled up alongside the dock.

Their dangerous work finished at Wonsan, the minesweepers still had to continue it in the Hungnam area. There they swept a channel 32 miles long and 1,600 yards wide, as well as an anchorage in the inner harbor. Actually, the minesweepers were busy as long as X Corps was in northeast Korea. Floating mines were common sights at this time off the east coast of Korea in the Wonsan-Hungnam area. One of the worst mine disasters occurred on 16 November, when an Army tug with a crane barge in tow struck a mine off the entrance to Wonsan Harbor and sank, with approximately thirty men lost out of forty.[24]

While the minesweeping was progress-

[23] *Ibid.*, pts. V, IV, and I–D–1 to I–D–3; Karig, *et al.*, *Battle Report, The War in Korea*, pp. 317–18; Lt. Comdr. R. N. Hartmann, USNR, "Minesweepers Go In First," *Armed Forces Chemical Journal*, vol. V, No. 2 (October, 1951), pp. 19, 46.

[24] Act Rpt, TF 90, Amphib Group 1, Hungnam Redeployment, 9–25 Dec 50, Status of Sweep Operations, 7 Dec 50; 3d Inf Div Comd Rpt, CofS Jnl, entry at 1637 17 Nov 50.

ing offshore, Lt. Col. William J. McCaffrey, Deputy Chief of Staff, X Corps, on 16 October brought the X Corps Advance Command Post to Wonsan by air, flying from Kimpo Airfield. He immediately established communications with the ROK I Corps and the commander of the minesweeping operations. McCaffrey's staff set to work at once with the ROK I Corps G–2 to learn who had laid the mines in the harbor and to find the warehouses where they had been stored. This was done successfully by the ROK I Corps intelligence section. The ROK's found a villager who had worked in the mine depot. After his fears were quieted, he guided a party to a depot north of Wonsan where the mines had been stored and assembled. He also provided information enabling the investigators to take into custody one of the sampan captains who had helped plant the mines. The information gained from these sources indicated that thirty Russians had been in Wonsan until 4 October assembling the mines and supervising laying the mine fields. Working almost entirely at night, from about thirty-two small boats, North Korean crews and their supervisors had laid approximately 3,000 mines.[25]

The North Koreans and their helpers had not confined laying mines at Wonsan to the waters in the harbor. The beaches were also heavily planted with land mines. This had been expected, and as soon as the ROK I Corps had secured Wonsan it cleared the beaches of mines.

An unusual incident growing out of this work occurred the night of 16 October. At the north end of the Wonsan Harbor ROK troops had stacked about 1,000 20-pound box mines they had just lifted from the beaches. A ROK lieutenant and five enlisted men decided to have a private celebration, and, moving off about 200 yards, the lieutenant fired into the stacked mines. The mines exploded, shattering panes of glass in the provincial capital building two miles away. Unfortunately, they also killed the six ROK's.

### *The X Corps Ashore*

The ships of Amphibious Group One and the LST's of the tractor group sailed from Inch'on late in the afternoon of 16 October. At 0800 on the 17th, the main body of the Attack Force with the 1st Marine Division aboard departed Inch'on, moved into the Yellow Sea, and headed south to round the tip of Korea. From Inch'on it was 830 miles to Wonsan by the shortest sea route.[26]

After arriving off the objective area, the flotilla carrying the 1st Marine Division steamed slowly back and forth from 19 to 25 October in the Sea of Japan just outside the Wonsan channel. The restless marines called it "Operation Yo-yo." It was a great relief to everyone afloat when twenty-one transports and

[25] Act Rpt, JTF 7, I–D–1 to I–D–3; Karig, *et al.*, *Battle Report, The War in Korea*, pp. 327–30; New York *Times*, October 21, 1950, dispatch from Hanson Baldwin on USS *Missouri* off Wonsan, 20 Oct 50; Ltr, Col William J. McCaffrey to Almond, 1 Dec 54, and forwarded by latter to author.

[26] X Corps WD, Diary of CG, 17 Oct 50; *Ibid.*, Notes of CofS X Corps, 17–19 Oct 50; 1st Mar Div SAR, 8 Oct–13 Dec 50, vol. I, p. 20 and vol. II, an. C, p. 4; 2d Log Comd Rpt, Oct 50, G–4 Sec, pp. 3–6; 3d Log Comd Hist Rpt, Oct 50; Karig, *et al.*, *Battle Report, The War in Korea*, pp. 301–05.

LANDING CRAFT APPROACHING BEACH *at Wonsan on 26 October.*

fifteen LST's came into the harbor on 25 October and dropped anchor off Blue and Yellow Beaches. The X Corps began a quiet, administrative landing at 0730 on 26 October. At 1000 27 October the command post of the 1st Marine Division closed aboard the USS *Mt. McKinley* and opened in Wonsan. By the close of 28 October all combat elements of the division were ashore.

Meanwhile, the 7th Division had remained idly afloat at Pusan for ten days. Finally, on October it received orders to proceed to Iwon, 150 miles above Wonsan, and to unload there across the beaches.

Because the X Corps mission by now had been changed to advancing northward instead of westward from Wonsan, General Almond decided to land the 7th Division as close as possible to its axis of advance inland toward North Korea's northern border. This was to be the Pukch'ong-P'ungsan-Hyesanjin road to the Yalu. On receipt of the changed orders, the 17th Regimental Combat Team, which was to be first ashore, had to unload its unit equipment from its transports at Pusan and reload combat equipment on LST's, in order to be prepared to land on a possibly hostile beach. This done, seven LST's with the 17th Regimental Combat Team aboard left Pusan on 27 October and headed up the coast for Iwon. The landing proved to be without danger for the minesweepers found no mines there, and the ROK Capital Division had captured and passed through the town several days earlier. The 17th Infantry landed over the beaches at Iwon unopposed on the morning of the 29th. Except for most of its tanks, the 7th Division completed unloading there on 9 November.[27]

---

[27] 7th Div WD, 17, 29 Oct 50 and G–3 Jnl, 26–28 Oct; *Ibid.*, POR, 9 Nov 50; X Corps WD, Summ, 29 Oct 50; Barr Notes, 26, 29 Oct and 9 Nov 50.

# CHAPTER XXXI

# The Capture of P'yongyang

To the devil with history and principles! After all, what is the problem?
Verdy du Vernois at battle of Nachod, anecdote told by Foch

### *The Logistical Situation*

The Eighth Army advance into North Korea had begun under great logistical difficulties and was supported only on the narrowest margin. On 10 October, the day after the attack began, General Milburn expressed himself as being disturbed by the logistical situation of I Corps. He felt that at least 3,000 tons of balanced stocks should be in the Kaesong ammunition supply points. But Col. Albert K. Stebbins, Jr., Eighth Army G–4, informed him that this could not be accomplished unless all the truck companies were diverted to that task. The unfavorable supply situation largely grew out of the fact that during the first half of October (1–17 October) unloading activities at Inch'on for Eighth Army were negligible. Practically all the port capabilities at that time were engaged in mounting out the 1st Marine Division for the Wonsan operation. Levels of some supplies for I Corps were at times reduced to one day, and only selective unloading enabled the supply sections to meet troop requirements. Most combat vehicles, such as tanks, operated in the forward zone without knowing whether they would have enough fuel at hand to continue the attack the next day.

Because it could not support any more troops north of the Han River at this time, Eighth Army had been compelled to undertake the movement north of the 38th Parallel with only one corps (I Corps), leaving IX Corps below the river. As rapidly as the logistical situation permitted, General Walker intended to move IX Corps into North Korea to help in the drive to the border. On 23 October, General Walker informed General Coulter that the ROK III Corps (5th and 11th Divisions) would relieve IX Corps in its zone as soon as practicable for this purpose, and not later than 10 November.[1]

On 19 October the army forward distributing point was at Kaesong. Hence, for most units supplies had to be trucked more than a hundred miles—a most difficult logistical situation even with good roads, and those in Korea were far from that. During this time Eighth Army used about 200 trucks daily to transport food,

[1] IX Corps WD, bk. I, sec. II, Oct 50; EUSAK WD, 23 Oct 50; Ltr of Instr, CofS to CG IX Corps, 23 Oct 50.

gasoline, and lubricants to dumps 50 miles north of Seoul.

A pipeline, completed in October, carried aviation gasoline from Inch'on to Kimpo Airfield and helped immensely in supplying the planes with fuel.[2]

The 3d Logistical Command at Inch'on was assigned to Eighth Army on 7 October with the primary mission of providing it with logistical support in North Korea. Eighth Army in turn attached the 3d Logistical Command to the 2d Logistical Command. From Pusan the 2d Logistical Command continued of necessity to forward by rail and truck supplies for Eighth Army.

The solution to Eighth Army's logistical problems rested in the last analysis on the railroads. Airlift and long-distance trucking were emergency measures only; they could not supply the army for an offensive operation several hundreds of miles from its railhead.

At the end of September, rail communications for Eighth Army did not extend beyond the old Pusan Perimeter. Yet the army itself was then at the Han River, 200 miles northward. Because of the resulting logistical strain, the repair of the rail line north of Waegwan was of the greatest importance.

The reconstruction of the railroad bridges over the major rivers north of Taegu constituted the greatest single problem. To rebuild these bridges Eighth Army marshaled all available bridging equipment and matériel. Engineer construction troops, aided by great numbers of Korean laborers, worked to the limit of their endurance to restore the rail lines northward. The Koreans assumed responsibility for repairing minor bridges, I Corps most of the highway bridges, and Eighth Army the rail bridges and the largest highway bridges.

The first great task was to repair the 165-foot break in the Waegwan rail bridge over the Naktong. Working fifty feet above the water, the engineers, after some preliminary work, in 7 days completed the major repairs. Rail traffic crossed the bridge on 5 October. At first all effort was concentrated on opening single track communications over the 200 miles of rail from the Naktong to the Han River. This was accomplished on 10 October, 17 days after reconstruction work started at the Naktong River bridge. It was not until 11 days later that a shoofly bridge carried rail traffic across the Han into Seoul.[3]

But even after trains crossed into Seoul they could proceed only as far as Munsan-ni on the south bank of the Imjin River. This was still 200 miles below the Eighth Army front at the Ch'ong-ch'on River in late October. Thus, at that time the railhead was still as many miles south of the Eighth Army front

[2] 3d Log Comd Hist Rpt, Oct 50; 2d Log Comd Rpt, G–4 Sec, Oct 50, pp. 3–6; EUSAK WD, G–4 Sec Rpt, 10 Oct 50; *Ibid.*, G–3 Jnl, 15 Oct 50; Interv, author with Eberle (FEC UNC G–4, 1950), 12 Jan 54; Interv, author with Maj Gen Leven C. Allen, 15 Dec 53; ORO, *An Evaluation of Service Support in the Korean Campaign*, ORO–T–6 (FEC), 1 Mar 51, p. 8.

[3] EUSAK WD, Engr Off Rpt, 30 Sep and 15 Oct 50; *Ibid.*, Trans Sec, 26 Oct, G–4 Staf Sec, 12 Nov, and G–1 Daily Hist Rpt, 20 Nov 50; Dept of State Pub 4051, *United Nations Command Eighth Report to the Security Council, United Nations, 16–31 October 1950*, p. 6; Col. Paschal N. Strong, "Army Engineers in Korea," *Military Engineer*, vol. 44, No. 302 (November–December, 1952), 405–10 and "Engineers in Korea—Operation Shoestring," vol. 43, No. 291 (January–February, 1951); Interv, author with Strong (Eighth Army Engr Off), 17 Sep 51.

as it had been a month earlier when the front was in the Seoul area and the railhead was at Waegwan. At Munsan-ni the supplies were unloaded, trucked across the Imjin, and reloaded on trains on the north side. Meanwhile, Engineer troops were at work repairing the Imjin River rail bridge. The water span was 1,600 feet long, with a length of several thousand feet of earth fill required in its approaches. As a generalization, it may be said that the railhead lagged 200 miles behind the Eighth Army front in October 1950.

The daily "must" trains from Pusan at this time were (1) a train of 9 cars to Taejon for the 25th Division, (2) a ration train of 20 cars (200,000 rations) to Yongdungp'o, (3) 2 ammunition trains of 20 cars each, (4) 1 hospital train, (5) 1 POL train of 30 cars, and (6) 1 train of 20 cars every other day in support of ROK troops based in the Seoul area.[4]

Repair of the major highway bridges presented a problem just as pressing as repair of the rail bridges. In some respects it was an even more immediate problem because, in general, the highway bridges could be repaired more quickly, and they were the first used to keep supplies moving forward to the troops. The 207-foot span break in the Naktong River highway bridge at Waegwan was closed with pile bents and a 100-foot triple single-panel Bailey bridge. The first traffic crossed the repaired bridge on 30 September. To provide a vehicular bridge across the Han River at Seoul quickly, the FEAF Combat Cargo Command, using seventy C-119 flights, flew in a ponton bridge from Japan. This 50-ton floating bridge was 740 feet long. On 30 September, 3,034 vehicles crossed it, and thereafter traffic passed over it day and night. A second bridge was completed across the Han on 7 October. The next afternoon two-way traffic resumed across the river.

At every turn in the operations in North Korea during October, Eighth Army's effort was limited by an adverse logistical situation. And it must be borne in mind that Eighth Army's troops had almost reached the North Korean capital of P'yongyang before it could get any supplies through the port of Inch'on, where facilities were still devoted exclusively to outloading the X Corps.

### *Sariwon Scramble*

With action in the Kumch'on Pocket ended, in the first phase of Eighth Army's drive into North Korea, the 2d Battalion, 7th Cavalry, marched from Hanp'o-ri on Namch'onjom. (*Map 19*) Air strikes on that town at 0700, 15 October, preceded the attack. The 2d Battalion then launched its assault, supported by artillery, against fiercely defending North Koreans. After hard fighting the 2d Battalion overcame the enemy force and entered Namch'onjom at noon, losing ten men killed and thirty wounded in the battle. North Korean prisoners said that strafing attacks on Namch'onjom during the morning had destroyed the *19th Division* command post and killed the division chief of staff.[5]

---

[4] EUSAK WD, G-3 Jnl, 15 Oct 50 and Surgeon's Rpt, 12-13 Oct 50.

[5] 7th Cav Regt Opn Ord 28, 141015 Oct 50; 5th Cav Regt WD, 15 Oct 50; 1st Cav Div WD, 14-16 Oct 50; 7th Cav Regt WD, 15-16 Oct 50; EUSAK WD, G-3 Jnl, 1130 15 Oct 50; Webel, MS review comments, 13 Apr 54; Ltr, Harris to author, 7 Apr 54; Crombez, MS review comments, 12 Jan 56; Interv, author with Crombez, 12 Jan 56.

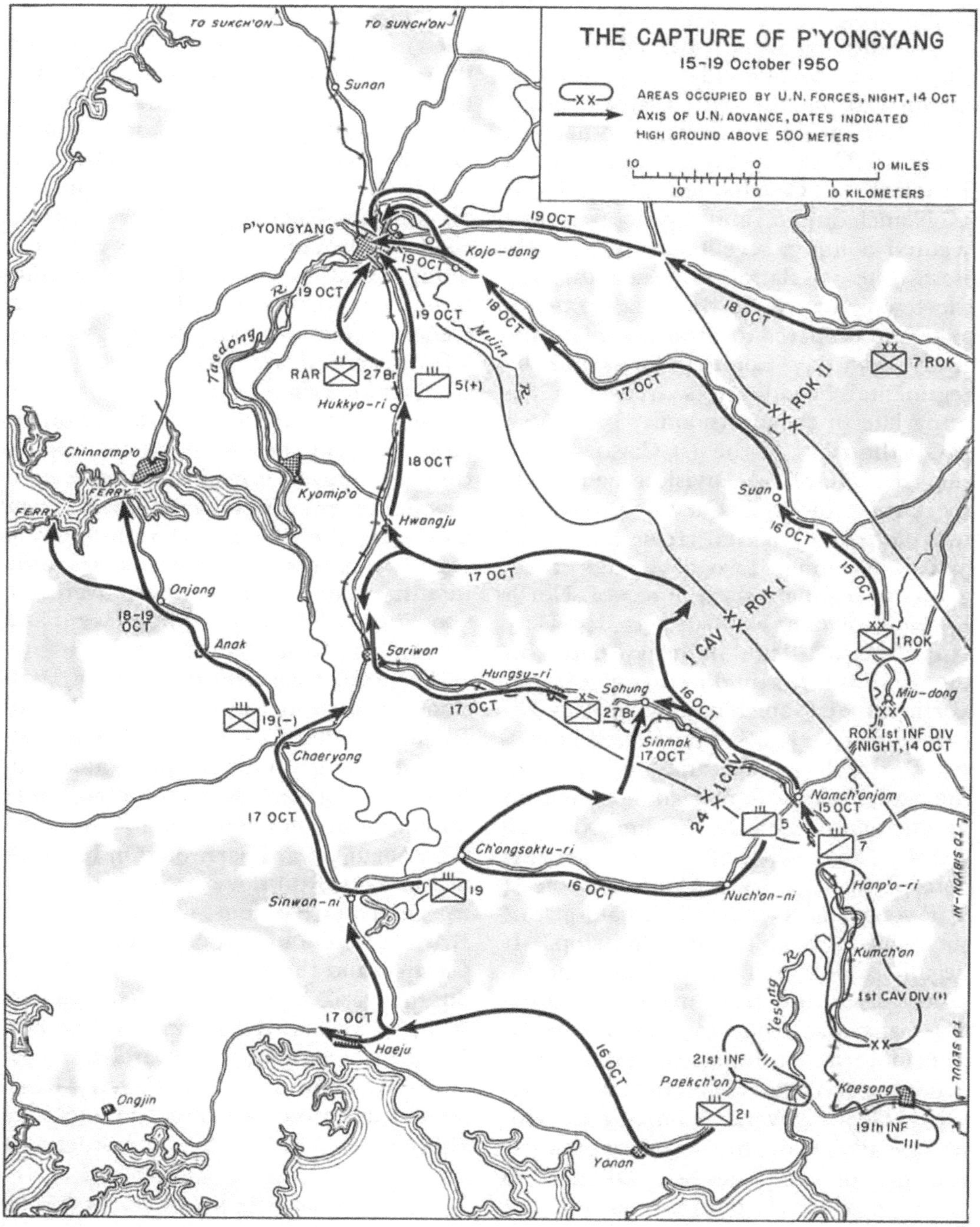
THE CAPTURE OF P'YONGYANG
15-19 October 1950
AREAS OCCUPIED BY U.N. FORCES, NIGHT, 14 OCT
AXIS OF U.N. ADVANCE, DATES INDICATED
HIGH GROUND ABOVE 500 METERS
10 MILES
10 KILOMETERS
TO SUKCH'ON
TO SUNCH'ON
Sunan
P'YONGYANG
Kojo-dong
19 OCT
18 OCT
17 OCT
16 OCT
15 OCT
7 ROK
ROK II
1 ROK
RAR
27 Br
5(+)
Hukkyo-ri
Taedong R.
Mujin R.
Chinnamp'o
FERRY
Kyomip'o
Hwangju
Suan
Onjang
18-19 OCT
Anak
Sariwon
Hungsu-ri
Sohung
ROK I
1 CAV
Miu-dong
ROK 1st INF DIV NIGHT, 14 OCT
19(-)
Chaeryong
Sinmak
24
Namch'onjom
Ch'ongsoktu-ri
Sinwon-ni
Nuch'on-ni
Hanp'o-ri
Kumch'on
Yesong R.
1st CAV DIV (+)
TO SIBYON-NI
TO SEOUL
Haeju
21st INF
Paekch'on
Kaesong
19th INF
Ongjin
Yonan
L. Booth

*MAP 19*

Torrential rains now turned the dusty roads into seas of mud, and maneuvers planned to put the 5th Cavalry in front of the retreating enemy came to naught.

On 16 October, Colonel Lynch's 3d Battalion, 7th Cavalry, led the attack out of Namch'onjom, and by noon it had secured Sohung, seventeen miles northwest. The 1st Battalion passed through the town, turned north on a secondary road, and prepared to advance on Hwangju the next day. Colonel Harris and the regimental headquarters arrived at Sohung late in the afternoon.

On the right of the 1st Cavalry Division the ROK 1st Division had made spectacular progress. On the 13th it entered Sibyon-ni, a vital crossroads northeast of Kaesong. Two days later it engaged a regiment-sized force of North Koreans, which was supported by six tanks and artillery, in heavy battle in the vicinity of Miu-dong, twelve miles northeast of Namch'onjom. Air strikes helped the ROK's. With his men following the high ground and his tanks on the road, Paik moved ahead. His division fought another battle the next day, 16 October, after which its leading elements entered Suan, forty air miles southeast of P'yongyang. General Paik said at this time that his tactics were "no stop." It began to look as if his division, the infantry afoot and traveling over secondary roads, was going to beat the American motorized columns to P'yongyang.[6]

On 15 October General Milburn reflected General Walker's impatience with what Walker thought was a slow advance. Milburn ordered the 24th Division to move into attack position on the left (west) of the 1st Cavalry Division and to seize Sariwon from the south, and then attack north toward the North Korean capital. On the same day General Gay ordered the 27th British Commonwealth Brigade to assemble behind the 7th Cavalry Regiment and be prepared to pass through it and seize Sariwon. Thus the stage was set for a continuation of the I Corps drive for P'yongyang. General Gay has said of that period, "The situation was tense, everybody was tired and nervous." [7]

Colonel Stephen's 21st Infantry of the 24th Division met just enough opposition as it moved from Paekch'on toward Haeju to prevent the infantry from mounting the trucks and rolling along rapidly as a motorized column. Its tank-infantry teams on 17 October overcame 300 North Koreans defending Haeju and secured the town that afternoon.[8]

The 19th Regiment of the 24th Division, meanwhile, trailed the 5th Cavalry Regiment. Both of them turned westward off the main highway at Namch'onjom. The 19th Infantry was to continue westward beyond Nuch'on-ni and then turn north toward Sariwon. On the 16th a bad traffic jam developed on the road up to Namch'onjom where the 27th British Commonwealth Brigade, the 5th Cavalry, and the 19th Regiment were all on the road. For long periods the vehicles moved slowly, bumper to bumper. From Namch'onjom westward, the 19th Infantry, behind the 5th Cavalry Regiment, was powerless to accelerate its pace although General Church had ordered it

[6] EUSAK WD, POR 279, 13 Oct and POR 289, 16 Oct 50; *Ibid.*, G–3 Jnl, 1130 15 Oct 50.

[7] Ltr, Gay to author, 23 Jan 54; I Corps Opn Dir 12, 151000 Oct 50; 24th Div WD, 15 Oct 50; Linklater, *Our Men in Korea*, p. 22.

[8] 24th Div WD, 16–19 Oct 50; EUSAK WD, G–3 Sec, 17 Oct 50.

to do so. Word came at this time that General Milburn had told Generals Gay and Church that whichever division—the 1st Cavalry or the 24th Infantry—reached Sariwon first would thereby win the right to lead the corps attack on into P'yongyang. The 24th Division was handicapped in this race for Sariwon, as it had a roundabout, longer route over inferior roads and poorer supply routes.[9]

A dominant characteristic of all units in the advance at this time was the strong rivalry prevailing between divisions, and even between units within a division, to gain the most ground and be the first to reach the North Korean capital. Flare-ups between units were frequent and nerves were taut.

One such flare-up occurred before dawn of 17 October. On the preceding afternoon two battalions and the regimental headquarters of the 7th Cavalry reached Sohung. The 3d Battalion held the town and together with F Company established roadblocks there. The 27th British Commonwealth Brigade was to pass through it the next morning in attack along the main highway to Sariwon. Holding a roadblock south of Sohung was Capt. Arthur H. Truxes, Jr., with F Company. Colonel Harris in posting his roadblock forces gave them orders to shoot at anything moving in front of the perimeter during the hours of darkness. He says he had no information that the 5th Cavalry was making a night approach toward his position. Captain Webel, S–3 of the 7th Cavalry Regiment, says that he told the 5th Cavalry liaison officer with the regiment of the roadblock forces and their orders to shoot, and asked him to return to the 5th Cavalry and inform it of the situation. This officer did not do that, however, but stayed in the 7th Cavalry command post overnight. The leading elements of the 1st Battalion, 5th Cavalry Regiment, approached the 7th Cavalry outpost one mile south of Sohung at 0300 the morning of 17 October, and a fire fight broke out between them, each thinking the other an enemy force. Before the mistake could be corrected, 7th Cavalry fire wounded seven men of the 5th Cavalry.[10]

On 17 October, with the 1st Battalion in the lead, the 7th Cavalry Regiment followed the secondary "cow path" road north from Sohung in a circuitous route toward Hwangju where it would strike the main P'yongyang highway north of Sariwon. The 27th British Commonwealth Brigade passed through the lines of the 7th Cavalry that morning at Sohung and took up the advance along the main highway toward Sariwon.

Sariwon lay some thirty miles up the highway almost due west from Sohung. At Sariwon the highway and railroad debouched from the mountains, turned north and ran through the coastal plain

[9] 24th Div WD, 16 Oct 50; 1st Cav Div WD, 16 Oct 50; 7th Cav Regt WD, 16 Oct 50; Crombez, MS review comments, 12 Jan 56; Ltr, Crombez to author, 12 Oct 54; Interv, author with Crombez, 12 Jan 56.

[10] 5th Cav Regt Unit Jnl, msg 181, 0630 17 Oct 50; 5th Cav Regt WD, 16–17 Oct 50; Ltrs, Harris to author, 23 Dec 53 and 7 Apr 54; Gay, MS review comments for author, 13 Mar 54; Ltr, Crombez to author, 12 Oct 54; Interv, author with Crombez, 12 Jan 56; Crombez, MS review comments, 12 Jan 56; Interv, author with Maj Gen Frank A. Allen, Jr., 28 Jan 54; Webel, MS review comments, 15 Nov 57. This episode is confused and the principals do not agree on all details. Captain Truxes' account of this incident was unobtainable as he was killed in action when the Chinese entered the war.

to P'yongyang, thirty-five miles away. Only occasional low hills lay across the road between Sariwon and P'yongyang. It was generally expected that the North Koreans would make their stand for the defense of P'yongyang, short of the city itself, on the heights before Sariwon.

A platoon of Maj. David Wilson's A Company of the Argyll 1st Battalion, mounted on American Sherman tanks, formed the point as the Argylls led the attack. Brig. Gen. Frank A. Allen, Jr., Assistant Division Commander, 1st Cavalry Division, accompanied the Argylls. Groups of haggard and hungry North Korean soldiers stood along the roadside waiting for a chance to surrender, and Russian-made trucks, their gas tanks empty, stood abandoned. Four miles short of Sariwon, on the hills guarding the approach to the town, it looked for a while as if the anticipated big battle had started. Enemy rifle fire suddenly burst on the column from a hillside apple orchard, 200 yards away. The column stopped and the men sought cover.

Behind the lead tanks, General Allen jumped from his jeep, stamped along the road, waved a map and shouted, "They're in that orchard, rake 'em, blast them out of there!" The general's aide, 1st Lt. John T. Hodes, climbed on one of the tanks and trained his glasses on the orchard to give fire direction. The pilot of a spotter plane above the ridge dipped his wings to indicate the presence of the enemy in force. A few North Koreans started running from the orchard when the tanks began firing into it. Suddenly, a mass of North Koreans broke from the orchard, rushed for the ridge line, and vanished over the top. Wilson's A Company of the Argylls moved on the orchard and swept it clean of remaining enemy troops. They killed about 40 and captured others in this brief action. The fleeing North Koreans left behind ten machine guns and, in the pass, they abandoned a battery of antitank guns. The British now entered Sariwon, a large town, which they found to be badly damaged by bombing. Their loss thus far for the day was 1 man killed and 3 wounded.[11]

About 1700 in the afternoon the Australian 3d Battalion passed through the Argylls in the town and advanced five miles north of it toward Hwangju. There the Australians went into a perimeter blocking position in front of a range of hills strongly held by the enemy, and prepared to attack in the morning.

Now began a succession of weird events in what proved to be a chaotic night in Sariwon. A British reconnaissance group south of the town encountered a truckload of North Korean soldiers driving north. The North Koreans shot their way through and continued into the town, but, finding the northern exit closed, they turned back and met the reconnaissance group again. In this second encounter, the reconnaissance party killed about twenty of the enemy troops.

A little later, Lt. Col. Leslie Nielson, commanding officer of the Argyll 1st Battalion, driving in the gloom near the southern end of Sariwon, was suddenly amazed to see coming toward him on either side of the road a double file of

[11] Maj Gen B. A. Coad, "The Land Campaign in Korea," *op. cit.;* Linklater, *Our Men in Korea*, pp. 22–23; Bartlett, *With the Australians in Korea*, pp. 27–28; Ltr, Gay to author, 23 Jan 54; 1st Cav Div WD, 17 Oct 50; Charles and Eugene Jones, *The Face of War*, pp. 150–51; New York *Herald Tribune*, October 17, 1950.

North Korean soldiers. The leading soldiers fired at him but missed. Nielson shouted to his driver, "Put your foot on it!" The driver did, and raced four miles through the marching North Koreans. Clearing the last of them, Nielson and his driver took to the hills and stayed there until morning. This enemy force, fleeing in front of the 19th Infantry, 24th Division, and approaching Sariwon from the south, did not know the town had already fallen to U.N. units.

There were many times during that wild night in Sariwon when U.N. soldiers thought the North Koreans were South Koreans coming up from the south with the 24th Division, and the North Koreans thought the British were Russians. There were several instances of mutual congratulations and passing around of cigarettes. One group of North Koreans greeted a platoon of Argylls with shouts of "Comrade!" and, rushing forward in the dim light, slapped the Scots on the back, offered cigarettes, and gave them the red stars from their caps as souvenirs. The ensuing fight was at very close quarters.

Lt. Robin D. Fairrey, the Argylls' mortar officer, walked around a corner into a group of North Koreans. Maintaining his composure, he said to them, "Rusky, Rusky," and after receiving several pats on the back, turned another corner and got away.

During this scrambled night at Sariwon about 150 North Koreans were killed; strangely enough, the British lost only one soldier. Most of the North Koreans passed through the town. North of it the Australian 3d Battalion reaped a harvest, capturing 1,982 North Korean soldiers at its roadblock. Maj. I. B. Ferguson played a leading role in capturing this large number of enemy troops. When the first of them came up to the Australian outpost a night battle seemed imminent. Ferguson mounted a tank and called out in the gloom for the North Koreans to surrender, telling them they were surrounded. After some hesitation, the leading enemy unit dropped its arms and surrendered, and most of the others followed its example.[12]

During the day, while the 27th British Commonwealth Brigade advanced on Sariwon along the main highway, the 7th Cavalry Regiment, with Colonel Clainos' 1st Battalion in the lead, hurried along the poor secondary roads through the hills north of it. This column was about three miles from Hwangju and the main highway above Sariwon when at 1600 in the afternoon it received a message General Gay dropped from a light plane. The message said that the roads out of Sariwon were crowded with hundreds of North Korean soldiers, and it directed Colonel Clainos to have one battalion of the 7th Cavalry turn south at Hwangju on the main highway to meet the British and help trap the large numbers of enemy soldiers in the Sariwon area, while another battalion turned right and held the town of Hwangju. Clainos and the two battalion commanders agreed that the 1st Battalion would turn to meet the British and the 2d Battalion would hold Hwangju.[13]

Soon after turning south on the Sariwon-P'yongyang highway the leading

---

[12] Coad, "The Land Campaign in Korea," *op. cit.;* Linklater, *Our Men in Korea;* 1st Cav Div WD, 17–18 Oct 50; Bartlett, *With the Australians in Korea,* p. 29; New York *Herald Tribune,* October 20, 1950.

[13] Ltrs, Gay to author, 23 Jan and 13 Mar 54; Clainos, MS review comments, 24 May 54.

elements of the 1st Battalion captured an enemy cavalry detachment and thirty-seven horses.

A little later the battalion came under fire from the enemy on the hill barrier ahead and separating it from the Australians. The battalion's motorized point had a short skirmish with an enemy group during which its South Korean interpreter, although wounded, tried and indeed succeeded in reaching the North Korean forward position. He told the North Koreans that the column they were fighting was Russian. The enemy platoon thereupon came up to the 7th Cavalry's point, which Colonel Clainos had just joined. Clainos turned the enemy group over to a squad leader who proceeded to disarm it. Finding that they had been tricked, some of the enemy tried to resist. This ended when the squad leader knocked one of the North Koreans into a ditch.

The enemy platoon's surrender took place in clear daylight and was observed by hundreds of North Korean soldiers in the nearby hills. Almost immediately, enemy soldiers from the eastern side of the position began pouring in to surrender. On the western side, however, small arms fire continued until dark when many there also came out to surrender. Altogether, more than 1,700 North Korean soldiers and thirteen female nurses surrendered to the 1st Battalion that evening.

Colonel Clainos had established radio communication with the Australians about 1800. At 2230, he radioed Colonel Green of the Australian battalion that, with vehicle lights on, he was coming through the pass with his battalion and prisoners. An hour before midnight the 1st Battalion, 7th Cavalry, reached the Australian perimeter. There, Colonel Clainos overheard one Australian soldier saying to another, "Now what do you make of this? Here we are all set for a coordinated attack in the morning, and the bloody Yanks come in at midnight from the north, with their lights burning, and bringing the whole damned North Korean Army as prisoners." [14]

### *Into P'yongyang*

It had become clear by the time the U.N. troops reached Sariwon that the remaining North Korean forces could not attempt a strong defense of P'yongyang without incurring total destruction or capture. The North Koreans by this time not only had to contend with the U.S. I Corps, approaching the capital city along the main Seoul axis from the south, but also the enveloping movements of the ROK Army forces from the southeast and east. Some of these forces, if they continued their rapid advance for a few days more, would almost certainly cut on the north the highways and exits from the doomed city. P'yongyang would then be surrounded and any forces retained in and around the city for its defense would face either destruction or surrender.

The flanking operation originally conceived by General MacArthur for the X Corps after it had landed on the east coast at Wonsan had, in fact, been carried out by ROK Army units under Eighth Army control before a single soldier of X Corps landed in the east. By

[14] Clainos, MS review comments for author, 24 May 54; Coad, "The Land Campaign in Korea," *op. cit.;* Linklater, *Our Men in Korea;* 7th Cav Regt WD, 17 Oct 50; 1st Cav Div WD, 17 Oct 50; Ltrs, Gay to author, 23 Jan and 13 Mar 54.

evening of 17 October four ROK divisions were racing each other, as well as the American and British units of the U.S. I Corps, to be first in reaching P'yongyang. The ROK 1st Division, only fifteen miles away to the southeast, was closest of all U.N. units to the city. On its right flank, the ROK 7th Division was swinging toward P'yongyang from the east. Still farther east the ROK 8th Division had almost reached Yangdok in the central mountains where it would turn west on the P'yongyang-Wonsan lateral road. And, finally, the ROK 6th Division was just short of Yangdok on this road, fifty air miles east of P'yongyang, after having turned west on 15 October from Wonsan on the coast, which it had reached by the road from Hwach'on. Thus, the U.S. I Corps was nearing P'yongyang from the south and southeast, the ROK 7th Division from the southeast, and the ROK 8th and 6th Divisions from the northeast. With approximately seven U.N. divisions converging on P'yongyang, obviously the North Korean Army in its state of depletion, disorganization, and demoralization could not hold the city.[15]

The Eighth Army G-2 estimated on 17 October that less than 8,000 effectives of the N.K. *32d* and *17th Divisions* were available for defense of P'yongyang. The estimate concluded that the enemy would undertake a token defense of the city while the main force withdrew northward across the Ch'ongch'on River for further operations.[16]

The 1st Cavalry Division had won the honor of leading the attack into P'yongyang when the British 27th Brigade, attached to it, beat the 24th Division into Sariwon. Leading elements of the 19th Infantry Regiment, 24th Division, were still several miles south of Sariwon when orders came at 1700 on 17 October to stop and hold up the attack because U.N. troops were already in the town. Morale in the 1st Cavalry Division was high. Most of the soldiers heard and passed on a rumor that the city was their final objective in the war, and once it was taken the American troops would leave Korea. Most of them expected to eat Thanksgiving Day dinner in Japan.[17]

Since the 7th Cavalry Regiment was the unit farthest north, General Gay ordered it to resume the advance on P'yongyang at daylight 18 October. The 3d Battalion at Hwangju became the assault battalion even though its men were tired from their long night movement to the town. At daylight on the 18th the battalion crossed the ford in Hwangju and began the advance. Resistance was light until the leading elements of the battalion arrived in front of the high ground south of Hukkyo-ri, halfway to P'yongyang. There enemy high velocity gun and heavy 120-mm. mortar fire struck the column. Captain Webel, the regimental S-3, estimated that a reinforced battalion of about 800 men held the prepared enemy defensive positions.

Twenty tanks of C Company, 70th Tank Battalion supported the battalion, but they had to contend with fire from three or four dug-in enemy tanks and a mined roadway. In the midst of the fighting, enemy small arms fire shot down an F-51 fighter plane. General Milburn, the corps commander, watched the ac-

[15] See EUSAK WD and POR's, 12-17 Oct 50, for movements and positions of ROK units.

[16] EUSAK PIR's 95, 15 Oct, and 97, 17 Oct 50.

[17] 1st Cav Div WD, 18 Oct 50; Interv, author with Crombez, 12 Jan 56; 24th Division WD, 17 Oct 50.

TANK-SUPPORTED CONVOY *winds through the safety channel of a mined area en route to P'yongyang.*

tion from an apple orchard at the side of the road, and about midafternoon General Gay came up and joined him. Dissatisfied with the progress of the attack, Gay ordered the regimental commander, Col. James K. Woolnough, who had temporarily replaced Colonel Harris, to start the other two battalions on flank movements against the enemy-held ridge. Captain Webel protested to General Gay that the enemy position was all but taken and that commitment of the other two battalions was unnecessary. But Gay let the order stand when he learned from Woolnough that the latter had already started to implement it. The two battalions upon coming up moved off toward the enemy flanks in what proved to be a nightlong movement. The next morning they found the enemy positions abandoned.

After giving the order on the 18th for a full regimental attack on the Hukkyo-ri position, General Gay informed Colonel Woolnough that the 5th Cavalry Regiment would pass through the 7th Cavalry the next morning and take up the attack on P'yongyang. He then went back and found Colonel Crombez and gave him the order. The 5th Cavalry Regiment was still strung out on the mountainous secondary road it had been traveling behind the 7th Cavalry from Sohung to Hwangju. Crombez did not have the last battalion in bivouac until 2300 that night.[18]

At 0500 on 19 October Lt. Col. Paul Clifford's 2d Battalion, 5th Cavalry Regiment, led north out of Hwangju. When it arrived at the 7th Cavalry lines at Hukkyo-ri those troops had just repulsed an enemy counterattack. At this point three enemy tanks rumbled up. A 5th Cavalry bazooka team led by a young Italo-American boy knocked out these tanks. Questioned about the exploit a little later, the boy explained, "Me and my two buddies were sitting over there behind that rock. These tanks came up toward us and stopped right out there on the road. They raised their turrets and started talking to each other. One of my buddies said, 'Christ, them ain't GI's, them are Gooks,' and I said, 'Let's shoot the S.O.B.'s' and that is what we did." [19]

---

[18] Webel, MS review comments, 13 Apr 54; Interv, author with Lynch, 9 Jun 54 (Lynch commanded the 3d Bn at Hukkyo-ri); Interv, author with Crombez, 28 Jun 55; Ltr, Harris to author, 8 Dec 53; Ltrs, Gay to author, 23 Jan and 19 Apr 54; Interv, author with Clainos, 30 Apr 54; 1st Cav Div WD, 17–18 Oct 50.

[19] Ltr, Gay to author, 23 Jan 54; Ltr, Capt James H. Bell (CO F Co, 5th Cav Regt Oct 50) to author, 11 Apr 56; Interv, author with Crombez, 28 Jun 55; 5th Cav Div Regt WD, 19 Oct 50.

The author has been unable to identify this boy, who reportedly was killed later.

F Company, led by 1st Lt. James H. Bell, reinforced with five tanks, a platoon of engineers, and a section of heavy machine guns, now passed through the 7th Cavalry and led the 5th Cavalry Regiment toward P'yongyang. Just as Bell was passing the first of the burning enemy tanks a friendly plane swooped down and rocketed it. The concussion almost made him a casualty.

Flights of jet planes coursed overhead in advance of F Company and, on at least two occasions, they helped supporting artillery reduce enemy forces that threatened to delay its advance. The regimental commander, Colonel Crombez, and a small command group followed immediately behind F Company most of the morning and pushed it hard.

At 1102, Lieutenant Bell's F Company reached the 20-yard-wide Mujin-ch'on River, a tributary of the Taedong at the southern edge of P'yongyang. North Korean troops from behind a 20-foot embankment on the north side defended the highway bridge over it with three antitank guns. Bell's troops were delayed there for about half an hour until their mortar fire caused the North Korean gun crews to abandon the antitank guns. Bell's F Company then crossed the Mujin-ch'on and entered the southwestern edge of P'yongyang just after 1100.[20]

P'yongyang is the oldest city in Korea, and for a long time was its capital. Its population at the outbreak of the war was approximately 500,000. The city is situated astride the Taedong River, one of the larger streams of Korea, forty miles from where it empties into the Yellow Sea. The main part of the city with the important public buildings lay on the north side of the river. A large, relatively new industrial suburb sprawled opposite on the south side. Two railroad bridges of the Pusan-Seoul-Mukden railroad cross the Taedong River here. Upstream from them about two miles was the main highway bridge. The Taedong at P'yongyang averages about 400–500 yards in width. As the current is swift, it constitutes a major military obstacle to north-south movement.

BURNING ENEMY TANK *knocked out by 5th Cavalry bazooka team at Hukkyo-ri, 19 October.*

Bell received orders to turn west and seize certain factory buildings, the railroad bridges, and a bridgehead on the north bank of the Taedong. In about

[20] Ltr, Bell to author, 8 Mar 54; Ltr, Crombez to author, 12 Oct 54; Interv, author with Crombez, 28 Jun 55; 5th Cav Regt WD, 19 Oct 50; 1st Cav Div WD, 19 Oct 50; EUSAK WD, Br for CG, 19–20 Oct 50.

Bell estimates the time he entered the south edge of P'yongyang as 1330. The official records, based on an aerial observer's report, give it as 1102.

5TH CAVALRY TROOPS *at the southern edge of P'yongyang, 19 October.*

half an hour he reached the river's southern bank and found that only one span of each of the two railroad bridges (each 3-span) was intact. After a hasty examination of the eastern bridge, Bell decided that infantry could cross on one of its spans to an island in the river. Leaving some riflemen and the Engineer platoon at its southern end to guard the tanks which gave supporting fire, he led the rest of F Company across to the island and secured it by midafternoon. While F Company was crossing to the island, enemy on the north bank destroyed a section of the bridge still intact there. During the afternoon the 3d Battalion, 5th Cavalry, crossed to the island and relieved F Company, which then moved back to the airfield on the south bank.

While F Company was trying to seize the railroad bridges over the Taedong, the rest of the 2d Battalion crossed the Mujin-ch'on and turned right toward the main highway bridge which crossed the Taedong River about midway on the city waterfront. This was the only bridge still intact on 19 October when U.N. troops entered P'yongyang. When the leading elements of E and G Companies neared the bridge the North Koreans blew up the center span.

Almost simultaneously with the 1st Cavalry Division's arrival at P'yongyang the ROK 1st Division entered the city on the Sibyon-ni–P'yongyang road at a point northeast of the 1st Cavalry Division. On the night of 18 October the chances had appeared excellent for the ROK 1st Division to be first into P'yongyang. After a day of very heavy fighting in which it gained two miles, it was only eight miles away. The leading elements of the 1st Cavalry Division were about 30 miles away. But the North Koreans made a stronger fight against the ROK 1st Division than against the 1st Cavalry Division, possibly because it was closer to the city and the more immediate threat. Also, the road on which the ROK's approached P'yongyang was heavily mined with both antipersonnel and antitank mines. Paik's division fought throughout the rainy night and finally overcame an enemy strongpoint an hour or two after daybreak. Enemy emplacements and automatic fire stopped the ROK infantry again about six miles from the city near Kojo-dong. Tanks of C Company, 6th Tank Battalion, in the ensuing ROK attack enveloped the enemy positions from both flanks, destroyed self-propelled guns, and overran the North Korean entrenchments, physically crushing machine guns and enemy soldiers. It was estimated that the tanks in this action killed nearly 300 North Koreans.

According to General Paik, extensive mine fields in the street behind the overrun enemy positions delayed the tanks, but the infantry of the ROK 2d Battalion, 12th Regiment, kept moving and General Paik affirms that they arrived at the edge of the Taedong River just before 1100 and deployed along the south bank northeast of the highway bridge. Leading elements of the 2d Battalion, 5th Cavalry Regiment, arrived at the traffic circle 100 yards east of the highway bridge almost at the same time. The leading tanks of C Company, 6th Tank Battalion, were in the southern edge of the city, according to their own records, at 1245. Tanks of D Company, 6th Medium Tank Battalion, entered the city along the same approach a little later, turned north, and together with troops of the ROK 11th Regiment secured the airfield at 1440. Other ROK units earlier had secured a smaller airfield a few miles to the east.[21]

CAPITOL BUILDING IN P'YONGYANG. *The men having coffee are members of Task Force Indianhead.*

After the North Koreans blew the highway bridge across the Taedong, elements of the 2d Battalion, 5th Cavalry, continued northeast along the river searching for a ford reported to be located there. When they found it a few miles east of the city they discovered that elements of the 15th Regiment, ROK 1st Division, already had crossed the river there, and others were then in the act of crossing into the main part of the city. Later, Colonel Crombez asked General Paik how his troops found the ford so quickly. Paik answered, "I am a native of P'yongyang. I know the fords." [22]

By dark most of the ROK 1st Division was in the main part of P'yongyang north of the Taedong River. Nor was that all. The 8th Regiment of the ROK 7th Division swung into north P'yongyang from the east and was in possession of Kim Il Sung University in the northern part of the city by 1700.[23]

The next day, 20 October, the ROK

[21] 6th Med Tk Bn WD, 19 Oct 50; EUSAK PIR 99, 19 Oct 50; EUSAK WD, 19 Oct 50, and G-3 Jnl, 1300-1600 19 Oct 50; 5th Cav Regt WD, 19 Oct 50; 10th AAA Group WD, 19-20 Oct 50; 1st Cav Div WD, 19 Oct 50; I Corps WD, Oct 50, p. 18; Gen Paik Sun Yup (CofS ROKA), MS review comments, 11 Jul 58.

[22] Ltr, Bell to author, 11 Apr 56; Interv, author with Crombez, 12 Jan 56.

[23] Interv, author with Schwarze (KMAG adviser with ROK 7th Div Oct 50), 3 Feb 54; 5th Cav Regt WD, 19-20 Oct 50; EUSAK WD, Br for CG, 190001-200800 Oct 50; EUSAK POR 299, 20 Oct 50.

1st Division advanced into the heart of the city and took the strongly fortified administrative center with ease. The enemy troops posted there were too demoralized to fight and they abandoned both guns and entrenchments. At 1000 the ROK 1st Division reported the entire city had been secured, including the City Hall, the Provincial Government offices, and the N.K. People's Committee offices. The ROK 8th Regiment aided the 1st Division by sweeping through the northwest section of the city and clearing it of the enemy. As soon as Engineer assault boats could be brought up, the 3d Battalion, 5th Cavalry, began crossing to the north side of the Taedong, and by noon that regiment, with the 3d Battalion, 7th Cavalry, attached to it, was across the river. Bells in Christian churches pealed a welcome. The people appeared friendly and there were no disturbances.[24]

When the operations of Eighth Army had progressed to the point where it appeared probable that P'yongyang would fall in the near future, the army on 16 October had organized a special task force known as Task Force Indianhead. Its name derived from the shoulder patch of the 2d Infantry Division. This task force was to enter the North Korean capital with the advance units of the 1st Cavalry Division. Its mission was to secure and protect specially selected government buildings and foreign compounds until they could be searched for enemy intelligence materials. Lt. Col. Ralph L. Foster, Assistant Chief of Staff for G–2, 2d Division, commanded the task force, which was built around K Company, 38th Infantry Regiment, and six tanks of C Company, 72d Medium Tank Battalion, and included Engineer demolition troops, automatic weapons vehicles of the 82d AAA Battalion, and counterintelligence troops. The task force secured most of its assigned objectives in P'yongyang on 20 October. It obtained a considerable amount of intelligence material, both military and political, which was turned over to a special team from GHQ, Far East Command, and transported by air to Tokyo.[25]

Twenty American prisoners escaped or were rescued from the North Koreans in the capture of P'yongyang. Most of the large number of prisoners held there, however, had been taken northward several days before the U.N. forces entered the city.

General Gay established his 1st Cavalry Division headquarters in the granite buildings of the North Korean Military Academy ten miles southwest of P'yongyang on the Chinnamp'o road. He was responsible for the internal security and order of P'yongyang after its capture. On 23 October he appointed Colonel Crombez civil assistance officer in the city because of the latter's special knowledge of the country and its people. Colonel Johnson, a veteran of Bataan, replaced Crombez in command of the 5th Cavalry Regiment until 14 December.

[24] EUSAK WD, G–3 Jnl, 1200 20 Oct 50; *Ibid.*, Br for CG, 20–21 Oct 50; I Corps WD, 20 Oct 50; Ltr, Crombez to author, 12 Oct 54.

[25] Ltrs, Foster to author, 11 and 21 May 54; Ltr, Gay to author, 13 Feb 54; EUSAK POR 292, 17 Oct 50; EUSAK WD, Br for CG, 190001–200800 Oct 50; 2d Div WD, Summ, 1 Sep–31 Oct 50, vol. II, pp. 47–49.

The 5th Cavalry Regiment was disposed in the southern outskirts of P'yongyang, the 8th Cavalry Regiment in the northern outskirts, and the 7th Cavalry Regiment at Chinnamp'o, P'yongyang's port. After the fall of P'yongyang, Colonel Harris had led the 7th Cavalry Regiment in a forced night movement from the city thirty-five miles southwest to Chinnamp'o. The regiment entered the port city in the dead of night, 22 October.

On 24 October, General Walker took personal command of his advance Eighth Army headquarters, established two days earlier by Colonel Collier of his staff, in the attractive and undamaged gray brick building in P'yongyang which had been the headquarters of Premier Kim Il Sung.[26]

KIM IL SUNG'S DESK. *Colonel Foster occupies the North Korean Premier's office in P'yongyang. Note portrait of Stalin.*

On 21 October a touching and revealing ceremony occurred on the P'yongyang airfield. General MacArthur had flown in from Tokyo to confer briefly with Generals Walker and Stratemeyer after the fall of the North Korean capital. In the course of his brief visit he reviewed F Company, 5th Cavalry Regiment, which had been the first American unit to enter P'yongyang. He asked all men in the company who had landed with it in Korea ninety-six days earlier, when it numbered nearly 200 men, to step forward. Only five men stepped forward; three of them had been wounded.[27]

---

[26] 5th Cav Regt WD, 22–23 Oct 50; GHQ UNC press release, 25 Oct 50. Ltr, Harris to author, 7 Apr 54; EUSAK WD, Br for CG, 22 Oct 50; 7th Cav Regt WD, 22–23 Oct 50.

[27] Ltr, Gay to author, 23 Jan 54; Crombez, MS review comments, 28 Jun 55.

## CHAPTER XXXII

# Up to the Ch'ongch'on

It is only common sense to say that we cannot hope to build up a true doctrine of war except from true lessons, and the lessons cannot be true unless based on true facts, and the facts cannot be true unless we probe for them in a purely scientific spirit.

Basil Henry LIDDELL HART, *The Ghost of Napoleon*

### *Airborne Attack: Sukch'on and Sunch'on*

When Eighth Army crossed the 38th Parallel and drove on P'yongyang, General MacArthur held the 187th Airborne Regiment, commanded by Col. Frank S. Bowen, Jr., in GHQ reserve at Kimpo Airfield near Seoul. He planned to employ the airborne troops in a drop north of P'yongyang in an attempt to cut off North Korean officials and enemy troops, and to rescue American prisoners of war who it was assumed would be evacuated northward when the fall of the North Korean capital seemed imminent.

After changing the date a time or two, General MacArthur set the airdrop for the morning of 20 October. There were to be two drop zones 30 air miles north of P'yongyang, the principal one at Sukch'on and the other at Sunch'on.[1] Two highways run north from P'yongyang like the sides of a narrow capital letter V, each roughly paralleling a rail line. The main highway from P'yongyang to the Yalu River and the Manchurian border at Sinuiju forms the left-hand side of the V. Sukch'on on this highway is situated in a wide valley surrounded by low hills, about 35 road miles north of P'yongyang. The right-hand road passes through rougher terrain to reach Sunch'on on the Taedong River, 17 air miles east of Sukch'on. (*Map 20*)

The airborne regiment turned out in a heavy rain for reveille at 0230 in the after-midnight darkness of 20 October. The men ate breakfast and then went to the airfield where they waited in the downpour for the weather to improve. Shortly before noon the sky began to clear. The regiment loaded into 113 planes, C–119's and C–47's of the 314th and 21st Troop Carrier Squadrons based in Japan. The planes were crowded—a typical C–119 carried 46 men in 2 sticks of 23 men each, 15 monorail bundles, and 4 door bundles. Each man had a main parachute, a .45-caliber pistol, and a carbine or M1 rifle.

The first aircraft, carrying Colonel

[1] 187th Abn RCT Unit Hist, pt. II, 17–19 Oct 50; I Corps WD, 20 Oct 50; EUSAK WD, 20 Oct 50.

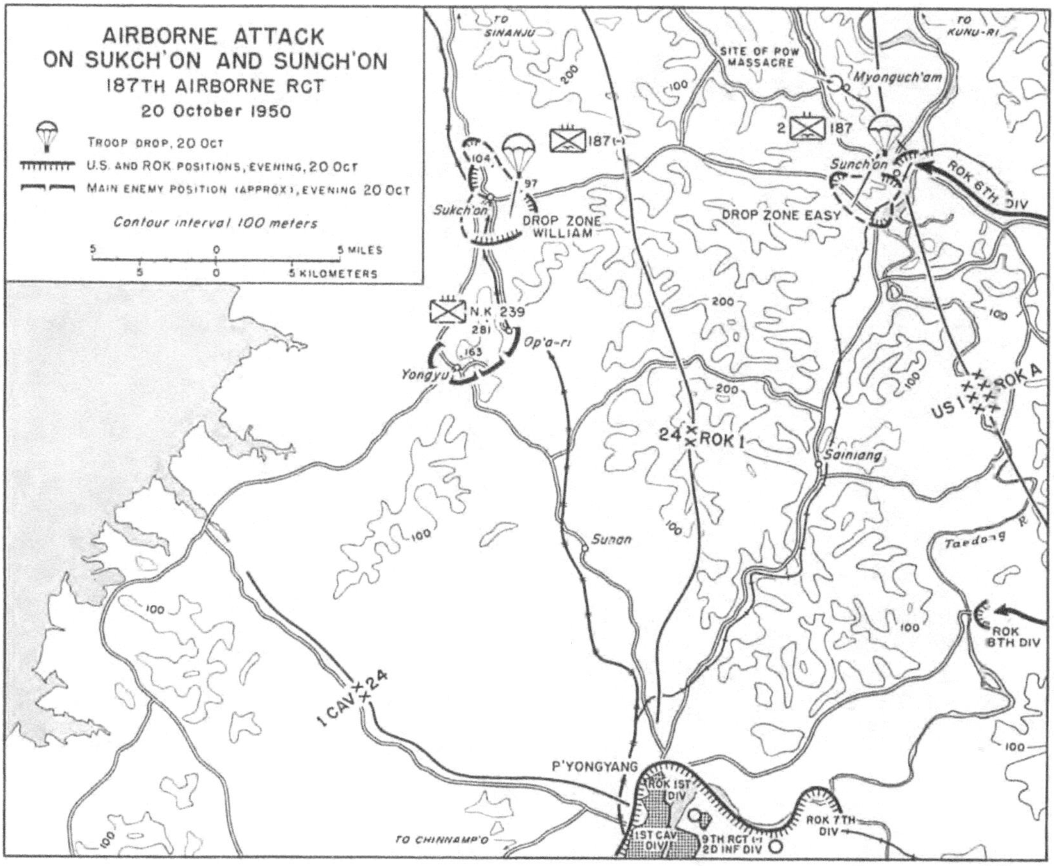


F. Temple

*MAP 20*

Bowen, was airborne at noon. When all the planes had assembled over the Han River estuary, they turned north along the west coast of Korea. This flight carried about 2,800 men. Recent intelligence had informed the airborne force that a trainload of American prisoners, traveling only at night and then slowly, was on its way north from P'yongyang. Colonel Bowen's men hoped to intercept this train and rescue the prisoners.

As the troop carriers approached the drop zone, fighter planes preceded them rocketing and strafing the ground. At approximately 1400 the first troops began dropping from the lead planes over Sukch'on. There was no enemy antiaircraft fire and only occasional sniper fire came into the drop zone. This first drop put Colonel Bowen and 1,470 men of the 1st Battalion, regimental Headquarters and Headquarters Company, and supporting engineer, medical, and service troops on the ground in Drop Zone William, southeast of Sukch'on. Twenty-five men were injured in this jump. One group landed a mile and a half east of the drop zone and lost one man killed

MASS AIRDROP NEAR SUNCH'ON

in his parachute by attacking enemy soldiers. Seventy-four tons of equipment were dropped with the men.[2]

After the troop drop came that of the heavy equipment—equipment organic to an airborne infantry regiment, including jeeps, 90-mm. towed antitank guns, 105-mm. howitzers, and a mobile radio transmission set equivalent in weight to a 2½-ton truck. Seven 105-mm. howitzers of the 674th Field Artillery Battalion and 1,125 rounds of ammunition were in the drop. Six of the howitzers were recovered in usable condition. About 90 percent of the shells were undamaged and none exploded. This was the first time heavy equipment had been dropped in combat, and it was the first time C–119's had been used in a combat parachute operation.

The 1st Battalion, against only light resistance, seized Hill 97 east of Sukch'on, where Colonel Bowen established his command post, and Hill 104 north of the town, cleared the town of Sukch'on itself, and set up a roadblock north of it.

In the meantime, the 3d Battalion had jumped in the same zone, turned south, taken up defensive positions on low hills two miles south of the town, and established roadblocks across the

[2] 187th Abn RCT Unit Hist, pts. I and II, 20 Oct 50.

ARTILLERY AIRDROP NEAR SUKCH'ON

highway and railroad at that point. It seized its objectives by 1700, killing five enemy soldiers and capturing forty-two others without loss to itself.

In the second jump area the 2d Battalion at 1420 began parachuting onto Drop Zone Easy, two miles southwest of Sunch'on. Twenty men were injured in this jump. The battalion secured its objective by night against virtually no resistance. Two companies established roadblocks south and west of Sunch'on. A third advanced to the town and established contact there with elements of the ROK 6th Division which had reached Sunch'on from the southeast in its push toward the Ch'ongch'on River.

During this and succeeding days, a total of approximately 4,000 troops and more than 600 tons of equipment and supplies were dropped at Sukch'on and Sunch'on. Included in the equipment were 12 105-mm. howitzers, 39 jeeps, 38 ¼-ton trailers, 4 90-mm. antiaircraft guns, 4 ¾-ton trucks, and 584 tons of ammunition, gasoline, water, rations, and other supplies.

On the morning after the airdrop, the 1st Battalion, 187th Airborne Regiment gained the dominant terrain it needed directly north of Sukch'on to carry out its mission of blocking the main highway running north. Strong enemy rear guard forces held the next line of hills northward. That afternoon elements of the 1st Battalion established contact with the 2d Battalion at Sunch'on.

General MacArthur, accompanied by Generals Stratemeyer, Wright, and Whitney, had flown from Japan to watch the airdrop. After seeing the parachute troops land and assemble successfully, he flew to P'yongyang. There he commented to reporters that the airborne landing seemed to have been a complete surprise to the enemy. He estimated that 30,000 North Korean troops, perhaps half of those remaining in North Korea, were caught in the trap between the 187th Airborne Regiment on the north and the 1st Cavalry and ROK 1st Divisions at P'yongyang on the south, and that they would be destroyed or captured. He termed the airdrop an "expert performance" and said, "This closes the trap on the enemy." The next day in Tokyo MacArthur predicted that "the war is very definitely coming to an end shortly." [3]

General MacArthur's optimism was not supported by the events of succeeding days. The airborne troops had not cut off any sizable part of the North Korean forces. The main body of the enemy had already withdrawn north of Sukch'on and Sunch'on and were either north of the Ch'ongch'on River or in the act of crossing it. No important North Korean Army or government officials were cut off and killed or captured. Civilians in P'yongyang said that the principal North Korean government officials had left P'yongyang on 12 October for Manp'ojin on the Yalu. The best information indicated, however, that the North Korean Government had moved to Kanggye in the mountains twenty air miles southeast of Manp'ojin. Most of the American and South Korean prisoners had been successfully removed into the remote part of North Korea.[4]

### *The Enemy Blocking Force Destroyed*

The most important action growing out of the airdrop occurred on 21–22 October in the zone of the 3d Battalion, 187th Regimental Combat Team, about eight miles south of Sukch'on in the vicinity of Op'a-ri. At 0900, 21 October, the 3d Battalion started south from its roadblock position toward P'yongyang in two combat teams: one (I Company) along the railroad, the other (K Company) along the highway. Following the railroad, I Company at 1300 reached Op'a-ri. There an estimated enemy battalion, employing 120-mm. mortars and 40-mm. guns, attacked it. After a battle lasting two and a half hours, the North Koreans overran two platoons and forced I Company, with ninety men missing, to withdraw to Hill 281 west of the railroad. The North Koreans did not press their advantage but withdrew to their own defensive positions on the high ground around Op'a-ri.[5]

Meanwhile, K Company, advancing south along the highway, encountered an estimated enemy battalion a mile

[3] *Stars and Stripes* (Pacific), October 21, 1950, p. 1, col. 6; New York *Times*, October 21, 1950 (including editorial); EUSAK WD, 22 Oct 50, Daily News Bul.

[4] GHQ FEC, History of the N.K. Army, pp. 41, 77–78; ATIS Interrog Rpts (N.K. Forces), Issue 17, p. 1, Rpt 2200, Bak Tong Hyon; *Ibid.*, Issue 19, p. 111, Rpt 2449, Jr Lt Chong Kil Hwan; EUSAK PIR's 99, 100, and 101, 19–21 Oct 50.

[5] 187th Abn RCT Unit Hist, 21 Oct 50.

north of Yongyu. After a sharp fight this enemy force withdrew south and east of the town to defensive positions on high ground, and K Company continued on into Yongyu and to Hill 163, just north of the town. Yongyu on the highway and Op'a-ri on the railroad are three miles apart and almost opposite each other.

The 3-mile gap separating the railroad and the highway here is the greatest distance between them at any point between P'yongyang and Sukch'on. Extending on a southwest to a northeast axis, and cutting across both the highway and railroad at Yongyu and Op'a-ri, is a line of high hills offering the best defensive ground between P'yongyang and the Ch'ongch'on River. Here, the N.K. *239th Regiment,* about 2,500 strong, had taken up defensive positions. This regiment had been the last force to leave P'yongyang. Its mission was to fight a delaying action against U.N. troops expected to advance north from P'yongyang. Now, suddenly, it found itself attacked by two separate forces from the rear.

At midnight the N.K. *239th Regiment* attempted to break out to the north. In its first attack a small group got into the K Company command post. In the close-quarter fight there Capt. Claude K. Josey, K Company commander, although wounded twice by an enemy burp gun, sprang on the gunner and wrested the gun from him before collapsing. The company executive officer was also wounded. Eventually, the enemy soldiers were either killed or driven off.

In two other attacks after midnight enemy soldiers forced the men at the roadblock near Hill 163 to withdraw after they had expended their ammunition. Aware of this withdrawal, the North Koreans attacked again at 0400. Then, at 0545, they ran blindly into the 3d Battalion command post and the L Company perimeter, and suffered very heavy casualties from direct and enfilading fire. In spite of these heavy losses the enemy renewed his attack, about 300 men striking L Company and 450 men assaulting Headquarters Company. At this point the airborne troops sent a radio message describing their situation and requesting help. Pfc. Richard G. Wilson, a medical aide, gave his life in heroic action in trying to reach and care for the wounded.[6]

Help was to come from close at hand as a result of a general advance northward of the U.S. I Corps. On 20 October, the day P'yongyang was secured, General Milburn had ordered the corps to continue the attack to the MacArthur Line, a line roughly thirty-five miles south of the Yalu River. The 24th Division, with the 27th British Commonwealth Brigade attached, was to lead this attack. On the right of the 24th Division three ROK divisions—the 1st, under I Corps, and the 6th and 8th under ROK II Corps, in that order eastward—were ready to join in the attack northward.[7]

---

[6] Department of the Army General Order 36, 4 June 1951, awarded the Distinguished Unit Citation to the 3d Battalion, 187th Airborne Regiment, and the 2d Section, Antitank Gun Platoon, Support Company, for this action. Department of the Army General Order 64, 2 August 1951, awarded the Medal of Honor posthumously to Pfc. Richard G. Wilson, Medical Company, 187th Airborne Regiment, for action near Op'a-ri, 21 October 1950. Eighth Army General Order 135, 12 March 1951, awarded the Distinguished Service Cross to Capt. Claude K. Josey for action near Yongyu.

[7] 24th Div WD, 20 Oct 50; EUSAK WD, 21 Oct and Br for CG, 210001–220800 Oct 50; EUSAK WD, G–3 Jnl, 0700 22 Oct 50.

At noon on 21 October, in this general Eighth Army advance, the British brigade crossed the Taedong River at P'yongyang and headed north on the main highway running toward Sukch'on, with the immediate mission of reaching the Ch'ongch'on River. Approaching Yongyu that evening, Brigadier Coad decided to halt for the night.

The British could hear the heavy night battle taking place a mile or two north of them. At first light on the 22d, two companies of the Argyll 1st Battalion advanced into Yongyu. There the Australian 3d Battalion passed through them, with Capt. A. P. Denness and his C Company in the lead riding tanks of D Company, U.S. 89th Tank Battalion. The tankers had orders not to fire because of the known proximity of the 187th Airborne troops.

Just north of Yongyu enemy rifle fire suddenly came from an orchard that spread out on both sides of the road. Captain Denness and his men jumped from the tanks and charged with fixed bayonets into the apple orchard. They went into it with a dash that brought forth admiration from all who witnessed it. One American officer present told of seeing a big, red-haired Australian jump into an enemy trench and come out later, his hands streaming blood from many cuts and his clothes slashed from head to foot. An inspection of the trench later revealed eight dead North Koreans there.

Colonel Green deployed a second company to seize high ground on the right of the road. Soon he had to send a third company to follow the second as the enemy fired on it from the rear. Then he sent his fourth company on the left of the road to follow C Company. The enemy was now using mortar as well as rifle and automatic fire. This action for the Australians was one of rifle, grenade, and bayonet. After committing all his rifle companies, Colonel Green moved his small headquarters into the orchard. There he was immediately attacked by a sizable group of North Koreans. In this fight his group killed thirty-four enemy soldiers. Among his own wounded were three men of his personal staff. One platoon of Australians crossed a rice field, kicked over stacks of straw, and shot the North Korean soldiers they found hiding in them.

In this hand-to-hand infantry fight the North Koreans lost about 270 killed and more than 200 captured; incredibly, the Australians had only approximately 7 wounded. Enemy survivors fled westward. The Middlesex 1st Battalion now passed through the Australians and, with the tanks, joined the 187th Airborne force at 1100.[8]

The 3d Battalion, 187th Airborne Regiment, reported that it had killed 805 of the enemy and captured 681 prisoners in the Yongyu battle. Caught between the airborne troops and the British 27th Brigade, the N.K. *239th Regiment* was practically destroyed at Yongyu. That afternoon the 3d Battalion returned to Sukch'on with the

[8] Linklater, *Our Men in Korea*, pp. 24–25; Bartlett, *With the Australians in Korea*, pp. 30–31; Barclay, *The First Commonwealth Division*, p. 23; 187th Abn RCT Unit Hist, 21–22 Oct 50; *Ibid.*, POR 10, 21 Oct 50; Interv, author with 1st Lt Francis Nordstrom (tk plat ldr, D Co, 89th Tk Bn), 31 Aug 51; Interv, author with Maj James W. Deloach (I Corps liaison off), 28 Jul 51. Both Nordstrom and Deloach witnessed the C Company bayonet attack. GHQ FEC General Order 54, 1 November 1950, awarded the Silver Star to Lt. Col. Charles H. Green.

British following it. There the British brigade relieved the 187th Airborne Regiment in its positions.

While the Yongyu battle was in progress, the 2d Battalion, 187th Airborne Regiment, remained relatively inactive in its drop zone at Sunch'on. The ROK 6th Division performed most of the work in clearing the town and its vicinity of enemy stragglers.

The 187th Airborne Regimental Combat Team returned to P'yongyang on 23 October, traveling by the secondary road through Sunch'on. This left the main highway free for the movement of the British 27th Brigade and the 24th Division. Altogether, the 187th Airborne Regiment suffered 46 jump casualties and 65 battle casualties in the Sukch'on-Sunch'on operations. It captured 3,818 North Korean prisoners in this operation.[9]

*Death in the Evening*

After the airdrop a new task force, formed around the 1st Battalion, 8th Cavalry Regiment, and a company of tanks, 70th Tank Battalion, started from P'yongyang to make junction with the airborne troops at Sunch'on. Lt. Col. William M. Rodgers of the tank battalion commanded the task force. It arrived at Sunch'on at 0900 21 October, picking up on the way five American prisoners who had recently escaped their North Korean captors. At the bridge just south of Sunch'on a few enemy troops hiding in holes under it opened fire as Task Force Rodgers came up and killed two men of the 8th Cavalry. The North Koreans had remained unobserved even though some airborne troops were on the bridge.

General Gay and Brig. Gen. Frank A. Allen, Jr., from an L-5 plane had watched Task Force Rodgers successfully establish contact with the airborne troops. Upon returning to P'yongyang, General Allen climbed into his jeep and accompanied by his aide, his driver, and two war correspondents (Don Whitehead of the Associated Press and Richard Tucker of the Baltimore *Sun*), started for Sunch'on, arriving there about noon.

Allen had been in the command post of the 2d Battalion, 187th Airborne Regiment, only a short time when a Korean civilian came in and excitedly told a story of North Koreans murdering about 200 Americans the night before at a railroad tunnel northwest of the town. Allen determined to run down this story at once.

His group set out with the Korean civilian and, on the way, stopped at the ROK 6th Division command post in Sunch'on. There a ROK colonel, an interpreter, and a driver in a second jeep joined Allen and drove with him to a railroad tunnel just beyond the village of Myonguch'am, five air miles northwest of Sunch'on. They arrived there at 1500. The railroad ran along a hillside cut and entered the tunnel some distance above the dirt road the men had followed. While the rest waited on the road, the ROK colonel climbed the hillside and entered the tunnel. He came back and said he had found seven dead Americans inside. Allen and the others now climbed to the tunnel. Inside it near the far end they found the seven emaciated bodies on straw mats beside

[9] 187th Abn RCT Unit Hist, 23–24 Oct 50; EUSAK WD, G–3 Jnl, Msg 1945 22 Oct 50.

NORTH KOREAN ATROCITY SITE *is marked by graves of American soldiers who were shot down as they waited for their evening meal.*

the rail track. These men had either starved to death or died from disease. Some had old wounds, apparently battle wounds.

The ROK colonel had walked on through the tunnel. He reappeared at the end and called out that he could see five Americans on the ridge top. Everyone hurried outside and started down the track. A little distance beyond the tunnel, a thin, wounded American soldier staggered from the brush. He was Pfc. Valdor John. Allen placed his coat around the shivering boy, who broke into tears and protested that he was too dirty to wear it. He then stammered out, "They are over there," and pointed into the brush. Seventeen dead Americans, all shot, lay there in a gulley. John had escaped by feigning death. Allen started climbing the ridge to the Americans who could be seen on top. Whitehead, sickened by the sight he had just seen, walked off alone across the railroad track into a cornfield on the other side. There he accidentally stumbled upon a semicircle of fifteen more dead Americans. They had been shot as they sat on the ground with rice bowls in hand expecting to receive food. Whitehead turned back to report to Allen; on his way back three American survivors came from among some bushes to him. Allen brought six more Americans who had escaped down off the ridge.

These survivors told the story of what had happened. Two trains, each carrying about 150 American prisoners of war, had left P'yongyang Tuesday night, 17 October, making frequent stops to repair the tracks, and crawling north at a snail's pace. Each day five or six men died of dysentery, starvation, or exposure. Their bodies were removed from the train. A few men escaped as the train traveled north. On the afternoon of the 20th, while the parachute jump was in progress, the second of the two trains stayed in the tunnel northwest of Sunch'on to escape the air activity in the vicinity. The group of 100 prisoners of this train, crowded into open coal gondolas and boxcars, was the remnant of 370 whom the North Koreans had marched north from Seoul more than a month earlier. That evening, the prisoners had been taken from the train in three groups to receive their evening meal. They were shot down as they waited for it. The train and the North Korean guards left that night.

From this story it appeared that there was another group of murdered men yet to be found. A search revealed a fresh burial place, and, upon removal of a thin covering of earth, the men discovered 34 more bodies. Altogether there were 66 dead (exclusive of the seven found in the tunnel) and 23 survivors, some of the latter critically wounded. Two of these died during the night, leaving only 21 who survived. A ROK detachment safely convoyed the rescued Americans and the dead to P'yongyang, where C-54's carried them to Japan.[10]

*The Advance Continues*

Even as the airborne troops came to ground at Sukch'on the Eighth Army G-2 was preparing his estimate that the North Koreans would be incapable of making more than a token defense of the Ch'ongch'on River barrier, forty-five air miles north of P'yongyang. He predicted that the enemy withdrawal would continue on to the north along the axes of two rail and highway routes, the first bending to the right and leading northeast from Sinanju and Anju on the Ch'ongch'on through Huich'on to Kanggye deep in the rugged mountains of central North Korea, twenty-two air miles from the Yalu River; and the second, the west coastal route, bending left and running northwest from the Ch'ongch'on River to Sinuiju near the mouth of the Yalu River at the Manchurian border.[11]

The Communist radio on 21 October announced that Premier Kim Il Sung's government had established a new capital at Sinuiju, on the south bank of the Yalu and opposite the Chinese city

[10] Interv, author with Maj Gen Frank A. Allen, Jr., 28 Jan 54; Interv, author with Whitehead, 27 Apr 56; Memphis *Commercial Appeal*, October 23, 1950 (detailed account by Whitehead, dateline Sunch'on, Korea, 22 Oct 50); Ltr, Lt Col Harry Fleming to author, 9 Mar 54 (Fleming was KMAG adviser with ROK 7th Regt, 6th Div, and joined Allen's party at the tunnel); 187th Abn RCT Unit Rpt, 21 Oct 50; EUSAK WD, G-3 Jnl, Msg 1930 22 Oct 50; Interim Hist Rpt, War Crimes Div, JAG, cumulative to 30 Jun 53. The author has relied principally on interviews with Allen and Whitehead and Whitehead's detailed account written at the time from notes made on the spot.

[11] EUSAK PIR 101, 21 Oct 50.

THE MIDDLESEX 1ST BATTALION *starts across the Ch'ongch'on River at Sinanju.*

of An-tung on the north bank.[12] But the fugitive North Korean capital soon moved on to Kanggye, and it was there in the mountains that the remnants of the North Korean Government and military power assembled. The Kanggye-Manp'ojin area, mountainous in the extreme and heavily wooded, was an ideal area in which to fight defensive delaying actions. It had been a stronghold of Korean guerrilla operations during Japanese rule. Many crossings of the Yalu were near at hand, it was centrally located, and it had lateral road communications to both northeast and northwest Korea.

On 22 October, C Company, 6th Medium Tank Battalion, designated Task Force Elephant, started from P'yongyang by way of Sunch'on for Kujang-dong to block the railroad there. Passing through Sunch'on, the task force arrived at its objective at 2200 and then turned west to Kunu-ri (Kaech'on on some maps), twenty miles downstream in the valley of the Ch'ongch'on. The ROK 1st Division followed behind the task force. (*Map 21*) The ROK's recovered 40 escaped American prisoners whom they evacuated at once to P'yongyang. Two more escaped prisoners came in at Kunu-ri the next morning, 23

October. That afternoon, a sergeant of the ROK 6th Division found the bodies of 28 American prisoners on the railroad track, and 3 men still alive, four miles north of Kujang-dong.[13]

On 23 October General Paik led his division from Kunu-ri down the valley of the Ch'ongch'on. Near Anju, D Company tanks knocked out two T34 tanks and two self-propelled guns, and captured one tank intact. Just before noon a platoon of tanks seized the damaged wooden bridge over the Ch'ongch'on River three miles northeast of Anju. A tank patrol continued downstream to Sinanju, which it found deserted by the enemy and the bridges there across the Ch'ongch'on destroyed.

Repair of the Anju bridge began at once and continued through the night. By 0900 24 October wheeled traffic, including 2½-ton trucks, could cross on it. During that morning a reconnaissance party found a tank ford three miles east of the bridge, and the 6th Medium Tank Battalion crossed the river there. All three regiments of the ROK 1st Division crossed the Ch'ongch'on on 23–24 October. The division then attacked northeast toward Unsan.[14]

Complying with I Corps' order to continue the advance beyond P'yongyang, advance elements of the 24th Division arrived in an assembly area north of the city the evening of 22 October, and there the division assumed control of the 27th British Commonwealth Brigade, the 89th Medium Tank Battalion, and the 90th Field Artillery Battalion. Meanwhile, the British Brigade had hurried on northward from Sukch'on. On 23 October it arrived at Sinanju only a few hours after the ROK 1st Division tank patrol entered the town. It also secured the airstrip five miles to the southwest. By this time the 24th Division completed its move to Sunan, twelve miles north of P'yongyang.[15]

The Ch'ongch'on River at Sinanju, not far from the sea, is wide, has 12-foot tides, and deep mud along its edges. On the 24th the British Middlesex 1st Battalion started crossing in assault boats. The rest of the brigade and the vehicles crossed that night over the ROK 1st Division bridge at Anju. The 3d Engineer Combat Battalion now worked to clear the highway to Sinanju, and to improve it for carrying the main part of Eighth Army's logistical support in the projected drive to the Manchurian border.[16]

While the U.S. I Corps on the U.N. left advanced to the Ch'ongch'on, two divisions of the ROK Army on its right also took up the advance. The ROK 6th Division turned northeast from Kunu-ri up the Ch'ongch'on River on the road that led through Huich'on to Kanggye.

---

[12] *Ibid.;* EUSAK WD, 23 Oct 50, Daily News Bul, dispatch of 21 Oct 50.

[13] 6th Med Tk Bn WD, 22–23 Oct 50; 10th AAA Group WD, Oct 50; Interv, author with Maj Roy M. Gramling (KMAG adviser with the ROK 6th Div), 17 Feb 54.

[14] 6th Med Tk Bn WD, 22–24 Oct 50; EUSAK WD, G–3 Sec and Br for CG, 24 Oct 50; EUSAK POR's 306 and 307, 22 Oct, and 309, 23 Oct 50.

[15] I Corps WD, 22 Oct 50; 24th Div WD, 22–25 Oct 50; EUSAK WD, G–4 Stf Sec, 22 Oct and G–3 Sec, 23 Oct 50; EUSAK PIR 103, 23 Oct 50. At Sunan staff officers investigated and confirmed a civilian report that General Dean had been held a prisoner in the town before being moved farther north.

[16] 24th Div WD, 23–24 Oct 50; British 27th Brig Unit Rpt, Sitrep 241800–261800 Oct 50; EUSAK WD, G–3 Sec, 24 Oct 50; 3d Engr Bn WD, Narr Summ, 29 Sep–Oct 50.

East of it the ROK 8th Division reached Tokch'on at midnight of 23 October. There it turned north and struck the Ch'ongch'on at Kujang-dong two days later. Both the ROK 6th and 8th Divisions (ROK II Corps) were now in exceedingly mountainous country. Near Kunu-ri the ROK 6th Division captured two trains, one carrying 8 tanks, and, farther on, near Kujang-dong, it captured 50 boxcars of ammunition. The division had a hard fight with an estimated regiment of North Koreans south of Huich'on but dispersed this force and entered Huich'on on the night of the 23d. There it captured 20 T34 tanks needing only minor repairs. At Huich'on the ROK 6th Division turned west, and later north, its objective being Ch'osan on the Yalu River. It was now far in front of any U.N. division.[17]

On 24 October, when Eighth Army troops crossed the Ch'ongch'on River and the ROK 6th Division passed through Huich'on and headed for the Yalu, less than six weeks had passed since that army had battled desperately to hold its lines 320 air miles to the south along the Pusan Perimeter. The Inch'on landing likewise was less than six weeks in the past. The capture of Seoul was about four weeks in the past. Since then, the Eighth Army, moving up from the south after breaking out of its embattled Perimeter, had penetrated 160 air miles north of Seoul and 130 air miles into North Korea. In doing this it had overrun the enemy's capital and breached the last important river barrier south of the northern border of the country.

At the same time the ROK I Corps under its command had fought its way northward equally far on the east coast to capture Wonsan. And in the closing days of this period the U.S. X Corps had moved amphibiously around the length of Korea to appear off Wonsan for an imminent landing and subsequent operations in that part of Korea. This Eighth Army–ROK–X Corps attack which moved the front northward more than 300 miles in less than six weeks had virtually destroyed the North Korean Army.

---

[17] EUSAK WD, Br for CG, 22–25 Oct 50; *Ibid.*, POR 307 and PIR 102, 22 Oct 50; Gen Paik Sun Yup, MS review comments, 11 Jul 58.

## CHAPTER XXXIII

# The Chinese Intervene

O divine art of subtlety and secrecy! Through you we learn to be invisible, through you inaudible; and hence hold the enemy's fate in our hands.

SUN TZU, *The Art of War*

The Far East Command near the end of October changed its policy with respect to Koreans attached to United States Army units, authorizing a strength of only twenty-five Koreans to an American infantry company or unit of similar size, instead of the 100 per company previously authorized. The resulting release of several thousand Korean soldiers who had been assigned to U.S. Army divisions since August made possible the activation of a new ROK division. On 25 October the ROK Army activated the 9th Division, composed of the 28th, 29th, and 30th Regiments, of two battalions each. On 30 October, three battalions of the 1st Anti-Guerrilla Group became the third battalion of each of these regiments. By 7 November, 8,272 Korean soldiers had been released and turned back to the ROK Army and several thousand more were on the point of being released. On that day the ROK Army reactivated in Seoul the 2d Infantry Division which had been shattered in the early days of the war. This division at first had only two regiments, the 17th and the 31st, but on 13 November the 32d Regiment was activated at Seoul as the division's third regiment.[1]

Other new military forces began making their appearance in Korea at this time, just when it began to look as if they would not be needed. The scheduled arrival of several United Nations troop organizations in Korea made necessary arrangements to equip and train them so that they could become effective parts of the Eighth Army command. In an attempt to accomplish this, General Walker on 8 October ordered the 2d Logistical Command to establish a United Nations Reception Center (UNRC) at Taegu University as soon as EUSAK moved from it. Its mission was "to clothe, equip, and provide familiarization training with U.S. Army weapons and equipment to U.N. troops as determined essential for operations in Korea by the Reception Center Commander." Not more than 6,200 troops were expected to be in training at the center at any one time. The first unit to

---

[1] EUSAK WD, G-1 Daily Hist Rpt, 26 Oct 50; *Ibid.*, POR 389, 19 Nov 50; *Ibid.*, Br. for CG, 9 Nov 50; GHQ UNC, G-3 Rpts, 27 and 30 Oct 50; 25th Div WD, 30 Oct 50.

make use of it was the 1st Turkish Armed Forces Command which arrived there on 18 October.[2]

The first of the new forces to arrive in Korea was the Thailand Battalion whose advance party arrived at Pusan on 3 October; the main party arrived more than a month later on 7 November. Following closely after the Thailand advance party came the advance party of the Turkish Brigade which arrived at Pusan on 12 October. The main body of the brigade (5,190 troops) arrived at Pusan five days later and began unloading on the 18th. The Turkish troops were fully equipped except for certain weapons. On 24 October, the advance parties of the Netherlands Battalion and the British 29th Brigade arrived in Korea. In Canada a special force of 10,000 men had volunteered and trained for combat service in Korea, and on 7 November an advance party of 345 of them arrived at Pusan to prepare the way for the main body. But with the war seemingly near an end, only a battalion followed; the main body was held in Canada.[3]

It should be emphasized that at this time when Eighth Army was making ready to continue the pursuit north of the Ch'ongch'on River, its logistical situation was not good. The breakdown of rail transportation in October, coincident with Eighth Army's rapid advance northward, caused an extraordinarily heavy burden to fall on truck transport operating over bad roads with long hauls from ports and railheads. At the end of October the 24th Division railhead was still at Yongdungp'o on the south side of the Han River, while the division service elements were in the vicinity of Pakch'on, 205 miles farther north. The longer the trucks ran over the rough Korean roads, the greater grew the number that became inoperable. It was a type of logistical support that promised soon to wear itself to destruction since the spare parts needed for repairs were not available.

During September, October, and on into November, 76 percent of Eighth Army's trucks operated on a 24-hour basis. In order to supply I Corps north of the 38th Parallel, Eighth Army had to take away from the 2d and 25th Divisions large numbers of their trucks, thereby virtually immobilizing these divisions. The 2d Division at one time furnished 320 trucks that were organized into a Red Ball Express to supply I Corps from the Han River. The need for trucks was so critical, and normal methods of delivering them to the tactical units so uncertain, that divisions and corps sent men back to Pusan by air and rail to drive the trucks 400 miles north over atrocious roads to the fronts.[4]

As soon as P'yongyang fell to Eighth Army an airlift of supplies to the airfield there got under way from Ashiya Air Base in Japan and from Kimpo Airfield near Seoul. The Kimpo airlift sought to attain a goal of transporting 1,000 tons daily to P'yongyang or north-

[2] Maj William F. Fox, History of the Korean War, Inter-Allied Co-operation During Combat Operations, vol. III, pt. 2, sec. B, pp. 10–11, MS in OCMH; Ltr Ord, Hq, EUSAK to CG, 2d Log Comd, 8 Oct 50, sub: Establishment of UNRC.

[3] 2d Log Comd Monthly Act Rpt, G–3 and G–4 Secs, Oct and Nov 50.

[4] Interv, author with Maj Gen Leven C. Allen, 15 Dec 53; EUSAK WD, Trans Sec, 18 Nov and G–4 Jnl, Msg 9, 241430 Nov 50; 24th Div WD, G–4 Daily Summ, 29–30 Oct 50; 2d Div WD, Narr Summ, Nov 50, p. 11.

ward. A large part of this airlift at the end of October carried ammunition. On 28 October, for instance, cargo planes carried 1,037 tons of ammunition from Kimpo to P'yongyang, and, as early as 31 October, planes carried ammunition to the hastily repaired fighter strip near Sinanju for the ROK units fighting along and above the Ch'ongch'on.[5]

### *American Optimism at End of October*

Offsetting the bad logistical situation at the end of October was the general belief among U.S. commanders that the war in Korea was all but ended. Viewed in this light, the situation looked so favorable that the Department of the Army and the Commander in Chief, Far East, made plans for the redeployment of Eighth Army units, including the return of the 2d Infantry Division to the United States or to Europe, and of other organizations later. On 25 October the Department of the Army notified General MacArthur that it planned to cancel shipment of enlisted reserve corps troops to the Far East scheduled for October and November, except 17,000 noncommissioned officers. All this was in accordance with general agreements reached at the Wake Island Conference earlier in the month.[6]

Even in Korea this cutback fever had taken hold. On 22 October General Walker requested authority from General MacArthur to divert to Japan all bulk-loaded ammunition ships arriving thenceforth in Korea from the United States, as he felt there was enough ammunition in Korea to satisfy future needs. MacArthur approved this request, and he also took steps to have six ammunition ships, en route to the Far East carrying 105-mm. and 155-mm. shells and Air Force bombs, diverted to Hawaii or returned to the United States. And General Weible, Commanding General, Japan Logistical Command, requested the Commanding General, San Francisco Port of Embarkation, to cancel all outstanding requisitions for ground ammunition and to unload any ships still in port.[7]

Morale was high in the U.N. forces as they crossed the Ch'ongch'on and set out on what most of them though would be the last, brief phase of the war. In the 1st Cavalry Division many men thought they would parade on the Plaza in Tokyo wearing yellow cavalry scarves on Thanksgiving Day. The division even started turning in its equipment in expectation of being the first organization to return to Japan. Others throughout the army threw away handbills listing prices of gifts available at post exchanges, saying they were going to do their Christmas shopping in Japan.

In the United States the New York *Times* probably expressed the prevailing opinion there at this stage of the war when it stated editorially, "Except for unexpected developments along the frontiers of the peninsula, we can now be easy in our minds as to the military outcome." [8]

---

[5] EUSAK WD, Ord Daily Act Rpts, 27, 28, 31 Oct and 6 Nov 50; *Ibid.*, Trans Sec, 25 Oct 50; 3d Log Comd Act Rpt, Nov 50, p. 8.

[6] Schnabel, FEC, GHQ Support and Participation in the Korean War, ch. VII, pp. 3–5, citing Msg 94651, JCS to CINCFE, 21 Oct 50, Msg C67065, CINCFE to DA for JCS, 21 Oct 50, and Msg S94985, DA to CINCFE, 25 Oct 50.

[7] *Ibid.*, ch. VII, pp. 6–7, citing Msg CX67506, CINCFE to CG Eighth Army and CG JLC, 26 Oct 50, and Msg CX7702, CINCFE to DA, 28 Oct 50.

[8] New York *Times*, October 29, 1950.

Looking ahead to the task of rehabilitating the people of the Republic of Korea, General MacArthur took steps to establish a Civil Assistance Command in Eighth Army. On 30 October General Walker activated this command with an authorized staff of 161 officers and 117 enlisted men, to become effective 1 November.[9]

Until 17 October General MacArthur's orders, based on the Joint Chiefs of Staff directive of 27 September, had restrained U.N. ground forces other than ROK troops from operating north of a line extending from Ch'ongju on the west through Kunu-ri and Yongwon to Hamhung on the east coast. On 17 October General MacArthur, in his UNC Operations Order 4, lifted this restriction and advanced northward the line below which all U.N. ground forces could operate. This new line, confirmed in a message to all commanders on 19 October, extended generally from Son-ch'on through Koin-dong–P'yongwon–P'ungsan to Songjin on the east coast. *(See Map 21.)* It was generally thirty to forty miles south of the Manchurian border across the greater part of the peninsula, and was within the spirit and meaning of the Joint Chiefs of Staff directive of 27 September, which was still in effect. In the policy laid down in this directive only ROK forces were to be used in the provinces of Korea bordering on the Yalu River.[10]

But on 24 October, as the leading U.N. forces crossed the Ch'ongch'on River, General MacArthur issued an order to his ground commanders in Korea which changed all earlier orders drastically. He now removed all restrictions on the use of U.N. ground forces south of the border, and instructed his commanders to press forward to the northern limits of Korea, utilizing all their forces.[11] Thus, when Eighth Army began what it thought would be the last series of maneuvers to end the war it did so under orders radically different from those that had so far guided its operations in Korea.

The day it was issued, this order brought a message from the Joint Chiefs of Staff to MacArthur stating that it was not in accord with the directive of 27 September and asking for an explanation. General MacArthur's reply the next day justified lifting the restriction as a matter of military necessity. He said that the ROK forces could not handle the situation by themselves, that he felt he had enough latitude under existing directives to issue the order, and that, furthermore, the whole subject had been covered in the Wake Island Conference.[12]

While it is clear that the Joint Chiefs of Staff felt that MacArthur had violated their basic 27 September directive, they did not countermand his orders to go to the Yalu. When the 27th British Commonwealth Brigade crossed the Ch'ong-

[9] EUSAK WD, G–3 Sec, 30 Oct 50; *Ibid.*, G–1 Sec, Civil Assistance Stf Sec Rpt, 1 Nov 50.

[10] Schnabel, FEC, GHQ Support and Participation in the Korean War, ch. VI, pp. 31–32, citing Msg CX66705, CINCUNC to all comdrs, 17 Oct 50, and CX66839, CINCUNC to all comdrs, 19 Oct 50.

[11] *Ibid.*, p. 34, citing Msg CX67291, CINCUNC to all comdrs, 24 Oct 50.

[12] C67397, CINCFE to JCS, 25 Oct 50; Senate MacArthur Hearings, MacArthur's testimony, pp. 97–98; Gen. of the Army Omar N. Bradley's testimony, pt. 1, p. 757, and Gen Collins' testimony, pt. 2, pp. 1216–17, 1229–30, 1235, 1239–41, 1312–13.

ch'on, that unit, the U.S. 24th Infantry Division which followed, and all the other U.N. troops deployed in Korea, were authorized to go to the Yalu River—to the extreme northern limits of the country.

General Walker on 25 October was quoted as saying, "Everything is going just fine." [13] And so it was—just then.

### *Continuation of the Pursuit*

The Ch'ongch'on River and its tributaries, the Kuryong and Taeryong Rivers, all flowing from the north, together form the last major water barrier in the western part of North Korea short of the border. The Ch'ongch'on valley is a wide one for Korea, varying in width from 3 to 20 miles. The Ch'ongch'on, like the Yalu, flows from the northeast to the southwest and it generally parallels the Yalu at a distance of approximately 65 air miles. The Ch'ongch'on River, the principal terrain feature in the field of operations for Eighth Army during late October and November 1950, largely dictated the army's deployment and tactical maneuvers.

The main P'yongyang highway crosses the Ch'ongch'on at Sinanju and runs west and northwest in the coastal area to Sinuiju at the North Korean border. Inland from the west coast, mountainous spines run down from the Yalu to the valley of the Ch'ongch'on and the terrain becomes ever rougher and more forbidding. These mountains reach their greatest heights and become almost trackless wastes in central Korea between the Changjin (Chosin) [14] Reservoir and the Yalu. The Yalu itself, save for its lower west coast reaches, runs through a gorgelike channel rimmed by high mountains on both sides. The great Suiho hydroelectric dam on the middle Yalu impounds a reservoir of the same name that extends upstream for sixty air miles, pushing water into hundreds of little lateral fjordlike mountain valleys.

Above the reservoir there is a major crossing of the Yalu at Manp'ojin. Twenty air miles southeast of Manp'ojin, situated in the very heart of the mountain fastness, is Kanggye. There the North Korean governmental officials and high military commanders assembled. From there, if necessary, they could retreat across the Yalu at Manp'ojin to the sanctuary of Manchuria.

From the valley of the Ch'ongch'on the principal road to Kanggye and Manp'ojin ran northeast from the Sinanju–Anju–Kunu-ri area through Huich'on. A railroad followed the same passageway. From the lower valley of the Ch'ongch'on, fifty air miles inland from the west coast, an important secondary road network ran north through Unsan to the Yalu. The events of the next few weeks were to give this particular road net special importance.

The configuration of the valley of the lower Ch'ongch'on in relation to the mountain ridges that approach it from the Yalu must be noted. North of the lower Ch'ongch'on for a distance of approximately fifteen air miles the ground is flat or only slightly rising with occasional low hills. A lateral road extending eastward from Yongsan-dong and generally paralleling the river marks the

[13] EUSAK WD, Daily News Bul, 25 Oct 50.

[14] See p. 729, n. 1, below.

cleavage line between this low ground, which in a broad sense can be described as the valley of the Ch'ongch'on, and the mountain spurs that rise rather abruptly from it and extend to the Yalu. The southern extremities of these mountain ranges with their limited corridors of passage form a natural defensive barrier to a military advance northward. The towns of Taech'on, Unsan, and Onjong stand at the entrances to these mountain corridors. There, the logical implications of the terrain were soon to be translated by an enemy into harsh and unwonted military reality.

The Eighth Army operation above the Ch'ongch'on began essentially as a continuation of the pursuit that had started with the breakout from the Pusan Perimeter; the U.S. I Corps was on the left, the ROK II Corps on the right. Within Milburn's I Corps, the 24th Division (British 27th Brigade attached) was on the left, the ROK 1st Division on the right. The U.N. Command expected little organized opposition from the enemy and emphasized a speedy advance to the northern border. Several columns were to strike out northward with little or no physical contact between them. The advance was not to be closely co-ordinated; each column was free to advance as fast and as far as possible without respect to gains made by others.

### *ROK Troops Reach the Yalu*

As Eighth Army resumed its general advance toward the North Korean border, the ROK 6th Division of the ROK II Corps appeared to have the greatest success of any front-line U.N. division. (*Map 22*) Meeting virtually no opposition and traveling fast up the valley of the Ch'ongch'on, it reached Huich'on the night of 23 October. There it left the valley of the Ch'ongch'on and turned west, the 7th Regiment leading. Its advanced battalion marched northwest over a cart trail, but the remainder of the regiment had to turn west from Huich'on on a road to Onjong. The night of 24–25 October, the 7th Regiment passed through Onjong, then turned north and joined its advanced battalion.[15] Finding the road clear, it headed north for its objective, the town of Ch'osan, fifty air miles away on the Yalu. Late in the afternoon the regiment stopped at Kojang, a sizable town eighteen air miles south of Ch'osan, and bivouacked there for the night.

The next morning, 26 October, Maj. Harry Fleming, KMAG adviser with the ROK 7th Regiment, accompanied the Reconnaissance Platoon, reinforced, into Ch'osan. The remainder of the regiment stayed at its overnight position. In Ch'osan the Reconnaissance Platoon found North Koreans retreating into Manchuria across a narrow floating footbridge that spanned the Yalu. Fleming and the ROK officers directed the setting up of machine guns to halt this foot traffic into Manchuria, but so placed the weapons that the impact area of their fire would not be in China across the river. After a thorough reconnaissance of the town, Fleming and the main body of the Reconnaissance Platoon returned to the regimental position. They left a small party in Ch'osan because the next morning the main force of the ROK 7th

---

[15] Interv, author with Gramling, 17 Feb 54; Ltrs, Fleming to author, 9 and 18 Mar 54; EUSAK WD, G–3 Jnl, 23–24 Oct 50.

Regiment was to come into the town.[16]

The Reconnaissance Platoon from the 7th Regiment, ROK 6th Division, was the first U.N. unit to reach the northern border of North Korea, and, as events turned out, it was the only element operating under Eighth Army command ever to get there during the war.

Following behind the 6th Division, the ROK 8th Division had reached the valley of the Ch'ongch'on at Kujang-dong the night of 25–26 October, marching from Sunch'on through Tokch'on. On the 26th, the day the advanced elements of the 6th Division reached the Yalu, the 8th turned up the Ch'ong-ch'on Valley toward Huich'on for the purpose of joining the 6th Division.

### *Chinese Strike the ROK II Corps*

The day before, on 25 October, the 3d Battalion, 2d Regiment, ROK 6th Division, had started northwest from the little crossroads village of Onjong, ten air miles northeast of Unsan, headed for Pukchin. There the 2d Regiment expected to turn north to Pyoktong on the Yalu. Eight miles west of Onjong the 3d Battalion came under enemy fire. The troops dismounted from their vehicles to disperse what they thought was a small force of North Koreans. But the roadblock turned out to be a Chinese trap. In the action that followed the Chinese destroyed the battalion as an organized force. Approximately 400 of 750 ROK's in the battalion escaped, however, and in the afternoon infiltrated back to Onjong. Among those captured in this action was Lt. Glen C. Jones, KMAG adviser to the battalion, who later died in a prison camp.[17]

Meanwhile, back at Onjong the 2d Battalion of the ROK 2d Regiment learned that the 3d Battalion had become heavily engaged, and moved out to support it. On the way, members of the battalion saw enemy troops moving about on the hills to the north. Patrols sent out to investigate came back with a Chinese prisoner. He said that Chinese forces had been waiting in the mountains around Pukchin since 17 October. Another Chinese soldier, badly wounded, was captured on the road ahead. That evening Chinese troops cut off the 2d Battalion from Onjong, but it escaped southward cross-country and succeeded in rejoining the 1st Battalion and regimental headquarters in the town.

At 0330 that night, the Chinese attacked Onjong. The ROK troops there broke in panic, but officers succeeded in stopping them at the southeast edge of town. When the Chinese penetrated this position at 0600 the ROK's started withdrawing eastward. They had gone only three miles when they came to a roadblock—the Chinese had cut them off. At this time not a single company of the ROK 2d Regiment was intact. The ROK's now scattered into the hills. Maj. Roy M. Gramling, KMAG adviser to

[16] This account of the ROK's at Ch'osan is based largely on Ltrs, Fleming to author, 9 and 18 Mar 54; also, Interv, author with Lt Col Willard G. Pearson, 1 Aug 51; EUSAK POR 319, 26 Oct 50; EUSAK WD, Br for CG, 26 Oct 50.

[17] Interv, author with Gramling, 17 Feb 54; EUSAK POR 319, 26 Oct 50. Unless otherwise noted, the Onjong narrative is based on the author's interview with Gramling. Army records have only a few fragmentary references to the ROK II Corps action at Onjong.

the regiment, and another KMAG officer escaped to Huich'on, but a third, Capt. Paul V. S. Liles, fell captive to the Chinese. That the 2d Regiment apparently did little determined fighting in its first encounter with the Chinese is indicated by the fact that about 2,700 men out of approximately 3,100 in the regiment eventually escaped to the Ch'ongch'on.

When the ROK 2d Regiment came under Chinese attack at Onjong, the 19th Regiment, except for one battalion, was in Huich'on. The 10th Regiment of the ROK 8th Division also was there. Maj. Gen. Yu Jae Hung, commanding the ROK II Corps, ordered these troops, less the 1st Battalion, 10th Regiment, which was to remain in Huich'on, to attack west in an attempt to recover the abandoned vehicles and artillery pieces of the 2d Regiment. These two regiments reached a point on 28 October from which advanced units could look down on Onjong and see some of the abandoned equipment, but they never got any farther. The next day these two regiments suffered the same fate that had overtaken the 2d Regiment. Heavily defeated, they lost their vehicles and three batteries of artillery—all they had.[18]

These startling developments in the Onjong area cut off the 7th Regiment of the ROK 6th Division to the north. At its headquarters at Kojang on the evening of the 26th, the regiment was making plans to occupy Ch'osan on the Yalu in the morning when it received a radio message from the ROK 6th Division. This message said that the 2d Regiment had been defeated and scattered, and ordered it, the 7th Regiment, to start south to rejoin the division. Major Fleming replied by radio that the regiment could not move unless it was resupplied with gasoline, food, and ammunition. An airdrop of supplies was successfully accomplished two days later at 1100.

The following morning, 29 October, the ROK 7th Regiment started south. Before noon, when approximately twenty miles south of Kojang, it ran into an enemy roadblock. In a very short time the entire regiment was committed against an enemy force. The tactical air control party called in strong air support and, according to Major Fleming, "with the tremendous help of the close air support we received, we were able to hold our own during the daylight hours, but after night fell and without the support of the fighter planes we could not hold the enemy off." [19]

During the night large numbers of ROK soldiers scattered into the hills in an effort to make their way south; some, however, stayed in position to the end. By daylight resistance had ended. It appears probable, from a hand-drawn operations map of the *373d Regiment*, CCF *125th Division*, captured in March 1951, that one battalion of this regiment set the trap and fought the action that destroyed the ROK 7th Regiment. Major Fleming, wounded in fifteen places and the only American to survive the battle, was captured at 0630 by the Chinese. Almost three years later Major Fleming returned to the United States in the pris-

[18] Interv, author with Gramling, 17 Feb 54; EUSAK WD, 28 Oct 50, Memo to CofS, G-2, *et al.*, from Acting CofS, G-3 (Rpt of Lt Col F. J. Lagasse after visiting Hq ROK II Corps); *Ibid.*, 29 Oct 50.

[19] Ltrs, Fleming to author, 9 and 18 Mar 54; EUSAK WD, Br for CG, 28 Oct 50; Lagasse Rpt, cited n. 17.

oner exchange in the fall of 1953.[20]

Eventually, about 875 officers and men of the 3,552 in the regiment escaped to Kunu-ri and rejoined the 6th Division. Col. Lim Bu Taik, the regimental commander, and two of his battalion commanders escaped, but the other principal regimental staff officers and the KMAG advisers were either killed or captured.[21]

The collapse of the ROK II Corps on the right of Eighth Army and the frightening, but confused, reports of Chinese troops in the action caused Eighth Army on 29 October to order the ROK 7th Division released from U.S. I Corps' control to revert to the ROK II Corps. It ordered the ROK II Corps to place the 8th Division in a defensive position north of the Ch'ongch'on, extending from Yongbyon eastward to the river at Kujang-dong, and then for the 7th Division to extend the line south toward Tokch'on.[22]

By 31 October Chinese forces were pressing against the ROK II Corps defensive line north and east of Kunu-ri. That morning they broke through the 16th Regiment of the 8th Division, near its boundary with the ROK 1st Division, causing one battalion to scatter.

On the south side of the Ch'ongch'on, Chinese forces by 1 November had pushed the ROK 7th Division back to the vicinity of Won-ni. The ROK II Corps of necessity by this time had pivoted to face generally east. This resulted in a gap between its left flank and Eighth Army. The U.S. 2d Division, attached by Eighth Army to I Corps, was assembled hurriedly in the vicinity of Sunch'on to meet a possible emergency in this gap.[23]

Thus matters stood on 1 November. Within a few days after its first action on 25 October, the CCF had driven back the ROK II Corps, crippling it disastrously, and was south of the Ch'ongch'on on the open right flank of Eighth Army. And disaster was also threatening in the center of the Eighth Army front at Unsan.

### *Unsan—Prelude*

In its part of the general advance, the ROK 1st Division on 25 October was strung out on the road running from the Ch'ongch'on River to Unsan. Its 15th Regiment passed through Yongbyon and continued on without opposition toward Unsan, fifteen air miles northward. Elements of D Company, 6th Medium Tank Battalion, led the way and passed through Unsan. A mile and a half northeast of the town, just before 1100, enemy mortar fire suddenly interdicted a bridge as the American tanks were approaching it. ROK troops deployed and engaged the enemy force. Half an hour later they reported 300 Chinese troops in the hills just north of Unsan. A little later they captured the first Chinese soldier taken prisoner by U.N. forces in the Korean War. American tank crewmen at the

[20] Ltrs, Fleming to author, 9 and 18 Mar 54; EUSAK WD, G-3 Jnl, Msg 0130 30 Oct 50; Interv, author with Gramling, 17 Feb 54; ATIS Enemy Documents, Korean Operations, Issue 42 (11 Jun 51), item 52, opposite p. 162.

[21] Interv, author with Lt Col Thomas E. Bennett (KMAG adviser to ROK 7th Regt in early Nov 50), 11 Dec 53; Paik Sun Yup, MS review comments, 11 Jul 58.

[22] EUSAK WD, G-3 Sec, 29 Oct 50; US I Corps WD, Oct 50, p. 45.

[23] EUSAK PIR 111, 31 Oct 50; EUSAK WD, G-3 Sec, 31 Oct and 1 Nov 50.

interdicted bridge learned at 1144 of the captured Chinese. The prisoner said there were 10,000 Chinese Communist troops in the hills north and northwest of Unsan and another 10,000 eastward toward Huich'on.[24]

During the afternoon the fighting north of town gradually intensified and just after 1400 the ROK troops in contact with the Chinese estimated them to be two reinforced companies. The TACP controller, who had been pinned down by enemy fire for more than an hour, finally established radio communication with the Mosquito plane overhead and informed its pilot of the Chinese prisoner and his story that 10,000 to 20,000 Chinese soldiers were in the vicinity of Unsan. The pilot related the alarming story at Eighth Army headquarters. That evening I Corps headquarters received a message from a G–2 liaison officer with the ROK 1st Division reporting the capture of the Chinese prisoner. Special arrangements were made to take the captive to the Eighth Army advance command post at P'yongyang for interrogation. The prisoner was interrogated there the next morning. There could be no doubt that he was Chinese. By midafternoon three more Chinese were brought into P'yongyang. They, too, looked Chinese, spoke Chinese, and understood neither Korean nor Japanese.[25]

The ROK 12th Regiment, second in the division column, turned west when it arrived at Unsan. Just beyond the town it, too, found Chinese troops blocking the way. The ROK 11th Regiment, bringing up the division rear, halted for the night a few miles below Unsan. The report spread rapidly among the ROK's during the afternoon that the enemy troops on their front were Chinese.[26]

Ironically, that same afternoon, 25 October, the U.S. I Corps headquarters at 1600 published its order directing its forces to go all the way to the Yalu. For its part, the ROK 1st Division was "to continue the destruction of North Korean forces." [27]

The 25th of October had been a cold day and carried promise of the bitter North Korean winter that lay ahead. All night the fight above Unsan continued with the sound of small arms and machine guns and the booming of supporting 155-mm. howitzers echoing through the darkness. A small flurry of snow fell early in the morning, the first snow of the winter for these troops. But it was a minor worry compared to the startling fact that during the night enemy forces had nearly surrounded Unsan. Morning brought more information that the forces were Chinese. One report told of thirty-three Chinese dead found north of the town.

Northeast of Unsan, ROK infantry of the 15th Regiment during the morning of the 26th fell back under enemy attack. At 1030 Lt. Col. John S. Growden, commanding the 6th Medium Tank Battalion, thinking that his tanks holding the road northeast of Unsan were in danger of being overrun, ordered D Company to fall back to high ground south-

[24] I Corps WD, 25 Oct 50; I Corps Intel Summ 122, 25 Oct 50; I Corps PIR 40, 25 Oct 50; 6th Med Tk Bn WD, 25 Oct 50.

[25] 6th Med Tk Bn WD, 25 Oct 50; EUSAK WD, G–3 Jnl, Msg at 1715 25 Oct 50; Ltr, Col Percy W. Thompson (G–2, I Corps Oct 50) to author, 9 Apr 54; Collier, MS review comments, 10 Mar 58.

[26] EUSAK POR 316, 25 Oct 50.

[27] I Corps Opn Dir 14, 251600 Oct 50.

east of the town. West of Unsan the ROK 12th Regiment held fast. The 11th Regiment moved up to join the 12th Regiment in the battle, but almost at once had to move back south of Unsan to combat an enemy force that cut the main supply road there in an envelopment from the west. Instead of driving off this enemy roadblock force, the 11th Regiment itself was pushed north to the edge of Unsan. An entry in the supporting U.S. 10th Antiaircraft Artillery Group War Diary at this time states, "Due to the seriousness of the situation around Unsan, the Group was prepared to move out on a moment's notice." The ROK's estimated that a full enemy division confronted them.[28]

The reaction of Eighth Army intelligence to this development was that the Chinese troops in the Onjong and Unsan areas indicated "some further reinforcement of North Korean units with personnel taken from the Chinese Communist Forces, in order to assist in the defense of the border approaches." The estimate stated there were "no indications of open intervention on the part of Chinese Communist Forces in Korea." [29]

On the 27th the situation at Unsan improved somewhat. An airdrop shortly after 1100 by ten C–119 planes flying from Ashiya Air Base eased the critical supply situation within the ROK 1st Division, the two supporting tank companies of the 6th Tank Battalion, and the 10th AAA Group. Freshly supplied with ammunition, the ROK 15th and 12th Regiments attacked and made slight gains north and west of the town. To the south of the town two battalions of the 11th Regiment cleared the road, and in the late afternoon reported the enemy there had withdrawn to the northwest. In these attacks the ROK's found the Chinese well dug-in, exceptionally well camouflaged, and very hard to locate.

When the ROK 1st Division first encountered the Chinese above Unsan on 25 October, General Paik, the division commander, was at P'yongyang attending a celebration. He had by now returned to his command post at Yongbyon. Going forward to the scene of fighting, Paik examined enemy dead. He said they were all Chinese. Paik had served with the Japanese Manchurian Army in World War II and he knew the Chinese well. He estimated there was a Chinese division of 10,000 soldiers—a solid organization and not just Chinese mixed with North Koreans—in front of him. He told General Milburn, the I Corps commander, there were "many, many Chinese." [30]

On the 28th the fighting at Unsan quieted down, although the ROK's captured two more Chinese prisoners. They repeated the same story told by previous

[28] 10th AAA WD, 25–31 Oct 50; U.S. units supporting the ROK 1st Division at this time were the 17th Field Artillery Battalion (105-mm.), 10th Antiaircraft Artillery Group (included 155-mm. howitzers and 90-mm. guns), and two companies of the 6th Medium Tank Battalion. EUSAK WD, Br for CG, 26–27 Oct 50; EUSAK PIR 106, 26 Oct 50; EUSAK WD, G–3 Jnl, Msgs at 1340, 1530, 1700, and 2115, 26 Oct, and Jnl, 27 Oct 50; I Corps WD, 26 Oct 50; 6th Med Tk Bn WD, 26–27 Oct 50; Arty Rpt 12, 27 Oct 50.

[29] I Corps Opn Dir 15, 26 Oct 50; EUSAK PIR 106, 26 Oct 50.

[30] Interv, author with Col William H. Hennig (CO, 10th AAA Group, Oct 50), 23 Mar 54; Interv, author with Milburn, 4 Jan 52; I Corps WD, 27 Oct 50; New York *Times*, October 27, 1950.

SUPPLY BY AIR *to the ROK 1st Division in the Unsan area.*

prisoners—that they were members of large Chinese organizations that had entered the war.

General Walker and the Eighth Army staff at P'yongyang had, of course, followed closely the many reports that came in concerning the puzzling and disturbing news from north of the Ch'ongch'on, particularly the information given by the first prisoners alleged by the ROK's to be Chinese. But, as would be natural in such a newly developing situation, the intelligence officials did not accept at face value all the information the prisoners related about the Chinese troop organizations they said were in Korea. As the extent of reverses north of the Ch'ongch'on mounted quickly within a day or two, General Walker and his staff, however, were forced to question the correctness of their initial reaction that the Chinese troops there represented only reinforcement of North Korean units.

By the morning of 28 October General Walker had become sufficiently concerned over events to order the 1st Cavalry Division relieved of its security mission at P'yongyang and to move north, pass through the ROK 1st Division at Unsan, and attack to the Yalu. General Gay ordered the 8th Cavalry Regiment to begin the division movement. It departed P'yongyang the next morning,

29 October. During the day it crossed the Ch'ongch'on at Anju and went into an assembly area at Yongsan-dong that evening.[31]

The ROK attack that began at first light on the morning of 29 October quickly developed into a stubborn fight against dug-in enemy using mortars, automatic weapons, and small arms. Even with the help of the artillery barrages and Fifth Air Force strafing attacks, the ROK's could not dislodge the Chinese.

Because Chinese forces had engulfed the ROK II Corps to the east in the Onjong and Huich'on areas, the ROK 1st Division now constituted a northern salient in the U.N. line. On its left there was a gap of fifteen air miles between it and elements of the U.S. 24th Division, the nearest Eighth Army unit on the west.

The next morning Lt. Col. Harold K. Johnson's 5th Cavalry Regiment arrived at Yongsan-dong. Johnson's mission was to protect the rear of the 8th Cavalry Regiment, which that morning had continued on north to Unsan where it was to relieve part of the ROK 1st Division. The 1st Battalion, 8th Cavalry, under Maj. John Millikin, Jr., arrived at Unsan that afternoon, 30 October. In conferring with KMAG officers attached to the ROK 12th Regiment, Millikin and his company commanders learned that the ROK line, about 8,000 yards north of Unsan, was under attack and being pushed back.[32]

On 31 October, the 2d and 3d Battalions, 8th Cavalry, relieved the ROK 12th Regiment. But on the right an enemy attack during the night had driven back the ROK 2d Battalion more than a mile. Its commander wanted his troops to regain the lost ground before they were relieved. Millikin's 1st Battalion, however, moved into a defensive position behind this part of the ROK line north of Unsan. That afternoon, General Milburn, U.S. I Corps commander, visited the 8th Cavalry regimental command post and was told everything was all right.

On the morning of 1 November the ROK's tried to regain their lost ground. Though assisted by elements of the 6th Tank Battalion they made only a slight gain. In this situation, elements of B Company of Millikin's battalion and an attached platoon of tanks from B Company, 70th Tank Battalion, attacked north along the west bank of the Samt'an River. In this fight tank fire greatly assisted the ROK's, but three of the tanks were damaged. By noon the 2d Battalion of the ROK 12th Regiment had regained something more than half a mile of ground. But it seemed that it would make no further gain as heavy enemy 120-mm. mortar fire started falling, forcing the tanks to withdraw. Eastward across the river, at a distance of about two miles, the ROK 15th Regiment could be seen under very heavy attack.

The ROK 2d Battalion commander gave notice in the afternoon that if he was not relieved by 1600 his battalion would leave its position anyway. Apparently Col. Raymond D. Palmer, the 8th Cavalry commander, refused to effect the relief of the ROK battalion while heavy

[31] I Corps WD, 28 Oct 50; EUSAK POR 325, 28 Oct 50; Ltr, Gay to author, 19 Feb 54.

[32] Ltr, Gay to author, 19 Feb 54; Ltr, Millikin (CO, 1st Bn, 8th Cav Regt, Oct 50) to author, 6 May 54; I Corps WD, 29–30 Oct 50; EUSAK WD, Br for CG, 29 Oct 50; EUSAK WD, G–3 Sec, 30 Oct 50; 6th Med Tk Bn WD, 30 Oct 50; 8th Cav Regt Unit Jnl, entry at 301915 Oct 50.

fighting was in progress. The 1st Battalion, 8th Cavalry, actually relieved the ROK's north of Unsan at 1600 when the latter fell back through its lines, thus requiring Millikin's men to hold where they were. Watching the action across the river in the area held by the ROK 15th Regiment, while this change was taking place in his own front, Colonel Millikin saw through his field glasses that the hillside seemed alive as waves of enemy troops moved along the ridge leading into the ROK lines.[33]

The positions held in the late afternoon of 1 November by the 8th Cavalry were anchored on the right, about a mile northwest of Unsan, on the road below the village of Maebong-dong near the west bank of the Samt'an River. The line extended from there in an arc southwest across the mountain to a point three miles west of Unsan. There it crossed the east-west road out of Unsan and curved southeast to strike the main supply road, the Yongsan-dong and Yongbyon road, about three miles below Unsan. The 1st, 2d, and 3d Battalions were on this line in that order from east to west and southeast. This semicircle had a radius of about three miles from Unsan at the center, except on the north where the 1st Battalion was only a mile distant from the town. The 8th Cavalry regimental command post was on the main supply road south of Unsan between the town and the 3d Battalion command post. The day before, General Gay had established the 1st Cavalry Division command post at Yongsan-dong, twelve miles to the south.[34]

The arrival of the U.S. 8th Cavalry Regiment at Unsan had set in motion a redeployment of the ROK 1st Division. Upon being relieved west of Unsan on 31 October, the ROK 11th Regiment had shifted southeast to establish contact with the ROK 8th Division on the corps boundary. The ROK 12th Regiment moved to a rest and reserve assembly area at Ipsok south of the Kuryong River, six air miles from Unsan. Still engaged in the battle at Unsan, the ROK 15th Regiment was desperately trying to hold its position across the Samt'an River east of the 8th Cavalry Regiment. In short, the U.S. 8th Cavalry Regiment was to the north, west, and south of Unsan; the ROK 1st Division to the northeast, east, and southeast of it.

On the morning of 1 November, pursuant to I Corps and division orders, Colonel Johnson, the 5th Cavalry regimental commander, made ready to move a battalion of the 5th Cavalry eastward to bolster the disintegrating ROK II Corps lines near the I Corps boundary. Johnson alerted the 3d Battalion to move at 1230. At noon, Lt. Col. Hallett D. Edson, the 8th Cavalry regimental executive officer, arrived at the 5th Cavalry command post from Unsan. He told Johnson that about halfway between the two regiments, near the Nammyon River where it flows into the Camel's Head Bend of the Kuryong, he had encountered a great number of Korean refugees. They told him that a large force of Chi-

[33] Ltr, Millikin to author, 6 May 54; Interv, author with Hennig, 23 Mar 54; 70th Tk Bn WD, 1 Nov 50, Msg file, 1210, 1435; 8th Cav Regt Jnl file, 31 Oct–1 Nov 50, and 8th Cav POR 181, Msgs 311730, 011413, 011645, 31 Oct–1 Nov 50.

[34] Ltr, Millikin to author, 6 May 54 and attached sketch map; Ltr, Lt Col William Walton (CO, 2d Bn, 8th Cav, Nov 50) to author, 27 Aug 54 and attached sketch map; I Corps POR's 150 and 151, 1 Nov 50; 8th Cav Regt Unit Jnl, 011430 Nov 50.

nese was approaching behind them from the west. Since it was the 5th Cavalry's mission to protect the rear of the 8th Cavalry, Edson asked Johnson what he would do about the report. Johnson at once directed the 1st Battalion to send at least a platoon-sized patrol to investigate it.[35]

Johnson then departed with the 3d Battalion and placed it in a defensive position six miles northeast of Yongbyon on a low line of hills astride the Yongbyon–Kujang-dong road, facing east. This accomplished, he and the battalion commander proceeded on eastward some miles until they encountered a mass of retreating troops of the ROK II Corps. Johnson said of them: "They were a solid mass of soldiers on the road—indifferent to vehicles moving, indifferent to all that was around them. They were a thoroughly defeated outfit at this particular time." Johnson had served in the Philippines during the fall of Bataan in World War II and he likened the appearance and behavior of these ROK's to what he had seen on Bataan just before the American surrender there.[36]

Johnson turned back, passed through the picturesque walled city of Yongbyon, and that evening, shortly after dark, arrived at his command post at Yongsan-dong. There he learned that the 1st Battalion patrol in the early afternoon had found Chinese soldiers astride the road opposite the Turtle Head Bend of the Kuryong River, four air miles south of the 3d Battalion, 8th Cavalry, positions. During the afternoon two companies of the 1st Battalion had moved to the scene and were now engaged in battle there with a large enemy force.

### *On the West Coastal Road*

Westward in the coastal area, the pursuit seemed to be going forward with success. On the evening of 25 October the 27th British Commonwealth Brigade crossed the Taeryong River at and near Pakch'on. On the west side it met enemy opposition. On the 27th the Middlesex 1st Battalion led the brigade attack, and three miles west of the river engaged an enemy force in a severe battle. In the course of it, air strikes and artillery preparations helped the infantry by knocking out ten North Korean T34 tanks and two self-propelled guns. After this battle, Brigadier Coad was convinced that the days of "rolling" were over, and he adopted a brigade formation better suited to heavy combat. On the 28th, after a 15-mile advance, the brigade stopped three miles from Ch'ongju.[37]

The next morning, 29 October, the Australian 3d Battalion attacked toward Ch'ongju. Aerial observers reported at least four enemy tanks with infantry on the ridge overlooking the road at the pass. In strikes against these positions with napalm and rockets, the Air Force destroyed four tanks. The Australian attack then gained the pass and the adjacent ridge lines. That evening the North Koreans attacked the Australians there in the two hours preceding midnight,

[35] Interv, author with Col Harold K. Johnson, 28 Apr 54; Johnson, MS review comments, recd Aug 54.

[36] *Ibid.;* Ltr, Millikin to author, 6 May 54.

[37] 27th British Commonwealth Brig Sitrep, 24–28 Oct 50; 24th Div WD, 26–28 Oct 50; EUSAK POR 316, 25 Oct 50; Bartlett *With the Australians in Korea*, pp. 32–34.

employing self-propelled gun and tank support. Australian bazooka teams destroyed three enemy T34's. Supporting American tank fire helped repel the attack. The Australians lost nine killed and thirty wounded in the battle before Ch'ongju. The next morning the Argylls entered Ch'ongju.

That evening, 30 October, North Korean high velocity shells landed in the vicinity of the town, six of them in the headquarters area of the Australian 3d Battalion. One of the six shells cleared a crest, hit a tree, and exploded outside Colonel Green's tent. Colonel Green was asleep inside on a stretcher. Strangely enough, although there were numerous soldiers in the area at the time, no one was injured except Colonel Green, who was struck in the stomach by a shell fragment. The seriously wounded officer was taken to the surgical hospital at Anju. There three days later the much admired Lt. Col. Charles H. Green, the Australian 3d Battalion commander, died. Lt. Col. I. B. Ferguson succeeded him in command of the battalion. The same night that North Korean fire struck down Colonel Green, similar artillery or tank fire killed Major Reith of the 27th Brigade.

Brigadier Coad on the 30th asked General Church, commanding the 24th Division, to pass a regiment through his British troops at Ch'ongju because they were very tired. Acceding at once, Church ordered the 21st Infantry Regiment to lead the advance.

At dark that evening, Lt. Col. Gines Perez' 2d Battalion, 21st Infantry, passed through the British lines and headed north past the burning houses of Ch'ongju. The moon was up and the silvery light promised to aid the night attack. Beyond Ch'ongju the men could hear the rumble of withdrawing North Korean tanks. At 0200 on high ground two and a half miles west of the village of Kwaksan seven enemy tanks and about 500 North Korean infantry troops tried to ambush the battalion column. The nearest enemy tank opened fire with its cannon at 300 yards' range. Other enemy tanks joined the fire, their shells looking like big orange balls as they came streaking down the road. Several of them hit American tanks, but all bounced off. These tanks returned fire at the enemy gun flashes. Colonel Stephens, the regimental commander, and Colonel Perez, the battalion commander, from their radio jeeps directed the battle that was now joined. By dawn the North Koreans abandoned their position, leaving behind fifty dead, 5 knocked-out tanks, 1 self-propelled gun, and 7 antitank guns. After daylight an air attack destroyed 2 more enemy tanks, and the infantry captured 2 on flatcars.[38]

After this night of battle, the regiment encountered only light resistance. Its advanced troops, Lt. Col. Charles B. Smith's 1st Battalion, by noon of 1 November reached the outskirts of Chonggo-dong, eighteen air miles from Sinuiju and the Yalu River.

There, acting on orders from the 24th Division, Colonel Stephens ordered the battalion to halt, consolidate its position, and be prepared to defend in depth. The order from the 24th Division for the

[38] Interv, author with Col Gines Perez, 6 Aug 51; Interv, author with Maj Charles R. Alkire (S-2, 21st Inf Regt), 6 Aug 51; 24th Div WD, 29-31 Oct 50; 21st Inf Unit Rpt 115, 31 Oct 50. Eighth Army General Order 244, 26 April 1951, awarded the Distinguished Service Cross to Colonel Perez.

regiment to halt, which in turn had come from I Corps, hit the 1st Battalion, 21st Infantry, "like a bolt out of the blue." [39]

At midafternoon seven enemy tanks and an estimated 500 infantry attacked the 1st Battalion at Chonggo-dong. Capt. Jack G. Moss, commanding A Company, 6th Medium Tank Battalion, led his tanks out to meet the enemy armor. In a blazing tank battle that lasted half an hour, they destroyed it. Two of the American tanks were slightly damaged. The fire of infantry, tanks, and artillery turned back the North Korean infantry and inflicted on it an estimated 100 casualties.[40]

This was the northernmost action fought by a unit of the United States Eighth Army in the Korean War. By a strange coincidence, the infantry element engaged was Colonel Smith's 1st Battalion, 21st Infantry, part of which had fought the first American ground battle near Osan nearly four months earlier. In a geographical sense at least, the action near Chonggo-dong on the afternoon of 1 November was the high-water mark of Eighth Army's effort to reach the Manchurian border and consolidate all of Korea for a unified government.

Advancing behind the 27th British Commonwealth Brigade, the 5th Regiment had been the first unit of the 24th Division to cross the Ch'ongch'on River. It continued on and crossed the Taeryong River above Pakch'on on 28 October. From there it marched northward on the right of the British troops toward Taech'on. On the 29th in a heavy battle it and supporting air units destroyed nine enemy tanks and four self-propelled guns. The 5th Regimental Combat Team then entered Taech'on. Two of 89 prisoners taken were Chinese, the first captured by American troops in the Korean War. They turned out to be deserters or stragglers from their units. There were no CCF units in contact with the 5th Regimental Combat Team at Taech'on.[41]

From Taech'on, Colonel Throckmorton's 5th RCT turned northwest toward Kusong. An enemy force estimated to number 5,000 to 6,000 men, supported by tanks, self-propelled guns, artillery, and mortars stubbornly opposed the advance. But the 5th RCT, strongly supported by tactical air, captured Kusong just after noon on the 31st. The regimental attack the next day secured the road junction a few miles north of Kusong. In this action the regiment killed an estimated 300 to 400 enemy soldiers, and destroyed 2 self-propelled guns, 8 76-mm. howitzers, 8 mortars, 6 antitank guns, and 5 machine guns. Advanced elements of the 5th RCT were about ten miles north of Kusong at midday when a liaison plane came over and dropped a message, as it had to the 21st Infantry along the coastal road. This ordered the regiment to stop and hold in place.[42]

---

[39] 21st Inf WD, 1 Nov 50; EUSAK WD, G-3 Sec, 1 Nov 50; 24th Div WD, 1 Nov 50.

[40] 21st Inf Unit Rpt 116, 31 Oct-1 Nov 50; Interv, author with Lt Col Charles B. Smith, 6 Nov 51; *Armor* (May-June, 1951), article by 1st Lt. Robert D. Wilcox (tk plat ldr in the action), p. 28.

[41] Interv, author with Col John L. Throckmorton, 16 Apr 54; Ltr, Thompson to author, 9 Apr 54.

[42] 24th Div WD, 30 Oct-2 Nov 50; EUSAK WD, Br for CG, 30 Oct 50; Interv, author with Maj Gen Garrison H. Davidson, 28 Jan 54.

The uncertainty of the 21st and 5th Regiments during the afternoon and evening of 1 November over their future courses of action was resolved an hour before midnight when the 24th Division ordered them to withdraw toward the Ch'ongchon.[43] Mystified and disappointed, the men of the two regimental combat teams traveled back toward the Ch'ongch'on the night of 1–2 November. They were to learn later that the explanation for this, to them, puzzling development lay in events that had taken place in the east. The day before, 31 October, General Walker verbally had ordered General Milburn to limit the 24th Division attack in keeping with the situation in the Unsan area. There lay the controlling events.

### *The X Corps' Changing Mission*

Contemporary events unfolding at the X Corps front in northeast Korea complete the view of the whole situation at the time the Chinese were first appearing in the suddenly changed picture of the almost ended war. As late as 16 October General Almond had received orders from General MacArthur's headquarters in Tokyo that, upon landing, X Corps would attack west along the Wonsan-P'yongyang axis. But the next day he received an alert order from General MacArthur stating that if Eighth Army had captured P'yongyang before X Corps landed, the corps would advance north instead of west. On the 18th came an alternate order from MacArthur which provided that, if P'yongyang was captured before D-day, X Corps would advance north in a zone parallel to Eighth Army, with the watershed of the Taebaek Range as the boundary between the two forces. The next day, 19 October, X Corps received the final and definite order to advance north.[44]

On 20 October General Almond flew from the USS *Missouri* by helicopter to the Wonsan Airfield. At noon he assumed command of troops in the X Corps area north of latitude 39° 10′ north and east of the Taebaek Range. The X Corps command post was now officially in Wonsan. By the end of October, X Corps had teletype communications with Eighth Army and the 2d Logistical Command, and radio communications with GHQ in Tokyo.[45]

The troops which General Almond found in his corps area on 20 October consisted of his own small command group of approximately 10 officers and 30 men, ROK I Corps troops numbering more than 23,000 men, and the few hundred troops of the 1st Marine Air Wing already at the airfield. His own X Corps troops still afloat would bring the total to nearly 84,000 men. In addition, there was the U.S. 3d Division which he expected would soon join the corps. Of the approximately 84,000 men then in the corps, more than a third—32,000—were South Korean soldiers. The major tactical organizations were the U.S. 1st Marine and 7th Infantry Divisions, and the 3d and Capital Divisions of ROK I

[43] 24th Div WD, 1–2 Nov 50; 21st Inf Unit Jnl, Msg 31, 012211 and Msg 35, 012300 Nov 50; EUSAK WD, G–3 Sec, 31 Oct 50.

[44] X Corps WD, Oct 50 Summ, p. 3; *Ibid.*, Opns, p. 19, citing CINCUNC Msg CX66705, 17 Oct 50, and CINCUNC Msg CX66739, 19 Oct 50; *Ibid.*, Diary, CG X Corps, 16 and 18 Oct 50.

[45] X Corps WD, Summ and Diary, CG X Corps, 20 Oct'50; *Ibid.*, Oct 50 Summ, Sig Sec, pp. 29–30.

Corps.[46] The only known organized enemy groups in the X Corps area at this time were north and northeast of Hungnam.

Rough terrain characterized the area in northeast Korea assigned to the X Corps. (*See Map 21.*) Even the coastal plain hardly deserved that name; the only level or semilevel land there consisted of isolated pockets extending inland generally for a distance of from three to five miles. These were separated from each other by hill spurs that came down to the sea. The Wonsan-Hamhung pocket is by far the largest of these northeast coast semilevel areas. Wonsan and Hamhung, and the latter's port of Hungnam, were the largest centers of population in the X Corps zone of operation. Wonsan in 1949 had a population of 150,000, but it had fallen to an estimated 90,000 in October 1950. Hamhung's population in October 1950 was placed at 80,000, at least 40 percent of it Communist or Communist-inclined. Chemical, dye, medical, gunpowder, and fertilizer plants in the Hungnam-Hamhung area made it the outstanding industrial area of Korea. At the ports of Wonsan and Hungnam ice is unusual, and when it occurs it is so thin as to be unimportant.

Back of the coastal strip lies the northern Taebaek Range with its steep slopes and narrow, twisting valleys. The peaks in the highest parts of the range reach an altitude of 6,000 feet or more. In the interior part of the northern Taebaeks the winter temperatures often reach 20° to 30° below zero. Snow in October and November is normally infrequent, and in December not usually heavy enough to form deep, permanent drifts. But rivers in the Taebaek Range usually freeze over by mid-December. Beginning forty air miles northward from Hamhung and extending another forty miles north in the heart of the Taebaek Range lies the Changjin Reservoir. Fifteen air miles east of it lies the smaller Pujon (Fusen)[47] Reservoir.

The principal road north from the Wonsan-Hamhung plain climbs the Taebaek Range to the Kot'o-ri plateau and then continues on to Hagaru-ri at the southern end of the Changjin Reservoir. From the Hamhung area a second important road, the east coast road, curves northeast toward the border of the Soviet Union. Inland from this coastal road the communication routes were poor—in places scarcely more than mountain trails.

Almond's general plan of deployment in this mountainous waste of northeast Korea was for the ROK I Corps to advance to the northeast border along the coastal and adjacent roads; the U.S. 7th Infantry Division, southwest of the ROK's, to advance to the northern border over the Iwon-Pukch'ong-Hyesanjin corridor; southwest of the 7th Division,

[46] X Corps POR 24, 20 Oct 50. The actual strengths of the various organizations were as follows:

| *X Corps* | *U.S.* | *ROK* |
|---|---|---|
| Total | 51,489 | 33,692 |
| Hq and Hq and Service Group | 3,870 | 80 |
| X Corps Combat Troops | 1,479 | |
| 1st Marine Div | 24,031 | |
| Army Attached Marine Div | 93 | |
| Engr Special Brig | 1,393 | |
| X Corps Tactical Air | 1,786 | |
| 7th Div | 18,837 | 7,804 |
| 1st and 5th ROK Marines | | 2,159 |
| Hq and Hq Co. ROK I Corps | | 664 |
| Capital Div (ROK) | | 11,626 |
| 3d Div (ROK) | | 11,359 |

[47] See p. 729, n. 1, below.

the U.S. 1st Marine Division to advance northward from Hamhung to the Changjin Reservoir, with its specific route beyond that point dependent on tactical developments in its front; and the U.S. 3d Division, when it arrived, to secure the Wonsan-Hamhung area, keep open the corps lines of communication, and protect the corps rear and left flank from guerrilla interference. Until the 3d Division arrived, the 1st Marine Division would have the responsibility of securing the Wonsan-Hungnam area. Accordingly it would not be entirely free to concentrate for the advance northward. As General Almond himself said a little later of X Corps, "We are scattered all over the landscape." But, generally, the deployment was controlled by the road net of the area in which the corps was to operate.[48]

On 26 October General Almond issued orders for his plan of operation. In its zone, the 1st Marine Division was split into three regimental combat teams: (1) the 1st Marines to relieve ROK I Corps elements in the Kojo and Majonni areas south and west of Wonsan; (2) the 5th Marines to secure the Wonsan area, the Yonp'o Airfield south of Hungnam, and the X Corps west flank; and (3) the 7th Marines to relieve the ROK 3d Division along the Hamhung-Changjin Reservoir corridor, and to secure the power installations of the Changjin and Pujon Reservoirs.[49]

### *The CCF Block the Way to Changjin Reservoir*

Acting upon its orders, the ROK I Corps had attacked north from the Hamhung area—the 3d Division north toward the Changjin Reservoir and the Capital Division northeast up the coastal road. The 26th Regiment led the advance for the ROK 3d Division. On the morning of 25 October two battalions of the regiment approached the first and second hydroelectric plants of the Changjin Reservoir area, about thirty road miles inland from Hungnam, and halfway to the reservoir itself. A message from Maj. Malcolm Smith, KMAG adviser with the regiment, to Colonel Emmerich that evening informed him that the regiment had captured a prisoner definitely identified as a Chinese soldier who said he belonged to the *5th Regiment* of the Chinese *8th Army*. This prisoner said there were 4,000 to 5,000 Chinese in the immediate vicinity.

During the next two days the ROK regiment moved ahead very slowly against increasing resistance. On the morning of 28 October the ROK's attacked in the vicinity of Sudong in what proved to be a very costly action, and suffered heavy casualties. ROK patrols to the Sinhung-ni and Koto-ri areas brought back news that they had seen at both places what they believed to be Chinese soldiers. That day two Chinese

---

[48] X Corps WD, 4 Nov 50, G–1 Rpt, Notes on Conference between CG X Corps and Partridge, 4 Nov 50; X Corps WD, 26 Oct 50; *Ibid.*, Catalogue of Plans and Orders, p. 50, Opn Instr 13, 261000 Oct 50.

The corps boundary was changed slightly on 28 October to run from the Sea of Japan at the 38th Parallel to longitude 128° east, thence northwest to Poftong-ni, longitude 127° 5′ east, latitude 38° 58′ north, thence north to the 39th Parallel, thence west on that parallel to longitude 126° 45′ east. See EUSAK WD, G–3 Sec, 28 Oct 50.

[49] X Corps WD, Oct 50, Opns, p. 20; 1st Mar Div SAR, 8 Oct–13 Dec 50, an. C, pp. 9–10.

soldiers were captured one mile west of Sudong.

All day of the 29th small arms close combat continued in the large fields around the second hydroelectric plant. In the afternoon enemy 120-mm. fire increased. The ROK troops at the same time began to show signs of demoralization as their supply of grenades ran low. In the fighting on this day, the ROK's captured sixteen more Chinese soldiers and learned from them that the *370th Regiment*, CCF *124th Division, 42d Army, XIII Army Group,* blocked the way north, with the rest of the division nearby. North Korean tanks supported these Chinese. The Chinese division and regimental headquarters reportedly were at Hagaru-ri at the southern end of the Changjin Reservoir. On the 30th, after a heavy battle with the CCF, the ROK 26th Regiment withdrew a short distance to a stronger defensive position.[50]

The capture of the sixteen Chinese on the 29th was a considerable prize, and General Kim, the ROK I Corps commander, telephoned the news to General Almond. The next day, 30 October, General Almond went to the ROK I Corps command post at Hamhung and personally inspected the captives and interviewed them through an interpreter. The Chinese told him they had not eaten for three days. They said they had crossed the Yalu River at Manp'ojin on 16 October (later they said they had crossed on the 14th) and had marched from there on foot at night, their mortars being carried on packhorses and mules. Most of the sixteen soldiers were members of the *Mortar Company, 370th Regiment.* At the time of their capture they said three of their four mortars had been destroyed and the fourth had been withdrawn. The men were well-clothed, healthy, and averaged twenty-eight to thirty years in age. They asserted that their entire division had crossed into Korea and marched to the front. Most of the men in this division had been in Chiang Kai-shek's Nationalist Army, stationed near Peiping, until about a year earlier, they said. Their division had surrendered there to the Communists and was immediately taken into the Red Army. General Almond at once sent a personal radio message to General MacArthur informing him of the presence of CCF units in northeast Korea and giving such details as he had learned in the course of his interview with the prisoners.[51]

The *370th Regiment* apparently arrived at its positions near Sudong on 23 or 24 October and first encountered ROK troops on the 25th. Behind it came the other two regiments of the *124th Division,* the *371st* and *372d,* one a few days behind the other. When General Almond visited General Kim again on 31 October, he learned that seven more CCF prisoners had been captured to make a total of twenty-five now in the X Corps zone. Some of them said a second CCF division was near the Changjin Reservoir.[52]

---

[50] X Corps WD, Oct 50, p. 15; X Corps PIR 33, 29 Oct, and 35, 31 Oct 50; Emmerich, MS review comments, 30 Nov 57.

[51] X Corps WD, Diary, CG X Corps, 30 Oct 50; *Ibid.*, G-1 Rpt, Notes on Conference between CG X Corps and Partridge, 4 Nov 50; X Corps PIR 34, 30 Oct 50; Interv, author with Almond, 13 Dec 51; Ltr, McCaffrey to Almond, 1 Dec 54, forwarded to author.

[52] X Corps WD, Diary, CG X Corps, 31 Oct 50; *Ibid.*, PIR 36, 1 Nov 50; ATIS Interrog Rpts (N.K. Forces), Issue 18, 2324, p. 57.

A search of enemy dead showed they carried no official identification, although a few had written their names and units in ink on the left inside of their blouses. The officer's uniform differed from that of the rank and file only by a vertical red piping on trousers, on the left side of the jacket, around the collar, and a diagonal across the sleeve cuff. The uniforms were heavily quilted cotton, usually a mustard brown color, although some of the Chinese wore dark blue. The quilted uniform was warm until it became water-soaked; then it was difficult to dry. Underneath it the soldiers wore the summer uniform and any other clothing they owned. Their laceless shoes were of cloth, low-cut, rubber-soled, and worn with sets of cotton socks. Heavy cotton caps had ear flaps that gave neck protection. These soldiers were armed mostly with Japanese rifles confiscated in Manchuria at the end of World War II. The greater part of their mortars, machine guns, and Thompson submachine guns were American-made, having been captured from the Nationalists. Approximately 70 percent of the *124th Division* had formerly been Nationalist soldiers in Chiang Kai-shek's army. Having left its artillery behind because of the mountains, the *124th Division* in the battle below the reservoir used nothing larger than 82-mm. mortars.[53]

North Koreans as well as the newly arrived CCF fought in the Sudong area. On the 29th, for instance, in addition to the sixteen Chinese, the ROK 26th Regiment captured sixty North Korean prisoners. There, as in front of Eighth Army at the same time, the North Koreans fought delaying actions while they backed up to a point where they met the approaching Chinese or the latter lay in wait. It was perhaps a coincidence that the Chinese entered action on both fronts, first against ROK troops as it chanced, at almost the same moment. In both east and west that fateful day was 25 October.

---

[53] X Corps PIR 43, 8 Nov and 46, an. 2, 11 Nov 50; 1st Mar Div SAR, vol. I, pp. 15, 19, 20, 30, an. B, 6–8 Nov 50; SSgt Robert W. Tallent, "New Enemy," *Leatherneck*, vol. XXXIV, No. 2 (February 22, 1951), pp. 12–14.

# CHAPTER XXXIV

# Unsan

So ends the bloody business of the day.

HOMER, *Odyssey*

In the Unsan and Onjong area at the end of October, great smoke clouds hung in the skies. What did these smoke clouds portend? Everyone in the area noticed them. Capt. Jack Bolt, commanding officer of C Battery, 99th Field Artillery Battalion, counted ten different forest fires burning in the mountains when his unit moved up on the 30th to support the 3d Battalion, 8th Cavalry Regiment, south of Unsan. The next day Colonel Johnson witnessed much the same thing during his visit to the 8th Cavalry regimental command post. And General Allen, the 1st Cavalry Division assistant commander, likewise saw them on 1 November when he drove to Unsan. These great smoke clouds north and northeast of Unsan came from forest fires set by the enemy. They obscured U.N. aerial observation and masked enemy troop movements.

The evidence on 1 November particularly indicated that some large enemy movement was in progress. That morning a Korean civilian reported that 2,000 Chinese soldiers were in a valley nine miles southwest of Unsan and that their mission was to move eastward and cut the road below the town. A member of a Home Guard unit reported there were 3,000 Chinese on Obong-san, six miles southwest of Unsan. Colonel Edson, in talking to Colonel Johnson, apparently referred to this force. At noon, air and artillery had dispersed an enemy column eight miles southeast of Unsan, killing approximtaely 100 horses and an unknown number of men. This column was approaching the ROK 11th Regiment positions which were then near the ROK II Corps boundary. In the afternoon aerial observers reported sighting large columns of enemy troops in motion northeast and southwest of Unsan. An air strike hit one of these columns, containing twenty-one vehicles loaded with troops, nine miles northeast of Unsan.[1]

At his command post at Yongsan-dong in the afternoon, General Gay and Brig. Gen. Charles D. Palmer, the division artillery commander, were listening to the chatter on the artillery radio set. Suddenly the voice of an observer in an L–5 plane directing fire of the 82d Field

[1] I Corps WD, Intel Summs, 142–43, and POR 150, 1 Nov 50.

Artillery Battalion (155-mm. howitzers) came in: "This is the strangest sight I have ever seen. There are two large columns of enemy infantry moving southeast over the trails in the vicinity of Myongdang-dong and Yonghung-dong. Our shells are landing right in their columns and they keep coming." The two places mentioned were about seven and five air miles respectively southwest and west of Unsan. General Palmer broke in on the radio to order the 99th Field Artillery Battalion to join in the fire on these enemy columns. General Gay, who had become uneasy about the dispersion of the 1st Cavalry Division, telephoned I Corps headquarters to request that the 7th Cavalry Regiment, which I Corps was holding south of the Ch'ongch'on, be ordered to join him at Yongsan-dong and that he be allowed to withdraw the 8th Cavalry Regiment a distance of several miles from Unsan. He also protested the use of the 3d Battalion, 5th Cavalry, at the corps boundary on the east. His requests were denied.[2]

While the record indicates general reluctance on the part of the American command to accept the accumulating evidence of Chinese intervention, at least one responsible staff officer seems to have agreed with the ROK interpretation of events at an early date. Col. Percy W. Thompson, G-2 of I Corps, briefed troops of the advanced party of the 1st Cavalry Division at I Corps headquarters when the division was committed in the Unsan area. He pointed out that they might be fighting Chinese forces. Their reaction was one of disbelief and indifference. This same attitude was apparent in the staff of the 8th Cavalry Regiment and some of the division officers when Colonel Hennig, who had been with the ROK 1st Division throughout the Unsan fighting, tried to tell them that they were up against Chinese forces. General Gay maintained that his first information on Chinese intervention came on 1 November when he visited General Paik at the latter's ROK 1st Division headquarters at Yongbyon. This is hard to reconcile with the fact that in the last two days of October officers and men of the 8th Cavalry Regiment at Unsan heard a great deal about the Chinese from the ROK 1st Division troops and the attached KMAG officers. Apparently most of the officers and men of the 8th Cavalry Regiment received this information with skepticism or disbelief.[3]

In the early afternoon of 1 November General Walker telephoned to General Milburn and told him the ROK II Corps had ceased to be an organized fighting organization, and that his right flank was unprotected. Walker told Milburn to take measures to protect his flank and to assume command of any ROK units that came into the U.S. I Corps area. General Milburn set out immediately for Kunu-ri to see the ROK corps commander, after giving orders to his chief of staff, Brig. Gen. Rinaldo Van Brunt, to organize a blocking force to take a position on the Kunu-ri–Anju road southwest of Kunu-ri. This blocking force was composed principally of Engineer and Ordnance troops. Its mission was to protect the I Corps right flank and the ponton bridges over the Ch'ongch'on River.

[2] Ltr, Gay to author, 19 Feb 54.

[3] Ltrs, Gay to author, 19 Feb, 15 Mar, and 24 June 54; Interv, author with Hennig, 23 Mar 54; Ltr, Thompson to author, 9 Apr 54.

When Milburn arrived at the ROK II Corps headquarters he found it in the act of moving to Sunch'on. The ROK corps commander told him that he had lost contact with and did not know the location of most of his subordinate units, that they were disorganized, and that so far as he knew he had only three battalions of the ROK 7th Division in the vicinity of Kunu-ri capable of fighting. Milburn told the ROK commander that he must hold Kunu-ri, and that a blocking force of U.S. troops west of the town would support him.[4]

Meanwhile, other disquieting events were taking place south of Unsan and behind the 8th Cavalry Regiment. When the platoon-sized combat patrol from the 1st Battalion, 5th Cavalry, in the early afternoon of 1 November moved north from Yongsan-dong it found its way blocked at a point four air miles, or six to seven road miles, below the position of the 3d Battalion, 8th Cavalry Regiment. (*Map 23*) As radio reports told him of the enemy's strength, the battalion commander rapidly reinforced the platoon with the full strength of A and B Companies. The enemy force held a position on the ridge extending across the road just south of the Turtle Head Bend of the Kuryong River.

Upon Colonel Johnson's return to the 5th Cavalry command post in the evening, the 1st Battalion commander requested him to release the third rifle company. Johnson approved the request, and C Company moved north. In the meantime, shortly after dark, the Chinese at the roadblock attacked the two companies in front of them and drove B Company from its position with the loss of four 81-mm. mortars and other equipment. Colonel Johnson then directed the withdrawal of A Company to the defensive position C Company established near midnight. There, A and B Companies assembled for reorganization. Johnson alerted the 2d Battalion at 2300, and two hours later ordered it north to support the 1st Battalion.[5]

By noon of 1 November, therefore, the Chinese had cut and blocked the main road six air miles south of Unsan with sufficient strength to turn back two rifle companies which had been strongly supported by air strikes during daylight hours. The CCF had set the stage for an attack that night against the 8th Cavalry Regiment and the ROK 15th Regiment. When dusk fell that evening enemy soldiers were on three sides of the 8th Cavalry—the north, west, and south. Only the ground to the east, held by the ROK 15th Regiment, was not in Chinese possession.

### *North of the Town*

The Chinese attack north of Unsan had gained strength in the afternoon of 1 November against the ROK 15th Regiment on the east, and gradually it extended west into the zone of the 8th Cavalry Regiment. The first probing attacks there, accompanied by mortar barrages, came at 1700 against the right flank unit, the 1st Battalion. There was also something new in the enemy fire support—rockets fired from trucks. These

[4] MS review comments, Milburn, Nov 57.

[5] Interv, author with Johnson, 28 Apr 54; Johnson, MS review comments, Aug 54; I Corps WD, 1 Nov 50 Intel Summ 142; 5th Cav WD, 1 Nov 50; I Corps POR 153, 2 Nov 50.

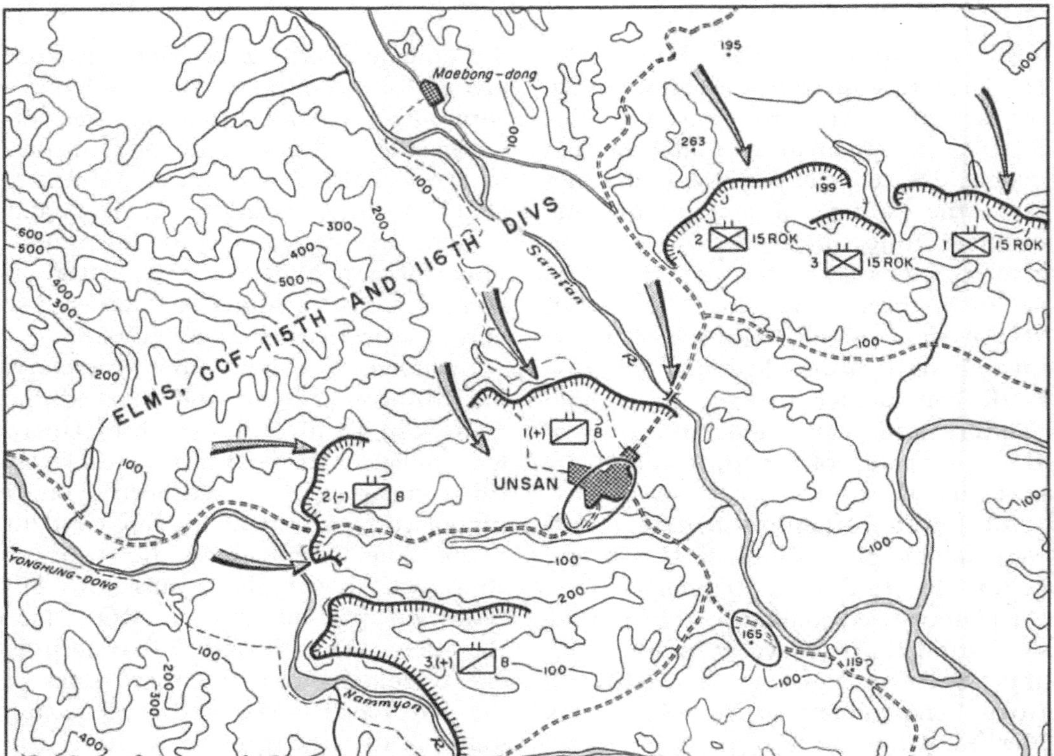
THE UNSAN ENGAGEMENT
8th CAVALRY REGIMENT
Night, 1–2 November 1950
8TH CAV AND 15TH ROK POSITIONS (APPROX) EVENING, 1 NOV
CCF BLOCKING POSITION, 1 NOV
CCF ATTACK, NIGHT, 1–2 NOV
PRINCIPAL CCF BLOCKING POSITIONS, NIGHT, 1–2 NOV
Elevations in meters
MILES
KILOMETERS
ELMS, CCF 115TH AND 116TH DIVS
Maebong-dong
Samtan R
UNSAN
YONGHUNG-DONG
Nammyon R
CAMEL'S HEAD BEND
OBONG MT
593
FORD
TO IPSOK 2 MI
TO MYONGDANG-DONG
"BUGLE HILL"
TURTLE HEAD BEND
Kuryong R
1/5 CAV
AS OF 2400, 1 NOV
15 ROK
L. Booth

MAP 23

rocket vehicles were to the northeast, across the Samt'an River. Supporting artillery soon located and forced the rocket vehicles to move, but not before their rockets had struck an ammunition truck at the battalion command post. Major Millikin's men recovered one of the rocket shells and found that it was of the Russian Katushka 82-mm. type fired from four multiple tubes, truck-mounted.

At dusk Millikin's 1st Battalion controlled the river approaches from the north except for portions in the ROK 15th Regiment zone on the east side. Millikin's position was weak on the left, however, where troop strength did not permit him to extend far enough to reach the main ridge leading into Unsan. He had physical contact with the 2d Battalion in that direction only by patrols. Neither battalion held this ridge except for outposts.[6]

East of the river the ROK's were under heavy attack. In this action the ROK's captured two 57-mm. recoilless rifles and two automatic rifles with Chinese markings. At 1900 the 10th AAA Group, supporting the ROK's, issued a march order and in a tense atmosphere began packing its equipment. An hour and a half later the group closed its fire direction center, and at 2100 its motor convoy moved south under blackout conditions. The 78th AAA Battalion's 90-mm. guns, which were tractor-drawn and could be moved quickly, remained behind and continued to fire in support of the ROK's for an hour or two longer; then they too withdrew on corps orders. After about 2300 the ROK 15th Regiment disintegrated rapidly, and shortly after midnight ceased to exist as a combat force. Very few of these ROK troops escaped; they were either killed or captured.[7]

A lull in the fighting at Millikin's position ended at 1930 when the Chinese struck his battalion all along its line. They drove the right flank back 400 yards. The left flank then withdrew half that distance. Millikin rushed fifty men from the Engineer platoon and the Heavy Mortar Company to the right flank, and with this reinforcement he held there. Heavy action continued. About 2100 the Chinese found the weak link on the ridge line and began moving through it down the ridge behind the 2d Battalion.

At 2200 the tanks holding the bridge northeast of Unsan in the right rear of Millikin's 1st Battalion reported large groups of men across the Samt'an River, moving south. The 4.2-inch mortars supporting the ROK 15th Regiment in that area had now quit firing. Radio reports from the ROK's made it clear that they were being defeated and pushed back. In order to ascertain what the situation was there, Millikin sent his assistant S-3 across the river in a jeep to locate the mortarmen. That officer, after crossing the river, was fired on but escaped and reported back to Millikin. The moon was now coming up. Night visibility was good.

Since it was apparent that enemy groups were passing him on the east, Millikin ordered the battalion trains and all noncombat vehicles to move south through Unsan to the road fork and be

[6] Ltr, Millikin to author, 6 May 54.

[7] Intervs, author with Hennig, 23-24 Mar 54; 10th AAA Group WD, 1-2 Nov 50.

prepared to move from there southeast across the Kuryong River ford in the ROK 1st Division zone to Ipsok. About the same time, Lt. Col. William Walton, commanding officer of the 2d Battalion, ordered his motor officer to take all vehicles in the motor pool across the river by this same route. These vehicles from the two battalions arrived safely at Ipsok.

With much sounding of bugles and whistles the Chinese extended their strong attacks westward to the 2d Battalion, and in a short time penetrated its right and encircled its left. At the same time the fight with the 1st Battalion went on. Near the battalion boundary, A Company reported that it was engaged in hand-to-hand combat with the enemy in two directions, had pulled back its left flank, and was withdrawing to the next ridge. On Millikin's right the tanks holding the ground along the river were being pressed back. By 2300 both the 1st and 2d Battalions had been forced back and their positions penetrated. The 1st Battalion had expended its basic load of ammunition and most of the reserve ammunition the regiment had sent forward. Millikin reported by radio to the regimental commander the increasingly desperate situation of the two front-line battalions and the fact that he was almost out of ammunition.[8]

While this night battle was increasing in intensity, an important conference was in progress at I Corps headquarters. During the afternoon, General Milburn and his staff had become more and more disturbed at reports of what was happening to the ROK II Corps eastward and of the increasing tempo of the action near Unsan. At noon on 1 November I Corps had ordered the 24th Division to halt its advanced units, then only a few miles from the North Korean border. Some hours later, about 1800 in the afternoon, General Milburn sent out a call for a meeting at corps headquarters that night to be attended by the commanding generals and certain staff members of the 24th Division, the 1st Cavalry Division, and the ROK 1st Division. Before General Gay left his command post for the I Corps meeting at Anju, he ordered Col. Ernest V. Holmes, his chief of staff, to send a warning order to the 8th Cavalry Regiment to be prepared to withdraw from Unsan.

The meeting at I Corps headquarters got under way about 2000. In this meeting General Milburn directed the corps to go from the attack to the defensive immediately. This was the first time I Corps had gone on the defensive since its breakout from the Pusan Perimeter. Milburn returned the 8th Cavalry Regiment to division control, and ordered that it and the ROK 15th Regiment withdraw at once from Unsan to positions above the Yongsan-dong–Yongbyon–Unhung east-west road. This would amount to a general withdrawal of approximately twelve air miles. Generals Gay and Paik were to co-ordinate the withdrawals of their advanced regiments, the ROK 15th Regiment to be the last to withdraw. General Gay telephoned Colonel Holmes from Anju, instructing him to issue the withdrawal order to the 8th Cavalry Regiment. Then he and General Paik and their parties

[8] Ltr, Millikin to author, 6 May 54; Ltr, Walton to author, 27 Aug 54; 8th Cav Unit Hist Rpt, 1 Nov 50; 70th Tk Bn Jnl, 10 Nov 50 (a report from CO B Co, 70th Tk Bn, for period 1–9 Nov 50).

started back to their headquarters. A little after midnight they reached the 1st Cavalry Division command post. There they learned the bad news from the Unsan front.[9]

Colonel Palmer received the withdrawal order from the 1st Cavalry Division about 2300. Fifteen minutes before midnight he issued a warning order alerting all battalions and the regimental trains for a withdrawal south. At midnight he issued the withdrawal order. The withdrawal route indicated was the only one possible—east from the road fork south of Unsan, across the ford of the Kuryong River, and then by the main supply route of the ROK 1st Division to Ipsok and Yongbyon. Major Millikin telephoned Colonel Walton that he would try to hold Unsan until the 2d Battalion cleared the road junction south of it. Then he would withdraw. The 3d Battalion, south of Unsan, was to bring up the regimental rear.[10]

In the 2d Battalion, Colonel Walton had lost communication by this time with all his companies except H Company. He gave that company the withdrawal order with instructions to relay it to the rifle companies since it still had communication with them. The 2d Battalion headquarters group, under ineffective enemy small arms fire, began withdrawing eastward to the sound of heavy firing in Unsan.

In Major Millikin's 1st Battalion area just north of Unsan, A Company had been forced from its left flank position and Chinese were infiltrating south along the ridge line into Unsan behind the battalion. At the same time, the Chinese were pressing hard against B Company on the right and the tanks of B Company, 70th Tank Battalion, along the river where they guarded the battalion's right flank. Millikin soon received word that the tanks had been forced back to the road junction at the northeast edge of the town. The tankers reported they would try to hold there until the 1st Battalion could withdraw past that point. Millikin issued orders for A and B Companies each to leave one platoon behind as rear guard, and for them and D Company to withdraw through C Company to the tank-held road junction. When Millikin himself arrived at the road juncture he found there two tanks and the D Company mortar vehicles. Other tanks had already passed on into Unsan. A din of small arms fire from Unsan indicated that the enemy held the town.

A few minutes later, about half an hour after midnight, elements of A and B Companies arrived at the road fork at the northeast edge of Unsan. Enemy troops in the town began firing at them and caused some casualties. Millikin then sent these A and B Company men around to the east of Unsan with instructions to wait for him at the road fork and bridge south of the town. Mil-

[9] Ltr, Gay to author, 19 Feb 54; Interv, author with Holmes, 26 Feb 54; Interv, author with Col Robert T. Hazlett (KMAG adviser to ROK 1st Div and present with Paik at I Corps conference), 25 Feb 54; Ltr, Thompson to author, 9 Apr 54; I Corps WD, 1 Nov 50. The verbal orders given at the conference were confirmed by I Corps in Operation Directive 19, published at 2200, 1 November.

[10] Ltrs, Millikin and Walton to author, 6 May and 27 Aug 54; Ltr, Col Hallett D. Edson, 16 Apr 54, and attached Exhibit A, Maj. William S. Coleman (S–3, 8th Cav Regt), Operations of the 8th Cavalry Regiment, 1–2 November 1950, in the Vicinity of Unsan (15 Nov 50); I Corps WD, Intel Summ 146; 1st Cav Div POR 182, 011700–021700 Nov 50; 8th Cav Unit Hist Rpt, 1 Nov 50; 70th Tk Bn Jnl, 10 Nov 50 (1–9 Nov 50, 70th Tk Bn Rpt).

likin and most of his staff remained at the northeast edge of Unsan. They hoped to direct the rest of the battalion on the escape route, and to send the mortar carriers with wounded, escorted by the two remaining tanks, through Unsan to the road fork southward.

Four tanks of the 1st Platoon of B Company, 70th Tank Battalion, already had fought their way through the town and arrived at the road fork south of it. It was the noise of this conflict in Unsan that Millikin and his men heard from the northeast edge of the town. Fifteen minutes later Millikin ordered the last two tanks and the mortar vehicles with the wounded to try to get through Unsan. A burning truck at the first turn going west into the town halted the column. In trying to get around the truck the first tank slid into a shell crater and got stuck. Chinese soldiers killed the tank commander as he struggled to free the tank. Other Chinese placed a satchel charge on the tracks of the second tank and disabled it. Of the ten tank crewmen, two were killed and five wounded. Apparently none of the wounded on the mortar carriers escaped.[11]

A little later, about 0100, a miscellaneous assortment of men, including elements of C Company, South Koreans attached to the 1st Battalion, ROK stragglers from the 15th Regiment, and Chinese soldiers, arrived at the road junction northeast of town at about the same time. Millikin still waited there. In the confusion that now spread out of control the men tried to escape in groups. Millikin and a small group went westward north of Unsan and then circled to the southwest. At 0200, they encountered parts of H Company from the 2d Battalion also trying to reach the road fork south of Unsan.

### *Roadblock South of Town*

When Colonel Palmer ordered the regimental withdrawal he placed Colonel Edson, the regimental executive officer, in charge of co-ordinating it and sent him to the road junction a mile and a half south of Unsan and a mile north of the regimental command post. That road junction was the critical point to be reached and passed by the scattered elements of the command. Accompanied by Capt. Rene J. Guirard, the regimental S–2, and two squads of the I&R Platoon, Edson arrived at the road junction just before midnight. Capt. Filmore W. McAbee, S–3 of the 3d Battalion, took one platoon of I Company and the company commander to the road fork about midnight, and after conferring there with Edson, he personally placed the platoon in position to protect the junction from the north.

The regimental trains passed through, as did the trains of the 1st and 2d Battalions; numerous groups of the 1st Battalion and some from the 2d Battalion also came through. The four 1st Platoon tanks arrived there about 0030. Edson placed them in defensive positions at the road junction. They remained there until two tanks of the 2d Platoon arrived. Then Edson ordered the first group of tanks to go southeast to the ford over the Kuryong River, and to protect it for the last part of the withdrawal. The two tanks that had just come through Unsan remained at the road junction.

[11] 70th Tk Bn, S–3 Jnl, Msgs at 0015 and 0030 2 Nov 50.

It was now about 0130, 2 November.[12]

As yet there had been no enemy action south of Unsan in the 3d Battalion area. Artillery elements supporting that battalion began withdrawing north through the road junction at this time. Headquarters and Service Battery and B Battery of the 99th Field Artillery Battalion passed eastward through the road junction. Next came C Battery. Captain Bolt, the battery commander, reached the road fork about 0220 at the head of his column of twenty vehicles, which included six prime movers towing six 105-mm. howitzers. He stopped briefly to talk with Colonel Edson who told him everything was all right and to go on.

The withdrawal route ran generally east from the road fork for a mile before it turned southeast to the ford across the Kuryong River and hence to Ipsok, four miles south of the river. Immediately east of the road fork the road ran on an embankment built above rice paddies with ditches on either side. North of the road, a considerable expanse of paddy ground extended to the Samt'an River at a point just before it turned east in a sharp bend to flow into the Kuryong a mile away. On the south side the paddy ground gave way to high ground which culminated in Hills 165 and 119. They crowded close on the road beginning at a point about 200 yards east of the road fork.

Captain Bolt turned east on this road and had proceeded about 200 yards when, upon glancing back, he saw that the second vehicle was not behind him. He told his driver to stop the jeep; they waited. The second vehicle had continued on past the turn at the road fork, had had to back up, and in doing so had jammed the column and caused the delay.

As he waited, Captain Bolt happened to glance to his left across the paddy ground, and in the moonlight he saw a line of men coming toward the road. He thought they were retreating 8th Cavalry infantrymen and remarked about them to his driver. When the oncoming soldiers were about fifty to seventy-five yards away the entire group opened fire on the road. Bolt shouted to his driver to get going, and upon rounding a curve where the hill came down to the road they lost sight of the rest of the battery at the road fork. Just around the curve from 15 to 20 enemy soldiers stood in the road. They opened fire on the jeep as it raced toward them. Bolt returned it with his submachine gun. The enemy group scattered to the sides of the road. The jeep raced on and passed 2 other small enemy groups, the last one numbering no more than 3 or 4 soldiers. Bolt soon caught up with the end of the regimental column, which he found consisted of B Battery and the four tanks of the 1st Platoon, B Company, 70th Tank Battalion. He tried to get one of the tanks to go back and fire down the road, but the tank commander said he was out of ammunition.[13]

---

[12] Coleman, Opns of 8th Cav, 1–2 Nov; Ltr, Edson to author, 16 Apr 54; Interv, author with Guirard, 21 Aug 54; Ltr, Maj Filmore W. McAbee to author, 8 Feb 57; 70th Tk Bn WD, 10 Nov and S–3 Jnl Msg at 0030 2 Nov 50; 8th Cav Unit Hist Rpt, 2 Nov 50.

[13] Details of Bolt's encounter with the CCF east of the road fork are based on Interv, Capt Edward C. Williamson with Bolt, 11 Jul 51, as reported in Williamson, Ambush of Battery C, 99th Field Artillery Battalion, 29 Oct–2 Nov 50, MS, copy in OCMH.

The enemy force at the road apparently had followed the 1st Battalion from north of Unsan, coming down along the west bank of the Samt'an River, although it is possible that they had crossed the river from the ROK 15th Regiment zone on the east.

As Bolt's jeep disappeared around the turn of the road the enemy soldiers reached the road embankment and opened fire on the next vehicle when it approached. This caused the driver to lose control and the 2½-ton truck upset over the side of the embankment, dragging the 105-mm. howitzer it was towing crosswise on the road and blocking it. One of the two tanks at the road fork went forward to try to break the roadblock, but the upset truck and howitzer blocked the way and the tank came under attack. Crewmen abandoned the tank after disabling its weapons. There is some evidence that a Chinese satchel charge had already broken the tank treads. Bolt's jeep was the last vehicle to pass eastward from the road fork below Unsan. Thus, at 0230 the Chinese had effectively cut the only remaining escape road from Unsan.[14]

At the road fork confusion swept over bewildered and frightened men. No one, it seems, was able to gather together enough men to fight the enemy roadblock force. Colonel Edson apparently made such an effort but it failed, and in the end he and his group escaped by circling around and through the roadblock force eastward and then south into the hills. Captain Guirard had several personal encounters with Chinese in this escape. Most of the artillerymen caught in the roadblock disappeared into the hill mass south of the road. A few officers, including one from I Company, and some of the noncommissioned officers, tried to assemble the men who had abandoned their vehicles and equipment on the road. But the few men they were able to bring together disappeared as soon as they turned their backs on them to look for others. A few Chinese soldiers came down among the vehicles and threw grenades, but most of them stayed at their roadblock position. Soon enemy machine gun and mortar fire began falling on the road junction area from the adjacent high ground.[15]

After watching Colonel Edson and his party disappear to the east, Colonel Walton, who meanwhile had arrived from west of Unsan, returned to his own 2d Battalion group at the road junction and led them southward across the hills. He came in through ROK lines at Ipsok after daylight with 103 men. When Major Millikin and his 1st Battalion group met elements of H Company west of the town, Millikin placed his wounded in their vehicles, and the combined party came on to the road fork. They found it a shambles of wrecked and abandoned vehicles and equipment.

Behind Millikin and the H Company group, the rest of the 2d Battalion never succeeded in reaching the road fork south of Unsan. Half a mile west of it, at the edge of the town, an enemy force cut the east-west road. There the Chi-

[14] 70th Tk Bn, S-3 Jnl, Msg at 0230 2 Nov 50.

[15] Williamson, Ambush of Battery C; Interv, author with Guirard, 21 Aug 54; Ltr, Edson to author, 16 Apr 54.

nese stopped A Battery, 99th Field Artillery Battalion, and the 3d Platoon, B Company, 70th Tank Battalion. Soon abandoned vehicles clogged the road at this point. The congestion was so bad that even the tanks could not get through and their crew members abandoned them after destroying their weapons. A few of these men filtered through to the road fork, but most of them went south over the hills. The infantry elements of the 2d Battalion for the most part scattered into the hills. Many of them reached ROK lines near Ipsok. Others came in to the position of the 3d Battalion, 8th Cavalry, the next morning. The sound of the 9th Field Artillery Battalion at Ipsok firing in support of the ROK 1st Division served as a guide for most of the men caught in the Unsan roadblock, and they moved in that direction.

When Major Millikin and his group arrived at the road fork they found Maj. Robert J. Ormond, commanding officer of the 3d Battalion, 8th Cavalry, there with a platoon of infantry. This was the I Company platoon McAbee earlier had posted north of the road fork blocking the road from Unsan. Millikin queried Ormond to find out what the latest orders were, as he had been out of communication with everyone since directing the 1st Battalion withdrawal. Ormond replied that he had no recent information, that his last orders were to try and hold the road fork until the 1st and 2d Battalions had gone through, and that he believed large portions of them had already passed eastward. Ormond then turned back south to his own battalion to start its withdrawal. The whole general area of the road fork was now under enemy small arms fire, some of it coming from the south which at first had been free of enemy soldiers.

Millikin found scattered elements of the 1st Battalion near the road fork and he collected about forty men, including Capt. Robert B. Straight of B Company who was wounded. Straight had stayed behind with one platoon north of Unsan when the rest of his company had withdrawn. There was one operable tank still at the road fork. Using its radio, Millikin tried to communicate with elements of the regiment, but was able to reach only one tank which was then engaged in running a roadblock near the ford over the Kuryong. The 1st Battalion commander then ordered the tank to start toward the enemy roadblock. He was following it with his men when enemy fire scattered them. The small groups infiltrated the Chinese lines and headed south. Millikin and the men with him crossed the Kuryong just before daylight and reached Ipsok about 0800. There he found his battalion trains and about 200 men of the 1st Battalion, most of them from those parts of A and B Companies that he had sent southeast around Unsan at the beginning of the withdrawal.

About noon on 2 November practically all men of the 1st Battalion who were to escape had reached the Ipsok area, and a count showed that the battalion had lost about fifteen officers and 250 enlisted men to all types of casualties. About half the battalion's mortars and heavy weapons had been lost to the enemy. Most of the regimental headquarters; the regimental trains; four tanks of B company, 70th Tank Battalion; and five artillery pieces crossed the Kuryong River ford safely and assembled in the vicinity of Yongbyon. From there they rejoined the 1st Cavalry Division at Yongsan-dong.[16]

### *Ordeal Near Camel's Head Bend*

During the evening and first part of the night of 1 November the troops of the 3d Battalion, 8th Cavalry, south of Unsan and their supporting artillery and tanks had enjoyed undisturbed quiet. Some of them in the late afternoon had noticed airplanes strafing a few miles to the south and were aware that an enemy force in that vicinity was on their main supply road. Major Ormond just before midnight had passed on to his company commanders word of the impending withdrawal. Lt. Col. Robert Holmes, commanding officer of the 99th Field Artillery Battalion, gave instructions for the two batteries of artillery (B and C) to withdraw. Battalion headquarters and B Battery departed at 0115 and cleared the road fork south of Unsan. Last of the artillery to march was Captain Bolt with C Battery at 0200, and, as already noted, he encountered the first of the enemy roadblock force. A platoon of twenty-five men from K Company accompanied C Battery.[17]

The 3d Battalion had taken a position just north of the Nammyon River, where it flowed into the nose of the Camel's Head Bend of the Kuryong three air miles southwest of Unsan. Its mission was to guard the regimental rear. Major Ormond had established his command post in a flat plowed field with a tight perimeter formed by headquarters and M Company command groups. Two squads of M Company held the bridge immediately in the rear (south) of the battalion headquarters, and the 4th Platoon of B Company, 70th Tank Battalion, was disposed in position there on either side of the road north of the river. The tree line extending west along the Nammyon was held by L Company with one platoon on a high hill on the south side; I and K Companies in that order were on a ridge line running from northeast to southwest overlooking the stream northwest of the battalion command post. The communications switchboard and the S–2 and S–3 sections of the battalion headquarters found just off the road a ready-made dugout for their use in a 20-by-20-foot hole with a log and straw roof over it which the North Koreans had dug at some earlier date to hide vehicles from aircraft.[18]

Upon receiving the regimental order to withdraw, with the 3d Battalion assigned the mission of guarding the regimental rear, Major Ormond issued instructions for K and I Companies to withdraw from their positions to the battalion command post. Company L was to cover their withdrawal. None of the rifle companies was engaged with the

[16] Ltr, Millikin to author, 6 May 54; I Corps POR 153, 2 Nov 50; EUSAK WD, G–3 Jnl, 0840 2 Nov 50; 8th Cav Unit Hist Rpt, 2 Nov 50.

[17] Williamson, Ambush of Battery C; 99th FA Bn WD, 1 Nov 50.

[18] Ltr, Lt Col Veale F. Moriarty (Ex Off, 3d Bn, 8th Cav, Nov 50) to author, and attached sketch map, 11 Jun 54; Ltrs, McAbee (S–3 3d Bn, 8th Cav, Nov 50) to author, 20 Aug 54 and 8 Feb 57; Ltr, SSgt Elmer L. Miller to Capt Carlos L. Fraser, CO B Co, 70th Tk Bn, 6 Nov 50, from 4th Field Hospital (Miller was a tank commander at the 3d Bn command post); Ltr, Capt Walter L. Mayo, Jr. (Arty Liaison Off, C Btry, 99th FA Bn, with L Co) to author, 15 Jan 58, together with notes prepared by him for Unit Historian, 8th Cav Regt, in 1954. These sources form the principal basis for the following account of the 3d Battalion except as otherwise noted.

enemy, and no difficulty was expected. Major Ormond then drove northward to the regimental command post and subsequently to the road fork south of Unsan where Major Millikin saw him.

Ormond started back south just a few minutes before enemy troops cut the road below the fork. As it was, he returned to his command post without trouble. There he told certain members of his staff that the 3d Battalion could not withdraw northward through the road fork below Unsan as planned because that road was now held by enemy forces. Using a map Ormond showed Maj. Veale F. Moriarty, the battalion executive officer, the cross-country route he intended the battalion to follow and sent the motor officer off to find a ford by which the vehicles could cross the river. He then gave instructions to SSgt. Elmer L. Miller, in charge of a section of tanks near the command post, to cover the battalion withdrawal. Miller passed this word on to the 4th Platoon tank commander, and then went to examine the ford selected for the vehicular crossing. All the vehicles in the battalion area, except the tanks, were lined up on the road bumper to bumper ready to begin the withdrawal.

At this time, close to 0300, a company-sized column of men (one source said platoon-sized) from the south approached the bridge over the Nammyon River below the battalion command post. The two squads of M Company charged with security of the bridge let the column pass over the bridge thinking they were ROK's. When this column was even with the command post one of its leaders sounded a bugle. This was the signal for a deadly surprise assault on the battalion command post from all sides. At the same time, other enemy forces engaged L Company along the stream bank to the southwest, and still others crossed the stream directly south of the command post and attacked the tanks there. Sergeant Miller crawled back to his tank in time to help fight enemy troops off the decks with a pistol. The tanks on both sides of the road backed up to the road except one which was first damaged by a satchel charge and then, in a few minutes, blew up. At the road the tanks held off other enemy troops trying to cross the stream from the south.[19]

In the command post itself the greatest confusion reigned after the onset of the Chinese attack. Hand-to-hand encounters took place all over the battalion headquarters area as the Chinese soldiers who had marched across the bridge fanned out, firing on anyone they saw and throwing grenades and satchel charges into the vehicles, setting many of them on fire. Part of the men around the command post were still in their foxholes or shelters, some of them apparently asleep awaiting the order to start the withdrawal. One man later said, "I woke up when the shooting started." Another said, "Someone woke me and asked if I could hear a bunch of horses on the gallop . . . then bugles started playing taps, but far away. Someone blew a whistle, and our area was shot to hell in a matter of minutes." Still another man was awakened by an exploding hand grenade. Lt. W. C. Hill said, "I thought I was dreaming when I heard a bugle

[19] The 70th Tk Bn S-3 Jnl, Msg at 0300 2 Nov, reporting Miller's radio message on the destruction of this tank is the most reliable evidence on the time of the CCF attack against the 3d Battalion.

sounding taps and the beat of horses' hooves in the distance. Then, as though they came out of a burst of smoke, shadowy figures started shooting and bayoneting everybody they could find." [20]

When the shooting started, Major Ormond and Captain McAbee left the command dugout to determine the extent of what they thought was a North Korean attack. Major Moriarty, battalion executive officer, who was in the dugout at the time never saw Ormond again.

Once outside the dugout, Captain McAbee started for the roadblock at the bridge and Major Ormond veered off to the right to go to L Company by the river. As McAbee approached the bridge small arms fire knocked off his helmet and a few seconds later another bullet shattered his left shoulder blade. He turned back toward the command post and ran into a small group of enemy soldiers. He dodged around a jeep, with the enemy in pursuit. As they came around the jeep he shot them. In the field along the road he saw about thirty more enemy troops attempting to set a tank on fire. McAbee emptied his carbine into this group, and then, growing weak from loss of blood, he turned again toward the dugout. A few steps farther and three enemy soldiers stepped from the roadside ditch and prodded him with bayonets. Not trying to disarm him, they jabbered to each other, seemingly confused. McAbee pointed down the road, and after a little argument among themselves they walked away. Once more on his way to the dugout McAbee fell into the hands of a small group of Chinese, and repeated his earlier experience. After this second group walked off up the road, McAbee finally reached the command post.

Meanwhile, a few minutes after Ormond and McAbee had left the dugout, Capt. Clarence R. Anderson, the battalion surgeon, and Father Emil J. Kapaun, the chaplain, brought in a wounded man. The small arms fire continued unabated and Major Moriarty stepped outside to investigate. Visibility was good, and in the bright moonlight he saw Captain McAbee stagger toward him. Just beyond McAbee, Moriarty saw three or four uniformed figures wearing fur headgear. He grabbed McAbee and thrust him into the dugout. Close at hand someone called for help. Responding to the call, Moriarty clambered over the dugout ramp leading from the road and found the battalion S-4 rolling on the ground grappling with an enemy soldier. Moriarty shot this soldier with his pistol and another who was crouching nearby. For the next fifteen or twenty minutes he was one of the many in the command post area waging a "cowboy and Indian" fight with the Chinese, firing at close range, and throwing grenades.[21]

Seeing a center of resistance developing around Miller's tank, Moriarty ran to it and found about twenty other men crouching around it. When enemy mortar fire began falling near the tank, Moriarty took these men and with them crossed the stream to the south. They destroyed a small group of enemy troops at the stream bank. The south side appearing free of the enemy at that point,

[20] New York *Herald Tribune*, November 3, 1950, dispatches written at Ipsok, 2 November, by a correspondent who escaped from the Unsan area.

[21] Ltr, Moriarty to author, 11 Jun 54; Ltrs, McAbee to author, 20 Aug 54 and 8 Feb 57.

they proceeded southeast. During the night this party was joined by others from different units of the regiment. When they reached friendly ROK lines near Ipsok after daylight there were almost a hundred men in the group.

After perhaps half an hour of hand-to-hand fighting in the battalion command post area the Chinese were driven out. In the meantime, most of L Company had withdrawn from the stream's edge back to the command post. Making its way toward the command post, pursuant to the earlier withdrawal order, K Company ran into an enemy ambush and lost its command group and one platoon. The remainder reached the battalion area closely followed by the Chinese. There on the valley floor the disorganized men of the 3d Battalion formed a core of resistance around Sergeant Miller's three tanks and held the enemy off until daylight.

Another island of resistance had formed at the ramp to the command post dugout. Three men who manned the machine gun there in succession were killed by Chinese grenades. When daylight came only five of the twenty or more men who had assembled there were left. After a final exchange of grenades with these men, the Chinese in the nearby ditches withdrew. The group at the ramp then joined the others in the small perimeter around the three tanks.

Enemy mortar fire kept everyone under cover until an hour after daylight. Then a Mosquito plane and fighter-bomber aircraft came over and began a daylong series of strikes against the Chinese. This kept the enemy under cover during the rest of the day and gave the men at the command post a chance to take stock of their situation and to gather in the wounded. They found Major Ormond, the battalion commander, very badly wounded and the rest of the battalion staff wounded or missing. There were approximately 6 officers and 200 men of the battalion still able to function. Within 500 yards of the 200-yard-wide perimeter there were more than 170 wounded. As they were brought inside the small perimeter the wounded were counted; the dead apparently were not.

The beleaguered men also used the daylight respite gained from the air cover to dig an elaborate series of trenches and retrieve rations and ammunition from the vehicles that had escaped destruction. An L–5 plane flew over and dropped a mail bag of morphine and bandages. A helicopter also appeared and hovered momentarily a few feet above the 3d Battalion panels, intending to land and evacuate the more seriously wounded, but enemy fire hit it and it departed without landing. The battalion group was able to communicate with the pilot of a Mosquito plane overhead who said a relief column was on its way to them.[22]

The relief column the pilot of the Mosquito plane referred to was the 5th Cavalry force that, after having been repulsed during the previous afternoon and night, resumed its effort at daylight to break through to the 3d Battalion of the 8th Cavalry. Just before 0400, 2 No-

[22] Ltrs, McAbee to author, 20 Aug 54 and 8 Feb 57; I Corps WD, 2 Nov 50, Surg Sec Daily Rpt; 8th Cav Unit Hist Rpt, 2 Nov 50; EUSAK WD, G–3 Jnl, Msg at 0840 2 Nov 50; 70th Tk Bn WD, 10 Nov 50 (Rpt, B Co, 70th Tk Bn, 1–9 Nov 50); Ltr, Miller to Fraser, 6 Nov 50. Two helicopters did evacuate twenty-two critically wounded from Ipsok.

vember, the 2d Battalion, 5th Cavalry, arrived at the defensive position the 1st Battalion had held during the latter part of the night. On General Gay's order, the 1st Battalion, 7th Cavalry, now also became available to Colonel Johnson. Gay directed that it strike off across country in an effort to flank the enemy position on the left while the 5th Cavalry attacked frontally. For the frontal attack, Colonel Johnson placed the 1st Battalion, 5th Cavalry, on the left of the road and the 2d Battalion on the right. His plan called for these two battalions to capture the enemy-held ridge in their front on a sufficient frontage to allow the 3d Battalion—which had been released that morning to his control and was then moving to join him, spearheaded by a tank company—to move through to the relief of the 3d Battalion, 8th Cavalry. The 3d Battalion would be up and ready for this effort by afternoon.

Colonel Johnson had a special interest in rescuing the 3d Battalion, 8th Cavalry. He had brought it to Korea from Fort Devens, Mass., where only two months earlier it had been part of the 7th Regiment of the 3d Division. It became the 3d Battalion of the 8th Cavalry Regiment, and he had commanded it through the Pusan Perimeter breakout battles. By right of this earlier association it was "his own battalion."

The two lead attack companies of the 5th Cavalry failed to reach and seize their objectives on 2 November. The 1st Battalion of the 7th Cavalry really contributed nothing to the effort as it merely moved off into rough country and never entered the fight. The attack had almost no support from artillery, since only two 155-mm. howitzers could reach the enemy positions and higher headquarters would not authorize moving up the lighter artillery. The repeated strikes by strong air cover against the enemy ridge positions probably did little damage because the dense smoke haze hanging over the area obscured the objective. The 2d Battalion in the afternoon made the last effort after an air strike had strafed the enemy-held ridge. But again the smoke haze was so heavy that the pilots could not see any targets and it is doubtful whether their strikes caused much damage. The dug-in Chinese did not budge. A prisoner said that five Chinese companies of the *8th Route Army* were holding the ridge.

In this night and day battle with the Chinese at the Turtle Head Bend of the Kuryong River the two battalions of the 5th Cavalry suffered about 350 casualties, 200 of them in Lt. Col. John Clifford's 2d Battalion which carried the brunt of the fighting on 2 November. The 5th Cavalry Regiment always thereafter referred to this ridge where it first encountered the CCF as "Bugle Hill." The name was well chosen for during the night and on into the day the Chinese had used bugles, horns, and whistles as signaling devices. No doubt they also hoped that these sounds would terrorize their enemy during the eerie hours of night battle.

With the battle still in progress against this Chinese force, General Milburn, the corps commander, after conferring with General Gay, at 1500 verbally instructed the latter to withdraw the 1st Cavalry Division. The two had agreed that with the forces available they could not break the roadblock. Approximately two hours later Gay received confirmation of the order from corps. General Gay at dusk made what he has described as the most difficult decision he was ever called on to make—to order the 5th Cavalry Regiment to withdraw and leave the 3d Battalion, 8th Cavalry, to its fate.[23] Thus, at dark on 2 November the 3d Battalion, 8th Cavalry, had no further hope of rescue.

At the 3d Battalion perimeter Chaplain Kapaun and Captain Anderson had risked their lives constantly during the day in attending the wounded. Many men not previously injured had been hit by sniper and machine gun fire in carrying wounded into the perimeter. Although wounded several times, and seriously, Major Ormond had refused treatment until all other wounded had been cared for. At dusk Chaplain Kapaun left the perimeter and went to join the fifty to sixty wounded who had been placed in the old dugout battalion command post. This dugout, initially at the southeast corner of the original perimeter, was now approximately 150 yards outside the new one. The three tanks moved inside the infantry position.

Just before dusk a division liaison plane flew over the 3d Battalion perimeter and dropped a message ordering it to withdraw under cover of darkness. Over his tank radio Miller received from a liaison pilot a similar message stating that the men were on their own and to use their own judgment in getting out. But, after talking over the situation, the tankers and the infantry in the little perimeter decided to stay and try to hold out during the night.[24]

As dusk settled over the beleaguered group and the last of the protecting air cover departed, the Chinese bombarded the little island of men with 120-mm. mortars which had been brought into position during the day. The tankers, thinking the mortar barrage was directed at them, moved the tanks outside the perimeter to divert it away from the infantry. The barrage followed them, but part of it soon shifted back to the infantry inside the perimeter. All the tanks were hit two or three times, and one of them started to burn. A crewman was killed in putting out the fire. His ammunition almost gone and his gasoline low, Miller decided that his tanks would not last out the night if they stayed where they were. He called the infantry over their SCR-300 radio and told them his conclusion that in the circumstances the tanks would be of no help to them. They agreed. Miller led the tanks off to the southwest. Three miles from the perimeter Miller and the other crew members had to abandon the tanks in the valley of the Kuryong. After some desperate encounters, Miller and a few of his men reached friendly lines.[25]

At the 3d Battalion perimeter the Chinese followed their mortar barrage with an infantry attack. To meet this, the men inside the perimeter fired bazooka rounds into the vehicles to start fires and light up the area. Attacking across the open field in successive waves and silhouetted against the burning vehicles, the Chinese made easy targets and were shot down in great numbers. Six times during the night the Chinese at-

[23] Ltr, Gay to author, 19 Feb 54; Interv, author with Johnson, 28 Apr 54; 5th Cav WD, 2-4 Nov 50; Milburn, MS review comments, Nov 57.

[24] Ltr, McAbee to author, 20 Aug 54; Ltr, Miller to Fraser, 6 Nov 50; 70th Tk Bn WD, Msg at 1620 2 Nov 50.

[25] Ltr, Miller to Fraser, 6 Nov 50; 70th Tk Bn WD, Msg (from S-2, 5th Cav, to G-2, 1st Cav Div) 081140 Nov 50, and Msg at 1050 3 Nov 50; EUSAK PIR 113, 2 Nov 50.

tacked in a strength of approximately 400 men, but each time they were beaten back from the perimeter. During the night about fifty men from the 2d Battalion who had been in the hills all day broke through to join those in the besieged 3d Battalion perimeter.

In this heavy action, the Chinese early in the evening, by mortar fire and grenades, knocked out the two machine gun positions at the old command post dugout. Then they overran it. Inside the dugout were between 50 and 60 badly wounded men. The Chinese took 15 of the wounded who were able to walk with some help, including Captain McAbee and Chaplain Kapaun, and removed them to the Nammyon River outside the range of fire. The others, unable to walk, were left inside the dugout. In getting out of the field of fire with their captors, the 15 men had to crawl over the dead. Major McAbee has stated that at the edge of the perimeter where he passed the enemy dead they were piled three high and he estimated there must have been 1,000 enemy dead altogether.[26] But this number seems excessive.

On the morning of 3 November a 3-man patrol went to the former battalion command post dugout and discovered that during the night the Chinese had taken out some of the wounded. That day there was no air support. Remaining rations were given to the wounded. Enemy fire kept everyone under cover. The night was a repetition of the preceding one, with the Chinese working closer all the time. After each enemy attack had been driven back men would crawl out and retrieve weapons and ammunition from the enemy dead. Their own ammunition was almost gone.

Daylight of 4 November disclosed that there were about 200 men left able to fight. There were about 250 wounded. A discussion of the situation brought the decision that those still physically able to make the attempt should try to escape. Captain Anderson, the battalion surgeon, volunteered to stay with the wounded. 1st Lt. Walter L. Mayo, Jr., and 1st Lt. Philip H. Peterson, accompanied by two enlisted men, left the perimeter to scout a way out. They crawled up the irrigation ditches to the old command post and talked with some of the American wounded the Chinese had left there. They found the ramp covered with dead Chinese and Americans. They then crawled up the roadside ditches to the small village farther north and found only some wounded Chinese in it. In reaching the village, Lieutenant Mayo has estimated that he crawled over the bodies of 100 Chinese. From there the four men scouted the ford across the river. That done, the two officers sent the two enlisted men back to the 3d Battalion perimeter with instructions to lead the group out, while they continued to scout the river crossing area. It was about 1430.

After the two enlisted men returned to the perimeter and reported on the escape route, Capt. George F. McDonnell of the 2d Battalion group and Capt. William F. McLain of E Company, together with 1st Lt. Paul F. Bromser of L Company and the able-bodied men, withdrew to the east side of the perimeter just as the Chinese let loose a terrific barrage of white phosphorus shells. These bursting shells completely covered the perimeter area and obscured it with

[26] Ltr, McAbee to author, 8 Feb 57.

smoke. There was no doubt that the Chinese were trying to screen an attack. Within five minutes the 200 men cleared the perimeter on the east side where an open field had prevented the enemy from taking positions. They left the wounded with Captain Anderson who was to surrender them. As they left the wounded behind, one who was present said none of the latter shed tears but, instead, simply said to come back with reinforcements and get them out. The wounded knew there was no alternative for those who still might escape.

The escaping group traveled all that night east and northeast and then south and southwest through a rain storm. In the morning from a mountainside they watched a few battalions of Chinese horse cavalry and infantry pass by on a road below them. Later in the day the battalion group went south through more hills and crossed the valley near Ipsok. The next day, within sight of bursting American artillery shells, Chinese forces surrounded them and the battalion group, on the decision of the officers, broke up into small parties in the hope that some of them would escape. At approximately 1600 on the afternoon of 6 November the action of the 3d Battalion, 8th Cavalry, as an organized force came to an end. Most of these men were either killed or captured that day, apparently in the vicinity of Yongbyon.[27]

The heroic 3d Battalion commander, Major Ormond, was among the wounded captured by the Chinese in the perimeter beside the Kuryong. He subsequently died of his wounds and, according to some reports of surviving prisoners, was buried beside the road about five miles north of Unsan. Of his immediate staff, the battalion S-2 and S-4 also lost their lives in the Unsan action. About ten officers and somewhat less than 200 enlisted men of the 3d Battalion escaped to rejoin the regiment. There were a few others who escaped later, some from captivity, and were given the status of recovered allied personnel.[28]

It is difficult to arrive at precise figures in totaling the losses at Unsan. In the night battle the troop loss in the ROK 15th Regiment was admittedly very heavy. The regiment's loss in weapons and equipment was virtually total, and included four liaison planes of the 9th Field Artillery Battalion and the 6th Tank Battalion which U.S. fighter planes subsequently demolished on the ground.

At first, more than 1,000 men of the 8th Cavalry Regiment were missing in action, but as the days passed, some of these returned to friendly lines along the Ch'ongch'on. Two weeks after the Unsan action tank patrols were still bringing in men wounded at Unsan and fortunate enough to have been sheltered and cared for by friendly Koreans. On 22 November the Chinese themselves, in a propaganda move, turned free 27 men who had been prisoners for two weeks or longer, 19 of them captured from the 8th Cavalry Regiment at Unsan. After all the stragglers and those who had

[27] The account of the 3d Battalion after the tanks left the perimeter is based on McAbee's and Mayo's letters to author. McAbee and Anderson lived to return to the United States in the prisoner exchange after the Korean armistice Chaplain Kapaun died in 1951 while a prisoner of war.

[28] Ltr, Moriarty to author, 11 Jun 54; Ltr McAbee to author, 8 Feb 57; Interv, author with Johnson, 28 Apr 54; Interv, Guirard, 21 Aug 54. The figures are from Moriarty, who remained as executive officer of the battalion.

walked south through the hills had reported in, the losses were found to total about 600 men. Enemy sources later indicated the Chinese captured between 200 and 300 men at Unsan. The principal officer casualties included a battalion commander and most of his staff, 5 company commanders, 2 medical officers, and 1 chaplain. In addition to the infantry losses, about one-fourth of the men of B Company, 70th Tank Battalion, were casualties. The Heavy Mortar Company also suffered heavily. The regiment's loss in weapons and equipment was very heavy indeed. It included 12 105-mm. howitzers and 9 tanks and 1 tank recovery vehicle. On 3 November the 8th Cavalry Regiment reported it had 45 percent of its authorized strength. The division G-4 considered the regiment inoperable until troops and equipment losses could be replaced.[29]

The Eighth Army announced on 5 November that "as a result of an ambush" the 1st Cavalry Division would receive all the new replacements until further notice. In the next twelve days, Eighth Army assigned 22 officers and 616 enlisted men as replacements to the 1st Cavalry Division. Nearly all of them went to the 8th Cavalry Regiment.[30]

To cover the withdrawal to the south side of the Ch'ongch'on of the 1st Cavalry Division and the ROK 1st Division, I Corps organized a special force known as Task Force Allen. The 2d and 3d Battalions, 7th Cavalry Regiment, and the 19th Engineer Combat Group were the principal organizations in the task force. Brig. Gen. Frank A. Allen, Jr., Assistant Division Commander, 1st Cavalry Division, commanded it. In addition to covering the withdrawal, it also had the mission of protecting the I Corps east flank in the Kunu-ri area.[31]

The Chinese force that brought disaster to the 8th Cavalry Regiment at Unsan was the *116th Division* of the *39th Army*. Elements of the *347th Regiment* imposed the roadblock east of the road fork south of Unsan that thereafter halted all vehicular traffic. The *115th Division* also fought in the Unsan action. It appears, therefore, that from first to last—from 25 October to 2 November—two Chinese divisions, or elements of them, engaged the ROK 1st Division and the U.S. 8th and 5th Cavalry Regiments in the Unsan area.[32]

---

[29] 1st Cav Div WD, 5, 6, and 17 Nov 50; 7th Cav Hist Rpt, 22 Nov 50; 8th Cav Unit Hist Rpt, 3 Nov 50; 70th Tk Bn WD, Jnl, 10 Nov 50; EUSAK WD, G-3 Daily Hist Rpt, 6 Nov 50; Ltr, Gay to author, 19 Feb 54.

The equipment loss figures are based on the following sources: EUSAK POR 339, 2 Nov 50; 70th Tk Bn WD, Summ, 2 Nov, and Jnl, Msg 4, 101000 Nov 50; 1st Cav Div WD, G-4 Jnl, 12 Oct 50, Div Arty and 70th Tk Bn battle losses, 27 Oct-4 Nov 50; 1st Cav Div POR 190, 10 Nov 50, an. A; 1st Cav Div WD, 5 Nov 50; EUSAK WD, 7 Nov 50, Ltr, CG EUSAK to CINCFE, sub: ROKA and U.S. Equipment Losses, 1-3 Nov 50; I Corps WD, POR 153, 2 Nov 50. These sources also gave losses in small arms, automatic weapons, and vehicles.

In a study of combat experience at Unsan prepared and distributed by *Headquarters, XIX Army Group, CCF,* the Chinese command, after recounting the large amount of equipment captured, apologized for what it considered relatively few prisoners. The study stated, "As a result of lack of experience in mopping-up operations in mountainous areas, only 200 odd were captured." See ATIS Enemy Documents, Issue 47, pp. 139ff, mimeographed booklet in Chinese, A Collection of Combat Experiences.

[30] EUSAK WD, G-1 Daily Hist Rpt, 5 and 17 Nov 50.

[31] Ltr, Gay to author, 19 Feb 54; Interv, Johnson, 28 Apr 54; 7th Cav Unit Hist Rpt, 1-4 Nov 50; I Corps WD, 2 Nov 50.

[32] ATIS Enemy Documents, Issue 47, pp. 139ff; I Corps, Armor Combat Bul 27, 15 Jun 51, quoting from captured Chinese notebook, Experiences in the Unsan Operation.

# CHAPTER XXXV

# Eighth Army Holds the Ch'ongch'on Bridgehead

The whole art of war consists in getting at what is on the other side of the hill.
DUKE OF WELLINGTON

### *Action North of the River*

Monday, 2 November, was a day of hectic activity and some confusion in the command posts of Eighth Army and its subordinate organizations. Orders and changes to orders came in an almost endless stream as steps were taken to withdraw U.S. I Corps below the Ch'ongch'on River. In the final instructions the 27th British Commonwealth Brigade and the 19th Infantry Regiment of the 24th Division were to remain north of the river in a bridgehead that encompassed and protected the bridges and tank fords over the Ch'ongch'on and Taeryong Rivers in the Anju-Pakch'on areas. General Walker wished to hold these river crossing sites for the purpose of resuming the offensive.[1]

By evening of 3 November the 27th Brigade had moved from Taech'on into its assigned defensive positions around Pakch'on at the northwest corner of the bridgehead. (*Map 24*) The fine fur boots on the feet of some of the enemy dead at Taech'on, and three Chinese deserters who entered their lines there, revealed to the British that they already had had minor brushes with the Chinese. Following close behind the departing British, Chinese forces entered Taech'on less than an hour after the British rear guard had left it.[2]

The other part of the bridgehead force, the 19th Infantry Regiment, also moved during 3 November to its defensive positions just north of the Ch'ongch'on and northeast of Anju. Once the 19th Infantry was in its defensive positions, the ROK 1st Division, still in contact with the enemy, withdrew through it to the south side of the Ch'ongch'on, completing its crossing before noon of 4 November.[3]

---

[1] 24th Div WD, 2 Nov 50; 27th British Commonwealth Brig Sitrep, 1–4 Nov 50; EUSAK POR 340, 2 Nov 50; I Corps WD, 1–2 Nov 50.

[2] 24th Div WD, G–2 Jnl, 4 Nov 50, Rpt from Home Guards of Taech'on; 27th British Commonwealth Brig Sitrep, 1–4 Nov 50; Coad, "The Land Campaign in Korea," *op. cit.*, p. 7; Linklater, *Our Men in Korea*, pp. 26–27; Bartlett, *With the Australians in Korea*, p. 39.

[3] 19th Inf Unit Rpt 114, 2–3 Nov 50; 24th Div WD, Nov 50 Summ; I Corps WD, 3 Nov 50.

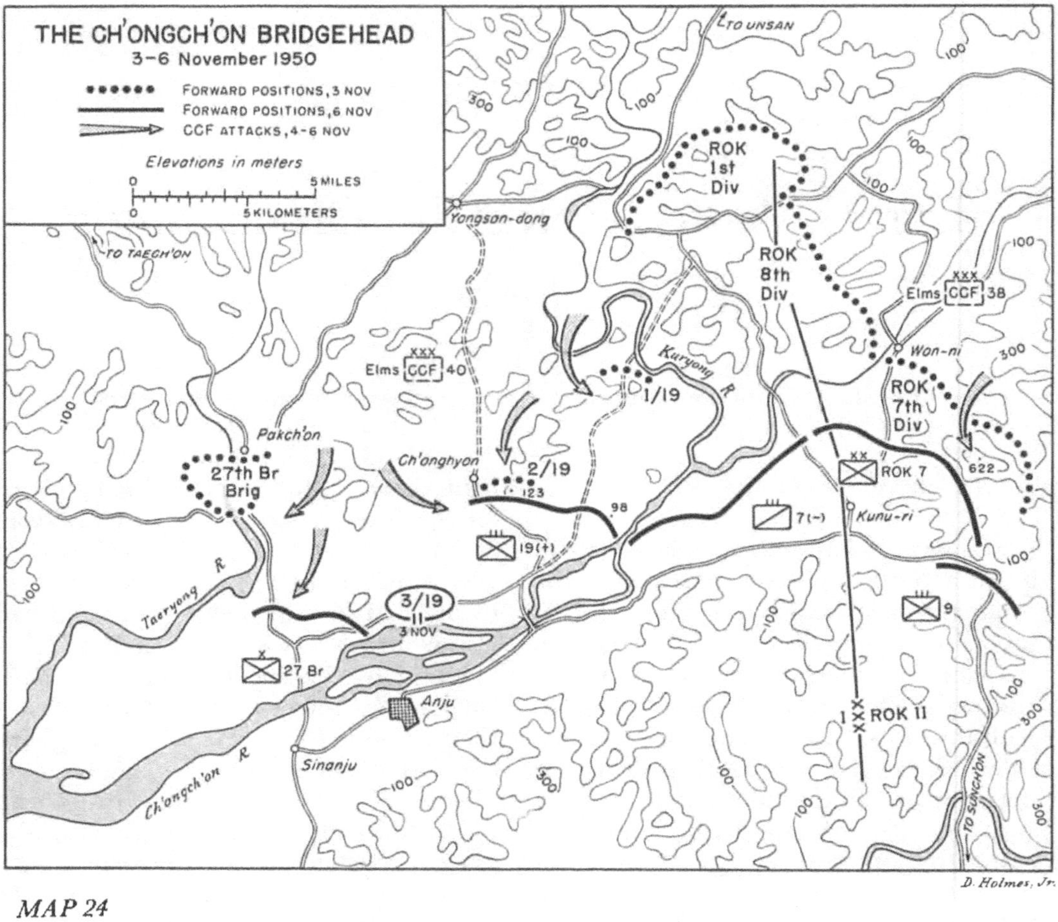


MAP 24

The boundary between the ROK Army and the U.S. I Corps crossed the Ch'ongch'on on a north-south line at Kunu-ri. Already the Chinese had crowded the ROK II Corps into a restricted defensive area near Won-ni about four miles northeast of Kunu-ri. On 3 November the 5th Regimental Combat Team of the 24th Division took a position at Kunu-ri behind the ROK II Corps. The 9th Regiment of the U.S. 2d Infantry Division was in position south of Kunu-ri protecting the road to Sunch'on. Enemy forces during the day advanced within two and a half miles of Kunu-ri.[4]

The next day, 4 November, enemy forces drove hard for Kunu-ri. Hill 622, a large mountain cresting three miles northeast of Kunu-ri, dominated the town, the valley of the Ch'ongch'on, and the rail and highway communication lines passing along it. The 3d and 5th

[4] 5th RCT WD, 3 Nov 50; 24th Div WD, 3 Nov 50; I Corps WD, 3 Nov 50; EUSAK WD, 3 Nov 50.

Regiments of the ROK 7th Division held this key terrain feature, with the 5th RCT, 24th Division, in blocking position just behind them. The ROK 8th Regiment was in reserve along the road east of Kunu-ri. That morning a strong CCF attack broke the ROK 3d Regiment position on the mountain and the South Koreans began streaming back through the 5th RCT. Capt. Hubert H. Ellis, commanding officer of C Company, stopped and reorganized these ROK troops and sent them back to retake the hill. The ROK 8th Regiment was now also committed to the battle. The hill changed hands several times throughout the day, but at dark ROK troops held its vital northwest ridge.

The 5th RCT itself had heavy fighting in this battle to hold Kunu-ri, and was forced to withdraw about 1,000 yards. Part of the fighting was at close quarters. Some men, like 1st Lt. Morgan B. Hansel of C Company who gave his life to prevent the disintegration of a platoon, charged Chinese machine gun emplacements alone. By evening the enemy attack in estimated division strength (elements of the CCF *38th Army*) had been repulsed. The ROK 7th Division and Col. "Rocky John" Throckmorton's 5th Regimental Combat Team had saved Kunu-ri and successfully protected the right flank of Eighth Army.[5]

Simultaneously with this attack south of the Ch'ongch'on against Kunu-ri, the enemy struck the bridgehead force north of the river. On 4 November both ground and aerial observers reported approximately 1,000 enemy soldiers crossing the Kuryong River two miles northwest of the 1st Battalion, 19th Infantry, and moving south through wooded terrain, evidently intent on getting into the rear of the battalion. The enemy maneuver succeeded. Chinese troops captured the battalion's radio while the operator was using it to report the situation to the regimental headquarters. The battalion did not make much of a fight, and, after destroying and abandoning its heavy equipment and vehicles, it withdrew eastward and infiltrated across the Kuryong and Ch'ongch'on Rivers to friendly positions. Nearly all the men escaped.[6]

Meanwhile, a task force of the 3d Battalion, 19th Infantry, subsequently reinforced by the entire battalion, tried to drive through to the 1st Battalion's position, but strong enemy forces on the road repelled it. With these difficulties developing in the bridgehead area, General Church ordered Brig. Gen. Garrison H. Davidson, the assistant division commander of the 24th Division, to assume command of all 24th Division troops north of the Ch'ongch'on and to co-ordinate the actions of the 27th British Commonwealth Brigade and the division troops. Davidson arrived at the 19th Infantry command post shortly after noon of the 4th to assume command of Task Force Davidson. The worsening situation caused General Church at 1630 also to order the 21st Infantry Regiment to

[5] I Corps WD, 4 Nov 50; 5th RCT WD, 4 Nov 50; Interv, author with Maj Grady R. Hamilton (S-3 5th RCT Aug 51), 10 Aug 51; ATIS Interrog Rpts (Enemy Forces), Issue 24, Interrog 2811, p. 29, Sgt Chon Song Hyon. Eighth Army General Order 397, 4 June 1951, awarded the Distinguished Service Cross posthumously to Lieutenant Hansel.

[6] 19th Inf Unit Rpt 115, 4-5 Nov 50; 19th Inf WD, Opn Summ, Nov 50; 24th Div WD, 4-5 Nov 50; 24th Div PIR 115, 3-4 Nov 50; Interv, Maj Gen Garrison H. Davidson (Asst Div Comdr 24th Div Nov 50), 28 Jan 54.

cross to the north side of the Ch'ongch'on River during the night and attack the next day, to clear the enemy from the 19th Infantry area and restore the bridgehead line.[7]

An enemy force made a further penetration in the 19th Infantry zone during the night, but the next morning, 5 November, the 2d and 3d Battalions, 21st Infantry, attacked and restored the position. Once again, as in the days on the Naktong, the 3d Engineer Combat Battalion assumed an infantry role in taking positions to protect the Anju bridges over the Ch'ongch'on. Fleeing the battle area, hordes of refugees crossed the Ch'ongch'on; 20,000 of them passed through the checking points on the south side of the river on 4–5 November.[8]

On the west, there was a 5-mile gap between the left flank of the 19th Infantry bridgehead position and the 27th British Commonwealth Brigade position. A large mountain mass lay in this no man's land, and over and through it enemy forces could move at will to the flank and rear of either the 27th Brigade or the 19th Infantry. On the 19th Infantry's extreme left flank at the edge of this gap the 2d Battalion held Hill 123 which overlooked a valley near the little village of Ch'onghyon, four miles above the Ch'ongch'on.

On the night of 5–6 November the enemy made a co-ordinated attack all along the bridgehead line. At Hill 123 the attack achieved surprise against E and G Companies, 19th Infantry. At least part of the enemy assault force came up to E Company's position from the rear, apparently following field telephone wire. The Chinese caught many men asleep in their sleeping bags and killed them where they lay. Others were shot in the back of the head. The Chinese virtually overran the battalion positions on Hill 123.

Cpl. Mitchell Red Cloud, an Indian from Wisconsin, gave the first alarm to E Company from his position on the point of the ridge where a trail climbed to the company command post. A group of Chinese suddenly charged him from a brush-covered area 100 feet away. Corporal Red Cloud sprang to his feet and fired his BAR into them. Enemy fire wounded and felled him, but he pulled himself to his feet, wrapped one arm around a small tree, and again delivered point-blank BAR fire until Chinese bullets cut him down. Later, American officers found a string of Chinese dead in front of his body.[9]

Another BAR man in E Company, Pfc. Joseph W. Balboni, was equally heroic. Chinese soldiers approached unnoticed within seventy-five feet of him. From this short distance they charged forward. Balboni met them with bursts from his BAR and stood in his tracks until killed. Two days later when a friendly patrol visited the spot seventeen enemy dead were found in front of Balboni's body.[10]

---

[7] 19th Inf WD, 4 Nov 50; 24th Div WD, 4 Nov 50; Interv, author with Davidson, 28 Jan 54.

[8] 19th Inf WD, 5 Nov 50; 19th Inf Unit Rpt 116, 4–5 Nov 50; 24th Div WD, 5 Nov 50; I Corps WD, 4–5 Nov 50; EUSAK WD, Br for CG, 5 Nov 50; EUSAK PIR 116, 5 Nov 50.

[9] Department of the Army General Order 26, 25 April 1951, awarded the Medal of Honor posthumously to Red Cloud. Interv, author with Davidson, 28 Jan 54; 19th Inf Unit Rpt 117, 5–6 Nov, and Rpt 119, 7–8 Nov 50; 24th Div WD, 6 Nov 50; 19th Inf Opn Summ, Nov 50.

[10] Eighth Army General Order 63, 10 February 1951, awarded the Distinguished Service Cross posthumously to Private Balboni.

By 0300 the battalion had withdrawn 1,000 yards. There it was only barely able to hold its new position. After daylight the enemy withdrew from contact with the 2d Battalion. Elsewhere the other battalions on the 19th Infantry front repulsed the attacks on them after hard fighting. Artillery firing from the south side of the Ch'ongch'on supported the 19th Infantry during the bridgehead battles.

After daylight the re-equipped 1st Battalion, 19th Infantry, which had recrossed to the north side of the Ch'ongch'on during the night, counterattacked and closed the gap between the 2d Battalion and the rest of the regiment on its right. The 19th Regiment then began restoring its original bridgehead line.

In these night battles of predawn 6 November the enemy had lost heavily. Two days later, the 2d Battalion alone counted 474 enemy dead in the vicinity of Hill 123, and it found evidence that many more dead had been buried. The 3d Battalion, 19th Infantry, counted more than 100 enemy dead. Interrogation of prisoners disclosed that the *355th Regiment*, CCF *119th Division;* the *358th Regiment*, CCF *120th Division;* and a North Korean regiment had attacked the 19th Infantry on the east side of the bridgehead.[11]

Simultaneously with its attacks against the Eighth Army right flank at Kunu-ri and the eastern side of the Ch'ongch'on bridgehead, the enemy struck with equal force against the western side of the bridgehead at Pakch'on on the Taeryong River. The Australian 3d and Argyll 1st Battalions were in defensive positions on the west side of the Taeryong River opposite Pakch'on, except A Company of the Argylls which was on the east side of the river south of Pakch'on. The Middlesex 1st Battalion was also on the east side in and around Pakch'on. Two and a half miles south of Pakch'on the 61st Field Artillery Battalion of the 1st Cavalry Division had emplaced to support the British.

During the night of 4–5 November an enemy force moved east around Pakch'on toward the support artillery where it could cut the road behind the brigade. By daylight these enemy troops were in position to launch their attack.

An estimated battalion of Chinese opened fire from the east with mortars and small arms on the 61st Field Artillery Battalion. Immediately, each battery commander placed all his men, except the gun crews who remained with their weapons, in a tight perimeter around the battery positions, manning every automatic weapon. Word of the attack reached the British, and A Company of the Argylls started south at once to the aid of the artillerymen. The brigade commander then ordered the remainder of the Argyll 1st Battalion to cross to the east side of the river.

The most severe attack fell on C Battery. Capt. Howard M. Moore, commanding officer of the battery, wheeled one 105-mm. howitzer around and fired point-blank into enemy troops in the rice paddies to the east. Forty-five minutes later he got a second howitzer turned around. The battery fired 1,400 rounds, some at a range of 50 yards, although the average range was about 300 yards. An aerial observer directed the fire of another battery in support of C Battery. Part of the enemy plan was to blow a

[11] 19th Inf Unit Rpts 117, 118, 119, 5–8 Nov 50; Interv, author with Davidson, 28 Jan 54.

critical bridge at the artillery position. If this had succeeded it is unlikely that the brigade would have saved a single tank or vehicle. The artillerymen killed one member of a demolition squad within twenty yards of the bridge.

At 0900 two tanks arrived at the C Battery perimeter. Before the Argylls arrived with more armored support, C Battery had lost 2 men killed and 18 wounded. The rest of the artillery battalion had 17 men wounded. Enemy action had destroyed 1 howitzer, 6 vehicles, 1 radio, and some other miscellaneous equipment. There were about 70 enemy dead in the vicinity of the gun positions. Speaking later before an English audience, Brigadier Coad said of C Battery, 61st Field Artillery Battalion, "I would like to say how magnificently these American gunners fought. Dead Chinese were lying thirty yards from the gun shields. . . . It was up to the very highest traditions of any artillery regiment." [12]

Upon arriving at the artillery position, A Company of the Argylls attacked a nearby hill, first winning, then losing it. Air reports now indicated that approximately one Chinese division had passed east around and below Pakch'on, virtually surrounding the brigade. While heavy air strikes went in to impede and disrupt the Chinese maneuver and attack, Brigadier Coad ordered the Australian 3d Battalion, still on the west side of the Taeryong, to withdraw to the east side.

Once on the east side, the Australians passed through the Argylls and in the early afternoon attacked north toward Pakch'on. In a bayonet charge they regained high ground the Argylls had lost two miles below the town. In this desperate battle the Australians lost twelve killed and sixty-four wounded. It now became apparent that the brigade could not hold the Pakch'on bridgehead. Under cover of the Australian attack, and a simultaneous one by the Middlesex 1st Battalion which cleared the road southward, the rest of the brigade and the artillery withdrew under fire toward the Ch'ongch'on River.

That evening the Australians joined the rest of the brigade in a defense perimeter astride the Pakch'on road on the first line of hills, 4,000 to 6,000 yards north of the Ch'ongch'on River. The Australians occupied the most advanced and exposed position. Soon after dark the Chinese struck them in an attack which continued unabated for four hours, forcing two companies to withdraw. By dawn of 6 November, however, the Chinese themselves had withdrawn from contact. The British estimated the Chinese lost 300 men to ground action and 600 to 1,000 to air action during the day and that night.

After daylight on Monday, 6 November, Australians in their positions could see enemy forces withdrawing northward up a valley. The predawn attacks of 6 November against the bridgehead area and Kunu-ri proved to be the last heavy engagements of the Chinese in their First Phase Offensive. During the day they withdrew generally out of contact. Aerial

---

[12] Department of the Army General Order 33, 31 March 1952, awarded the Distinguished Unit Citation to the 61st Field Artillery Battalion, 1st Cavalry Division. 61st FA Bn WD, 5 Nov 50; 1st Lt Hal W. Chaney (C Btry, 61st FA Bn), Debriefing Rpt 64, Arty School, Ft. Sill, Okla., 22 Jan 52; 27th British Commonwealth Brig Sitrep 4-Nov 50; Coad, "The Land Campaign in Korea," *op. cit.;* Linklater, *Our Men in Korea*, pp. 26–27; Bartlett, *With the Australians in Korea*, pp. 39–40.

observers reported many sightings of large enemy forces moving northward. Limited objective attacks by the 24th Division on 7 November all reached their goals against light opposition, and seemed to indicate that the violent enemy attacks of the past twelve days had ended for the present.[13]

### *MIG's and Jets Over the Yalu*

Coinciding with the appearance of the Chinese in the Korean War came a new turn in air action over Korean territory. Except for its opening weeks, the Korean War had been characterized by the U.S. Fifth Air Force's complete dominance of the skies. Now, suddenly, Russian-built MIG's began crossing over into North Korea from Manchurian bases and challenging the Fifth Air Force when its planes approached the Yalu border. In the closing days of October, American planes had for the first time been allowed to approach the border.

At the beginning of the month, Fifth Air Force planes were prohibited from flying within fifty miles of the border. On 17 October this restriction was eased, and finally on 25 October it was lifted to allow close support missions, under control of a tactical air control party or a Mosquito observer, as near the border as necessary. Pilots of these missions were carefully selected and flew under experienced leaders, for there was to be no bombing within five miles of the border.[14]

On 31 October the famous Russian-built MIG–15 jet first entered combat over Korean territory when a number of them attacked American propeller-driven aircraft in the Sinuiju area. The Fifth Air Force reportedly destroyed several MIG's at this time. It was not known then whether the enemy plane build-up was North Korean or some "volunteer" air force.[15]

On 5 November General MacArthur gave a new turn to the course of the war by ordering aerial bombing to destroy the Yalu River bridges from the Korean side halfway into the stream. He felt this action necessary to stop or greatly reduce the flow of Chinese troops and equipment into Korea. The order came at the time the CCF attack threatened to overrun Eighth Army's Ch'ongch'on River bridgehead and to capture Kunu-ri on the south side. The Joint Chiefs of Staff in Washington received from MacArthur a radio report of the order. They immediately countermanded it, repeating their directive not to bomb targets within five miles of the border.[16]

General MacArthur replied at once to the JCS message in one of the strongest protests he ever dispatched to Washington. He said the only way to halt the reinforcement of the enemy was to destroy the bridges. Paraphrased, his message continued, "Under the gravest protest that I can make I am carrying

[13] EUSAK WD, G–3 Sec, 6 Nov 50; I Corps WD, 6 Nov 50; 24th Div WD, 6 Nov 50; 5th Cav WD, Narr Rpt, Nov 50; Ltr, Gay to author, 19 Feb 54. Task Force Allen was dissolved at 1700, 6 November.

[14] USAF Hist Study 71, p. 80.

[15] FEAF Opn Hist, II, 19; Dept of State Pub 4051, *United Nations Command Ninth Report to the Security Council, United Nations, 1–15 November 1950*, pp. 11–12; EUSAK WD, 4 Nov 50, EUSAK Daily News Bul, UP dispatch, 1 November, quoting General Partridge.

[16] Msg, JCS 95878, JCS Personal for MacArthur, 5 Nov 50, quoted in Schnabel, FEC, GHQ Support and Participation in the Korean War, ch. VII, pp. 16–17; Senate MacArthur Hearings, pt. 1, p. 20, testimony of MacArthur, 3 May 51.

out your instructions and suspending this strike." He argued that what he proposed to do was within the rules of war and the resolutions and directions he had received, and that it constituted no act of belligerency against Chinese territory. He asked that the matter be brought immediately to the attention of the President, "as I believe your instructions may result in a calamity of major proportions for which, without his personal and direct understanding of the situation, I cannot accept responsibility." [17]

This message produced the result MacArthur desired. The Joint Chiefs of Staff on 6 November authorized MacArthur to proceed with the planned bombing of the Korean end of the Yalu bridges, provided that he still considered such action essential to the safety of his forces. MacArthur was expressly forbidden, however, to bomb any dams or power plants on the Yalu River or to violate Manchurian property and airspace.[18]

Along the Manchurian border there were known to be at least seven major bridges across the Yalu River and three across the Tumen River. The most important Yalu structures were the 3,000-foot-long rail and highway bridges connecting Sinuiju and An-tung, the 2,000-foot-long rail and highway bridges thirty-five air miles northeast of Sinuiju near Sakchu, and the 1,500-foot-long rail bridge at Manp'ojin.

The first bomber strike against the Yalu River bridges at Sinuiju went in on 8 November, seventy-nine B-29's under fighter escort executing the mission. The date is memorable also for the first battle between jet planes in aerial warfare. Lt. Russell Brown, pilot of an F-80 in the fighter escort, shot a MIG-15 down in flames.

Thereafter, land-based and carrier-based planes attacked the Yalu bridges almost daily during the rest of the month, and the doctrine of "hot pursuit" into the enemy's Manchurian sanctuary soon became a burning issue, not only in the battle zone but in the diplomatic capitals of the world.

[17] Msg C68396, CINCFE to DA, 6 Nov 50.
[18] Msg, JCS 95949, JCS to CINCFE, 6 Nov 50.

CHAPTER XXXVI

# The Chinese Appraise Their First Phase Korean Action

Those who wage war in mountains should never pass through defiles without first making themselves masters of the heights.

MAURICE DE SAXE, *Reveries on the Art of War*

What were the Chinese Communist Forces that fought the battles in late October and early November, and just when and how did they come to Korea? What did they think of their own efforts in their first actions against American troops?

The Chinese troops that appeared suddenly in the Korean fighting near the end of October had crossed the Yalu River from Manchuria in the period from 13 or 14 to 20 October. Hiding from aerial observation during the day and marching at night, they had reached their chosen positions on the southern fringe of the high mountain mass fifty air miles south of the Yalu. There they lay in wait overlooking the corridors that entered these mountains from a point near Huich'on on the east along a line extending sixty air miles westward through Onjong and Unsan.

Three Chinese armies, the *38th*, *40th*, and *39th*, each composed of three infantry divisions, were deployed in that order from Huich'on westward, on this line of battle in front of Eighth Army and the ROK II Corps. The *38th Army* was northwest of Huich'on, the *40th* in the Onjong area, and the *39th* above Unsan. Two more armies, the *66th* and the *50th*, of three divisions each, remained hidden in reserve and out of action in the west during the Chinese First Phase Offensive. This made a total of fifteen divisions. Elements of another division, the *125th*, apparently were the troops that cut off and dispersed the 7th Regiment of the ROK 6th Division below Ch'osan.[1]

The CCF *39th Army* arrived at Antung on the north side of the Yalu River in Manchuria in mid-October and immediately crossed the river into Korea

[1] ATIS Enemy Documents, Issue 47, pp. 139ff, booklet, A Collection of Combat Experiences, issued by *Hq XIX Army Group*, CCF, 29 Mar 51 (also partially reproduced by Hq U.S. I Corps, G–2 Sec, Aug 51); *Ibid.*, Issue 11, pp. 74–82, Primary Conclusions of Battle Experiences at Unsan, issued by *Hq 66th Army*, 20 Nov 50; *Ibid.*, Issues 6, 2 Sep 51, A–9 and 35, 1–15 Nov 52, p. 45; FEC Intel Digest, vol. I, Nr 4, p. 26 (17 Feb 53), *XIII Army Group*, CCF; FEC, Order of Battle Information, CCF, 15 Jun 51.

at Sinuiju. There it came under the operational control of the *XIII Army Group.* The *38th* and *40th Armies* crossed into Korea about the same time. Later in the month, the *50th* and *66th Armies* crossed into Korea in support of the others. It may be concluded that if the U.S. I Corps had not withdrawn when it did, elements of these two CCF armies would have engaged the 5th and 19th Regiments of the 24th Division above Kusong. As it was, small parts of the *66th Army* did accidentally encounter the 19th Regiment there on 1 November and exchanged shots with it.[2]

While the CCF *39th Army* stopped the ROK 1st Division at Unsan, the *40th* engaged and virtually destroyed the ROK 6th Division at Onjong. On the second day of the action, 26 October, the CCF *38th Army* joined the *40th Army* in the battle against the ROK 6th and 8th Divisions between Onjong and Huich'on. These two Chinese armies then rolled the ROK II Corps back southwest along the valley of the Ch'ongch'on to the edge of Kunu-ri.

Simultaneously with the deployment of these Chinese divisions in front of Eighth Army, others deployed in front of the X Corps in northeast Korea. Three CCF divisions, the *124th, 125th,* and *126th,* forming the *42d Army,* entered Korea at Manp'ojin. The first to cross the border was the *124th Division.* On 14 October it left T'ung-hua in southern Manchuria, about fifty air miles from the North Korean border, and proceeded by train to Chi-an on the Yalu opposite Manp'ojin, where it crossed the river the same day. On the 16th it started on foot from Manp'ojin, marching southeast through Kanggye and Yudam-ni to Hagaru-ri. From there its advanced elements proceeded to the point south of the Changjin Reservoir where they met the ROK 26th Regiment on 25 October. The remainder of the division moved up to the point of contact and joined in the battle near Sudong against the U.S. 1st Marine Division troops that replaced the ROK 26th Regiment. The other two CCF divisions that followed the *124th* into Korea at Manp'ojin remained out of action behind the *124th* in this phase of the fighting, but assumed defensive positions in the Changjin Reservoir area, blocking the roads to Kanggye.

Except for the action against the *124th Division,* it would appear that the ROK and U.S. troops had not reached the points in northeast Korea where the CCF intended to block their advance. While the *124th Division* at first drove back the ROK troops it encountered, and then slowed the advance of the U.S. Marine troops that replaced them on the road to the reservoir, it did not have the success that attended the CCF action against the ROK II Corps and part of the U.S. I Corps in the west. In fact, this CCF division was virtually destroyed.

In the First Phase Offensive, highly skilled enemy light infantry troops had carried out the Chinese attacks, generally unaided by any weapons larger than mortars. Their attacks had demonstrated that the Chinese were well-trained disciplined fire fighters, and particularly adept at night fighting. They were masters of the art of camouflage.

[2] FEC Intel Digest, vol. I, Nr 4, 1–15 Feb 53, Histories of CCF Army Groups Active in Korea: XIII Army Group, pp. 30–37; EUSAK PIR 109, 29 Oct 50.

Their patrols were remarkably successful in locating the positions of the U.N. forces. They planned their attacks to get in the rear of these forces, cut them off from their escape and supply roads, and then send in frontal and flanking attacks to precipitate the battle. They also employed a tactic which they termed Hachi Shiki, which was a V-formation into which they allowed enemy forces to move; the sides of the V then closed around their enemy while another force moved below the mouth of the V to engage any forces attempting to relieve the trapped unit. Such were the tactics the Chinese used with great success at Onjong, Unsan, and Ch'osan, but with only partial success at Pakch'on and the Ch'ongch'on bridgehead.

The Chinese soldiers engaging in the First Phase Offensive were well-fed, in excellent physical condition, well-clothed, and well-equipped. As the British had noted, some of them even wore fur-lined boots. Despite numerous American intelligence conjectures at the time that the first Chinese were integrated and scattered throughout the North Korean units, this was not the fact. From the very beginning the Chinese fought in Chinese organizations and were never mixed as individuals into North Korean organizations.

In the offensive against the Eighth Army and the ROK II Corps at the end of October and the first week of November the action was almost entirely by Chinese troops. The delaying action along the west coast against the 24th Division, however, was by North Koreans. Only in rare instances, such as in the final action to the east against the 19th Infantry and the 27th British Commonwealth Brigade, were North Korean infantry troops involved, and then they apparently constituted only a relatively small part of the total attacking enemy force. The North Korean *105th Armored Division,* partly reconstituted since the Naktong battles, was committed to help the Chinese and did so with tank fire in a few instances, but it played a negligible role in the fighting. The Fifth Air Force destroyed most of its tanks back of the battle front. On 7 November, for instance, U.N. aircraft reportedly destroyed 6 tanks, 3 armored cars, and 45 vehicles in Pakch'on and the area eastward.[3]

At the same time in northeast Korea the ROK I Corps encountered Chinese forces at only one point on the road to the reservoir. Within a few days elements of the 1st Marine Division relieved the ROK's there, and they had a stubborn fight for nearly a week in gaining the Kot'o-ri plateau south of the Changjin Reservoir. The Chinese intervention in northeast Korea had not been on the same scale as in the west central part, nor had it had the same success. It is to be noted that the Chinese appeared only in the mountainous, inland part of the peninsula; on neither coast were there any Chinese contacts at this time.

Since the First Phase Offensive was the first engagement in the Korean War between American and Chinese troops, the opinion the Chinese formed of their adversary as a result of it may be of interest. On 20 November, less than three weeks after the CCF *39th Army* had driven the U.N. forces from the Unsan area, the headquarters of the

[3] 24th Div WD, 7 Nov 50; GHQ FEC, History of the N.K. Army, p. 80.

*66th Army,* "Chinese Peoples' Volunteer Army," published a pamphlet entitled, "Primary Conclusions of Battle Experiences at Unsan." In it the Chinese listed what they considered the strengths and weaknesses of the American forces, based on their experience with the 8th Cavalry Regiment. On the favorable side the pamphlet described in some detail the American method of making an attack and said:

> The coordinated action of mortars and tanks is an important factor. . . . Their firing instruments are highly powerful. . . . Their artillery is very active. . . . Aircraft strafing and bombing of our transportation have become a great hazard to us . . . their transportation system is great. . . . Their infantry rate of fire is great and the long range of fire is still greater.[4]

Not so favorable was the Chinese estimate of the American infantry. The pamphlet said that American soldiers when cut off from the rear

> . . . abandon all their heavy weapons, leaving them all over the place, and play opossum. . . . Their infantrymen are weak, afraid to die, and haven't the courage to attack or defend. They depend on their planes, tanks, and artillery. At the same time, they are afraid of our fire power. They will cringe when, if on the advance, they hear firing. They are afraid to advance farther. . . . They specialize in day fighting. They are not familiar with night fighting or hand to hand combat. . . . If defeated, they have no orderly formation. Without the use of their mortars, they become completely lost . . . they become dazed and completely demoralized. . . . At Unsan they were surrounded for several days yet they did nothing. They are afraid when the rear is cut off. When transportation comes to a standstill, the infantry loses the will to fight.

After analyzing the Americans' strength and weakness, the Chinese set forth certain principles for future operations:

> As a main objective, one of the units must fight its way rapidly around the enemy and cut off their rear. . . . Route of attack must avoid highways and flat terrain in order to keep tanks and artillery from hindering the attack operations. . . . Night warfare in mountainous terrain must have a definite plan and liaison between platoon commands. Small leading patrol groups attack and then sound the bugle. A large number will at that time follow in column.

The Chinese admitted they did not have an effective weapon against the American tank, but said that 20-pound TNT charges placed on the tracks or under the tank would disable it. Antitank sections consisted of four men carrying two 20-pound and two 5-pound charges.

The Chinese summed up their viewpoint on the first phase of their intervention:

> Our Army [39th] was the first expeditionary force ordered to hurry to the I-Ung-pong area of Unsan to relieve the North Korean Army and intercept the enemy advancing northwards at Unsan. We deployed our main force to encircle and annihilate the enemy at Hichon [Huichon], Onjong, and Chosan. At that time, we did not fully comprehend the tactical characteristics and combat strength of the enemy, and we lacked experience in mountain warfare. Moreover, we engaged the enemy (first, in the form of interdiction, then in that of attack) without sufficient preparation; yet the result was satisfactory.

[4] ATIS Enemy Documents, Issue 11, pp. 74–82, reproduces this document, captured by ROK 1st Div, 26 Nov 50. See also ATIS Enemy Documents, Issue 47.

## CHAPTER XXXVII

# Guerrilla Warfare Behind the Front

> Do not now go into such dangerous and difficult places, but order your guides to lead you by the easiest road, unless it is much too long; for the easiest road is the shortest for an army. XENOPHON, *Cyropaedia*

In late October and during November, guerrilla warfare was increasing in intensity behind the United Nations front lines in North Korea. Remnants of the North Korea People's Army retreating into North Korea from the Pusan Perimeter and, to a lesser degree, miscellaneous guerrillas carried on this action.

While the U.S. I Corps advanced beyond the 38th Parallel into North Korea together with the bulk of the ROK Army, the U.S. IX Corps in early October assembled in the Taejon-Chonju area, well south of the Han River. There it prepared to secure the supply routes from the old Pusan Perimeter, repair the railroad, and hunt down and destroy bypassed enemy groups. The IX Corps command post opened at Taejon on 5 October. The 2d Infantry Division assumed responsibility generally for the area west and southwest of Taejon, and the 25th Division for the Taejon area and that south and east of it. The newly activated ROK 11th Division was attached to IX Corps on 5 October to help with security in rear areas. On 10 October the Eighth Army Ranger Company was attached to IX Corps, and it joined with the 25th Division Reconnaissance Company in antiguerrilla activity in the Poun area, northeast of Taejon.[1]

During October the 25th Division was the American organization most actively employed in antiguerrilla warfare. It had 6,500 square miles of mountainous country in its zone of responsibility. This lay athwart the escape routes from the old Pusan Perimeter of the larger part of the enemy units cut off or bypassed. On occasion, large groups of enemy soldiers were cornered and either destroyed or captured. On 7 October the 1st Battalion, 35th Infantry, largely as a result of unusually effective artillery fire, killed or wounded about 400 of an enemy force estimated to number approximately 500 men. On another occasion the 3d Battalion of the same regiment captured 549 prisoners in one day.[2]

Enemy guerrilla action, in fact, extended all the way to the southern tip

---

[1] IX Corps WD, Sep 50, Personal Recollections of General Coulter; *Ibid.*, bk. I, sec. II, Oct 50; 2d Inf Div Hist, vol. II, Sep 50, p. 44; 25th Div WD, 7 Oct 50; IX Corps WD, 10 and 15 Oct 50.

[2] EUSAK WD, Br for CG, 24 Oct 50; 25th Div WD, Hist, bk. I, 7–8 Oct 50.

of Korea. Reportedly it was co-ordinated and directed by Kim Chaek, the commander of the North Korean *Front Headquarters.* The guerrillas harassed isolated villages, ambushed patrols, fired on trains, cut telephone lines, and attacked South Korean police stations.

In South Korea bands of guerrillas ambushed vehicles on both the Green Diamond (east coast) and Red Diamond (mountainous western) routes between Pusan and Taegu, and at night raided villages for food, clothing, and hostages. In the latter part of November an estimated 20,000 guerrillas operated in the southwest corner of Korea. The Chirisan area there between Hadong and Koch'ang continued to be a center of guerrilla activity despite the efforts of the newly organized ROK 11th Division and the National Police to suppress it.[3]

Near the 38th Parallel in the central part of Korea there was serious guerrilla action. A long series of incidents occurred in the Ch'unch'on-Hwach'on area north of Wonju. On 22 October an enemy force attacked and dispersed sixty ROK soldiers and police guarding the Hwach'on Dam. The North Koreans then opened the dam's control gates and the Pukhan River rose four feet by the next day and one pier of a railroad bridge many miles below was washed out. On the 25th when a ROK police battalion reoccupied Hwach'on and the hydroelectric dam east of it they found the turbine and control panels at the dam seriously damaged. In November the 1st, 3d, 5th, and 7th ROK Anti-Guerrilla Battalions were almost constantly engaged with North Korean forces in the Hwach'on area. Most of them apparently were elements of the N.K. *10th Division.* At one time the North Koreans held Ch'unch'on, cutting off all its communications with Seoul only forty-five miles to the southwest.[4]

In the east, in the high Taebaek Mountains southwest of Samch'ok, in mid-November there were 4,000 bypassed soldiers of the N.K. *3d, 4th,* and *5th Divisions,* with miscellaneous guerrillas, according to the report of a North Korean regimental commander who surrendered there. Guerrillas even operated at the edge of Seoul. On 15 October a group of them attacked a radio relay station four and a half miles north of the Capital building.

In northeast Korea there was guerrilla action behind the X Corps front from the very moment troops came ashore at Wonsan and it continued until the corps left that part of the country.

Within a few hours after it landed there on 26 October, the 1st Battalion, 1st Marine Regiment, loaded onto gondola cars and just after noon started for Kojo, thirty-nine miles down the coast. There it relieved ROK troops guarding a supply dump. That first night passed quietly, and the next afternoon the ROK's loaded into a train and departed northward.

In a few hours the deceptive quiet at Kojo exploded into a wild night battle

[3] 2d Log Comd Act Rpt, G-2 Sec, Nov 50; EUSAK WD, Br for CG, 23 Nov 50; IX Corps WD, bk. 1, 17-18 Oct 50.

[4] EUSAK WD, G-3 Jnl, Msg 2100 23 Oct 50; *Ibid.*, G-3 Jnl, Msg at 1200 12 Nov 50; *Ibid.*, G-3 Sec, 13 Nov 50; *Ibid.*, Br for CG, 25 Oct 50; GHQ FEC, History of the N.K. Army, p. 69 (N.K. *10th Div*); ATIS Res Supp Interrog Rpts, Issue 104 (N.K. *10th Div*), p. 50; ATIS Interrog Rpts, Issue 27 (Korean Opns), p. 81ff, Lt Col Yun Bong Hun, CO *19th Regt,* N.K. *13th Div.*

that began soon after dark and continued after daylight of the 28th. An organized force of North Koreans, later determined by prisoner interrogations to have been three battalions of the N.K. *5th Division,* had silently crept close to the hill positions of the marines south and northwest of Kojo. The enemy attack came so swiftly at one platoon position that fifteen marines were killed there—seven in their sleeping bags. For a time two Marine companies were cut off, but they eventually fought off the enemy and re-established contact with the battalion.[5]

Upon receiving word of the enemy attack, General Almond requested the dispatch of two destroyers to Kojo where they delivered naval gunfire in support of the marines, and on his orders the 1st Marine Air Wing delivered strikes on Kojo that almost completely demolished and burned it. In the afternoon the 1st Marine Division started the 2d Battalion, 1st Marines, from Wonsan for Kojo, and that evening a tank company began loading into an LST at Wonsan, its destination, Kojo. Before noon of the 28th, however, the North Koreans moved off west into the hills and the fight was over. The marines recovered their dead and counted their losses—27 killed, 39 wounded, 3 missing.[6]

Other guerrilla actions south and west of Wonsan, apparently by elements of the N.K. *5th Division,* followed the Kojo incident. On 2 November, an enemy force ambushed a supply convoy behind the 3d Battalion, 1st Marines, near Wonsan, killing nine and wounding fifteen men.

Some of the most destructive of the attacks behind the lines occurred along the main rail and highway route between Wonsan and Hungnam. On the night of 6–7 November, after dark, an enemy force ambushed a convoy south of Kowon, thirty-five miles north of Wonsan, and, after the drivers fled into the adjacent rice paddies, burned three jeeps and twenty trucks. North of town another group ambushed a military police patrol and then attacked a signal unit, inflicting casualties on both.

At the town of Kowon itself, before midnight, North Koreans attacked a northbound Marine Corps supply train. Thirty marines were riding the train as guards. The night was dark, and snow flurries increased the poor visibility. After the train had stopped at a water tank, a North Korean soldier violently flung open the door of the front coach and burst inside. A marine killed him instantly. Enemy burp gun and rifle fire now ripped into the coaches. Outside, enemy soldiers blew the track ahead and killed the engineer when he started to back the train. The North Koreans persisted in their efforts to enter the coaches despite heavy losses until they finally succeeded, shooting and clubbing every marine they thought still alive. But 2 wounded marines inside the coach successfully feigned death and escaped later. In this action the train engineer and 6 marines were killed, and 8 marines wounded.[7]

---

[5] 1st Mar Div SAR, 8 Oct–13 Dec 50, vol. I, p. 23; *Ibid.*, vol. II, an. C, pp. 6–8, 26–28 Oct 50; Geer, *The New Breed,* pp. 196–203.

[6] X Corps WD, G–3 Jnl, Msg 39, 28 Oct 50; Diary of CG, X Corps, 28 Oct 50; 1st Mar Div SAR, vol. II, an. C, pp. 10–15, 29 Oct–2 Nov 50.

[7] 1st Mar Div SAR, vol. II, an. C, p. 27, 7 Nov 50; X Corps WD, 7–8 Nov 50; *Ibid.*, PIR 42, 7 Nov 50; Geer, *The New Breed,* pp. 248–51.

Later in the night, at 0230, an estimated 500 North Koreans attacked a battalion of the 65th Infantry Regiment, 3d Division, which just the previous day had occupied Yonghung, eight miles farther north and midway between Wonsan and Hungnam, and inflicted 40 casualties. This same enemy force then struck a detachment of the 4th Signal Battalion nearby, and, in moving northward from it, stumbled into the camp of the 96th Field Artillery Battalion. There the North Koreans penetrated the battalion positions and by mortar fire destroyed six 155-mm. howitzers and the battalion trains containing all the ammunition supply, inflicting about 40 casualties. The enemy withdrew after daylight. Aerial observers later discovered an estimated 2,000 enemy troops moving on secondary roads into the mountains northwest of Yonghung. Two heavy air strikes on them caused an estimated 300 casualties.[8]

To the west, elements of the N.K. *15th Division* held the upper Imjin River valley and centered their harassing action against Majon-ni. From the beginning this area was one of the worst centers of North Korean guerrilla action in the X Corps zone of operations. The Wonsan–Majon-ni road was the eastern end of the main lateral road across the peninsula to the North Korean capital city of P'yongyang. From Wonsan, ridge after ridge of the Taebaek Range rose westward for an airline distance of 60 miles. The road twisted along narrow gorges and climbed zigzag over mountain passes. Majon-ni, only 16 air miles from Wonsan, is 28 by road. The little village of 300 population is located at the junction of the lateral Wonsan-P'yongyang road and the north-south road following the Imjin River from Sibyon-ni. For 60 air miles this latter road followed the cramped, twisting gorge of the Imjin River through the heart of the Taebaek Range. It was a favorite route of travel for cutoff North Koreans and traversed one of the most important centers of guerrilla activity in all Korea.

The 3d Battalion, 1st Marines, commanded by Lt. Col. Thomas Ridge, drew the task of holding the Majon-ni road junction, and relieved ROK 3d Division troops there on 28 October. Ridge established a tight perimeter defense line around the village with a roadblock on each of the three roads coming into the Y-shaped junction. On 7 November the 3d Battalion came under general attack at Majon-ni and lost some outpost positions. Air strikes after daylight helped repel the North Koreans. That same day enemy forces on the Majon-ni–Wonsan road ambushed a convoy in Ambush Alley, as it came to be known, on the east side of the 3,000-foot pass. E Company, 1st Marines, accompanying the convoy as guard, lost eight men killed and thirty-one wounded.[9]

On 12 November the 1st Battalion, 15th Infantry, left Wonsan to relieve the marines at Majon-ni. Because of enemy roadblocks it did not arrive there until the next day. On the 14th, the ROK 3d Korean Marine Corps (KMC) Battalion, which had landed at Wonsan four days earlier, arrived at Majon-ni and proceeded fourteen air miles westward to

[8] X Corps PIR 42, 7 Nov 50 and PIR 43, 8 Nov 50; X Corps WD, Diary of CG, 7 Nov 50.

[9] 1st Mar Div SAR, vol. II, an. C, pp. 8–26, 28 Oct–7 Nov 50.

Tongyang where it established a blocking position at another north-south road intersection with the Wonsan-P'yongyang road, near the U.S. X Corps–Eighth Army boundary. Relieved at the isolated Majon-ni mountain crossroads, the 3d Battalion, 1st Marines, joyfully departed on 14 November for Wonsan.[10]

On the night of 20–21 November, an estimated 200 enemy soldiers with armored support again attacked the Majon-ni perimeter. After penetrating it at one point the enemy was repulsed just before daylight.

That day, 21 November, a motorized patrol consisting of two officers and ninety-one men of the 1st Battalion, 15th Infantry, moved west from Majon-ni to establish contact with the 3d KMC Battalion at Tongyang. In an 8-mile gorge about midway between the two places a hidden enemy force ambushed the column. In this action the patrol lost twenty-eight men and most of its vehicles and heavy weapons. A platoon of tanks and a company of infantry from the 2d Battalion arrived to reinforce the Majon-ni garrison. A second attempt to get through to Tongyang on the 22d also failed when a tank-led column turned back because, overnight, the enemy had badly cratered and mined the road. It now became necessary to supply the 3d KMC Battalion at Tongyang by airdrop.

Again, on the 23d, a reinforced rifle company left Majon-ni with the task of repairing the cratered road and proceeding west to contact the ROK marines. The ROK's at the same time were instructed by an airdropped message to send a force eastward on the road to meet the 15th Infantry force. The westbound column had just passed the point of the ambush two days earlier when an estimated 300 enemy troops opened fire on it. Fortunately, the ROK force arrived from the opposite direction and joined in the fight. But both the ROK and the 15th Infantry forces withdrew after the latter had lost sixteen wounded and three missing in action.

Finally, on 25 November the 1st Battalion, 15th Infantry, and the 3d KMC Battalion made a co-ordinated attack against the enemy force that had cut the road between them and decisively defeated it, killing 150, capturing four 120-mm. mortars, and destroying a large ammunition cache. This action temporarily reopened the supply road to Tongyang. But determined and destructive enemy activity in this area never ceased as long as U.N. forces were there.[11]

Between the U.S. 65th Regiment and the ROK 26th Regiment to the north, a large gap existed in the U.S. 3d Division zone that was also a center of guerrilla activity. Near the Eighth Army–X Corps boundary there the 5,600-foot-high mountain mass of Paek-san interposed its trackless waste between the two regiments and the lateral roads they had to use for communication and supply. Between 18 and 22 November B Company, 65th Infantry, tried to establish contact with the ROK 26th Regiment to the north but was driven back after two engagements with enemy forces on Paek-san. The 3d Division estimated that the bulk of 25,000 North Korea

[10] 1st Mar Div SAR, Oct–Dec 50, vol. II, an. C, pp. 26–40, 7–14 Nov 50; 3d Inf Div Comd Rpt, sec. III, Narr of Opns, pp. 5–14, Nov 50; 1st Lt. Charles R. Stiles, "The Dead End of Ambush Alley," *Marine Corps Gazette* (November, 1951), p. 39; Geer, *The New Breed*, pp. 206–15.

[11] 3d Div Comd Rpt, sec. III, pp. 5–9.

People's Army guerrillas in its zone were within a 10-mile radius of Paek-san.[12]

Because there were no usable roads in the northern part of its zone, the U.S. 3d Division could supply the ROK 26th Regiment only with great difficulty. Because of this, Maj. Gen. Robert H. Soule on 21 November requested General Almond to move the boundary between the 3d Division and the 1st Marine Division northward a few miles so as to place the Huksu-ri–Sach'ang-ni road within the 3d Division zone. The next day the X Corps granted the request and changed the boundary. General Soule thereupon ordered the ROK 26th Regiment to establish blocking positions at Huksu-ri and Sach'ang-ni. In the course of capturing these places, the ROK's fought several engagements. An enemy battalion defending Sach'ang-ni was driven away only after it had lost more than 100 men killed or captured.

The ROK's then attacked north and west out of Sach'ang-ni for several miles, encountering enemy delaying forces. On 24 November in this continuing action, the ROK 26th Regiment captured twenty-six CCF soldiers in the vicinity of Sach'ang-ni, the first Chinese soldiers encountered in the X Corps zone with the exception of those in the reservoir area. The road on which this action occurred ran south from Yudam-ni and the west shore of the Changjin Reservoir. Their presence here indicated that Chinese forces were working southward along the X Corps left flank. Information obtained from prisoners indicated that Chinese forces had been in position there for approximately three weeks to check any U.N. attempt to advance north over this road. It chanced that the ROK 26th Regiment on 24 November was the first organization to reach the point where Chinese of the *125th Division* stood guard. This was the newly developed situation when on 25 November the 1st Battalion, 7th Infantry, relieved the ROK 26th Regiment at Sach'ang-ni.[13]

A survey of 110 enemy prisoners taken in one 24-hour period early in November in the X Corps rear areas showed they came from sixty-two different organizations. In the latter part of November, X Corps intelligence estimated that 25,000 North Korean Army guerrillas operated south and west of Hungnam in the X Corps rear areas. The pattern of guerrilla activity gradually shifted northward during November, from below Wonsan to the mountains west of Hungnam. A special report prepared by X Corps listed 109 separate guerrilla attacks in the corps zone during the month, an average of better than three and a half a day.[14]

In the west coastal area of North Korea in the Eighth Army zone, guerrilla activity at this time was not as great as in northeast and central Korea. Patrols, nevertheless, were necessary.

In the hills north and east of P'yongyang these patrols uncovered large quantities of enemy war supplies in caves,

[12] 3d Div Comd Rpt, sec. II, p. 2, and sec. III, p. 7, Nov 50; 65th Inf Comd Rpt, Nov 50, p. 3.

[13] 3d Div Comd Rpt, sec. III, pp. 10–13; X Corps PIR 60, 25 Nov 50; FEC Intel Digest, vol. I, Nr 4, 1–15 Feb 53, Histories of CCF Army Groups Active in Korea, pt. III, *XIII Army Group*, pp. 32–33.

[14] 1st Mar Div SAR, vol. I, pp. 23, 28; X Corps PIR's 38, 3 Nov, 43, 8 Nov, and 46, 11 Nov 50; 3d Inf Div Comd Rpt, Nov 50, sec. II, Intel Summ, p. 2; X Corps, Guerrilla Activities, X Corps Zone, Nov 50, with location map of attacks.

tunnels, and other hiding places. Much of this was found in the mining area east of P'yongyang in old mine shafts and tunnels accessible to spur rail lines. The most important find of all was made by Lt. Doric E. Ball's patrol from M Company, 23d Infantry Regiment, 2d Division on 6 November. With the aid of a local chief of police, the patrol found a huge arsenal in an old lead mine nine miles northeast of Kangdong. In it were about 400 lathes of mixed American, English, and Russian manufacture; all the machinery needed to make Russian-type burp guns and 120-mm. mortars was there. The arsenal consisted of nineteen large rooms used variously as machine shop, storage room, office space, and an auditorium. This armament factory, known as Arsenal No. 65, had originally been above ground in P'yongyang, but aerial bombing had caused the North Koreans to move it to the mine.

East of the main supply road strong guerrilla bands were active in the Sibyon-ni area. One of the worst incidents in this area occurred on 6 November about midway between Kumch'on and Sibyon-ni when an enemy force ambushed the I&R Platoon and a platoon of L Company, 27th Infantry, reinforced by C Battery, 8th Field Artillery Battalion. The 1st Battalion hurried to the scene and at midnight rescued several wounded and missing men, but the enemy had disappeared. Three days later the bodies of fifteen men from the ambushed patrol were found in a shallow grave. According to a survivor, the North Koreans had murdered these men when they followed the example of 1st Lt. Harold G. Parris, an officer captured with them, in refusing to give their captors information beyond their name, rank, and serial number.[15]

At the same time, an enemy force estimated at 1,000 men held control of the Ich'on area some miles eastward. It required a combined attack of the U.S. 27th Infantry from the southwest and of the ROK 17th Regiment from the east on 7 November to drive this force into the hills. A week earlier, on 28 October, North Koreans had killed all the ROK wounded and hospital personnel in a field hospital at Ich'on.[16]

It was inevitable that the Iron Triangle of central North Korea would become the scene of enemy guerrilla and harassing action since it was a principal assembly area for North Korean soldiers retreating northward. The U.N. effort to open the railroad from Seoul through the Iron Triangle to Wonsan encountered almost unceasing guerrilla attacks. Most of them occurred just north of P'yonggang at the apex of the Triangle. ROK troops rode all trains as guards, and pitched battles often were fought between them and the guerrillas.

On the morning of 2 November one of the largest of these actions occurred when a force of about 1,000 guerrillas ambushed a work train ten miles north of P'yonggang. Fortunately, 200 soldiers of the ROK 17th Regiment were riding the train as guards and they succeeded in beating off the attackers, killing 40 and capturing 23 of them. These prison-

---

[15] 27th Inf WD, 6, 9 Nov 50; 8th FA Bn WD, 5–9 Nov 50. Eighth Army General Order 484, 1 July 1951, awarded the Distinguished Service Cross posthumously to Lieutenant Parris. The sixteen captured men were from C Battery, 8th Field Artillery Battalion.

[16] EUSAK WD, Br for CG, 7 Nov 50; Ministry of National Defense, ROK, *Korea in War*, vol. I, 1 May 1950–June 1951, p. A–67.

ers said there were 4,000 guerrillas in the vicinity and that they planned to attack every train going toward Wonsan. Enemy activity became so intense in this area that the entire ROK 17th Regiment was sent there, and gradually elements of the U.S. 25th Division also entered the fight in the Iron Triangle.[17]

On 9 November about 1,400 enemy troops of the *18th Regiment,* N.K. *4th Division,* attacked Ch'orwon in the Iron Triangle, driving away 800 South Korean police. The next day about 500 men of the *5th Regiment* of the same division occupied Yonch'on, ten miles to the south on the road to Seoul. Elements of the 24th Infantry Regiment, 25th Division, tried but failed to regain the town. That evening a North Korean roadblock force ambushed an I&R Platoon and A Company, 24th Infantry, column near there. The next day when the regiment retook Yonch'on it found the bodies of thirty-eight American soldiers at the ambush site and all vehicles burned.

Co-ordinated with the enemy action at Ch'orwon and Yonch'on, an enemy regiment, supported by mortars, attacked the 1st and 2d Battalions, ROK 17th Regiment, at P'yonggang on 10 November, and had surrounded them by daylight the next morning. Maj. Robert B. Holt, a KMAG adviser with the regiment, radioed a request for the 3d Battalion of the ROK regiment at Ich'on to attack eastward, and he also arranged for an airdrop of ammunition. During the day a Mosquito plane landed at P'yonggang and picked up Holt, who directed air strikes on the enemy positions with reportedly heavy casualties to the North Koreans. The enemy force withdrew from around P'yonggang during the night of 11–12 November. But two nights later part of it returned and again attacked the airstrip and the town. The ROK's cleared both places of enemy troops before noon of the 14th at a cost of 11 men killed and 23 wounded, as against 141 enemy killed and 20 captured. The enemy pressure in this area never ceased and the situation there became more precarious when the 25th Division troops moved north to participate in the 24 November Eighth Army attack.[18]

[17] EUSAK WD, G–3 Jnl, Msg at 2315, 2 Nov, and Off for Co-ordination of Protection of Lines of Comm, Rear Areas, Daily Rpt, 28 Oct and 2 Nov 50.

[18] EUSAK WD, G–3 Jnl, Msg 0700 9 Nov 50; EUSAK PIR's 121–25, 10–14 Nov 50; EUSAK POR 370, 12 Nov 50; 25th Div WD, 11–14 Nov 50; GHQ FEC, History of the N.K. Army, pp. 59, 68; ATIS Res Supp Interrog Rpts, Issue 94 (N.K. *4th Div*), p. 50; ATIS Res Supp Interrog Rpts, Issue 100 (N.K. *9th Div*), p. 54; ATIS Interrog Rpts, Issue 18 (Enemy Forces), p. 94 and p. 152; Interv, author with 1st Lt Robert J. Tews (Plt Ldr A Co, 24th Inf, at time of incident and present at scene of ambush, 11 Nov), 3 Sep 51.

CHAPTER XXXVIII

# The X Corps Advances to the Yalu

> Water shapes its course according to the ground over which it flows; the soldier works out his victory in relation to the foe whom he is facing. Therefore, just as water retains no constant shape, so in warfare there are no constant conditions.
>
> SUN TZU, *The Art of War*

### *ROK I Corps Attacks up the Coastal Road*

After the landing of the X Corps at Wonsan on 26 October, the ROK Capital Division, already north of Hungnam, continued its attack northward in three regimental combat teams. (*Map 25*) The ROK Cavalry Regiment of the division, a motorized organization, constituted what General Almond called the "flying column." It was to advance as rapidly as possible toward the border. Almond made arrangements for supplying this flying column from an LST at sea, and he provided it with a tactical air control party from the 7th Infantry Division. Two days after the X Corps landing at Wonsan, the ROK Cavalry Regiment against strong opposition captured Songjin, 105 air miles northeast of Hungnam. At the same time, the 1st Regiment of the Capital Division approached P'ungsan, inland halfway to the border on the Iwon–Sinch'ang-ni–Hyesanjin road. Two days later the third regiment of the division, the 18th, reached the south end of Pujon Reservoir.[1]

In front of the ROK Cavalry Regiment on the coastal road an estimated North Korean battalion retreated northward toward Kilchu, a sizable town twenty miles north of Songjin. Kilchu is fourteen air miles inland from the coast, the farthest point inland for a town of any size along the whole length of the east coastal road. Situated beyond the reach of effective naval gunfire, it was a favorable place for the North Koreans to fight a delaying action. The ROK attack before daylight of 3 November developed into a day-long battle which

---

[1] X Corps PIR 34, 30 Oct 50; X Corps WD, Summ, 28 Oct 50 and WD, 4 Nov 50, G–1 Rpt, Rpt on Almond-Partridge Conference.

Like both AMS maps of Korea, scales 1:250,000 and 1:50,000, the text uses the Korean place names Pujon and Changjin for the two reservoirs that figure prominently in the narrative hereafter. Hagaru-ri is used for the combined areas of the town of Changjin and the village of Hagaru-ri. Because the American forces were using Japanese maps (some early maps gave dual names), they used the Japanese names Chosin and Fusen rather than the Korean Changjin and Pujon for the reservoirs.

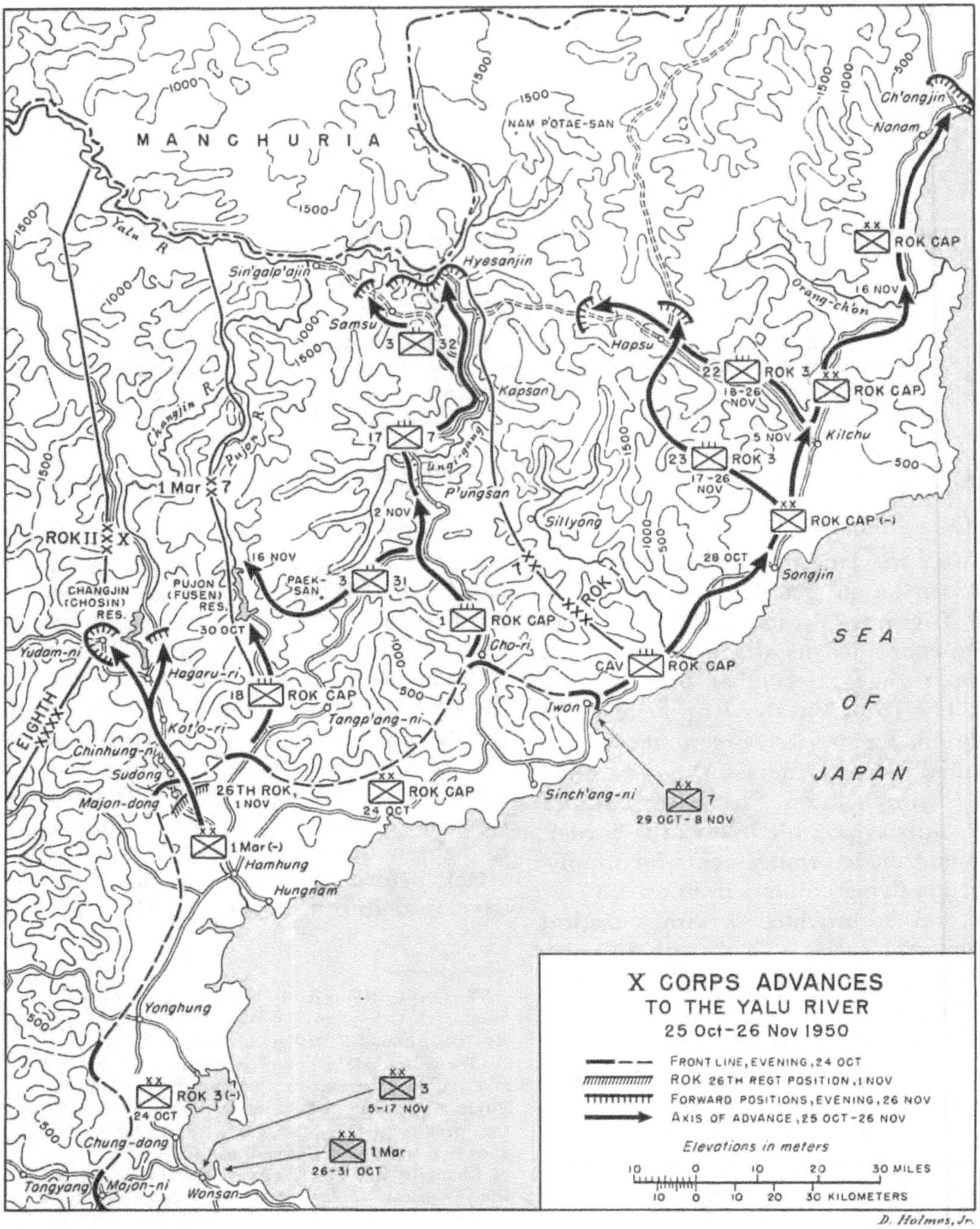
X CORPS ADVANCES
TO THE YALU RIVER
25 Oct–26 Nov 1950
FRONT LINE, EVENING, 24 OCT
ROK 26TH REGT POSITION, 1 NOV
FORWARD POSITIONS, EVENING, 26 NOV
AXIS OF ADVANCE, 25 OCT–26 NOV
Elevations in meters
10 0 10 20 30 MILES
10 0 10 20 30 KILOMETERS
MANCHURIA
SEA OF JAPAN
Yalu R
Hyesanjin
Sin'galp'ajin
Samsu
Kapsan
Hapsu
Ch'ongjin
Nanam
Orang-ch'on
Kilchu
P'ungsan
Sillyong
Songjin
Cho-ri
Iwon
Sinch'ang-ni
Tangp'ang-ni
PAEK-SAN
CHANGJIN (CHOSIN) RES.
PUJON (FUSEN) RES.
Yudam-ni
Hagaru-ri
Kot'o-ri
Chinhung-ni
Sudong
Majon-dong
Hamhung
Hungnam
Yonghung
Chung-dong
Tongyang
Majon-ni
Wonsan
NAM P'OTAE-SAN
Changjin R
Pujon R
ROK II
EIGHTH
ROK CAP
ROK 3
ROK 3(-)
ROK CAP(-)
1 Mar
1 Mar(-)
26TH ROK, 1 NOV
CAV
24 OCT
16 NOV
5 NOV
28 OCT
2 NOV
30 OCT
18-26 NOV
17-26 NOV
29 OCT-8 NOV
5-17 NOV
26-31 OCT
D. Holmes, Jr.

MAP 25

failed to win the town. The ROK 1st Regiment joined the Cavalry Regiment in the battle. By daylight of 5 November the two ROK regiments had encircled Kilchu, and they captured it before noon. On the day of Kilchu's capture Corsair air strikes from the 1st Marine Air Wing were credited with destroying 2 enemy tanks, 4 artillery pieces, and 350 counted enemy dead. The next day a count of all the North Korean dead reached 530. In the Kilchu battle the ROK's captured 9 45-mm. antitank guns, 6 82-mm. mortars, and 10 heavy machine guns. The ROK Cavalry Regiment lost 21 killed and 91 wounded. Prisoners said the N.K. *507th Brigade* had defended the town. The local North Korean commander reportedly ordered the execution of a battalion commander whose unit had retreated.[2]

After the Kilchu battle, aerial reports indicated that fresh enemy troops were moving south along the coastal road from the Ch'ongjin-Nanam area. Supported by tanks, this force, estimated at six to seven battalions, met the Capital Division on 12 November just north of the Orang-ch'on, thirty miles above Kilchu. In the resulting battle it forced the ROK 18th Regiment to withdraw south of the stream. Bad weather prevented effective close air support, and, since the scene of action was beyond the range of destroyers' guns, the heavy cruiser *Rochester* was sent to provide naval gunfire support. Clearing weather enabled Corsairs to join in the battle on the 13th, and that afternoon an air strike destroyed two tanks, damaged a third, and forced another to withdraw. Six inches of snow covered the Orang-ch'on battlefield.[3]

The enemy attack resumed and made further penetrations in the positions of the 18th Regiment on 14 November. The next day it compelled the 18th and the 1st Regiments to withdraw again. Close air support for the ROK's prevented the North Koreans from exploiting this success and during the day destroyed 3 more tanks, 2 self-propelled guns, and 12 trucks. At the same time, thirty B-29's dropped 40,000 incendiary bombs on Hoeryong, a rail and road communication center of 45,000 population on the Tumen River at the Manchurian border, 100 miles southwest of Vladivostok. By 16 November the four days of ground battle and three of aerial attack had so weakened the enemy force that it faltered, and the ROK 18th Regiment once more advanced slowly.

A delayed report covering the three days from the evening of 14 November to that of 17 November listed 1,753 enemy killed, 105 prisoners, and the capture of 4 rapid-fire guns, 62 light machine guns, 101 burp guns, and 649 rifles. On the 19th, air attacks destroyed 2 more tanks and 2 artillery pieces. The USS *St. Paul* now gave naval gunfire support to the ROK's. The enemy force that fought the battle of the Orang-ch'on consisted of about 6,000 troops of the N.K. *507th Brigade* and a regiment of the N.K. *41st Division*, supported by a battalion of 8 tanks.[4]

---

[2] X Corps WD, Summ of Opns, 3-6 Nov 50; *Ibid.*, PIR's 39, 4 Nov, 40, 5 Nov, and 41, 6 Nov 50.

[3] X Corps POR 48, 13 Nov 50; X Corps PIR 47, 12 Nov, and 48, 13 Nov 50.

[4] X Corps POR 50, 15 Nov, and 57, 22 Nov 50; X Corps PIR 49, 14 Nov, 51, 16 Nov, and 54, 19 Nov 50; X Corps WD, 25 Nov 50, app. 1, Intel Est to an. AO 7, p. 2; GHQ UNC, G-3 Opn Rpt, 14 Oct 50; EUSAK WD, 16 Nov 50, EUSAK Daily News Bul.

The ROK troops fought these battles in northeast Korea under great physical hardship from the cold. On 16 November the temperature in their zone already had dropped to 16° below zero. Their uniforms in most cases consisted of one old suit of fatigues, worn-out shoes, one half blanket per man, and an old U.S. Army overcoat. The situation was about to improve, however, for the next day at Songjin an LST unloaded 26,000 sets of ROK winter clothing. Another LST loaded with shoes, wool socks, underwear, shirts, and field jackets was to sail from Pusan that same day.

By 17 November the ROK 3d Division had moved up behind the Capital Division on the coastal road and had started its 23d Regiment inland from Songjin toward Hapsu. The next day its 22d Regiment started for the same objective from farther north at Kilchu. The larger part of the ROK 3d Division, therefore, was now deployed on the left of and inland from the Capital Division. Six LST's of the Korean Navy supplied the ROK I Corps.[5]

Able at last to resume its advance after the battle of the preceding week, the Capital Division on 20 November crossed the Orang-ch'on and resumed its drive toward Ch'ongjin, the big industrial center 30 miles north of the river and 65 air miles southwest of the Siberian border. A little more than a month earlier a naval task force had heavily bombarded Ch'ongjin with 1,309 rounds of 6-inch, 400 rounds of 8-inch, and 163 rounds of 16-inch shells. ROK troops, following behind a rolling barrage of naval gunfire, enveloped Ch'ongjin on 25 November. That evening the 1st Regiment moved around the city to a point 5 miles north of it; the Cavalry Regiment seized the airfield on its western edge; and the 18th Regiment was on its southern outskirts. The next day, Ch'ongjin fell to the Capital Division. The ROK's now planned to turn due north and inland along the highway and railroad leading to Hoeryong at the Manchurian border.[6]

### *U.S. 7th Infantry Division Reaches Manchurian Border*

The U.S. 7th Infantry Division, in its zone of operations between the ROK I Corps to the north and the 1st Marine Division to the south, likewise made important gains in carrying out its part of the X Corps mission in northeast Korea. On 29 October the 17th Infantry Regiment of the division landed unopposed at Iwon. That same day the 1st Battalion of the regiment; the 49th Field Artillery Battalion; and A Company, 13th Engineer Combat Battalion moved from the beachhead to Cho-ri, a distance of 50 miles. From Cho-ri the 7th Division was to strike north for the Manchurian border at Hyesanjin, 70 air miles away. But over the poor dirt road that twisted its way through the mountains and the Korean upland the distance was much greater. On the last day of the month the 1st Battalion and regimental headquarters moved on to P'ungsan, 120 road miles from the Iwon beaches and approximately halfway between the coast and Hyesanjin. The 1st Regiment of the

[5] X Corps POR's 52–55, 17–20 Nov 50.

[6] X Corps POR's 55–61, 20–26 Nov 50; X Corps PIR's 59–61, 24–26 Nov 50; 2d Log Comd Hist Rpt, Water Div, Trans Sec, Nov 50.

ROK Capital Division had cleared the road of enemy troops that far. When the 7th Division got all its elements ashore its total strength would exceed 26,600 men. The division on 1 November counted 18,837 men, almost full-strength, and to this were added 7,804 attached South Korean soldiers.[7]

The U.S. 7th Infantry Division had its initial action in northeast Korea on 1 November when the 1st Battalion, 17th Infantry, helped the ROK 1st Regiment repulse a strong North Korean attack two miles north of P'ungsan. Col. Herbert B. Powell, commander of the 17th Infantry Regiment, ordered an attack by the 1st Battalion the next morning at 0800 to clear the enemy force from in front of the town. But the North Koreans in approximately regimental strength attacked first, at 0700, starting an action that continued throughout the day. Except for one company, all units of the 17th Regiment arrived at P'ungsan by the end of the day. Colonel Powell's regiment now relieved the ROK 1st Regiment, which turned back to join the Capital Division on the coastal road.[8]

Because the open beaches were wholly at the mercy of the weather and high seas, unloading of the 7th Infantry Division at Iwon went forward slowly. The relatively few vehicles ashore, the long haul, and the low stockpile on the beach combined to cause the 17th Infantry on 4 November to request an airdrop at P'ungsan of 4.2-inch, 81-mm., and 60-mm. mortar ammunition. An airdrop the next day had considerable breakage loss. Patrols on the 4th discovered the enemy had withdrawn from in front of P'ungsan, and the 17th Infantry advanced unopposed to the Ungi River. The temperature stood at 2° below zero.[9]

At the Iwon beachhead, the 3d Battalion, 31st Infantry Regiment, landed on 3 November and the rest of the regiment followed ashore the next day. The regimental mission was to move to the left (west) flank of the 17th Infantry. This would place it in the mountainous country extending to the Pujon Reservoir. ROK troops previously had advanced into that region.[10]

Carrying out its mission, the 31st Infantry Regiment advanced on the division left flank toward the reservoir. On 8 November it encountered Chinese soldiers on the eastern slopes of Paek-san, a 7,700-foot peak twelve miles east of the southern end of the reservoir. This was the U.S. 7th Division's first contact with the CCF. There, during the afternoon, elements of the regiment engaged in a battle with an estimated battalion of Chinese, later identified as part of the CCF *376th Regiment, 126th Division.* Before nightfall the 31st Infantry seized that part of Paek-san, and the Chinese force withdrew with at least 50 killed. On this same day a patrol of the regiment met a Marine patrol about midway between Hamhung and the Pujon Reservoir, thus establishing the first contact between the two divisions in northeast Korea.[11]

---

[7] 7th Inf Div WD, 29–31 Oct 50; X Corps POR 36, 1 Nov 50.

[8] 7th Inf Div WD, 1–2 Nov 50; 17th Inf WD, 1–2 Nov 50; Interv, author with Brig Gen Herbert B. Powell, 13 Apr 54.

[9] 7th Inf Div WD, G–4 Jnl, 4–5 Nov 50; *Ibid.*, 5–7 Nov 50; Barr Notes, 6–7 Nov 50.

[10] 7th Inf Div WD, 2–4 Nov 50.

[11] 7th Inf Div WD, 7–10 Nov 50.

THE 1ST BATTALION, 17TH INFANTRY, *enters Hyesanjin on the Yalu River on the morning of 21 November.*

On the division right flank the next day, 9 November, the 7th Reconnaissance Company moved to Sillyong, east of P'ungsan, to protect a power installation.

On 12 November the division received orders from X Corps to continue the advance northward. The 17th Infantry was to seize Kapsan, and then go to Hyesanjin on the Yalu; the 31st Infantry was to advance on the left of the 17th; and the 32d Infantry was to seize the southeast shore of the Pujon Reservoir. The 32d Infantry, which began unloading on 4 November and was the last of the regiments to come ashore at Iwon, moved southwest from the beach along the coast through Hamhung and there turned northeast to Tangp'ang-ni in preparation for its part in the operation.[12]

In accordance with the corps order, the 17th Infantry prepared to attack across the Ungi River on 14 November. To replace the bridge which the North Koreans had blown, Colonel Powell had ROK troops in the regiment construct a floating footbridge made of planking extending between empty oil drums. En-

[12] 7th Inf Div WD, 9–12 Nov 50; X Corps Opn Ord 9, 12 Nov 50.

LOOKING ACROSS THE YALU, *from left to right: Brig. Gen. Homer Kiefer, Generals Hodes, Almond, and Barr, and Colonel Powell.*

emy fire on the bridge site was at long range and ineffective. The 2d Battalion, 17th Infantry, crossed over this footbridge without difficulty on the 14th and proceeded to the attack.

The 3d Battalion was scheduled to cross the river at the same time over a shallow ford a few miles to the east. During the night of 13–14 November enemy forces apparently opened dams upstream. The water level rose two feet, making the river waist-deep. In the face of heavy small arms and some mortar fire, six men of L Company waded the stream in weather 7° below zero. A few other men entered the water, but it soon became apparent that all who crossed the stream would be frozen and out of action in a few minutes unless they were specially cared for. The battalion commander ordered the men who had crossed to the north side to return. Their clothes had to be cut from them. They were then wrapped in blankets and taken to the 3d Battalion command post tent to warm. Casualties from this abortive crossing attempt were 1 killed, 6 wounded, and 18 men suffering frostbite from exposure in the river. Colonel Powell agreed with the battalion commander that the 3d Battalion could not cross by wading the icy water. Both Generals Barr and Almond concurred in this decision. The battalion subsequently crossed over the oil drum footbridge.[13]

The need for shelters and warming areas for the front-line troops led the 7th Division the next day to request the immediate delivery of 250 squad tents and 500 oil-burning stoves. In order to keep vehicle gasoline lines and carburetors from freezing it was necessary to mix alcohol or alcohol-base antifreeze with gasoline.

On 15 November, the 1st Battalion crossed the Ungi River behind the 2d Battalion and moved up on its left, but the two battalions made only small gains. On the 16th, aerial observers reported the enemy forces separating into small groups and withdrawing toward Kapsan. That day the 17th Regiment gained about eight miles. On the 19th, the 1st Battalion seized Kapsan at 1030 after a co-ordinated infantry, tank, and artillery attack. In this action the 17th Tank Company overran enemy troops in their foxholes, while the heavy fire of the 15th AAA Battalion 40-mm. weapons drove other North Koreans from log-covered trenches and pillboxes and then cut them down. Under cover of the combined fire of the tanks and the antiaircraft weapons the infantry then crossed the river. That night the 1st Battalion was eight miles north of Kapsan, only twenty-three road miles from Hyesanjin on the Yalu. The 2d and 3d Battalions followed behind the 1st Battalion. The regimental command post set up in Kapsan for the night.[14]

[13] 17th Inf WD, 14 Nov 50; 7th Div WD, 14 Nov 50; Barr Notes, 14–15 Nov 50; Interv, author with Powell, 13 Apr 54; Interv, author with Lt Col Walter Gunthorp (2d Bn, 17th Inf, Nov 50), 13 Apr 54.

[14] 7th Div WD, 15–19 Nov 50; Barr Notes, 19 Nov 50; Capt George H. Worf, "Enroute to the Yalu," *Antiaircraft Journal* (March–April, 1951), p. 22.

The next day, 20 November, the 17th Regiment in a column of battalions—the 1st, 3d, and 2d, in that order—advanced on foot nineteen miles over icy roads through and over the mountains to a point only a few miles from the Yalu. Small enemy groups opposed the advance with only brief exchanges of fire, and then fled. On the morning of 21 November, without opposition, the 1st Battalion, 17th Infantry, led the way into Hyesanjin, and by 1000 had occupied the town and surrounding ground to the banks of the Yalu River. General Almond had flown into Kapsan on the 20th and, together with General Barr and Colonel Powell, accompanied the leading elements of the 17th Infantry Regiment into the town. A week earlier, on the 13th, Navy carrier planes had attacked the military camp at Hyesanjin, burning the barracks buildings and warehouses. The town was now about 85 percent destroyed by this and earlier aerial action.[15]

Upon receiving word that elements of the 7th Division had reached the Yalu, General MacArthur immediately sent a message to General Almond saying, "Heartiest congratulations, Ned, and tell Dave Barr that the 7th Division hit the jackpot." Almond added his own congratulations to Barr on the 22d, saying in part, "The fact that only twenty days ago this division landed amphibiously over the beaches at Iwon and advanced 200 miles over tortuous mountain terrain and fought successfully against a determined foe in subzero

[15] 17th Inf WD, 21 Nov 50; 7th Inf Div WD, 20–21 Nov 50; X Corps WD, Diary of CG, 21 Nov 50; X Corps POR 55–56, 20–21 Nov 50; X Corps PIR 56, 21 Nov 50; Barr Notes, 21 Nov 50.

weather will be recorded in history as an outstanding military achievement." [16]

The Yalu River at Hyesanjin is not the great river it becomes near its mouth at Sinuiju. At Hyesanjin, near its source on the southwest slopes of 8,000-foot-high Nam P'otae-san, the White Head Mountain famous in Korean history, the Yalu was approximately 50 to 75 yards wide. On the day the 17th Infantry first stood on its banks the river was frozen over except for a 6-foot-wide channel; four days later it was completely frozen over. The bridge across the stream at Hyesanjin had been destroyed before the 17th Infantry arrived there. Upstream about 300 yards on the north side of the Yalu in Manchuria there was an undamaged Chinese village larger than Hyesanjin. Officers and men of the 17th Infantry had mixed emotions, some apprehensive, as they looked across the ribbon of ice and water into Manchuria. There they saw CCF sentries walk their rounds and their officers come and go.[17]

Meanwhile, to the southwest, the 31st Infantry Regiment patrolled extensively and advanced in its zone. This mountainous waste was virtually roadless, and ox-drawn carts were used to transport supplies and evacuate wounded. On 15 November a patrol from the 3d Battalion, 31st Infantry, reached the eastern shore of Pujon Reservoir. The next day another patrol encountered about 200 CCF soldiers at the northern end of the reservoir and drove them away after a brief fight. On the 18th, patrols ranged both sides of the reservoir. Leaving strong detachments to guard the mountain passes from the reservoir eastward into the division's rear along the Cho-ri–P'ungsan road, General Barr on 20 November began moving the bulk of the 31st and 32d Regiments to the P'ungsan-Kapsan area behind the 17th Infantry. On the division right, ROK troops finally arrived at Sillyong shortly before midnight of the 20th to relieve I Company, 32d Infantry. Unfortunately, in the darkness each group at first thought the other the enemy and a brief fire fight resulted in the wounding of five ROK's.

The 32d Infantry Regiment, concentrating now at Kapsan, prepared to strike northwest through Samsu to Sin'galp'ajin on the Yalu. This would put it on the Manchurian border to the left or west of the 17th Regiment at Hyesanjin. Task Force Kingston, commanded by 2d Lt. Robert C. Kingston, a platoon leader of K Company, started for Samsu on 22 November and entered the town unopposed at midafternoon, followed later by the rest of the 3d Battalion, less I Company.[18]

The 17th Infantry at Hyesanjin was to co-operate with the 32d Infantry by attacking west to meet it. On 22 November, the first day that it attempted to move west to join the 32d Infantry, one of its combat patrols encountered a force of North Koreans about seven miles west of Hyesanjin, and a stubborn fight developed. This set a pattern of

[16] Schnabel, FEC, GHQ Support and Participation in the Korean War, ch. VII, p. 2, quoting Msg U/CX2867, CG X Corps to CG 7th Div, 21 Nov 50, and Msg U/CX2859, CG X Corps to CG 7th Div, 22 Nov 50.

[17] Interv, author with Almond, 23 Nov 54; Interv, author with Barr, 1 Feb 54; Interv, author with Powell, 13 Apr 54.

[18] 7th Div WD, 22 Nov 50; X Corps PIR 57, 22 Nov 50; 1st Lt Martin Blumenson, Task Force Kingston (32d Infantry Regiment, 22–29 November 50), copy in OCMH.

ON THE BANKS OF THE YALU, *two soldiers look across the valley into the mountains of Manchuria.*

action that occupied the 17th Infantry during the next week, as long as it was in that part of Korea—daily fights with small but stubborn enemy forces that blew bridges, cratered roads, all but immobilized the regiment, and kept it from making any appreciable gains. At the same time, in front of the 32d Infantry, enemy forces fought effective delaying actions north of Samsu so that not until 28 November did Task Force Kingston, reinforced, reach Sin'galp'ajin.[19]

The intense cold of northeast Korea in late November took its toll in frostbite casualties in the 7th Division. The worst to suffer was the 31st Infantry which operated in the remote mountain regions east of the Pujon Reservoir. A total of 142 men in the division were treated for frostbite up to 23 November; 83 of them were from the 31st Regiment. Of the 58 men evacuated because of frostbite, 33 were from that regiment.[20]

### *3d Infantry Division Joins X Corps*

During November the U.S. 3d Infantry Division joined the X Corps in Ko-

[19] X Corps PIR 59, 24 Nov 50; 7th Div WD, 22–28 Nov 50; Senate MacArthur Hearings, pt. 4, p. 2954, testimony of Gen Barr.

[20] 7th Div WD, G–1 Rpt, 23 Nov 50.

OX-DRAWN SLEDS *replace trucks for the movement of supplies west of P'ungsan across the icy wastes.*

rea. One of its regiments, the 65th, had been in South Korea for more than two months. It had embarked on two transports in Puerto Rico on 25 August, passed through the Panama Canal, and sailed directly for Korea. It arrived at Pusan on 22 September and disembarked the next day. The other two regiments, the 7th and 15th, and the division headquarters sailed from San Francisco between 30 August and 2 September. The last ship of the division transports arrived at its destination, Moji, Japan, on 16 September.

Because the division was greatly understrength it was scheduled to receive large numbers of Koreans (KATUSA) for integration into its units. The division, minus the 65th Infantry Regiment, had an actual strength of only 7,494 men on 15 September. Beginning the first of October and continuing through the month, the 3d Division in Japan received 8,500 Korean draftees. Squads often consisted of two American enlisted men and eight Koreans.[21]

The 65th Infantry was the first part of the 3d Division to join X Corps in

[21] EUSAK WD, G-1 Hist Rpt, 15 Sep 50; 3d Div Comd Rpt, Nov 50, sec. 1, p. 1; Maj. James A. Huston, Time and Space, ch. IV, p. 117, MS in OCMH; ORO, Utilization of Indigenous Manpower in Korea, ORO (FEC), pp. 54-56.

northeast Korea. Its 2d Battalion, the first to land, came ashore at Wonsan on 5–6 November and was thereupon attached to X Corps for operations. The regiment was composed of white Puerto Ricans, Virgin Islands Negroes, white soldiers from the United States, Negroes from the United States (tank company), Americans of Japanese descent, and, finally, integrated South Koreans. When it sailed from Puerto Rico all the enlisted men in the infantry regiment were Puerto Rican, as were sixty-four of the 206 officers.[22]

In Japan the main body of the 3d Division made ready to outload at the port of Moji the first week of November. A division advance party opened the 3d Division tactical command post at Wonsan on 10 November. The 15th Regimental Combat Team began unloading there on the 11th, and the 7th RCT finished landing on 17 November.

The 3d Division's primary mission was to relieve all 1st Marine Division troops in the Wonsan area and south of Hamhung, to block the main roads in the southern part of the corps zone against guerrillas and bypassed North Koreans, and to protect the Wonsan-Hungnam coastal strip. The 3d Division zone of responsibility measured approximately ninety by thirty-five miles, an area so large as to make centralized division control impracticable. Therefore, General Soule, the division commander, decided to establish four regimental combat teams and to assign sectors and missions to each. These were the 7th RCT, commanded by Col. John S. Guthrie; the 15th RCT, commanded by Col. Dennis M. Moore; the 65th RCT, commanded by Col. William W. Harris; and the ROK 26th Regiment of the ROK 3d Division (attached to the U.S. 3d Division for operations), commanded by Col. Rhee Chi Suh. The 10th Field Artillery Battalion supported the 7th RCT; the 39th Field Artillery Battalion, the 15th RCT; the 58th Armored Artillery Battalion (self-propelled guns) and C Company, 64th Heavy Tank Battalion, the 65th RCT; and A Battery, 96th Field Artillery Battalion, the ROK 26th Regiment.[23]

The 15th RCT had the mission of protecting Wonsan and the area south and west of the city, with the Wonsan–Majon-ni–Tongyang road the probable axis of major enemy activity. North of the 15th RCT, the 65th RCT was to hold the west central part of the division zone, with the Yonghung–Hadongsan-ni lateral road the principal route into the regimental sector from the coast. The northern sector of the division zone, west of Hamhung, fell to the ROK 26th Regiment; included among its missions was that of patrolling west to the Eighth Army–X Corps boundary. The 7th RCT was in 3d Division reserve with the mission of securing the coastal area from Chung-dong, a point about eight miles north of Wonsan, to Hungnam. The 64th Heavy Tank Battalion (–) was also in division reserve.[24]

---

[22] 3d Div Comd Rpt, Nov 50, sec. III, p. 3; *Ibid.*, G–3 Jnl, entry 1225, 10 Nov 50; X Corps POR 52, 17 Nov 50; X Corps PIR 46, 11 Nov 50; 2d Log Comd Hist Rpt, G–3 and G–4 Secs., Nov 50; MS review comments, Col John H. Chiles to author, Nov 54.

[23] 3d Div Comd Rpt, Nov 50, sec. III, p. 4; X Corps Opn Ord 6, 11 Nov 50.

[24] 3d Div Comd Rpt, Nov 50, sec. III, p. 5.

The 3d Division did not engage in any major military operation during the November period covered in this volume, but beginning on 12 November it did have a number of engagements with North Korean forces in ambushes and roadblocks along the regimental main supply routes, particularly in the sector of the 15th RCT west of Wonsan between Majon-ni and Tongyang. Several of these were serious and resulted in heavy losses of men and equipment. They grew progressively worse toward the end of November; apparently the North Korean actions were co-ordinated with Chinese intervention in the reservoir area of northeast Korea.

### *7th Marines Clear Road to Reservoir*

While the ROK I Corps and the U.S. 7th Infantry Division advanced toward the northeast border of Korea against scattered and ineffective North Korean opposition, the 1st Marine Division began moving up its assigned axis of advance toward the Changjin Reservoir to the southwest of them. Its rate of advance was not to be as rapid as theirs, nor was it to go as far.

At 0730 Sunday, 29 October, the 1st Battalion, 7th Marines, began loading into trucks at the Wonsan Agricultural College, and an hour later it started on the 83-mile trip to Hamhung. The next day X Corps ordered the 1st Marine Division to relieve the ROK I Corps in the reservoir area. At the end of the month the 7th Marines was in an assembly area north of Hamhung, and the 5th Marines was en route there from Wonsan.

From Hamhung to the southern tip of the Changjin Reservoir the road climbs for 56 miles. For slightly more than half the distance, to Chinhung-ni, the climb is easy and gradual over a two-lane road. From there a one-lane dirt road zigzags up precipitous slopes to the 4,000-foot-high plateau lying just south of the reservoir. In an air-line distance of 4 miles, and a road distance of 8 miles, north of Chinhung-ni, the road climbs 2,500 feet in elevation to the rim of the plateau, two and a half miles south of Kot'o-ri. A narrow-gauge railroad paralleled the road from Hamhung to Chinhung-ni, but from that point to the top of the plateau a cable car incline replaced it. Once on top of the plateau, the railroad track continued north to Hagaru-ri and the Changjin Reservoir. There were four mountain power plants on the road to the reservoir.

Six road miles below Chinhung-ni is the village of Sudong. There, just below the steep climb to the plateau, the CCF *124th Division* held its blocking position. Three miles south of Sudong the road, climbing northward, crosses from the west to the east side of a mountain stream. The bridge at this crossing was of importance; if it were destroyed the U.N. forces north and south of it would be separated and those on the north cut off from their source of supply. Hill 698 dominated this bridge from the west, as did Hill 534 from the east. These two hills were critical terrain features.

Early on 1 November the 7th Marines entrucked at its Hamhung assembly area and, together with its attached artillery, the 2d Battalion, 11th Marines, it moved north 26 road miles to defensive positions behind the ROK 26th Regiment, 3 miles below Changjin Power Plant 3. The ROK troops had fallen back more than 5 miles since they first met the Chinese.[25]

At 1030, 2 November, the 7th Marines relieved the ROK 26th Regiment in its position and the 1st Battalion, followed by the 2d Battalion, attacked north. By noon it had confirmed that Chinese troops opposed it, and during the day captured three of them—one from the *370th Regiment* and two from the *372d*. The 1st Battalion at 1630 began to prepare defensive positions for the night about one mile south of Sudong. The 2d Battalion, nearly a mile behind the 1st Battalion, meanwhile had engaged in a hard struggle for Hill 698 west of the road. Three miles behind the 2d Battalion the regimental headquarters and the 3d Battalion were at the Majon-dong road fork.[26]

Shortly after midnight the CCF launched a co-ordinated attack, calculated to separate the 1st and 2d Battalions from each other and from the regiment behind them. The enemy infantry cut in between the 1st and 2d Battalions and almost overran the 4.2-inch mortar company in position along the road. Fighting was close and at grenade range for both battalions. In the course of the battle the Chinese gained a position dominating the crucial bridge in the 2d Battalion area. With the coming of daylight, Marine aircraft went into feverish action, repeatedly attacking the Chinese and eventually forcing them from their roadblock positions. Sandbagged trucks successfully brought sixty-six wounded marines through CCF small arms fire to the rear. During the day combined ground and air action killed about 700 enemy soldiers. Identification on the dead showed that nearly all of them were members of the *370th Regiment*.[27]

After the heavy battle on the 3d, the CCF apparently withdrew, for the marines encountered only light opposition on the 4th as they entered and passed through Sudong and continued on to the higher ground around Chinhung-ni. At Samgo railroad station, just north of Chinhung-ni, the marines destroyed the last four tanks of the N.K. *344th Tank Regiment*.

Less than a mile beyond Chinhung-ni the steep climb began through the Funchilin pass to the Kot'o-ri plateau. The marines could see enemy troops on the heights flanking the road at the pass. Farther north, according to aerial observers, an estimated 400 soldiers and three tanks were moving south from the Changjin Reservoir. Strafing reportedly caused heavy casualties in this column. A critical terrain feature, Hill 750, or How Hill as it came to be called, a mile and a half beyond Chinhung-ni, dominated the road where it made a hairpin loop of 1,000 yards eastward in starting the climb. This hill was, in fact, the

---

[25] 1st Mar Div SAR, Oct–Dec 50, vol. II, an. C, 1–2 Nov, pp. 14–16; *Ibid.*, vol. III, an. RR, 2 Nov, p. 12; X Corps POR 34, 30 Oct 50.

[26] 1st Mar Div SAR, Oct–Dec 50, vol. III, p. 12, an. RR, 2 Nov 50; X Corps WD, 2 Nov 50, Msg at 1730 from Maj Sayre; X Corps PIR 37, 2 Nov 50; Lynn Montross and Capt. Nicholas A. Canzona, USMC, *U.S. Marine Operations in Korea, 1950–1953*, vol. III, *The Chosin Reservoir Campaign* (Washington: Historical Branch, G–3, Headquarters, U.S. Marine Corps, 1954), is an excellent detailed account of the 1st Marine Division in northeast Korea in 1950.

[27] 1st Mar Div SAR, Oct–Dec 50, vol. II, p. 19, an. C, 3 Nov 50; X Corps WD, PIR 39, 4 Nov 50; Geer, *The New Breed*, pp. 234–36; Montross and Canzona, *The Chosin Reservoir Campaign*, pp. 103–09.

southern knob of a long finger ridge that extended southward from the rim of the Kot'o-ri plateau, on the east side of the road.

The 3d Battalion on 5 November moved through the 1st Battalion to start the attack up the pass. From How Hill the CCF stopped its advance. A map taken from a dead CCF officer showed that reinforced battalions on either side of the road were holding the high ground. Marine aircraft repeatedly attacked How Hill but failed to force the enemy from his positions.

The marines had to take How Hill before they could advance farther. The next day H Company made a long flanking march to approach the hill from the southeast. At 1600 in the afternoon of the 6th, H Company reached the position from which it was to make its assault. After half an hour of air strikes and an artillery preparation, two platoons started for the top. Four times the CCF drove them back. When darkness fell the Chinese still held the hill, and H Company withdrew. All night artillery and mortars pounded How Hill, the 81-mm. mortars firing 1,800 rounds.[28]

Meanwhile, west of the road a Marine force had seized Hill 611 without difficulty. A prisoner taken there supplied the second report X Corps had received that two more Chinese organizations, the CCF *125th* and *126th Divisions,* were in the reservoir area.

The next morning, patrols from the 3d Battalion found that the enemy had withdrawn from the heights behind Chinhung-ni, including How Hill,

CHINESE COMMUNIST POW'S, *wearing quilted cotton winter uniforms and fleece-lined caps.*

leaving behind many dead and some wounded. Information gained later from prisoners disclosed that the artillery and mortar barrage against How Hill during the night had caused crippling casualties in the *372d Regiment* (possibly the *371st*) of the *124th Division* while it was moving up to reinforce the line. These losses had caused the CCF to withdraw. On the afternoon of 7 November the 3d Battalion moved ahead and reached the village of Pohujang and Power Plant 1.[29]

During the marines' six days of battle with the Chinese *124th Division* the 1st Marine Air Wing had inflicted great numbers of casualties on the Chinese. But according to prisoners, the support-

[28] 1st Mar Div SAR, vol. II, pp. 23–25, an. C, 5–6 Nov 50; X Corps PIR's 40–41, 5–6 Nov 50; Geer, *The New Breed*, pp. 239–41.

[29] 1st Mar Div SAR, vol. II, pp. 27–28, an. C, 7 Nov 50; ATIS Interrog Rpts (Enemy Forces), Issue 19, Nr 2400, p. 1, Wang Fu Tien, *372d Regt, 124th Div.*

ing artillery and mortars had taken an even greater toll. After 7 November the CCF *124th Division* reportedly was down to a strength of about 3,000 men. Except for its stragglers, the *124th Division* did not again enter the fighting in the reservoir area.

Thus far the action against the CCF *124th Division* from 2 through 7 November had cost the 7th Marines 46 men killed, 262 wounded, and 6 missing in action.[30]

For two days after reaching Power Plant 1, the 7th Marines sent out patrols which failed to contact the enemy. On 10 November the regiment moved up over the pass without opposition and occupied Kot'o-ri. Only seven miles now separated it from Hagaru-ri.

In its fight to reach the Kot'o-ri plateau the 7th Marines had captured 58 Chinese prisoners, 54 of them from the *124th Division* and 4 from the *126th Division*. It had taken its first prisoner from the *126th Division* on 7 November.[31]

During the time the 7th Marines was heavily engaged in combat with the CCF *124th Division,* a controversy between General Almond and General Partridge over the control of the 1st Marine Air Wing came to a head. Under existing procedure the Fifth Air Force Joint Operations Center at Seoul controlled the assignment of missions to the 1st Marine Air Wing. General Almond felt that, during a period of active ground combat when the local ground tactical situation could change drastically within an hour or two, he, the local commander, should have complete command over the air units supporting the ground troops. On 4 November General Partridge flew to Wonsan to hold a conference with General Almond on the subject. General Almond won his point; the Fifth Air Force ordered the 1st Marine Air Wing to assume direct responsibility for close support of X Corps without reference to the Joint Operations Center. Close support requests beyond the capabilities of the 1st Marine Air Wing were to be reported to the Fifth Air Force.[32]

The first night on the Kot'o-ri plateau, 10–11 November, was to be one painfully unforgettable to the men of the 7th Marines. During the afternoon and night the temperature dropped 40 degrees to 8° below zero—and with it came a wind of 30 to 35 miles an hour velocity. Although the weather later became colder, with the temperature dropping to from 20° to 25° below zero, it did not affect the men as did this first shock of subzero temperature. During the succeeding three or four days more than 200 men of the regiment collapsed from severe cold and were placed in sick bays for medical treatment. Stimulants had to be used to accelerate depressed respiration. Water-soluble medicines froze, and morphine could be maintained in satisfactory condition only when kept against the body. Plasma could be used only after a 60- to 90-minute preparation in a warm tent.[33]

[30] 1st Mar Div SAR, vol. III, an. RR, p. 34.

[31] X Corps PIR 47, 12 Nov 50.

[32] X Corps WD, 4 Nov 50, G–1 Rpt, Notes on Conference between CG X Corps and Gen Partridge, 4 Nov 50; USAF Hist Study 71, pp. 76–78.

[33] 1st Mar Div SAR, Oct–Dec 50, vol. III, p. 79, an. RR, app. 3, Cold Weather Rpt, and p. 95, app. H, Medical Rpt.

On the plateau and at the reservoir the men found that they needed more energy-giving food. Candy was at a premium. One veteran of the Changjin Reservoir operation said later, "I think we consumed more [candy] in one cold Korea week than we averaged in a Stateside year. . . . I saw many others do as I did—eat six or seven large Tootsie Rolls within a 10 to 15 minute period." [34]

### *The Gap Between Eighth Army and X Corps*

A glance at a map of North Korea which shows the locations across the breadth of the peninsula of forces under United Nations command at this time, in mid-November, would reveal to any student of war a situation at once startling and perhaps bewildering. In northeast Korea, forces under X Corps command were far to the north, and in some places stood at the northern boundary of the country. In west and central North Korea the forces under Eighth Army command were far south of these latitudes. A line drawn due east from the Eighth Army front after the battle to hold the Ch'ongch'on River bridgehead in early November would cross the X Corps rear areas far behind the corps front. Not only was the X Corps front far north of Eighth Army's, but it was also separated from it by a wide lateral gap. Virtually all of North Korea west or northwest of the X Corps front in November was in enemy hands.

This great gap, seldom penetrated even by army or corps patrols, extended a minimum distance of 20 air miles from the northernmost right flank positions of Eighth Army to the nearest left flank positions of X Corps. Farther south the gap was greater, being about 35 air miles on a line east of P'yongyang and west of Wonsan. This was the distance after the Korean Marine Corps 3d Battalion established its blocking position at Tongyang on 14 November; before that the distance was about 50 air miles when the X Corps' westernmost position was at Majon-ni. The road-mile distances of this gap over exceedingly bad mountain trails—they hardly could be called roads—were far greater. The 20 air miles, for instance, between Maengsan, the easternmost position of Eighth Army beyond Tokch'on, and that of the X Corps at Kwangch'on just across the X Corps boundary was about 50 road miles.

On any line projected westward north of Hamhung there were no Eighth Army troops opposite X Corps. The enemy held all the territory there from the X Corps boundary west to the Yellow Sea. Accordingly, physical contact between the two commands would have to be made in the southern part of the X Corps zone if it was to be accomplished at all.

This wide gap between the two major tactical organizations of the U.N. Command in Korea caused great concern to Eighth Army and some to the Joint Chiefs of Staff in Washington; but less concern in X Corps, and very little, apparently, to General MacArthur. He believed that the mountainous backbone of North Korea was so destitute of roads and usable means of communication that it would be impossible for the U.N. forces to maintain a continuous line

[34] 1st Lt. William J. Davis, "Lessons Learned Up North," *Marine Corps Gazette* (April, 1952), p. 45.

across the peninsula that far north, and that the enemy would be unable to use this mountainous spine for effective military operations. Indeed, it seems quite clear that it was principally because of this forbidding terrain and the lack of lateral communications between the western and eastern parts of North Korea that General MacArthur established the two separate commands in North Korea.[35]

Many times it has been alleged that the Chinese in late November and in December 1950 took advantage of this great gap between Eighth Army and X Corps to defeat the U.N. forces in Korea. But this concept can be refuted. First, a study of CCF troop movements and deployments in November and December 1950, and the subsequent military action, will show that the Chinese forces did not use the area of this gap for extensive or decisive military operations. Instead, they operated against the Eighth Army right flank where the ROK II Corps was on line, just as they had in their First Phase Offensive in late October and early November. In short, they crushed and rolled up the Eighth Army right flank; they did not attack it from the gap, nor did they move around and behind it through the gap between Eighth Army and X Corps. Second, the unified line formed later across Korea by Eighth Army and X Corps under Eighth Army command was farther south where the terrain and communication facilities were much more favorable for a continuous line than in the area which was the scene of operations in November 1950.

While General MacArthur never expected solid and continuous physical contact between Eighth Army and X Corps in North Korea, he did expect communication and co-ordination between them by radio and personal liaison to the extent possible. There was radio communication between the two commands, and there was a daily trip by air of a liaison officer from X Corps to Eighth Army and back. As early as 25 October, before X Corps troops had landed on the east coast, arrangements had been made by the Fifth Air Force Joint Operations Center for two reconnaissance flights daily between the ROK II Corps right flank and X Corps left flank, which were to report on front lines and enemy concentrations.[36]

The two commands made many attempts to establish physical contact between them by means of patrols scheduled to meet at designated points along the Eighth Army–X Corps boundary. The first of these efforts was made on 6 November when the 2d Infantry Division of Eighth Army sent a reinforced patrol from K Company, 23d Infantry, to the designated point (the village of Songsin-ni) on the boundary five miles east of Yangdok. The patrol reached this point the next day but there were no elements of the X Corps there to meet it. At Yangdok the K Company patrol found and destroyed 16 boxcars of 120-mm., 80-mm., and 47-mm. ammunition;

[35] Senate MacArthur Hearings, pt. 1, p. 246, testimony of MacArthur, and pt. 2, pp. 972–75, testimony of Bradley.

[36] EUSAK WD, G–3 Jnl, Msg 1700, 25 Oct 50; Interv, author with Col LeRoy Lutes, Jr. (Planning Off G–3, X Corps, 1950), 7 Oct 51; Interv, author with Maj Gen Frank A. Allen, Jr., 28 Jan 54.

6 self-propelled guns; 16 47-mm. antitank guns; 30 57-mm. antitank guns; 1 120-mm. mortar; 3 heavy machine guns; and 3 antitank rifles. The X Corps sent a radio message to Eighth Army saying that marines from the 3d Battalion, 1st Marines, at Majon-ni, 30 air miles to the east, could not meet the Eighth Army patrol because of the distance and intensive guerrilla action in the area to be traversed. It suggested other contact points on roads to the north—at Hadong-san-ni and at Sach'ang-ni.[37]

Upon receiving this message Eighth Army withdrew the 23d Infantry patrol and prepared to send another one from the 38th Infantry Regiment, 2d Division, to Hadongsan-ni on the next lateral road northward. General Almond meanwhile had ordered Colonel Harris, commanding officer of the 65th Infantry Regiment, 3d Division, to place one battalion near the boundary to establish contact there with elements of Eighth Army. For this purpose the 1st Battalion, 65th Infantry Regiment, on 10 November after some delay rolled west and established its patrol base at Kwang-ch'on, about four air miles from the boundary, but double that in road miles.[38]

On 9 November IX Corps of Eighth Army had ordered the 38th Infantry to send a patrol to Hadongsan-ni. But the patrol of the 2d Reconnaissance Company which tried to reach that point was turned back by craters and boulders in the road. The next day patrols discovered all roads leading east from the 38th Infantry area were cratered or blocked by boulders. From the X Corps zone a patrol of the 65th Infantry went to the boundary on 11 November, and the next day it went two miles beyond its boundary without meeting a patrol from Eighth Army. On the 12th the pilot of a liaison plane working with the patrol discovered a ROK force westward and dropped a message to it arranging a meeting for the next day. On the 13th, however, even though it went nine miles beyond the boundary to a point where the road became impassable, the 65th Infantry patrol failed to meet an Eighth Army patrol. Eighth Army had tried to keep the rendezvous at Hadongsan-ni, but its strong motorized patrol built around E Company, 38th Infantry, was stopped by road craters ten miles east of Maengsan. All the mountain roads and trails leading eastward from this area were examined by 38th Infantry patrols but none were found that would permit passage of motorized vehicles. Some of the craters were about fifteen feet deep and thirty-five feet in diameter. Although work was started on a bypass it appears that it was never completed.[39]

While the 38th Infantry patrol failed to get through to the contact point on

[37] EUSAK WD, G–3 Jnl, Msg at 1335 6 Nov, G–3 Sec, 7 Nov 50, and Br for CG, 10 Nov 50. Songsin-ni does not appear on the 1:250,000 revised map of Korea, but the village of Yakhyon-dong is at the map co-ordinates given in the records for Songsin-ni.

[38] X Corps WD, Summ, 10 Nov 50; *Ibid.*, Diary of CG; 65th Inf Comd Rpt, Nov 50, p. 5; Interv, author with Almond, 23 Nov 54.

[39] 2d Inf Div WD, Narr Summ, Nov 50; 2d Inf Div POR's 300, 13 Nov 50, and 303, 14 Nov 50; IX Corps WD, bk. I, sec. IV, Opns, Nov 50; 3d Inf Div WD, G–3 Jnl, 13 Nov 50; EUSAK WD, Br for CG, 13 and 16 Nov 50; 38th Inf WD, 12–18 Nov 50.

the 14th, there was success from another quarter. On the 13th an Eighth Army liaison plane dropped two messages to the X Corps patrol saying that a patrol from the 10th Regiment, ROK 8th Division, was working its way to the contact point along a different route. On 14 November at 1000 the two friendly patrols, a platoon from the 2d Battalion, 10th Regiment, ROK 8th Division, and a patrol from the 1st Battalion, 65th Infantry, did meet near the village of Songha-dong just west of the boundary. The ROK patrol had come on foot from its patrol base at Maengsan, forty-five miles to the west. En route it had encountered an estimated total of 400 North Korean guerrillas and had fought several minor engagements. The round trip of the ROK patrol to the boundary and back to its base took ten days. This should explain why there were not daily meetings between Eighth Army and X Corps patrols at the boundary contact point.[40]

On 18 November, just before noon, a patrol from the 3d Battalion, 38th Infantry, reportedly reached Hadong-san-ni on the boundary where it found a blown bridge which it could not bypass. There were no patrols there at that time from the 65th Infantry in the X Corps zone. Only once, therefore, on 14 November, did patrols from the Eighth Army (ROK II Corps) and the X Corps make physical contact with each other at the army-corps boundary.[41]

[40] 3d Inf Div WD, G-3 Jnl, Msg CG X Corps to Eighth Army, 1500, and entries at 1020, 2220 13 Nov 50; *Ibid.*, G-3 Jnl entry at 141330 Nov 50; X Corps PIR 49, 14 Nov 50; 65th Inf Comd Rpt, Nov 50, p. 4; EUSAK POR 379, 15 Nov 50; EUSAK WD, G-3 Sec, 14 Nov 50; Lt. Col. Robert C. Cameron, "The Lost Corps," *Military Review*, vol. XXXIII, No. 2 (May, 1953), p. 11. Cameron was KMAG adviser to the 10th Regt, ROK 8th Div.

[41] 2d Inf Div WD, POR 315, 18 Nov 50; 38th Inf WD, 16–18 Nov 50.

# CHAPTER XXXIX

# The Big Question

> A bold operation is one which has no more than a chance of success but which, in case of failure, leaves one with sufficient forces in hand to cope with any situation. A gamble, on the other hand, is an operation which can lead either to victory or to the destruction of one's own forces.
>
> IRWIN ROMMEL, "Account of the War in Africa," *Rommel Papers*

By the end of the first week of November it was clear that Chinese Communist Forces had intervened in the Korean War. This intervention, long feared and by some expected, had become a fact. The intervention came in sufficient force to drive Eighth Army back to the Ch'ongch'on River and to delay the advance of X Corps in the east toward the Changjin Reservoir. After accomplishing this, the Chinese Communist Forces withdrew from immediate contact with Eighth Army behind a screen of North Korean soldiers. The big question now loomed—what was the purpose and extent of the Chinese intervention? The U.N. command and intelligence agencies had to ponder and answer it in determining the future conduct of military operations.

### *The Chinese Communist Forces*

A look at the background of the Chinese Communist Forces seems necessary to an understanding of the problem. While the Chinese Communist Party itself dates from 1921, Chinese Communist fighting forces may be said to have come into existence with the outbreak of the Nanchang rebellion in China on 1 August 1927. This date is commemorated as the founding date of the Chinese Communist Forces by the Chinese ideograms in the upper lefthand corner of the CCF flag and insignia.[1] A month before the Nanchang rebellion, Generalissimo Chiang Kai-shek had expelled the Communist Party from the Kuomintang. From that time on the two groups became increasingly hostile toward each other and engaged in an intermittent civil war lasting for two decades, until at last Chiang Kai-shek's Nationalist Government in 1949 was driven from the China mainland to Formosa.

For a time in 1934 it had appeared that Chiang Kai-shek was on the point of destroying the Communist forces in China. But in a series of battles, be-

[1] GHQ FEC, MI Sec, Order of Battle Information, Chinese Communist Forces in Korea, 15 Jun 51 (hereafter cited as FEC MIS, Order of Battle Info, CCF), p. 1.

ginning on 21 October of that year and lasting through the 29th, the Communists broke through the Nationalist forces that had surrounded them in Kiangsi Province and started out on what has since been called the "Long March" to Yenan in Shensi Province, far to the northwest. Approximately 100,000 Communist soldiers began this march with Lin Piao leading the *First Army Corps.*

Finally, on 20 October 1935, a year after they had started from Kiangsi, 20,000 survivors of the Long March met units of the *25th, 26th,* and *27th Red Armies* in Shensi that had been there since 1933. The veterans of the Long March had traveled more than 6,000 miles; they had crossed 18 mountain ranges, 24 rivers, 12 different provinces, and had averaged nearly 24 miles a day for the 235 days and 18 nights of actual travel.[2]

In Shensi Province the Chinese Communists then began a reorganization and consolidation of their forces. They established there a secure base that became the foundation for their future operations. Survivors of the Long March were organized into the *8th Route Army,* commanded by Chu Teh. By 1947 the Chinese Communist Army was entrenched in North China and Manchuria with about 600,000 troops. The Soviet withdrawal from Manchuria had taken place in April of the preceding year. Chu Teh, commander of the Chinese Communist Army, at that time announced the primary mission of his force was the piecemeal annihilation of the Nationalist Armies by guerrilla-type action.

The Chinese Communist Army in Manchuria was called the North-East People's Liberation Army (NEPLA). Lin Piao commanded this army, which by the end of 1947 had cut Nationalist lines of communication to Manchuria and isolated that important area from the rest of China. In the spring of 1949 NEPLA was redesignated the *Fourth Field Army,* incorporating five army groups, the *XII* through *XVI,* which in turn comprised the CCF *38th* through *58th Armies,* a total of 60 divisions of about 10,000 men each. This gave a total of approximately 600,000 men in the *Fourth Field Army.* The Korean volunteers and Manchurian Korean veterans in this army numbered about 145,000.

Lin Piao's *Fourth Field Army* played a prominant role in the Chinese Communist Forces' great triumph of wresting control of the China mainland from the Nationalists in 1949. Some elements of his army marched all the way from Manchuria to South China where they made the amphibious attack against Hainan Island in the spring of 1950 and began preparations for a similar attack against Formosa. The *Fourth Field Army* had fought from Manchuria to Hainan Island in the China Civil War without a major defeat. In June 1950 these elements of the *Fourth Field Army* marched to Canton and entrained there for Antung, Manchuria, across the Yalu River from Korea. Lin Piao was now taking them back to the Korean border to stand ready for any eventuality arising from the impending Communist invasion of South Korea. Still other elements of the *Fourth Field Army* moved during the

[2] *Ibid.*, pp. 1–2; Edgar Snow, *Red Star Over China* (New York: Modern Library, 1944), pp. 207–16.

summer from other points in China back to Manchuria. A part of the army had always remained there.[3]

Following on the heels of the *Fourth Field Army,* elements of the *Third Field Army,* which consisted of the *20th* through the *37th Armies,* moved to Manchuria in the late summer and early autumn of 1950. By mid-October the Chinese forces of the *Third* and *Fourth Field Armies* had concentrated more than 400,000 troops in Manchuria close to Korea. A Chinese Communist army comprised normally three divisions, although a few of them had four. A full-strength Chinese division had approximately 10,000 soldiers. It was elements of the *Fourth Field Army,* the best field army of the Chinese Communist Forces, that first intervened in the Korean War.[4]

CHINESE COMMUNIST FLAG

### *Eighth Army Estimate of CCF Intervention*

On 25 October, Col. Percy W. Thompson, G–2 of U.S. I Corps, made special arrangements to transport the first Chinese prisoner, captured that day at Unsan by the ROK 1st Division, to the Eighth Army advanced command post at P'yongyang for interrogation. There could be no doubt that he was Chinese; he spoke neither Korean nor Japanese. His story seemed straightforward and credible.[5] With this first interrogation of a captured Chinese soldier in Korea by U.S. Army intelligence officials began the build-up of a large body of information on Chinese Communist units in Korea. The Chinese Communist prisoners captured in the Eighth Army zone of responsibility grew steadily in number from the 3 captured at Unsan and Onjong on 25 October. By 29 October 10 had been captured; by 2 November, 55; by 20 November, 84; and by 23 November, as Eighth Army assumed its final deployment for the attack designed to reach the Yalu, 96 CCF prisoners had been captured. They identified six Chinese Communist armies—the *38th, 39th, 40th, 42d, 50th,* and *66th,* of which they were members—as being in Korea. (Each army had three divisions, thus the six totaled eighteen divisions.) Eastward in its zone, X Corps had captured prisoners from the *42d Army.*[6] As events

[3] FEC MIS, Order of Battle Info, CCF, CCF *Fourth Field Army,* pp. 1–9; FEC Intel Digest, vol. 1, No. 4, 1–15 Feb 53, CCF *XIII Army Group,* pp. 30–37.

[4] FEC MIS, Order of Battle Info, CCF, pp. 9, 171; FEC Intel Digest 10, 2 Nov 51.

[5] Ltr and attached statement, Col Percy W. Thompson to author, 9 Apr 54.

[6] EUSAK PIR's 109, 29 Oct; 11, 31 Oct; 113, 2 Nov; 116, 5 Nov; 120, 9 Nov; 121, 10 Nov; 124, 13 Nov; and 134, 23 Nov 50.

CHU TEH

were to prove, the *42d Army* near the end of this period had sideslipped from the X Corps area southwest into the Eighth Army zone.

Within the first week of CCF intervention in Korea, prisoners had been taken from four different Chinese armies in the Eighth Army zone. Eighth Army interrogated these first Chinese prisoners intensively, even using a lie detector on three selected and flown to P'yongyang.[7] But in evaluating the interrogations, Eighth Army intelligence officials were skeptical of the stories of large Chinese forces in Korea and, lacking what they believed adequate confirmation, did not accept the substance of the prisoners' accounts. One intelligence item reproduced at different headquarters minimized the scale of Chinese intervention, stating that of 344 prisoners taken at Unsan in two days only two were Chinese. This statement was certainly inaccurate because that number of prisoners was never captured at Unsan. This figure could have been obtained only by counting the many disorganized and retreating North Koreans captured on the road from the Ch'ongch'on River to Unsan. After the Chinese entered the fight against the ROK 1st Division just above Unsan, very few prisoners were captured and these were wounded Chinese.[8]

The initial estimate at Eighth Army headquarters based on prisoners' reports was that the presence of Chinese troops at Unsan and Onjong indicated some reinforcement of North Korean units with troops taken from the Chinese Communist Forces in order to assist in defense of the border approaches, but that there were "no indications of open intervention on the part of Chinese Communist Forces in Korea."[9] On 30 October the 1st Cavalry quoted in one of its

[7] Interv, author with Col Robert G. Fergusson (Deputy ACofS G-2, Eighth Army Oct–Nov 50), 23 Jun 52.

[8] EUSAK PIR 107, 27 Oct 50; 1st Cav Div PIR 100, OB an., 30 Oct 50. Colonel Thompson says that at the time this report appeared he had objected to the statistical statement appearing in the 1st Cavalry Division PIR which allegedly was based on a I Corps report. Colonel Thompson, I Corps G-2; Colonel Hennig, Commanding Officer, 10th AAA Group, supporting the ROK 1st Division at Unsan; and Colonel Hazlett, KMAG adviser to the ROK 1st Division, all have told the author there was no such number as 344 prisoners taken at Unsan. And there is no indication in the combat records that that many prisoners were taken there.

[9] EUSAK PIR 106, 26 Oct 50; 1st Cav Div WD, 30 Oct 50; I Corps WD, Intel Summ 135, 30 Oct 50; New York *Times*, October 28, 1950.

reports from a I Corps estimate of a day or two earlier, "There are no indications at this time to confirm the existence of a CCF organization or unit, of any size, on Korean soil."

The first week of Chinese action in Korea, however, did cause General Walker to restrain I Corps, which, acting on his warning, ordered the 24th Division back to the Ch'ongch'on River from its advanced position. This order on 1 November disturbed the Far East Command. General Hickey telephoned General Allen about the withdrawal and the latter had to explain why General Walker and his staff considered it advisable.[10]

During the first phase of Chinese intervention the units involved tried to conceal their identity by using code names. This succeeded rather well at first. The *54th Unit,* for instance, was not suspected of being in reality the *38th Army,* which prisoners said it was, but was accepted as being only a small part of it. Similar evaluations were made of the *55th, 56th, 57th,* and *58th Units,* each of which represented a CCF army. In point of fact, the *54th Unit* was the *38th Army;* the *55th Unit,* the *39th Army;* and the *56th Unit,* the *40th Army.* This misconception was further deepened by the Chinese use of a battalion code to represent a full division; thus the *1st Battalion, 55th Unit,* was actually the *115th Division, 39th Army.* Chinese officials maintained from the first the fiction that the Chinese fighting in Korea were volunteers. Thousands of interrogations of Chinese prisoners later and scores of captured Chinese documents proved this contention false.[11]

LIN PIAO

Eighth Army intelligence noted on 31 October that out of a total of eleven Chinese prisoners taken in the Eighth Army zone by that date, six claimed to belong to the *56th Unit* and two claimed the *55th Unit.* The army intelligence estimate on that date spoke of these two units as "token units" and surmised that they probably were concerned with protecting the approaches to the Kanggye area. By 1 November, however, on the

[10] Interv, author with Maj Gen Leven C. Allen, 15 Dec 53; Ltr, Thompson to author, 9 Apr 54; Ltr, Lt Gen Doyle O. Hickey to author, 14 Feb 56.

[11] EUSAK WD, 31 Oct 50, G–2 Sec, Interrog of Yang Jun-tzu, 164–MISDI–1171; *Ibid.,* 28 Oct 50, G–2 Sec, Interrog of Yen Shu-cheng, 164–MISDI–1167; *Ibid.,* 1 Nov 50, Interrog of Huang Che-chan, 164–MISDI–1174; *Ibid.,* PIR 113, 2 Nov 50; FEC Intel Digest, vol. 1, Nr 4, 17 Feb 53, *XIII Army Group,* CCF; FEC MIS, Order of Battle Info, CCF.

eve of the 8th Cavalry disaster at Unsan, Eighth Army intelligence had changed its view so far as to say that the enemy forces in the vicinity of Unsan included "possibly at least two Chinese units of regimental size." [12]

By 4 November, Eighth Army intelligence had accepted two Chinese units of division size in Korea. The next day it raised this estimate to "three divisional sized Task Units," tentatively identified as the *54th, 55th,* and *56th Units*, totaling approximately 27,000 men. It will be noted that the Chinese code unit still was underestimated—now being accepted as of division size rather than army size, which it in fact was.[13]

On 6 November, at the very time as it chanced that the CCF withdrew from general contact with Eighth Army, General Walker wrote a letter to General MacArthur in which he expressed his views of the tactical situation on his front. He said in part:

> There has never been and there is now no intention for this army to take up or remain in passive perimeter or any other type of defense. Every effort is being made to retain an adequate bridgehead to facilitate the resumption of the attack as soon as conditions permit. All units continue to execute local attacks to restore or improve lines. Plans have been prepared for the resumption of the offensive employing all forces available to meet the new factor of organized Chinese Communist forces. These plans will be put into execution at the earliest possible moment and are dependent only upon the security of the right flank, the marshalling of the attack troops and the restoration of vital supplies.[14]

On 13 November Eighth Army received its first report of a prisoner from the CCF *42d Army*. The report came from the ROK II Corps on the right flank of the army front. In the next few days other Chinese prisoners from the *42d Army* were taken in this same general area around Tokch'on. All came from the *374th Regiment, 125th Division, 42d Army*.[15] This Chinese army at the time was sideslipping southwest from the Changjin Reservoir area in the X Corps zone to the right flank of Eighth Army, while elements of the CCF *Third Field Army* were replacing it in northeast Korea.

At the end of the third week in November, as U.N. forces made ready to resume their attack toward the border of Korea, Eighth Army intelligence estimated there were about 60,000 Chinese troops in Korea. Various field reports reaching the Department of the Army in Washington differed in their estimates, their figures ranging from 46,700 to 70,000.[16]

The opinions held by the ranking members of the Eighth Army staff on the extent of Chinese intervention, capability, and intention seem to have varied. General Walker apparently shared the view held by Lt. Col. James C. Tarkenton, his G–2, that the Chinese in Korea

[12] EUSAK PIR's 109, 29 Oct; 111, 31 Oct; and 112, 1 Nov 50.

[13] EUSAK PIR 115, 4 Nov, and 116, 5 Nov 50; Ltr, Thompson to author, 9 Apr 54.

[14] MacArthur, MS review comments, with Ltr to Maj Gen Richard W. Stephens, Chief, OCMH, 15 Nov 57. MacArthur's comments on the copy of the manuscript on file in OCMH are in pencil and each one is initialed by him.

[15] EUSAK PIR's 124, 13 Nov; 129, Incl 2, 18 Nov; and 131, 20 Nov 50.

[16] Interv, author with Fergusson, 23 Jun 53; DA *Intel Rev*, 175, Dec 50, p. 33; Ltr, Maj Gen John A. Dabney to author, 3 Feb 54.

numbered only a few divisions composed possibly of volunteers; that there were no organized CCF armies, as such, in Korea; and that China would not enter the war.[17] Colonel Dabney, Eighth Army G-3, was skeptical of this view. He arrived at the conclusion that the Chinese might well have crossed the Yalu River in great strength, but he too was still searching for final answers.

In the two weeks after the Chinese broke off their First Phase Offensive and withdrew from contact with Eighth Army, the impact of the Chinese menace on the American command gradually subsided. Among Eighth Army staff members, the motive generally ascribed for the first CCF intervention in October and early November was that the Chinese wanted to protect the power plants south of the Yalu River. Many now thought they would dig in on a defensive line to do this. As the days passed and the front remained quiet, fear of massive Chinese intervention dwindled.

A careful study of Eighth Army daily intelligence reports for the month of November 1950 reveals that, despite daily reference to the Chinese potential north of the Yalu River in Manchuria, there was a tapering off of concern about full Chinese intervention from about 10 November until 24 November, when Eighth Army resumed its offensive. In this connection it should be noted that the controlling Eighth Army viewpoint could scarcely avoid being influenced somewhat by that of the Far East Command, which seems to have been that China would not intervene with major forces.

### *The X Corps Estimate*

The capture of sixteen prisoners from the *370th Regiment* of the *124th Division* by the ROK 26th Regiment on 29 October convinced the X Corps G-2 that "integral CCF units have been committed against U.N. forces."[18] General Almond said at this time he intended to attack with sufficient strength to find out if the Chinese were in only regimental or in greater strength. The 7th Marines' attack on the road to the Changjin Reservoir disclosed that the Chinese were there in at least division strength.

Elements of X Corps encountered and captured only a few Chinese soldiers from another division, the *126th,* near the Changjin Reservoir. On the east side of the Pujon Reservoir, a few more were captured who said they were from the same division. These prisoners reported that a third division, the *125th,* was in the Changjin Reservoir area. Because this division guarded the road and rail line approaching Yudam-ni and the reservoir from the southwest—the central part of the peninsula—X Corps was not in contact with it until the third week of November. By then most of this enemy division had already moved southwestward into the Eighth Army zone.

Although X Corps did not have as harsh an experience with Chinese forces

---

[17] Interv, author with Lt Gen Frank W. Milburn, 4 Jan 52; Ltr, Dabney to author, 3 Feb 54; Interv, author with Fergusson, 23 Jun 53; Interv, author with Allen, 15 Dec 53; Interv, author with Brig Gen William C. Bullock (Asst G-3 EUSAK, Nov 50), 28 Jan 54. The author has considered carefully and taken into account the MS review comments of Col. Tarkenton (Eighth Army G-2 at the time), Dec 57, in making final revisions in this chapter.

[18] X Corps PIR 34, 30 Oct 50; Pacific *Stars and Stripes,* November 1, 1950.

in the First Phase Offensive as did Eighth Army, it nevertheless received intelligence which indicated large-scale Chinese intervention. In mid-November, for instance, two civilian draftsmen, formerly employed by the Traffic Department of the P'yongyang Railway Bureau, reported on what they saw and heard of Chinese intervention before they left Manp'ojin on 26 October. According to them, there had been a continuous flow of CCF soldiers through Manp'ojin beginning on 12 October. One of the men estimated 80,000 Chinese had passed south through the border town. Chinese officers had variously told the two men that 200,000 and 400,000 Chinese soldiers were to enter Korea.[19]

The lack of enemy activity in front of X Corps during the second and third weeks of November prompted the corps intelligence officer to state officially on 18 November that "the enemy's recent delaying operations are apparently concluded and he is once again withdrawing to the north. The speed of his movements has caused a loss of contact at most points." General Almond himself at this time did not think that the Chinese had intervened in the Korean War in force.[20]

As the date approached for the U.N. attack intended to complete the occupation of all Korea, X Corps, like Eighth Army, seemed to take the view that enemy forces in front of it would fight only a delaying defensive action. On 22 November, corps intelligence reported that the enemy was "apparently preparing to make a defensive stand in his present positions," and that there was "no evidence to indicate any considerable number of CCF units have crossed the border since the initial reinforcement." [21]

The next day, however, elements of the 7th Marines captured two Chinese soldiers seven miles west of Hagaru-ri who said they belonged to the *267th Regiment, 89th Division,* which had crossed the Yalu ten days earlier. The men had deserted from their unit the night before and walked toward the U.N. lines.[22] The most interesting and important thing about these two deserters was that, if their story was true, not only was there another new Chinese division in Korea, but it came from a new army, the *20th,* a new army group, *IX,* and, most important of all, from a new field army, the CCF *Third Field Army,* which in the summer had been in the Shanghai area.

In preparing for its part in the impending U.N. offensive, X Corps anticipated that when the Eighth Army advance reached Huich'on, and X Corps itself neared the Kanggye road above Huich'on in the U.N. double envelopment attack, the enemy would react violently, and the possibility of enemy ground and air reinforcements from Manchuria could not be overlooked.[23]

While X Corps recognized the capabilities of CCF strength beyond the border, it seems clear that on the eve of

---

[19] X Corps PIR 49, 14 Nov 50.

[20] X Corps PIR's 53, 18 Nov, and 54, 19 Nov 50; X Corps POR 54, 19 Nov 50; Interv, author with Almond, 13 Dec 51.

[21] X Corps PIR 57, 22 Nov 50.

[22] X Corps PIR 58, 23 Nov 50.

[23] X Corps PIR 59, 24 Nov 50; X Corps WD, 25 Nov 50, app., an. A to X Corps Order 7, pp. 2, 5–6.

the impending U.N. attack—set for 24 November in the west and 27 November in the northeast—corps believed that not more than one or two CCF divisions were on its front and that the enemy's efforts in the near future would be wholly defensive in character. General Almond, the corps commander, held this view. As in the case of Eighth Army, the controlling X Corps view was probably influenced by that of the Far East Command.

### *The Far East Command's and MacArthur's Estimates*

Of all the intelligence levels of the U.N. command and the American government, perhaps the most decisive in evaluating the intention and capability of Chinese intervention in the Korean War was that of the Far East Command in Tokyo. The evaluation by General MacArthur and his intelligence officers of Chinese intervention and Chinese military capability in Korea in October and November 1950 seems to have been the determining factor in shaping the future course of U.N. military action in that country.

Why this was so requires explanation, for normally the intelligence evaluation of whether a foreign power has decided to intervene in a war in national force involves political intelligence at the highest level. Field and theater commanders could expect such an evaluation to be made by the government in Washington with the advice of its Central Intelligence Agency. The intelligence responsibility of Eighth Army and X Corps was tactical; strategic intelligence responsibility rested with the Central Intelligence Agency, the Department of the Army, and the theater headquarters, with the ultimate political intelligence the responsibility of the President and his immediate advisers. But apparently the Central Intelligence Agency and the administration generally did not evaluate the available intelligence so as to reach a conviction on the question of whether the Chinese intended to intervene in the Korean War different from that held by General MacArthur. It must be inferred that either Washington was undecided or that its view coincided with that of the Commander in Chief, Far East, since it did not issue directives to him stating a different estimate. The conclusion, then, is that in the developing situation of November the views of the Far East Command were decisive on the military course to be taken in Korea at that time.

The Korean War had scarcely started when the Far East Command began to consider the threat of CCF intervention. On 28 June its daily intelligence summary stated that the possibility existed that North Korea might receive Chinese Communist reinforcements from Manchuria. In early July General MacArthur informed the Joint Chiefs of Staff in Washington that if Chinese combat forces did become involved in the war the assistance of the Strategic Air Command would be required to destroy communications into and through North Korea from China. At this time the Far East Command estimated there were 116,000 CCF regular troops in Manchuria. An increase in CCF troop strength there became perceptible during the month and continued steadily thereafter. Much of the information concerning CCF troop movements from south to north China came from Chinese

CHOU EN-LAI

Nationalist sources on Formosa. Chiang Kai-shek's government received a steady stream of intelligence from its agents on the China mainland, and it, in turn, provided General MacArthur's command with numerous reports.[24]

Many of the intelligence reports received at Tokyo, as is usual in such matters, were inaccurate and unreliable. Such were several reports in August and September that CCF troops had crossed the border into North Korea. One of these reports, on 29 August, alleged that four CCF armies had crossed the Yalu and were deployed in North Korea.[25]

On 8 September the daily intelligence summary included a report of the Chinese Nationalist Ministry of Defense G-2 that if the outcome of the war seemed doubtful, elements of Lin Piao's *Fourth Field Army* probably would be committed. This report further indicated that such troops would not be used as CCF units but would be integrated into the North Korea People's Army.[26]

The Far East Command learned in mid-September of an alleged conference in mid-July in Peiping where it was decided to support North Korea short of war. Chou En-lai was quoted, however, as having said that if the North Koreans were driven back to the Yalu, the CCF would enter Korea. Far East Command intelligence, in commenting on this report, said .hat the Chinese Communist authorities apparently were worried over Korea and would regard a U.N. advance to the Yalu as a "serious threat to their regime." [27] Two weeks later, on the last day of September, the daily intelligence summary reported on an alleged high-level conference in Peiping on 14 August, at which it had been decided to provide 250,000 CCF troops for use in Korea.[28]

A new note of more official character entered into the intelligence clamor on 3 October. The Chinese Communist Foreign Minister, Chou En-lai, summoned Ambassador Sardar K. M. Panikkar of India to his office in Peiping and told him that if the United States or United Nations forces crossed the 38th Parallel, China would send troops to de-

[24] FEC, Daily Intel Summ, 28 Jun 50 (hereafter cited as FEC DIS); *Ibid.*, Nr 6, 2 Sep 51, p. A-3; FEC telecon with DA, TT3467, 6 Jul 50.

[25] FEC DIS 2915, 2 Sep 50; X Corps PIR 22, an. 2, 10 Oct 50.

[26] FEC DIS 2921, 8 Sep 50.

[27] FEC DIS 2929, 16 Sep 50.

[28] FEC DIS 2943, 30 Sep 50.

fend North Korea. He said this action would not be taken if only South Korean troops crossed the 38th Parallel. This information was communicated quickly by the Indian Ambassador to his government, which in turn informed the United States and the United Nations. The government at Washington immediately dispatched the message to General MacArthur in Tokyo.[29] Representatives of other nations reported similar statements coming from Chinese officials in Peiping. Then, on 10 October, the Peiping radio broadcast as a declaration of Chinese Communist intentions a statement to the same effect. On 15 October the Department of the Army informed MacArthur's headquarters of another report from a reliable source that Moscow was preparing a surprise for American troops when they approached the northern border.[30]

In early October an escaped American officer informed American intelligence authorities that he had been interrogated in North Korea by three Soviet officers and that one of them, a senior colonel, told him on 22 September that if U.S. forces crossed the 38th Parallel new Communist forces would enter the war in support of North Korea.[31]

On 5 October for the first time Far East Command intelligence listed as number one priority in enemy capabilities "Reinforcement by Soviet Satellite China." But this estimate did not long remain in first priority; it dropped to second place the next day, to third place on 9 October, and remained there through 13 October. On 14 October the intelligence estimate again raised the reinforcement of North Korea to first priority. There it remained during the Wake Island Conference.[32]

The Far East Command daily intelligence summary for 14 October carried a lengthy analysis of the problem and presumably represented the official view of Maj. Gen. Charles A. Willoughby, Far East Command G-2. This intelligence estimate accepted a total strength of thirty-eight CCF divisions in nine armies in Manchuria. It expressed the view that Russia would find it convenient and economical to stay out of the conflict and let the Chinese provide the troops if there was to be intervention. It went on to say that the interest of all intelligence agencies was focused on the "elusive Lin Piao" and the Yalu River. One significant paragraph stated:

> Recent declarations by CCF leaders, threatening to enter North Korea if American forces were to cross the 38th Parallel, are *probably in a category of diplomatic blackmail.* [Italics supplied.] The decision, if any, is beyond the purview of collective intelligence: it is a decision for war, on the highest level; i.e. the Kremlin and Peiping. However, the numerical and troop potential in Manchuria is a fait-accompli. A total of 24 divisions are disposed along the Yalu River at crossing points. In this general deployment, the grouping in the vicinity of Antung is the most immediately available Manchurian force, astride a suitable road net for deployment southward.[33]

[29] FEC DIS 2947, 4 Oct 50; Senate MacArthur Hearings, pt. 3, p. 1833, testimony of Secy of State Dean Acheson.

[30] Schnabel, FEC, GHQ Support and Participation in the Korean War, ch. VII, p. 13, citing Msg 94214, DA to CINCFE, 15 Oct 50.

[31] FEC DIS 2950, 7 Oct 50.

[32] FEC DIS 2948, 5 Oct, and 2949-60, 6-17 Oct 50.

[33] FEC DIS 2957, 14 Oct 50.

This same report pointed to the recent fall of Wonsan as a serious loss to the enemy and one jeopardizing his entire defense structure. It went on to say, "This open failure of the enemy to rebuild his forces suggests that the CCF and Soviets, in spite of their continued interest and some blatant public statements, have decided against further expensive investment in support of a lost cause."

Meanwhile, President Truman on 10 October had announced his intention to fly to the Pacific for a meeting with General MacArthur over the coming weekend to discuss "the final phase of U.N. action in Korea." The conference between the President, General MacArthur, and selected advisers of each took place on Wake Island, Sunday, 15 October. Most of the talk concerned plans for the rehabilitation of Korea after the fighting ceased. General MacArthur said he expected formal resistance to end throughout North and South Korea by Thanksgiving Day and that he hoped to get the Eighth Army back to Japan by Christmas. In response to President Truman's question, "What are the chances for Chinese or Soviet interference?", notes of the conference indicate that General MacArthur replied substantially as follows:

> Very little. Had they interfered in the first or second months it would have been decisive. We are no longer fearful of their intervention. We no longer stand with hat in hand. The Chinese have 300,000 men in Manchuria. Of these probably not more than 100,000 to 200,000 are distributed along the Yalu River. Only 50,000 to 60,000 could be gotten across the Yalu River. They have no Air Force. Now that we have bases for our Air Force in Korea, if the Chinese tried to get down to Pyongyang there would be greatest slaughter.[34]

General MacArthur then discussed briefly the chance of Russian intervention, holding the view that it was not feasible and would not take place.

General MacArthur has challenged the accuracy of the notes of the conversations at the Wake Island Conference. He maintains that the question concerning possible Chinese or Soviet intervention was low on the President's agenda, and that while he replied that the chances of such intervention were "very little," he added that this opinion was purely speculative and derived from the military standpoint, while the question fundamentally was one requiring a political decision. His view, he states, was also conditioned by the military assumption that if the Chinese did intervene United States forces would retaliate, and in a peninsular war could work havoc with their exposed lines of communication and bases of supply. He says, in effect, that he took it for granted that Chinese knowledge of this capability would be a powerful factor in keeping them from intervening.[35]

---

[34] Substance of Statements Made at Wake Island Conference on October 15, 1950, compiled by General of the Army Omar N. Bradley, Chairman, JCS, from notes kept by the conferees from Washington. Copies of this Substance of Statements were forwarded to General MacArthur by the JCS on 19 October 1950, and receipted for by his aide on 27 October. Neither General MacArthur nor his headquarters advised the JCS of any nonconcurrence with this record of the conference. See Bradley, Letter of Submittal, 2 May 1951, to Senator Richard Russell, Chairman, Senate Armed Services Committee, included as part of the document (printed by the Committee on Armed Services and the Committee on Foreign Relations, U.S. Senate (Washington, 1951); New York *Times*, October 11, 1950; Dept of State Pub 4263, *United States Policy in the Korean Conflict, July 1950–February 1951*, p. 19.

[35] MacArthur, MS review comments, 15 Nov 57.

It must be noted that General MacArthur's opinion on the subject was not questioned by the President or any of the others present, who must be assumed to have had knowledge of the highest level of intelligence bearing on the matter. In fact, so thoroughly did they seem to agree with his opinion that the Chairman of the Joint Chiefs of Staff asked him when he could spare a division for European duty. So it would seem that General MacArthur in responding to the President's question merely voiced the consensus of the highest officials from the seat of government. This is how matters stood at mid-October.

On 20 October the Far East Command daily intelligence summary carried a report from a source it regarded as reliable that 400,000 CCF troops were at the border alerted to cross on the 18th or 20th. The Far East Command stated that precautionary measures had been taken of conducting daily air reconnaissance flights over all avenues of approach to the U.N. forces from the Yalu but that "so far no positive movements except intermittent though large-scale truck convoys have been picked up." [36] On this same date, the Far East Command issued CINCFE Plan 202, which was to be the basis for withdrawal of U.N. forces from Korea when the fighting ended. This plan assumed there would be no intervention either by Chinese or Soviet forces.[37]

Strangely enough, beginning on 25 October and continuing throughout the month, and at a time when the U.N. forces were actually fighting the Chinese Communist Forces in North Korea and capturing Chinese prisoners, the Far East Command daily intelligence summary placed Chinese intervention second in priority to guerrilla operations in enemy capabilities. The intelligence summary for 27 October carried the story of the first CCF prisoners captured two days earlier. The G–2 comment on the prisoners' account of Chinese intervention was that it was "based on PW reports and is unconfirmed and thereby unaccepted." [38]

The next day, 28 October, after discussing further the question of possible CCF intervention, the intelligence estimate said:

> From a tactical viewpoint, with victorious U.S. Divisions in full deployment, it would appear that the auspicious time for such [Chinese] intervention has long since passed; it is difficult to believe that such a move, if planned, would have been postponed to a time when remnant North Korean forces have been reduced to a low point of effectiveness.[39]

On 31 October the daily summary discussed the fact that ten Chinese prisoners had been taken by Eighth Army and that the ROK II Corps had suffered reverses. It then said that the situation "may signify the commitment of Chinese Communist Forces in the Korean conflict." The United Nations Command report to the Security Council covering the period 16–31 October, in mentioning the capture of Chinese prisoners, said there was no positive evidence

[36] FEC DIS 2963, 20 Oct 50.

[37] Schnabel, FEC, GHQ Support and Participation in the Korean War, ch. VII, pp. 1–2, citing CINCFE Opn Plan 202, 20 Oct 50, JSPOG files.

[38] FEC DIS 2968–2974, 25–31 Oct 50.

[39] FEC DIS 2971, 28 Oct 50.

that Chinese units as such had entered Korea.[40]

On 3 November, however, the Far East Command accepted the estimate that 16,500 CCF troops were in contact with U.N. forces in Korea and that possibly the total might be 34,000. This intelligence report listed CCF strength in Manchuria totaling 833,000 men, of whom 415,000 were Chinese Communist regular ground forces. On this same day the Joint Chiefs of Staff in Washington dispatched a message to MacArthur expressing their concern over what appeared to be "overt intervention in Korea by Chinese Communist units," and asked his views on the matter. He replied the next day that while it was a distinct possibility, "there are many fundamental logical reasons against it and sufficient evidence has not yet come to hand to warrant its immediate acceptance." [41]

On the same day, 3 November, that MacArthur received the inquiry from the Joint Chiefs of Staff, the Peiping radio broadcast a joint communiqué in Chinese by the Chinese Communist Party and various political parties participating in the Peiping government. It declared the Korean War was a direct threat to the safety of China and that the Chinese people should take the initiative and exert utmost efforts to resist the United States and assist North Korea. The Far East Command G–2, in commenting on this broadcast, said that preceding ones had sounded like "bombast and boasting. The above does not." [42] Two days later, on 5 November, the daily intelligence summary stated that the Chinese Communist Forces had the potential to launch a large-scale counteroffensive at any time and without warning.[43]

On the heels of this estimate came General MacArthur's well-publicized special communiqué on 6 November which charged the Communists with having "committed one of the most offensive acts of international lawlessness of historic record by moving without any notice of belligerency elements of alien Communist forces across the Yalu River into North Korea," and of massing a great concentration of possible reinforcements behind the sanctuary of the Manchurian border. Concerning the future, he said, "Whether and to what extent these reserves will be moved forward to reinforce units now committed remains to be seen and is a matter of the gravest international significance." [44]

The Far East Command intelligence report the next day raised its accepted number of CCF troops in Korea to 34,500; 27,000 in the Eighth Army zone and 7,500 in the X Corps zone.[45] On the 9th it carried a rather detailed analysis of the CCF in Korea. It accepted 8 CCF divisions from 4 armies with a strength of 51,600 men as being in contact with U.N. forces; it accepted 2 more

[40] FEC DIS 2974, 31 Oct 50; Dept of State Pub 4051, *United Nations Command Eighth Report to the Security Council, United Nations, 16–31 October 1950*, p. 2.

[41] FEC DIS 2977, 3 Nov 50; Schnabel, FEC, GHQ Support and Participation in the Korean War, ch. VII, p. 14, quoting Msg W95790 CSUSA to CINCFE, 3 Nov 50, and Msg C68285 CINCFE to DA, 4 Nov 50.

[42] FEC DIS 2980, 6 Nov 50.

[43] FEC DIS 2979, 5 Nov 50.

[44] GHQ FEC Communiqué 11, 6 Nov 50; see also New York *Times*, November 6, 1950; Dept of State Pub 4263, pp. 20–22.

[45] FEC DIS 2981, 7 Nov, and 2982, 8 Nov 50.

divisions with 12,600 men as probably in contact, and still another 2 divisions with 12,600 men as possibly being in the X Corps zone but not in contact with U.N. forces. This analysis gave a total of 76,800 CCF troops as probably being in North Korea.[46]

The report of a conference in Peiping on 17 October when Chinese officials allegedly decided to go to war received further consideration on 12 November, in the light of more information, and the G–2 comment was that "this information may be evaluated as probably true." Reports continued to reach Tokyo from the Chinese Nationalist government on Formosa that the Chinese Communists intended to throw their main forces against the United Nations in Korea and also to increase their participation in the Indochina fighting.[47]

At mid-month, the U.N. Command reported to the Security Council that elements of twelve CCF divisions had been identified in forward areas, nine in Eighth Army zone, three in X Corps. As the third week of November passed, the Korean front was relatively quiet. Far East Command intelligence took notice of the continuing propaganda actively being carried on in Peiping and elsewhere by radio broadcasts, letters to newspaper editors, rallies, and other devices, in what seemed to be a campaign to prepare the nation for a defensive intervention war in Korea. The Far East Command now apparently accepted Chinese Communist strength in Korea at a maximum of 70,051 and a minimum of 44,851. Apparently this estimate did not change up to the beginning of the U.N. attack on 24 November.[48]

The Department of the Army estimate of CCF strength in Korea was essentially the same as that of the Far East Command, Eighth Army, and X Corps. In the week preceding the resumption of the U.N. attack on 24 November it accepted the estimate of 51,600 CCF troops in Korea, and a probable total of 76,800 CCF troops in Korea. It credited these troops to four CCF armies (the *38th, 39th, 40th,* and *42d*) with twelve divisions, giving each division a strength of 6,300 men.[49]

It is obvious that the Far East Command was in possession of a great amount of intelligence concerning the Chinese and their relation to the Korean War. But the vital questions still remained. Just what did General Willoughby, MacArthur's G–2, think, and, most important, what did General MacArthur himself think, of the probability of full-scale Chinese intervention?

While General Willoughby frequently pointed out in his intelligence summaries the potentialities of CCF intervention, it appears on all the evidence

[46] FEC DIS 2983, 9 Nov, and 2988, 14 Nov 50.

[47] FEC DIS 2989, 15 Nov 50.

[48] Dept of State Pub 4051, *United Nations Command Ninth Report to the Security Council, United Nations, 1–15 November 1950,* p. 9; FEC DIS 2993, 19 Nov 50. The original CCF strength in Korea before casualties was given as between 50,400 and 76,600. This indicated that FEC intelligence believed that approximately 5,500 CCF soldiers had become casualties in the fighting up to that time. See also FEC DIS 2994–2998, 20–24 Nov 50.

[49] See DA Wkly Intel Rpts 91, 17 Nov 50, pp. 19–21, and 92, 24 Nov 50, p. 32. A full treatment of the Department of the Army and the Joint Chiefs of Staff levels in the Korean War may be found in Schnabel, Theater Command.

that he did not think it would take place. When Maj. Gen. Leven C. Allen passed through Tokyo in early September on his way to Korea to assume the post of Chief of Staff, Eighth Army, Willoughby in a conversation with him played down the possibility of Chinese intervention and said the Chinese were sensible and would keep out of the Korean affair.[50] When General Ruffner, Chief of Staff, X Corps, in talking with General Willoughby in November, expressed concern about the great number of CCF divisions identified in the Eighth Army and X Corps zones, Willoughby answered they may have been only elements of that many divisions—not that many full divisions.[51] And in November when General Hickey, Acting Chief of Staff, Far East Command, with the G-2, G-3, and G-4 of that command, visited the X Corps in Korea to form a first-hand estimate of the degree of Chinese intervention, he asked General Willoughby substantially the following question: "If, as General Almond states, Chinese forces have intervened, how many Chinese troops do you estimate are now in Korea?" General Willoughby reiterated that only volunteers had entered Korea and that probably only a battalion of volunteers of each division identified was actually in Korea. In this same conversation, in response to a question from General Almond about what had happened to the 8th Cavalry Regiment in the Eighth Army zone, General Willoughby reportedly replied that the regiment had failed to put out adequate security, been overrun by a small, violent surprise attack, and had scattered during the hours of darkness.[52]

The Far East Command intelligence reports themselves during October and November 1950, although filled with intelligence data and estimates of the CCF capabilities, seem never to reflect the evaluated opinion that the Chinese would intervene in full force.

General MacArthur's view seems to have paralleled closely that reflected in the Far East Command intelligence evaluations, but he may have been somewhat more apprehensive of massive Chinese intervention. On 7 November he sent a report to the Joint Chiefs of Staff, in response to their request for one, that reflects his views and outlook at the time. His message, paraphrased, said:

> Unquestionably . . . organized units of CCF have been and are being used against U.N. forces; that while it is impossible to determine accurately the precise strength, it is enough to have taken the initiative in the west and to have slowed appreciably our offensive in the east. The pattern seems established that such forces will be used and increased at will, probably without a formal declaration of hostilities. If this enemy build-up continues, it can easily reach a point preventing our resumption of the offensive and even force a retrograde movement. An attempt will be made in the west, possibly within 10 days, again to assume the initiative if the flow of enemy reinforcements can be checked. Only through such an offensive can an accurate measure of the enemy strength be taken.[53]

Despite this somewhat somber view, MacArthur two days later expressed himself optimistically in a message to the JCS with respect to future possible

[50] Interv, author with Gen Allen, 15 Dec 53.

[51] Interv, author with Maj Gen Clark L. Ruffner, 27 Aug 51.

[52] Ltr, Col William J. McCaffrey (X Corps Deputy CofS, Nov 50) to Lt Gen Edward M. Almond, 1 Dec 54, and forwarded by Almond to author.

[53] CINCFE to DA for JCS, C68465, 7 Nov 50.

military operations against the Chinese Communist Forces. He said:

> I believe that with my air power, now unrestricted so far as Korea is concerned except as to hydroelectric installations, I can deny reinforcements coming across the Yalu in sufficient strength to prevent the destruction of those forces now arrayed against me in North Korea.[54]

On the day the U.N. attack began, 24 November, General MacArthur gave further evidence of the degree to which this view guided his thinking. In a communiqué that day he announced:

> The United Nations massive compression envelopment in North Korea against the new Red Armies operating there is now approaching its decisive effort. The isolating component of the pincer, our Air Forces of all types, have for the past three weeks, in a sustained attack of model coordination and effectiveness, successfully interdicted enemy lines of support from the North so that further reinforcement therefrom has been sharply curtailed and essential supplies markedly limited.[55]

Perhaps even more revealing of MacArthur's state of mind was his special communiqué to the United Nations later the same day in which he said:

> The giant U.N. pincer moved according to schedule today. The air forces, in full strength, completely interdicted the rear areas and an air reconnaissance behind the enemy line, and along the entire length of the Yalu River border, showed little sign of hostile military activity. The left wing of the envelopment advanced against stubborn and failing resistance. The right wing, gallantly supported by naval air and surface action, continued to exploit its commanding position.
>
> Our losses were extraordinarily light. The logistic situation is fully geared to sustain offensive operations. The justice of our course and promise of early completion of our mission is reflected in the morale of troops and commanders alike.[56]

In these dispatches General MacArthur expressed a viewpoint which apparently dominated his thinking during most of the critical period of October and November 1950 while he and the United Nations wrestled with the problem of CCF intervention. He seems to have believed, first, that the Chinese would not intervene in full force, and, second, that should they do so, his air power would destroy them. General MacArthur very likely expected to fight a battle with the Chinese Communist Forces short of the Yalu, but he expected to win it through the decisive effect of the interdiction and close support capabilities of his air power. This reliance on air power in dealing with the CCF was perhaps the crucial factor in MacArthur's calculations.

### *Actuality*

The statement of Chou En-lai to the Indian Ambassador on 3 October, the announcements made over the Peiping radio, the timing of CCF troop movements as learned from prisoners, and other forms of intelligence, taken in connection with later events, make it seem reasonably clear that the Chinese Communist government had decided by early

---

[54] CINCFE to DA for JCS, C68572, 9 Nov 50.

[55] GHQ UNC Communiqué 12, 24 Nov 50, in EUSAK WD, 24 Nov 50, EUSAK Daily News Bul, 24 Nov 50.

[56] Senate MacArthur Hearings, pt. 3, p. 1834, testimony of Acheson, quoting MacArthur's message.

October on intervention in North Korea if United Nations troops other than ROK's crossed the 38th Parallel. Whether the Chinese Communists believed the United Nations Command would cross the Parallel is unknown, but there is at least one good reason to think the North Korean Government believed the U.N. Command would stop at the 38th Parallel. Kim Il Sung, Commander in Chief of the North Korea People's Army, in an order to the army dated 14 October 1950, stated in part, "Other reasons that we have failed are that many of us felt that the 38th Parallel would be as far as the US Forces would attack. . . ." [57]

Within a few days after the leading elements of the U.S. forces crossed the 38th Parallel at Kaesong on 9 October, elements of the CCF were crossing the Yalu River at the Manchurian border into North Korea. The first of these troops apparently crossed the boundary on 13 or 14 October, although it is possible that some may have crossed on the 12th.

Four CCF armies, each of three divisions, crossed the Yalu River between 14 and 20 October. Two of them, the *39th* and the *40th,* crossed from An-tung, Manchuria, to Sinuiju, North Korea; the other two, the *38th* and *42d,* crossed from Chi-an, Manchuria, to Manp'ojin, North Korea. All four armies were part of Lin Piao's *Fourth Field Army* and upon arrival in Korea were subordinated to the CCF *XIII Army Group.* The *1st Motorized Artillery Division,* two regiments of the *2d Motorized Artillery Division,* and a cavalry regiment also crossed into Korea at An-tung about 20–22 October in support of the four armies already across.[58]

Three of the four CCF armies entering Korea deployed in front of Eighth Army, the fourth deployed in front of X Corps. From west to east these armies took the following positions: the *39th Army* was in front of Unsan, the *40th Army* in front of Onjong. The *38th Army,* marching from Manp'ojin through Kanggye, reached a position on Eighth Army's right flank in the Huich'on area. Very likely it was troops from this CCF army that General Dean saw in the early dawn one morning in mid-October twenty miles north of Huich'on as he was being taken by his captors to Manp'ojin. The *39th* and *40th Armies* entered combat with U.N. forces for the first time on 25 October; the *38th* entered combat on the 26th.

The fourth army, the *42d,* moved from Manp'ojin through Kanggye to the Changjin Reservoir area in front of the X Corps main axis of advance in northeast Korea. It, like the *39th* and *40th Armies,* first entered combat against forces under U.N. command on 25 October. The west flank units of this army, elements of the CCF *125th Division,* overlapped into the Eighth Army zone and apparently constituted the enemy force that dispersed the ROK 7th Regi-

[57] X Corps PIR 55, 20 Nov 50, reproduces this captured document, Order 1–1, 14 Oct 50, signed by Kim Il Sung and Pak Kun Yon, Chief, Korean People's Supreme Political Bureau.

[58] FEC Intel Digest, vol. 1, Issue 4, 1–15 Feb 53 pp. 26–38; *Ibid.,* Issue 26, 16–30 Jun 52, Individual Histories, Chinese Communist Support and Service Units, pp. 46–47; FEC MIS, Order of Battle Info CCF, p. 9.

ment below Ch'osan at the end of October.[59]

At the time, then, that General MacArthur was expressing to President Truman and his advisers at Wake Island on 15 October his belief that there was very little likelihood that the Chinese Communist Forces would intervene, that, if they did, no more than 60,000 could get across the Yalu and that his air force would destroy them, approximately 120,000 CCF soldiers either had already crossed, were in the act of crossing, or were moving from their assembly and training areas to the crossing sites for the purpose of crossing.

Following the first four armies in approximately ten days, two more CCF armies crossed into North Korea at the end of October. These were the *50th* and *66th Armies* which, crossing from An-tung to Sinuiju, had completed their crossing into North Korea by 31 October. Each of these armies was composed of three divisions recently brought up to war strength. In these two armies approximately 60,000 more CCF troops came into North Korea, to make a total during the month of approximately 180,000 troops. The *50th Army* deployed southward on the CCF west flank and remained in reserve during the CCF First Phase Offensive. An element of this army did, however, exchange fire with the 19th Infantry Regiment on 1 November near Kusong.[60]

[59] FEC Intel Digest, vol. 1, Nr 4, 1–15 Feb 53, pp. 28–33; EUSAK WD, G–2 Sec, 12 Nov 50, interrog of Ma Yu-fu; *Ibid.*, 14 Nov 50, ATIS Interrog Rpts (Enemy Forces), Issue 17, 2279, p. 186, interrog of Lin Piao Wu (a company grade officer); 1st Mar Div SAR, 8 Oct–15 Dec 50, vol. 1, an. B, p. 23.

[60] FEC Intel Digest, vol. 1, Nr 4, 1–15 Feb 53, pp. 36–37.

KIM IL SUNG

Also, before the end of October the CCF *42d Truck Regiment* entered Korea at Sinuiju from An-tung, and the *5th Truck Regiment* and the *8th Artillery Division* entered Manp'ojin from Ch-ian. Both truck units supported the First Phase Offensive.[61]

In the X Corps zone, the *42d Army* had sent the *124th Division* south of the Changjin Reservoir where it fought the delaying battle in late October and the first part of November with elements of the ROK 3d Division and the U.S. 7th Marines. The *126th Division* remained in reserve in the reservoir area, but it had a number of minor engagements and patrol actions with the U.S. 7th Division in the Pujon Reservoir area. The *125th Division* moved southward from Yudam-ni to block the axis of approach from

[61] *Ibid.*, Issue 26, 16–30 Jun 52, pp. 48, 52.

TABLE 5—ORGANIZATION OF THE XIII ARMY GROUP

| Armies | Divisions | Regiments |
|---|---|---|
| 38th | 112th | 334th, 335th, 336th |
| | 113th | 337th, 338th, 339th |
| | 114th | 340th, 341st, 342d |
| 39th | 115th | 343d, 344th, 345th |
| | 116th | 346th, 347th, 348th |
| | 117th | 349th, 350th, 351st |
| 40th | 118th | 352d, 353d, 354th |
| | 119th | 355th, 356th, 357th |
| | 120th | 358th, 359th, 360th |
| 42d | 124th | 370th, 371st, 372d |
| | 125th | 373d, 374th, 375th |
| | 126th | 376th, 377th, 378th |
| 50th | 148th | 442d, 443d, 444th |
| | 149th | 445th, 446th, 447th |
| | 150th | 448th, 449th, 450th |
| 66th | 196th | 586th, 587th, 588th |
| | 197th | 589th, 590th, 591st |
| | 198th | 592d, 593d, 594th |

*Source:* FEC Intel Digest, vol. 1, Nr 4, 17 Feb 53; FEC MIS, Order of Battle Info, CCF, 15 Jun 51; *Ibid.*, Nr 35, 1–15 Nov 52, p. 45; *Ibid.*, Nr 6, 2 Sep 51, p. A–9.

Sach'ang-ni to the north. By 13 November it had moved southwest across the X Corps–Eighth Army boundary into the Eighth Army zone where it appeared before the ROK 8th Division in the Tokch'on area on the Eighth Army east flank.

These six armies, five in the Eighth Army zone and one in the X Corps zone, composed the CCF *XIII Army Group* with a total of 18 divisions, each division at the standard strength of about 10,000 men. In the Chinese Army the division and regiment numeration proceeds progressively in sequence. The organization of the *XIII Army Group* illustrates this. (*Table 5*)

A third major CCF entry into North Korea now took place. The *IX Army Group, Third Field Army,* entered Korea during the first half of November. This army group had come by rail directly to the border from Shantung Province, China, in late October and early November and had started crossing at once. It comprised three armies, the *20th, 26th,* and *27th,* each of three divisions. Each of these armies was reinforced by a division taken from the *30th Army,* this giving each army four divisions.[62] The *IX Army Group* entering Korea in the first part of November, therefore, added 12 infantry divisions to the 18 already there, for a total now of 30 divisions. In addition to these 30 infantry divisions, the Chinese Communist Forces also had in North Korea a number of artillery, cavalry, and support units.

The *IX Army Group* moved southeast to the Changjin Reservoir area, large units of it arriving there on or before 13 November. On that date, elements of the *IX Army Group* relieved the *42d Army* of its responsibility in that sector, and the *124th* and *126th Divisions* of the *42d Army* followed the *125th* southwest into the Eighth Army zone. By the end of the third week of November, therefore, the *XIII Army Group* of the CCF *Fourth Field Army,* with 18 divisions of infantry (180,000 men), was concentrated in front of Eighth Army, and the *IX Army Group* of the CCF *Third Field Army,* with 12 divisions of infantry (120,000 men), was concentrated in front of X Corps. A formidable total of approximately 300,000 CCF infantry troops

[62] The *89th Division* reinforced the *20th Army;* the *88th Division,* the *26th Army;* and the *90th Division,* the *27th Army.*

were now deployed in North Korea.[63]

The entry of the Chinese Communists into the Korean War necessarily brought changes in the enemy command. It appears that Peng Teh-huai, Deputy CCF Commander, established a joint *CCF-NKA Headquarters* in Mukden and there made basic decisions concerning enemy operations. A subordinate headquarters, called the *N.K. Army–CCF Combined Headquarters,* under Kim Il Sung, Commander in Chief of the North Korean Army, was publicly given credit for controlling military operations in Korea, but it seems certain that actual control rested in Mukden. There is some indication that the CCF *XIII Army Group* at first may have been under Kim Il Sung's North Korean command. The *IX Army Group* apparently was under complete CCF control from the beginning.

PENG TEH-HUAI

*Conclusion*

Starting with an acceptance of only a few Chinese "volunteers" mixed with North Korean units, the U.N. Command in the course of a month had gradually raised its estimate to accept about 60,000 to 70,000 Chinese troops in Korea by 24 November, less than one-fourth the number actually there. How was it possible for the U.N. Command to mistake so grossly the facts in the situation, even after it had met a considerable part of these Chinese forces in combat?

The answer seems clear enough. First, although the Chinese Communist government had several times openly stated it would intervene if U.N. forces other than ROK troops crossed the 38th Parallel, American authorities were inclined to disbelieve this and to consider these statements to be in the nature of threats and diplomatic blackmail. Second, the actual troop movements across the Yalu and deployment south were made at night and so in the main were not subject to aerial observation. During the day, aerial observation failed to discover

[63] The approximate figure of 300,000 for the Chinese was arrived at only after careful study of the intelligence information. Although the 30 regular divisions may not all have been precisely 10,000 strong, the round figure takes into account the miscellaneous service, cavalry, and artillery units. FEC Intel Digest, vol. 1, Nr 3, 16–31 Jan 53, *IX Army Group*, pp. 32–37; *Ibid.*, vol. 1, Nr 4, 1–15 Feb 53, *XIII Army Group*, p. 33; FEC Intel Digest, Nr 8, 16–30 Sep 51, and Nr 6, 16–31 Aug 51, p. A–12; FEC MIS, Order of Battle Info, CCF, pp. 9–10, 87–91, and *Third Field Army*, 1 Mar 51; ATIS Enemy Docs, Issue 29, p. 84, Chinese Notebook, 11–24 Nov 50; 1st Mar Div SAR, vol. 1, an. B, 8 Oct–15 Dec 50, pp. 25, 32.

the troops, who remained hidden in the hills under perfect camouflage discipline. Third, because they were not adequately confirmed, the reports from prisoners and Korean civilians of mass CCF movements across the border were not accepted by intelligence authorities. The intelligence system of Eighth Army, for various reasons, did not work as well in North Korea as it had in South Korea during the days of the Pusan Perimeter.

A word should be said about the CCF march discipline and capabilities, which in large part accounted for the secrecy with which the Chinese Communists entered and deployed in North Korea. This march capability and performance equaled the best examples of antiquity. In Xenophon's account of the retreat of the 10,000 Greeks, a day's march on the average came to a little less than 24 miles. The Roman military pace was set to cover 20 miles in 5 hours, the usual day's march for a Roman legion. In normal training exercises the Roman legions had to make three such marches every month. On occasion the legions were required to march 24 miles in 5 hours. When Caesar besieged Gergovia in Gaul, he marched 50 miles in 24 hours.[64]

In a well-documented instance, a CCF army of three divisions marched on foot from An-tung in Manchuria, on the north side of the Yalu River, 286 miles to its assembly area in North Korea, in the combat zone, in a period ranging from 16 to 19 days. One division of this army, marching at night over circuitous mountain roads, averaged 18 miles a day for 18 days. The day's march began after dark at 1900 and ended at 0300 the next morning. Defense measures against aircraft were to be completed before 0530. Every man, animal, and piece of equipment were to be concealed and camouflaged. During daylight only bivouac scouting parties moved ahead to select the next day's bivouac area. When CCF units were compelled for any reason to march by day, they were under standing orders for every man to stop in his tracks and remain motionless if aircraft appeared overhead. Officers were empowered to shoot down immediately any man who violated this order.[65]

These practices, especially the march and bivouac discipline, explain why United Nations aerial observation never discovered the CCF deployment into Korea. The Chinese Communist Forces moved 300,000 men into position in October and November and none of them was ever discovered by the U.N. Command prior to actual contact. While the planes were overhead searching for possible Chinese movement into Korea, the Chinese, perfectly camouflaged, lay hidden below. The aerial observers did not see them nor did the aerial photographs reveal their presence.

### *The Pregnant Military Situation*

On 6 November General MacArthur took official notice of the recent CCF offensive and summed up his estimate of the changing situation in Korea. He said

[64] See Xenophon, *The Anabasis*, and William Duncan's *Caesar* (includes Caesar's Commentaries), pp. 46–50.

[65] FEC Intel Digest 16, 16–31 Jan 52, March of a CCF Army, pp. 33–39; I Corps WD, 30 Oct 50, Intel Summ 135, 30 Oct 50. Although the march described actually occurred early in 1951, a few months after the initial CCF intervention, the initial CCF troops entering Korea apparently marched at an equal rate of speed at night to reach their assembly areas.

the defeat of the North Koreans had been decisive when the Chinese intervened in "one of the most offensive acts of international lawlessness of historic record." Speaking in high praise of General Walker, he said the possible trap "surreptitiously laid calculated to encompass the destruction of the United Nations Forces" was avoided, "with minimum losses only by the timely detection and skillful maneuvering of the United States commander responsible for that sector." General MacArthur announced his future intentions in these words, "Our present mission is limited to the destruction of those forces now arrayed against us in North Korea with a view to achieving the United Nations' objective to bring unity and peace to the Korean nation and its people." He intended, obviously, to destroy the Chinese forces in Korea as well as the remaining North Koreans. To accomplish this he considered it necessary to establish an integrated continuous front in western and central Korea for co-ordinated large-scale offensive action.[66]

That same day, 6 November, General Walker issued Eighth Army's operation plan for a renewal of the offensive. It called for an advance to the Korean border with three corps abreast—the U.S. I Corps on the west, the U.S. IX Corps in the center, and the ROK II Corps on the east in the army zone.[67] In preparing for the projected offensive, tentatively set for 15 November, Eighth Army had to bring the IX Corps into the line. Steps had already been taken to accomplish this. On 2 November the ROK III Corps had assumed responsibility for the then IX Corps zone, and IX Corps completed its move to Sunch'on on 4 November. The next day at noon the IX Corps became operational there with control of the 2d Infantry Division.[68]

GENERAL MACARTHUR

Colonel Stebbins, Eighth Army G–4, estimated that Eighth Army needed 3,000 tons of supplies daily for passive defensive operations and 4,000 daily for active combat. At no time in October or up to about 20 November was Eighth Army able to supply this minimum need for active offensive combat. The 4,000

[66] GHQ FEC Communiqué 11, 6 Nov 50; Dept of State Pub 4051, *United Nations Command Ninth Report to the Security Council, United Nations, 1–15 November 1950*, p. 10.

[67] EUSAK Opns Plan 14, 6 Nov 50; EUSAK WD, G–3 Sec, 6 Nov 50.

[68] EUSAK POR 342, 3 Nov 50; EUSAK WD, G–3 Sec, 4 Nov 50; IX Corps WD, Nov 50, bk. I, sec. IV, Opns.

tons daily was finally achieved by bringing forward approximately 2,000 tons daily to P'yongyang by rail from the south, by unloading 1,000 tons daily at Chinnamp'o, and by bringing approximately 1,000 tons daily into North Korea by airlift.[69]

While Eighth Army was striving to overcome the logistical difficulties that delayed its resumption of the attack in mid-November, X Corps in northeast Korea continued its headlong rush to the border against scattered and ineffective opposition except in the 1st Marine Division sector below the Changjin Reservoir. There the Marine division and regimental commanders, much to the X Corps commander's dissatisfaction, deliberately slowed their advance.

From the very beginning of X Corps operations in northeast Korea, General Smith had looked with disfavor on the wide dispersal of the subordinate units of the Marine division. On 7 November in a conference with General Almond, he again urged the concentration of the division. The recent experience of X Corps with the CCF *124th Division* and Eighth Army's encounter with the CCF in the west apparently caused General Almond to be more amenable to General Smith's arguments for concentrating the Marine division, and he agreed to it. Smith then went further and argued that the division should not advance to the Kot'o-ri plateau at the south end of the Changjin Reservoir with winter at hand, but General Almond felt that the marines should hold Hagaru-ri at the southern end of the reservoir. With the withdrawal of the CCF from contact both in the west and northeast Korea about 7 November, confidence soon reasserted itself in both Eighth Army and X Corps, and X Corps on 11 November reiterated its directive to proceed to the Yalu. Both Eighth Army and X Corps were still enjoined under General MacArthur's directive of 24 October to proceed to the Yalu. Apparently General Almond hoped that the troops could reach the border quickly, turn over the area to ROK troops, and withdraw before winter really set in.[70]

After the 7th Marines reached the Kot'o-ri plateau on 10 November, neither Colonel Litzenberg, the regimental commander, nor General Smith, the division commander, showed any inclination to hurry the advance. General Smith plainly indicated that he was apprehensive about his western exposed flank, that he wanted to improve the road up the pass from the division railhead at Chinhung-ni, that he wanted to develop a secure base at Hagaru-ri, and that he wanted to garrison key points on the main supply road back south. And most of all he wanted to concentrate the full strength of the Marine division in the Hagaru-ri area before trying to advance further toward the Yalu.

Winter struck early in the Changjin

[69] Interv, author with Stebbins, 4 Dec 53; Interv, author with Maj Gen Leven C. Allen, 15 Dec 53; Ltr, Allen to author, 27 Nov 54; EUSAK WD, G-3 Sec, 13 Nov 50, and G-4 Jnl, Msg 7, 241015 Nov 50.

[70] Interv, author with Ruffner, 15 Aug 51; Interv, author with Lynn Montross and Capt Nicholas Canzona, Marine Corps Hist Sec, 6 Apr 54; Interv, author with Almond, 13 Dec 51; Lt Gen Oliver P. Smith, MS review comments, 15 Nov 57; X Corps Special Rpt on Chosin Reservoir, Nov 50; Schnabel, FEC, GHQ Support and Participation in the Korean War, ch. VII, pp. 21-23.

Reservoir area of Korea in 1950. It arrived with violent force and subzero temperatures on 10 November, the day the marines reached the Kot'o-ri plateau. From that day on the troops there were involved in a winter campaign. Patrols sent out from Kot'o-ri on 11 and 12 November found only small scattered enemy groups in the hills, and the next day a Marine unit advanced to Pusong-ni, halfway to the reservoir. On 14 November the 7th Marines, wearing their heavy arctic parkas, trudged in subzero weather toward Hagaru-ri over a road now covered with an inch of snow. Vehicles froze up on the move, brakes grabbed, transmissions were stiff, and the men themselves had difficulty in moving forward. Entering Hagaru-ri, the marines found it burned out by previous bombing attacks and practically deserted. Natives told them that the 3,000 Chinese soldiers occupying the town had departed three days earlier, going north and west. A Chinese soldier from the *377th Regiment, 126th Division,* captured near Hagaru-ri during the day, said elements of his division were east of the reservoir. That night, 14–15 December, the temperature dropped to 15 degrees below zero.[71]

The next day the 7th Marines completed its movement into Hagaru-ri, and Colonel Litzenberg made arrangements for a perimeter defense. The 1st Battalion protected the northwest approaches, the 2d Battalion the southern, and the 3d Battalion the northeast approaches to the town. That same day the 2d Battalion, 5th Marines, arrived at Kot'o-ri, beginning the concentration of the rest of the 1st Marine Division in the Changjin Reservoir area behind the 7th Marines. The 5th Marines now guarded the main supply route back to Hamhung.

Two days after the first Marine units entered Hagaru-ri, General Smith and Maj. Gen. Field Harris (Commanding General, 1st Marine Air Wing) on 16 November looked over the ground there and selected the site for a C–47 airstrip. Smith felt that such an airstrip would be needed to supplement supply by road and for fast evacuation of casualties. Engineer troops began work on the airstrip on 19 November, and others continued work on improving the road up the pass from Chinhung-ni. The first trucks climbed through the pass to Hagaru-ri on the 18th. Smith held the Marine advance to Hagaru-ri while this work continued.

Thus it was, that with virtually no enemy opposition, the marines advanced at an average rate of only a mile a day between 10 and 23 November. But this caution on the part of General Smith in concentrating the division and his insistence on securing its supply lines and of establishing a base for further operations in the frigid, barren wastes of the Changjin Reservoir area were to prove the division's salvation in the weeks ahead.

Although the projected Eighth Army attack on the 15th had to be postponed because of logistical difficulties, the army on the 14th ordered an attack to be made, on a day and hour to be announced, to seize a line running generally from Napch'ongjong, on the west coastal road, eastward through Taech'on–

[71] 1st Mar Div SAR, vol. II, pp. 38, 40, an. C, 13–14 Nov 50; X Corps PIR 49, 14 Nov 50; ATIS Interrog Rpts (Enemy Forces), Issue 18, Nr 2304, Cheng Cheng Kwo.

Onjong–Huich'on to Inch'o-ri. This was to be the line of departure for the projected co-ordinated attack. The army was then to be prepared to continue the advance on order to the northern border of Korea. General Walker's order reflects an intention to proceed with a closely co-ordinated attack in order to have the army under control at all times. It also reflects a considerable degree of caution and a certain respect for the enemy forces. It appears on the weight of the evidence that General Walker wanted to make the attack. He expected opposition, but apparently believed he could reach the border. His chief of staff, General Allen, shared this view.[72]

On 17 November, with the logistical situation improved, Eighth Army announced to its subordinate organizations that the attack north would start on 24 November. General MacArthur notified the Joint Chiefs of Staff at the same time of the tentative attack date, emphasizing that the delay in mounting the offensive had been due to logistical difficulties. He optimistically reported that the intensified air attacks of the past ten days had isolated the battlefield from added enemy reinforcements and had greatly reduced the flow of enemy supplies.[73]

Up to the launching of the 24 November attack the U.S. Eighth Army and the X Corps had suffered a total of 27,827 battle casualties in the Korean War; 21,529 in Eighth Army and 6,298 in X Corps. Of the Eighth Army total, 4,157 had been killed in action, 391 more had died of wounds, and 4,834 were missing in action.[74]

On the afternoon of 21 November Eighth Army advised I and IX Corps and the ROK Army that H-hour for the army attack was 1000 24 November. Word of the attack hour had reached the front-line units by 23 November. That was Thanksgiving Day. The army front was generally quiet. Patrols went out several thousand yards in front of the line with little enemy contact. Nearly everywhere the enemy seemed to have withdrawn during the past week, leaving behind light outpost and covering positions. At no place did U.N. forces uncover what could be considered a main line of resistance.[75]

As Eighth Army units moved out in attack on 24 November they encountered only a few small enemy squad- and platoon-sized groups employing small arms fire. Even in the ROK II Corps zone of attack enemy opposition was unexpectedly light. In most places the U.N. advance was unopposed.

Generals MacArthur, Stratemeyer, Wright, Willoughby, and Whitney, to-

[72] EUSAK Opn Plan 15, 14 Nov 50; EUSAK WD, G–3 Sec, 11, 14 Nov 50; Interv, author with Allen, 15 Dec 53; Interv, author with Stebbins, 4 Dec 53. Colonel Stebbins, the Eighth Army G–4 at the time, however, told the author that General Walker did not want to make the attack.

[73] EUSAK WD, G–3 Sec, Msg 172100 to CG I and IX Corps and CG ROK; I Corps WD, 18 Nov 50; Schnabel, FEC, GHQ Support and Participation in the Korean War, ch. VII, p. 26, citing Msg C69211, CINCUNC to DA, 18 Nov 50.

[74] EUSAK WD, AG Sec, Statistical Casualty Rpt, 24 Nov 50; X Corps POR 59, 24 Nov 50; TAGO Rpt, May 52, shows Korean casualties through 15 Nov 50 as 28,159: 5,702 KIA; 18,909 WIA, of which 482 were DOW; and 3,548 missing or captured.

[75] EUSAK WD, G–3 Sec, Opn Instr, 211645 Nov 50; EUSAK PIR 133, 22 Nov 50; 24th Div WD, 23 Nov 50; 89th Med Tk Bn Unit Hist, Nov 50, p. 3; I Corps WD, 22 Nov 50.

Insignia of Major U.S. Ground Force Units in Korean War in 1950

*Eighth Army*

*I Corps*

*IX Corps*

*X Corps*

*1st Cavalry Division*

*2d Infantry Division*

*3d Infantry Division*

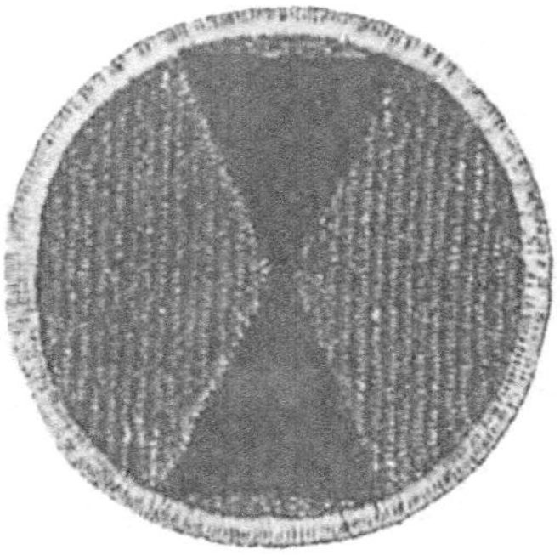

*7th Infantry Division*

*11th Airborne Division*
*(187th RCT)*

*24th Infantry Division*

*25th Infantry Division*

*1st Marine Division*
*(World War II patch)*

gether with several chiefs of U.S. press bureaus in Tokyo, had flown to Korea the morning of the 24th to witness the beginning of the attack. General Walker joined them in visits to I Corps, IX Corps, and 24th Division headquarters along the Ch'ongch'on. At I Corps headquarters General Milburn cast a momentary shadow over the bright picture being drawn when he told the party that his patrols had found the Unsan area heavily defended, and in his opinion the projected IX Corps attack there would not progress easily. General Church briefed the party at the 24th Division headquarters shortly after noon on the progress of the attack. Optimism and enthusiasm as to chances of the attack succeeding seemed to prevail.[76]

In the afternoon when General Leven Allen returned to Eighth Army headquarters from the airstrip where he had accompanied the party, he remarked to some of the staff members, "I think the attack will go. General MacArthur would not have come over if he did not think so." [77]

---

[76] EUSAK PIR 135, 24 Nov 50; EUSAK WD, CG memo for record (aide-de-camp), 24 Nov 50; IX Corps WD, bk. I, 24 Nov 50; I Corps WD, Nov 50; Interv, author with Maj Gen Edwin K. Wright, 7 Jan 54.

[77] Interv, author with Brig Gen William C. Bullock, 28 Jan 54; Ltr, Maj Gen Leven C. Allen to author, 4 Dec 54.

# The Sources

There is little need for extended comment on the sources used in preparing this volume. Their nature will be evident in the footnote citations. With the exception of a few passages, and these largely in the first chapters, the entire work is based on primary sources. These consist of the official records of the United States and United Nations armed forces bearing on the land, sea, and air action. ROK records were also consulted, but generally they were scant and sometimes nonexistent.

The U.S. Army ground organizations kept war diaries from army level down through infantry regiments and artillery and tank battalions. They also maintained daily periodic intelligence and operations reports and journal message files. Separate battalions kept similar records. Although not required by the Adjutant General for record purposes, infantry battalion war diaries in some instances were attached to regimental war diaries and in that manner found their way into the permanent records of the Department of the Army. Included in the daily Eighth Army War Diary was a useful G–3 Air report.

Some contemporary army records were lost in combat. The most important of these were the records of the 34th Infantry Regiment, lost in the battle of Taejon in July 1950, and the records of the 2d Infantry Division and of its 9th and 38th Regiments, lost at Kunu-ri in late November 1950. Substitutes were made up from memory of individuals and scattered sources bearing on the period. The records of the 5th Regimental Combat Team for its first month of combat, August 1950, were never located, although a thorough search for them was made in the Adjutant General's Office.

The Marine combat units kept records called special action reports that were similar to the Army war diaries, together with annexes covering subordinate units and staff sections. They also maintained G–2 and G–3 journal files.

For the Inch'on landing and the Wonsan-Hungnam operation, the naval records of Joint Task Force Seven were indispensable sources.

All the war diaries, special action reports, periodic reports, and journal files of combat units and headquarters at all levels were read and studied, as well as similar reports of logistical and service organizations.

Supplementing the records, and often more important, as indicated in the Preface, were the firsthand accounts of participants. These have come from army commanders, corps and division commanders, regimental and battalion commanders, company commanders, platoon leaders, staff officers, platoon sergeants, other noncommissioned officers, and privates in personal interviews, correspondence, and their review of early manuscript drafts on actions in which they fought. Each such personal account was checked against the others and the records, and all were used in the framework of the time sequence and map locations given in the operations journals and periodic reports of the various command echelons. Discrepancies abounded. Often it was not easy to reconcile conflicting evidence and arrive at a narrative in which one could have confidence.

Enemy sources used consisted of captured North Korean and Chinese Communist Forces documents and thousands of reports of interrogations of prisoners from both the North Korean Army and the Chinese Communist Forces. The value of the interrogation reports varied widely depending upon the rank and position in the enemy organization held by the prisoner. Many were entirely valueless, but one could never know in advance what little scrap of information any report might yield. The most useful body of enemy material was in the collections of the Allied Translator and Interpreter Section (ATIS) of General MacArthur's Far East Headquarters, although a sizable amount of similar material was included as attachments to the S–2 and G–2 periodic reports of regiments, divisions, corps, and army. Interrogation of prisoner reports found with the records of combat units had a particular value because they often represented the best information available to the troops on the enemy situation in their immediate front, although later interrogation reports might be more complete.

Names of enemy organizations, whether North Korean or Chinese Communist Forces, are italicized in the text.

The writer believes that good military history cannot be written without a clear knowledge of the terrain involved in action. He studied the terrain on the ground itself whenever possible, and gave to map study a proportionately large allotment of time. The Army Map Service map of Korea, scale 1:50,000, fourth edition, February 1951, was the standard terrain map used, although earlier editions and maps at both larger and smaller scales were also consulted. Special maps were employed in the case of actions involving towns and cities of any size.

# Glossary

| | |
|---|---|
| AAA | Antiaircraft Artillery |
| Abn | Airborne |
| Act | Action |
| ADCOM | Advance Command and Liaison Group in Korea |
| ADVATIS | Advanced Allied Translator & Interpreter Section |
| AFF | Army Field Forces |
| AGC | Amphibious force flagship equipped with special communication facilities |
| AK | Cargo ship |
| Amphib | Amphibious |
| AMS | Army Map Service |
| an. | Annex |
| AP | Armor-piercing |
| AP | Transport |
| APA | Attack transport |
| APD | High-speed transport |
| Armd | Armored |
| Arty | Artillery |
| ATIS | Allied Translator & Interpreter Section |
| AW | Automatic Weapons |
| BAR | Browning Automatic Rifle |
| BC | N.K. Border Constabulary (called *Bo An Dae*) |
| BLUEHEARTS | Code name for the original plan for an amphibious landing behind enemy lines, abandoned by 10 July 1950. Succeeded by CHROMITE |
| Bn | Battalion |
| Br | Briefing |
| Brig | Brigade |
| Btry | Battery |
| Bul | Bulletin |
| C | Combat |
| CCF | Chinese Communist Forces |
| CCS | Combined Chiefs of Staff |
| CG | Commanding General |
| Chmn | Chairman |
| CHROMITE | Code name for amphibious operations in September 1950, one of which was a landing at Inch'on |
| CINCFE | Commander in Chief, Far East |
| CINCUNC | Commander in Chief, United Nations Command |
| CM-IN | Classified message—In |
| CO | Commanding Officer |

| | |
|---|---|
| Comd | Command |
| Comdr | Commander |
| Comm | Communication |
| COMNAVFE | Commander, U.S. Naval Forces Far East |
| CP | Command post |
| CofS | Chief of Staff |
| CSGPO | Chief of Staff, G–3, Plans & Operations Div. |
| CSUSA | Chief of Staff, U.S. Army |
| DA | Department of the Army |
| Det | Detachment |
| Dir | Director |
| DIS | Daily Intelligence Summary |
| Div | Division |
| DOW | Died of wounds |
| DUKW | Amphibious truck |
| EDT | Eastern Daylight Time |
| Engr | Engineer |
| Est | Estimate |
| EST | Eastern Standard Time |
| EUSAK | Eighth United States Army in Korea |
| Ex Off | Executive Officer |
| FA | Field Artillery |
| FDC | Fire Direction Center |
| FEAF | Far East Air Forces |
| FEC | Far East Command |
| FO | Field Order |
| Fonecon | Telephone conversation |
| G–2 | Intelligence section of divisional or higher staff |
| G–3 | Operations and training section of divisional or higher staff |
| Gen | General |
| GO | General Orders |
| GS | General Staff |
| HE | High explosive |
| HEAT | High explosive, antitank |
| Hist | History, Historical |
| Hq | Headquarters |
| Hv | Heavy |
| IG | Inspector General |
| I&R | Intelligence & Reconnaissance |
| Incl | Inclosure |
| Inf | Infantry |
| Info | Information |
| Instr | Instruction |
| Intel | Intelligence |
| Interrog | Interrogation |
| Interv | Interview |
| JAG | Judge Advocate General |
| JCS | Joint Chiefs of Staff |

| | |
|---|---|
| JLC | Japan Logistical Command |
| Jnl | Journal |
| JOC | Joint Operations Center |
| JSPOG | Joint Strategic Plans and Operations Group |
| JTF 7 | Joint Task Force Seven |
| KIA | Killed in action |
| KMAG | U.S. Korean Military Advisory Group |
| KMC | Korean Marine Corps |
| LCVP | Landing craft, vehicle, personnel |
| Ldr | Leader |
| Log | Logistical |
| LSD | Landing ship, dock |
| LSMR | Landing ship, medium (rocket) |
| LST | Landing ship, tank |
| LSV | Landing ship, vehicle |
| Ltr | Letter |
| LVT | Landing vehicle, tracked |
| MATS | Military Air Transport Service |
| Med Stf | Medical Staff |
| Med Tk | Medium Tank |
| MIA | Missing in action |
| Mil | Military |
| MIS | Military Intelligence Service |
| (−) | Minus |
| Mort | Mortar |
| MS | Manuscript |
| Msg | Message |
| MSgt | Master Sergeant |
| Narr | Narrative |
| Natl | National |
| NAVFE | U.S. Naval Forces, Far East |
| NKPA | North Korea People's Army |
| Nr | Number |
| OCMH | Office of the Chief of Military History |
| Ofc | Office |
| Off | Officer, Officers |
| Opns | Operations |
| Ord | Order |
| OSAF | Office of the Secretary of the Air Force |
| PC | Patrol vessel |
| Pers | Personnel |
| PIR | Periodic Intelligence Report |
| Plat | Platoon |
| PLR | Periodic Logistics Report |
| POL | Petroleum, oil, and lubricants |
| POR | Periodic Operations Report |
| Pub | Publication |
| PW | Prisoner of War |

| | |
|---|---|
| Rad | Radio |
| RCT | Regimental Combat Team |
| Rec | Recoilless |
| Recd | Received |
| Recon | Reconnaissance |
| Regt | Regiment |
| Rep | Representative |
| Res | Research |
| Rev | Review |
| Rpt | Report |
| RTO | Rail Transportation Office |
| S–1 | Adjutant |
| S–2 | Intelligence Officer |
| S–3 | Operations and Training Officer |
| S–4 | Supply Officer |
| S. Comm. | Senate Committee |
| SAR | Special Action Report |
| SCAP | Supreme Commander for the Allied Powers |
| SCR | Set complete radio |
| Sec | Section |
| Secy | Secretary |
| Sep | Separate |
| SFC | Sergeant First Class |
| Sig | Signal |
| Sitrep | Situation Report |
| SO | Special Order |
| SP | Self-propelled |
| Stf | Staff |
| Summ | Summary |
| Supp | Supplement |
| Surg | Surgical |
| TACP | Tactical air control party |
| TAGO | The Adjutant General's Office |
| Telecon | Teletypewriter conference |
| TF | Task Force |
| Tk | Tank |
| T/O | Tables of Organization |
| Trans | Transport, Transportation |
| Transl | Translation |
| UNC | United Nations Command |
| UNRC | United Nations Reception Center |
| USAFIK | U.S. Army Forces in Korea |
| USSR | Union of Soviet Socialist Republics |
| VHF | Very High Frequency |
| VT | Variable Time Fuze |
| WD | War Diary |
| WIA | Wounded in action |
| Wkly | Weekly |

# Basic Military Map Symbols

Symbols within a rectangle indicate a military unit, within a triangle an observation post, and within a circle a supply point.

## Military Units—Identification

| | |
|---|---|
| Antiaircraft Artillery |  |
| Armored Command |  |
| Army Air Forces | 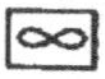 |
| Artillery, except Antiaircraft and Coast Artillery |  |
| Cavalry, Horse |  |
| Cavalry, Mechanized |  |
| Chemical Warfare Service | G |
| Coast Artillery | 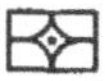 |
| Engineers | E |
| Infantry |  |
| Medical Corps |  |
| Ordnance Department |  |
| Quartermaster Corps | Q |
| Signal Corps | S |
| Tank Destroyer | TD |
| Transportation Corps |  |
| Veterinary Corps |  |

Airborne units are designated by combining a gull wing symbol with the arm or service symbol:

| | |
|---|---|
| Airborne Artillery |  |
| Airborne Infantry |  |

## Size Symbols

The following symbols placed either in boundary lines or above the rectangle, triangle, or circle inclosing the identifying arm or service symbol indicate the size of military organization:

| | |
|---|---|
| Squad | • |
| Section | •• |
| Platoon | ••• |
| Company, troop, battery, Air Force flight | I |
| Battalion, cavalry squadron, or Air Force squadron | II |
| Regiment or group; combat team (with abbreviation CT following identifying numeral) | III |
| Brigade, Combat Command of Armored Division, or Air Force Wing | X |
| Division or Command of an Air Force | XX |
| Corps or Air Force | XXX |
| Army | XXXX |
| Group of Armies | XXXXX |

### EXAMPLES

The letter or number to the left of the symbol indicates the unit designation; that to the right, the designation of the parent unit to which it belongs. Letters or numbers above or below boundary lines designate the units separated by the lines:

Company A, 137th Infantry

8th Field Artillery Battalion

Combat Command A, 1st Armored Division

Observation Post, 23d Infantry

Command Post, 5th Infantry Division

Boundary between 137th and 138th Infantry

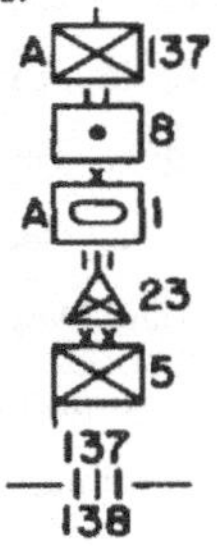


## Weapons

Machine gun

Gun

Gun battery

Howitzer or Mortar

Tank

Self-propelled gun

# Index

Made in United States
Cleveland, OH
10 July 2026

40161911R00243